NUMERICAL ANALYSIS WITH SLIDING POLYNOMIALS

WILLARD M. SNYDER
Athens, Georgia

RICHARD H. McCUEN
University of Maryland
College Park, MD 20742

NUMERICAL ANALYSIS WITH SLIDING POLYNOMIALS

Lighthouse Publications / Mission Viejo, California

Library of Congress Catalog Number 91-061003

First Lighthouse Publications Edition

NOTICE

No patent liability is assumed with respect to the use of the information contained herein. While every precaution has been taken in the preparation of this book, the publisher assumes no responsibility for errors or omissions. Neither is any liability assumed for damages resulting from the use of the information contained herein.

Library of Congress Cataloging-in-Publication Data:

Snyder, Willard M.
Numerical analysis with sliding polynomials/Willard M. Snyder,
Richard H. McCuen.
p. cm.
Includes bibliographical references and index.
ISBN 0-914055-09-7
1. Interpolation. 2. Numerical analysis. 3. Approximation theory. I. McCuen, Richard H., 1941- . II. Title. III. Title: Sliding polynomials.
QA281.S57 1991
511'.4--dc20 91-14630
CIP

CONTENTS

LIST OF PROGRAMS

PREFACE

NUMERICAL ANALYSIS WITH SLIDING POLYNOMIALS has two main thrusts of presentation. First, sliding polynomials are discussed with emphasis on structure, algebraic derivation, and wide scope of utility. Second, numerical analysis is treated as a cohesive set of operations. These operations extend from simple interpolation through smoothing and basic probability analysis, and well into the solution of differential equations. Cohesion is provided by consistent use of sliding polynomials in all operations.

The text is problem oriented, intended for the student who needs to use numerical analysis, rather than the student who requires the rigorous development of theory. Nevertheless, the text will provide a sound understanding of details of computational procedures for those students who wish to continue into advanced work.

Sliding polynomials were first presented as a simple interpolation technique in 1961. Since that time, the development of sliding polynomials has continued into many fields of numerics. The sliding polynomials are unique piece-wise operators, functionally continuous at their nodes where the pieces join. They may superficially resemble splines, but they are structurally different. They are symmetrical in the independent variable, and hence have the powerful characteristic of reversibility. They exhibit less polynomial over-swing than splines.

Sliding polynomials are, computationally, unbiased linear operators. This feature allows their use in an identical manner in treating experimental data through smoothing, or in performing numerical integration of a differential equation. By optimization with least squares one accomplishes some "best" functional approximation, whether an empirical function through smoothing or a process estimation through solution of differential or integral equations. Smoothing provides a useful alternative to conventional regression analysis, particularly in the treatment of parameter uncertainty. Differential equations can readily be solved across discrete variation of elements in the problem space.

The book is organized into three main sections. Section One, *Basics of Analysis*, includes Chapters 1 through 4. Section Two, *Analysis Based on Observations,* includes Chapters 5 and 6. Section Three, *Analysis Based on Process*, includes Chapters 7 and 8.

CHAPTER 1
Introduction to Sliding Polynomials

1.1 BACKGROUND

Numerical analysis could be called the art of approximate calculation. We must calculate approximate answers to numerical problems when we have no mathematically rigorous means to specify an exact answer. Usually, this means we do not have an exact and specific function, or equation, that we can evaluate. Sometimes we need to approximate a value intermediate to two or more given values. Sometimes we have only imprecise relationships of variables measured in experiments. We may have differential equations with no means to get an exact integral.

In all cases where we need approximate answers to problems, we must choose the system to use to get the approximation. This may be as simple as estimating on a straight line to interpolate a value between two given points. None-the-less, the straight line was chosen. Good approximation often involves subjective choices of the type of approximating system and the subjective design of this system to a particular application. Sometimes we must be guided by intuitive appreciation of a particular mathematical structure. Experience and good judgment are essentials. We lump all such non-rigorous elements under the art of approximation.

It is one of the fundamental premises of this text that we do a better job when we are familiar with the tools needed on the job. We will do a better job in approximation if we are completely familiar with some approximating system and comfortable with the results of using it. Therefore, this text discusses in detail one system of approximation, the sliding polynomials. Use of these polynomials in diverse applications from simple interpolation to solution of differential equations will establish familiarity for the user.

Sliding polynomials were first presented as continuously transforming parabolas (Snyder, 1961). The first objective in development of these special piece-wise arcs was for interpolation. However, least-squares smoothing was soon added (Snyder, 1962). The transforming parabola is a four-point structure, meaning that four check points, or nodal points, are needed for evaluation of each arc. A six-point procedure, which added mathematical continuity of the second derivative across adjacent arcs, was presented next (Snyder, 1967). The use of this six-point procedure to obtain estimates of the rates and gradients was pointed out. Because of continued development of interest an overview was prepared and the name "sliding polynomials" was introduced (Snyder, 1976a).

Sliding polynomials can be used as submodels (Snyder, 1976b). They can be used for the reduction of sets of limited data (Thomas et al., 1977). Coefficients derived from sliding polynomials can be used in the calculation of smoothly continuous moving averages (Snyder, 1979). They can also be used as substitutes for regression models with the added features of boundary controls and the calculation of localized uncertainties (Snyder, 1978, 1980). The four-point sliding polynomial has been used as a free-form probability distribution (Thomas and Snyder, 1983) and in significance testing (Harper and Snyder, 1987).

1.2 STRUCTURE OF SLIDING POLYNOMIALS

Sliding polynomials are one of a number of piece-wise mathematical forms. "Piece-wise" means an ensemble of connected segments that covers a problem space. The forms of the segments differ, the means of connections differ, but all piece-wise forms are intended to establish a free-form mathematical continuum through the problem space. This free-form continuum is a substitute for an unknown rigorous problem solution. Most students will be familiar with bar graphs or histograms, or with linear segmented functions. Some will be familiar with splines. We will develop sliding polynomials of three degrees of complexity for alternative levels of precision in approximation. These will be called two-point, four-point, and six-point sliding polynomials.

1.2.1 Two-Point Sliding Polynomials

The two-point sliding polynomial is a variant of linear segments. However, we will place these linear segments in a form that will be highly useful in later sections of this text. Development of this form is simple and will serve as preparation in later development of the four-point and six-point systems.

Consider the sequence of x - y pairs in Fig. 1-1a. Suppose that we consider a system to estimate values of y between two points Y_1 and Y_2. This system will be a straight line that could be written $\hat{y} = a_1 + b_1 x$, where $\hat{y}$ is an estimate of y. Given values of x at Y_1 and Y_2, a_1 and b_1 could be evaluated by solving the two simultaneous equations. However, in sliding polynomials we always use a local unitized variable z, placed as shown in Fig. 1-1a. The straight line becomes $\hat{y} = a_2 + b_2 z$. We can always go back and forth between x and z by some transform function $x = f(z; a_1, a_2, ..., a_n)$.

The equation in $\hat{y}$ and z can be evaluated by making it pass through the points $(-1/2, Y_1)$, and $(1/2, Y_2)$. Simple algebra gives the solution

$$\hat{y} = (Y_1 + Y_2)/2 + (Y_2 - Y_1)z \tag{1-1}$$

It will be found convenient to re-write Eq. 1-1 to the form:

$$\hat{y} = \left(\frac{1}{2} - z\right) Y_1 + \left(\frac{1}{2} + z\right) Y_2 \tag{1-2}$$

The terms in parenthesis now express how the points Y_1 and Y_2 are weighted to give a value $\hat{y}$. These terms can be evaluated for any value of z between -1/2 and 1/2. For example, at $z = 0$, half way between Y_1 and Y_2, $\hat{y} = (Y_1 + Y_2)/2$, which is the simple average of the two points defining the line. With the transform function $x = f(z_1; a_1, a_2, ..., a_n)$ we can always locate in the x-scale the place where $\hat{y}$ was evaluated in the z scale.

The value of the local unitized variable will become more evident as we study the four-point and six-point sliding polynomials. Basically, the expressions for the weighting coefficients in Eq. 1-2 need to be algebraically evaluated only once for the standardized z-variable. Notation can be simplified if we define a set of coefficients C as:

$$\begin{aligned} C_1(z) &= \frac{1}{2} - z \\ C_2(z) &= \frac{1}{2} + z \end{aligned} \tag{1-3}$$

When there is no confusion, we will simplify the notation further to the form:

$$\hat{y} = C_1Y_1 + C_2Y_2 \tag{1-4}$$

The functional dependence of the C_i and z, as in Eqs. 1-3, is always implied.

Referring again to Fig. 1-1a, it is evident that we need to connect more than just points Y_1 and Y_2. However, we can now do this simply by writing, for the next span:

$$\hat{y} = C_1Y_2 + C_2Y_3 \tag{1-5}$$

Equations similar to Eqs. 1-4 and 1-5 can be written for any span between the given values Y_i and Y_{i+1}. The C_i are always calculated by Eqs. 1-3. The only requirement is that we keep account of where we are on the x-scale.

The straight-line segments can be visualized as the track of a point sliding from the left to the right through the data. The point slides in a straight line from each Y_i to the next Y_{i+1}. The places where two segments join are usually called nodes.

1.2.2 Four-Point Sliding Polynomials

The graphical structure of the four-point sliding polynomial is shown in Fig. 1-1b. We will develop an approximating system based on four sequential values such as Y_1, Y_2, Y_3, and Y_4. The local standardized variable z now extends from -3/2 to 3/2; however, the approximating function is defined only for the central range $-1/2 \leq z \leq 1/2$.

The approximating function, $\hat{y}$, for the four-point sliding polynomial is based on the following cubic equation:

$$\hat{y} = a + bz + cz^2 + dz^3 \tag{1-6}$$

The cubic polynomial is used so that four parameters match in number the set of four nodes Y_i. Four mathematical criteria are necessary to evaluate the four parameters in Eq. 1-6. These are as follows. First, the cubic polynomial must pass through Y_2 and Y_3, establishing two criteria. Third, at the left side of the z scale the cubic is required to be tangent to a parabola, ${}_1\hat{y}$, which passes through Y_1, Y_2, and Y_3.

Fourth, at the right the cubic is required to be tangent to the parabola ${}_2\hat{y}$, which passes through Y_2, Y_3, and Y_4. Thus four criteria are established.

The first derivative of Eq. 1-6 is given by:

$$d\hat{y}/dz = b + 2cz + 3d\,z^2 \tag{1-7}$$

Parabola ${}_1\hat{y}$ can be written

$${}_1\hat{y} = e + fz + gz^2 \tag{1-8}$$

The first derivative of this parabola is

$$d_1\hat{y}/dz = f + 2gz \tag{1-9}$$

If we pass the parabola of Eq. 1-8 through the three points $(-3/2, Y_1)$, $(-1/2, Y_2)$, and $(1/2, Y_3)$, we get the simultaneous equations:

$$Y_1 = e - \frac{3}{2} f + \frac{9}{4} g \tag{1-10a}$$

$$Y_2 = e - \frac{1}{2} f + \frac{1}{4} g \tag{1-10b}$$

$$Y_3 = e + \frac{1}{2} f + \frac{1}{4} g \tag{1-10c}$$

Equations 1-10 can be solved for the parameters giving

$$e = \frac{1}{8} (-Y_1 + 6Y_2 + 3Y_3) \tag{1-11a}$$

$$f = Y_3 - Y_2 \tag{1-11b}$$

$$g = \frac{1}{2} (-Y_1 + 2Y_2 + Y_3) \tag{1-11c}$$

The first derivative, Eq. 1-9, becomes

$$d_1\hat{y}/dz = \frac{1}{2} (-Y_1 + Y_3) \tag{1-12}$$

The same procedure can be used to find the first derivative of the parabola through Y_2, Y_3, and Y_4:

$$d_2y/dz = \frac{1}{2}(-Y_2 + Y_4) \quad (1\text{-}13)$$

We can now form four simultaneous equations expressing the four criteria for evaluation of the parameters of Eq. 1-6. The equation passes through (-½, Y_2) giving

$$Y_2 = a - \frac{1}{2}b + \frac{1}{4}c - \frac{1}{8}d \quad (1\text{-}14a)$$

The equation also passes through (½, Y_3) giving

$$Y_3 = a + \frac{1}{2}b + \frac{1}{4}c + \frac{1}{8}d \quad (1\text{-}14b)$$

The derivatives of Eqs. 1-7 and 1-12 are equal at z = -½ which yields

$$\frac{1}{2}(-Y_1 + Y_3) = b - c + \frac{3}{4}d \quad (1\text{-}14c)$$

The derivative of Eqs. 1-7 and 1-13 are equal at z = ½ , which yields:

$$\frac{1}{2}(-Y_2 + Y_4) = b + c + \frac{3}{4}d \quad (1\text{-}14d)$$

Solving the four simultaneous equations, Eqs. 1-14, for the parameters yields

$$a = -\frac{1}{16}Y_1 + \frac{9}{16}Y_2 + \frac{9}{16}Y_3 - \frac{1}{16}Y_4 \quad (1\text{-}15a)$$

$$b = \frac{1}{8}Y_1 - \frac{11}{8}Y_2 + \frac{11}{8}Y_3 - \frac{1}{8}Y_4 \quad (1\text{-}15b)$$

$$c = \frac{1}{4}Y_1 - \frac{1}{4}Y_2 - \frac{1}{4}Y_3 + \frac{1}{4}Y_4 \quad (1\text{-}15c)$$

$$d = -\frac{1}{2}Y_1 + \frac{3}{2}Y_2 - \frac{3}{2}Y_3 + \frac{1}{2}Y_4 \quad (1\text{-}15d)$$

Equations 1-15 can be substituted for the respective parameters in Eq. 1-6. If we then re-arrange and collect all multipliers of the nodal ordinate Y_1, all multipliers of the nodal ordinate Y_2, and so on, we can write Eq. 1-6 in the form

$$\hat{y} = C_1(z)\ Y_1 + C_2(z)\ Y_2 + C_3(z)\ Y_3 + C_4(z)\ Y_4 \qquad (1\text{-}16)$$

The equations for $C_i(z)$ are given by:

$$C_1(z) = -\frac{1}{16} + \frac{1}{8}\ z + \frac{1}{4}\ z^2 - \frac{1}{2}\ z^3 \qquad (1\text{-}17a)$$

$$C_2(z) = \frac{9}{16} - \frac{11}{8}\ z - \frac{1}{4}\ z^2 + \frac{3}{2}\ z^3 \qquad (1\text{-}17b)$$

$$C_3(z) = \frac{9}{16} + \frac{11}{8}\ z - \frac{1}{4}\ z^2 - \frac{3}{2}\ z^3 \qquad (1\text{-}17c)$$

$$C_4(z) = -\frac{1}{16} - \frac{1}{8}\ z + \frac{1}{4}\ z^2 + \frac{1}{2}\ z^3 \qquad (1\text{-}17d)$$

When arranged in a nested polynomial form for efficient computer programming, these become

$$C_1(z) = (((-8z + 4)\ z + 2)\ z - 1)/16 \qquad (1\text{-}18a)$$

$$C_2(z) = (((24z - 4)\ z - 22)\ z + 9)/16 \qquad (1\text{-}18b)$$

$$C_3(z) = (((-24z - 4)\ z + 22)\ z + 9)/16 \qquad (1\text{-}18c)$$

$$C_4(z) = (((8z + 4)\ z - 2)\ z - 1)/16 \qquad (1\text{-}18d)$$

Equation 1-16 covers only the span between Y_2 and Y_3 in Figure 1-1b. Now the span between Y_3 and Y_4 can be considered. A cubic arc through nodal values Y_2, Y_3, Y_4, and Y_5 could be evaluated in a manner similar to that through Y_1, Y_2, Y_3, and Y_4. Particularly, if we define z on the nodal locations Y_2 to Y_5, then the coefficients will be identical to those in Eqs. 1-17. We can write directly for this second span

$$\hat{y} = C_1Y_2 + C_2Y_3 + C_3Y_4 + C_4Y_5 \qquad (1\text{-}19)$$

The cubic arc in the first span is tangent to the parabola through Y_2, Y_3, and Y_4 at its right edge at Y_3. But the second cubic arc is also tangent to the parabola through Y_2, Y_3, and Y_4 at its left edge. Therefore, the two arcs are common and tangent to each other at Y_3. The ensemble of the two arcs is continuous and has a continuous first derivative. More arcs could be added to cover additional spans. By the same arguments used above, the total assembly of arcs is continuous with a continuous first derivative.

In the two-point system, a point slides from node to node and generates a straight line. In the four-point system, the parabola slides from the three-point set Y_1, Y_2, and Y_3 forward to the three-point set Y_2, Y_3, and Y_4. This parabola flexes as it moves generating a cubic arc between Y_2 and Y_3. It flexes in the same way as it continues to slide forward to the set Y_3, Y_4, and Y_5, generating a second cubic arc that is common and tangent to the first arc. The sliding continues through the entire data set that defines the problem.

Table A-1.1 lists the four coefficients of Eqs. 1-18 for all values of z by increments of 0.01 from -0.5 to 0.5. These are convenient for small calculator use in approximating values at a small number of points. For example, suppose we have the values -1, 0, 2, 3, and 3 uniformly spaced with an interval dx, and we wish to approximate values at the quarter points of the intervals between 0-2 and 2-3. The calculations are shown in Table 1-1. The coefficients are taken from Table A-1.1 and the calculations follow Eq. 1-16.

The given nodal values and the approximations are shown in Figure 1-2. A curve has been drawn through the interior points and the approximations. This curve is made up of two cubic arcs and demonstrates the smooth transition through the node (xL + 2 dx, 2). This smooth joint is due to the continuous first derivative through the node. The nodes located at xL and xL + 4dx can be called outlier nodes. They are required by the four-point system, but the cubic arcs do not pass through them; however, the parabolas to which the cubic arcs are tangent pass through these outliers.

1.2.3. The Six-Point Sliding Polynomial

In the six-point sliding polynomial a fourth-order polynomial, which is two degrees greater than the parabola of the four-point system, slides through the data. As it slides it generates a fifth-order polynomial, which is two degrees greater than the cubic of the four-point system. The functions are sketched in Fig. 1-1c.

A general fifth-order polynomial is written as

$$\hat{y} = a + bz + cz^2 + dz^3 + ez^4 + fz^5 \tag{1-20}$$

At the left edge where z = -0.5 this polynomial has first and second derivatives that are equal to the first and second derivatives of the fourth-order polynomial

$$_1y = j + kz + lz^2 + mz^3 + nz^4 \tag{1-21}$$

The first and second derivatives of Eq. 1-20 are

$$d\hat{y}/dz = b + 2cz + 3dz^2 + 4ez^3 + 5fz^4 \tag{1-22}$$

and

$$d^2\hat{y}/dz^2 = 2c + 6dz + 12ez^2 + 20fz^3 \tag{1-23}$$

The first and second derivatives of Eq. 1-21 are:

$$d_1\hat{y}/dz = k + 2\ lz + 3mz^2 + 4nz^3 \tag{1-24}$$

and

$$d_1^2\hat{y}/dz^2 = 2\ l + 6mz + 12nz^2 \tag{1-25}$$

It is now necessary to evaluate the parameters in Eq. 1-21. This is most conveniently done by passing the polynomial through a set of points (w_i, Y_i) where w = z + 0.5. The set of points at the left in Fig. 1-1c, which are symmetrical in w, is (-2, Y_1), (-1, Y_2), (0, Y_3), (1, Y_4), and (2, Y_5). The solutions of the parameters are given by:

$$j = Y_3 \tag{1-26a}$$

$$k = \frac{1}{12} Y_1 - \frac{2}{3} Y_2 + \frac{2}{3} Y_4 - \frac{1}{12} Y_5 \tag{1-26b}$$

$$l = -\frac{1}{24} Y_1 + \frac{2}{3} Y_2 - \frac{5}{4} Y_3 + \frac{2}{3} Y_4 - \frac{1}{24} Y_5 \tag{1-26c}$$

$$m = -\frac{1}{12} Y_1 + \frac{1}{6} Y_2 - \frac{1}{6} Y_4 + \frac{1}{12} Y_5 \tag{1-26d}$$

$$n = \frac{1}{24} Y_1 - \frac{1}{6} Y_2 + \frac{1}{4} Y_3 - \frac{1}{6} Y_4 + \frac{1}{24} Y_5 \tag{1-26e}$$

We can now write the following from Eqs. 1-24 and 1-25:

$$(d_1y/dz)_{z=-0.5} = (d_1y/dw)_{w=0} = k = \frac{1}{12} Y_1 - \frac{2}{3} Y_2 + \frac{2}{3} Y_4 - \frac{1}{12} Y_5 \tag{1-27}$$

$$(d^2{}_1y/dz^2)_{z=-0.5} = (d^2{}_1y/dw^2)_{w=0} = 2l = -\frac{1}{12} Y_1 + \frac{4}{3} Y_2 - \frac{5}{2} Y_3 + \frac{4}{3} Y_4 - \frac{1}{12} Y_5 \tag{1-28}$$

By a similar analysis we can develop at the right-hand edge of the span where z = 0.5 and w = z - 0.5 the derivatives

$$(d_2y/dz)_{z=0.5} = (d_2y/dw)_{w=0} = \frac{1}{12} Y_2 - \frac{2}{3} Y_3 + \frac{2}{3} Y_5 - \frac{1}{12} Y_6 \tag{1-29}$$

$$(d_2^2y/dz^2)_{z=0.5} = (d_2^2y/dw)_{w=0} = -\frac{1}{12} Y_2 + \frac{4}{3} Y_3 - \frac{5}{2} Y_4 + \frac{4}{3} Y_5 - \frac{1}{12} Y_6 \tag{1-30}$$

The next step is to solve a set of six simultaneous equations based on the six criteria:

$$(\hat{y})_{z=-0.5} = Y_3 \tag{1-31a}$$

$$(\hat{y})_{z=0.5} = Y_4 \tag{1-31b}$$

$$(d\hat{y}/dz)_{z=-0.5} = (d_1y/dz)_{z=-0.5} \tag{1-31c}$$

$$(d\hat{y}/dz)_{z=0.5} = (d_2y/dz)_{z=0.5} \tag{1-31d}$$

$$(d^2\hat{y}/dz^2)_{z=-0.5} = (d_1^2y/dz^2)_{z=-0.5} \tag{1-31e}$$

$$(d^2\hat{y}/dz^2)_{z=0.5} = (d_2^2y/dz^2)_{z=0.5} \tag{1-31f}$$

The six simultaneous equations are:

$$Y_3 = a - \frac{1}{2} b + \frac{1}{4} c - \frac{1}{8} d + \frac{1}{16} e - \frac{1}{32} f \tag{1-32a}$$

$$Y_4 = a + \frac{1}{2} b + \frac{1}{4} c + \frac{1}{8} d + \frac{1}{16} e + \frac{1}{32} f \tag{1-32b}$$

$$\frac{1}{12} Y_1 - \frac{2}{3} Y_2 + \frac{2}{3} Y_4 - \frac{1}{12} Y_5 = b - c + \frac{3}{4} d - \frac{1}{2} e + \frac{5}{16} f \tag{1-32c}$$

$$\frac{1}{12} Y_2 - \frac{2}{3} Y_3 + \frac{2}{3} Y_5 - \frac{1}{12} Y_6 = b + c + \frac{3}{4} d + \frac{1}{2} e + \frac{5}{16} f \tag{1-32d}$$

$$-\frac{1}{12} Y_1 + \frac{4}{3} Y_2 - \frac{5}{2} Y_3 + \frac{4}{3} Y_4 - \frac{1}{12} Y_5 = 2c - 3d + 3e - \frac{5}{2} f \tag{1-32e}$$

$$-\frac{1}{12} Y_2 + \frac{4}{3} Y_3 - \frac{5}{2} Y_4 + \frac{4}{3} Y_5 - \frac{1}{12} Y_6 = 2c + 3d + 3e + \frac{5}{2} f \tag{1-32f}$$

The solution of these six equations gives the parameters of Eq. 1-20:

$$a = \frac{9}{768} Y_1 - \frac{75}{768} Y_2 + \frac{225}{384} Y_3 + \frac{225}{384} Y_4 - \frac{75}{768} Y_5 + \frac{9}{768} Y_6 \tag{1-33a}$$

$$b = -\frac{13}{384} Y_1 + \frac{81}{384} Y_2 - \frac{281}{192} Y_3 + \frac{281}{192} Y_4 - \frac{81}{384} Y_5 + \frac{13}{384} Y_6 \tag{1-33b}$$

$$c = -\frac{5}{96} Y_1 + \frac{39}{96} Y_2 - \frac{17}{48} Y_3 - \frac{17}{48} Y_4 + \frac{39}{96} Y_5 - \frac{5}{96} Y_6 \tag{1-33c}$$

$$d = \frac{9}{48} Y_1 - \frac{53}{48} Y_2 + \frac{57}{24} Y_3 - \frac{57}{24} Y_4 + \frac{53}{48} Y_5 - \frac{9}{48} Y_6 \tag{1-33d}$$

$$e = \frac{1}{48} Y_1 - \frac{3}{48} Y_2 + \frac{1}{24} Y_3 + \frac{1}{24} Y_4 - \frac{3}{48} Y_5 + \frac{1}{48} Y_6 \tag{1-33e}$$

$$f = -\frac{5}{24} Y_1 + \frac{25}{24} Y_2 - \frac{25}{12} Y_3 + \frac{25}{12} Y_4 - \frac{25}{24} Y_5 + \frac{5}{24} Y_6 \tag{1-33f}$$

If the terms in Eqs. 1-33 common to each Y_i, i = 1 to 6, are collected, the following equations for the six-point sliding polynomial coefficients are obtained:

$$C_1(z) = (9 - 26z - 40z^2 + 144z^3 + 16z^4 - 160z^5)/768 \tag{1-34a}$$

$$C_2(z) = (-75 + 162z + 312z^2 - 848z^3 - 48z^4 + 800z^5)/768 \quad (1\text{-}34b)$$

$$C_3(z) = (450 - 1124z - 272z^2 + 1824z^3 + 32z^4 - 1600z^5)/768 \quad (1\text{-}34c)$$

$$C_4(z) = (450 + 1124z - 272z^2 - 1824z^3 + 32z^4 + 1600z^5)/768 \quad (1\text{-}34d)$$

$$C_5(z) = (-75 - 162z + 312z^2 + 848z^3 - 48z^4 - 800z^5)/768 \quad (1\text{-}34e)$$

$$C_6(z) = (9 + 26z - 40z^2 - 144z^3 + 16z^4 + 160z^5)/768 \quad (1\text{-}34f)$$

These can be converted to the computer-efficient nested polynomial form:

$$C_1(z) = (((((-160z + 16)z + 144)z - 40)z - 26)z + 9)/768 \quad (1\text{-}35a)$$

$$C_2(z) = (((((800z - 48)z - 848)z + 312)z + 162)z - 75)/768 \quad (1\text{-}35b)$$

$$C_3(z) = (((((-1600z + 32)z + 1824)z - 272)z - 1124)z + 450)/768 \quad (1\text{-}35c)$$

$$C_4(z) = (((((1600z + 32)z - 1824)z - 272)z + 1124)z + 450)/768 \quad (1\text{-}35d)$$

$$C_5(z) = (((((-800z - 48)z + 848)z + 312)z - 162)z - 75)/768 \quad (1\text{-}35e)$$

$$C_6(z) = (((((160z + 16)z - 144)z - 40)z + 26)z + 9)/768 \quad (1\text{-}35f)$$

The foregoing development of the six-point polynomial covers one span, between Y_3 and Y_4, in Fig. 1-1. We must now slide the sub-function, Equation 1-21, forward through the span between Y_4 and Y_5. The same argument can be used as in the four-point system. The fifth-order polynomial, Equation 1-20, for the span between Y_3 and Y_4 is common to and has common first and second derivatives with the fourth-order polynomial, Equation 1-21, through nodes Y_2, Y_3, Y_4, Y_5, and Y_6, at node Y_4. The fifth-order polynomial for span Y_4 and Y_5 is also common to and has common first and second derivatives with the same fourth-order polynomial at Y_4. Therefore, the two fifth-order polynomial arcs are common and have common first and second derivatives with each other at node Y_4.

The same argument will apply to any number of adjacent spans. The fourth-order polynomial flexes as it slides forward from span to span, generating fifth-order polynomial arcs within each span. The ensemble of fifth-order arcs is mathematically continuous through the second derivative.

Table A-1.2 gives values of the six $C_i(z)$ of Eq. 1-34. Values are given for each 0.01 increment of z. These values are convenient for calculating a small number of points of the fifth-order arcs. Computer programs will be given in Chapter 2 for the calculation of numerous points. A brief example of the use of the coefficients in Table A-1.2 follows. Given the set of uniformly spaced nodal values -1, -1, 0, 2, 3, and 3, a value at the midpoint where $z = 0$ can be approximated. The calculation is (0.01172*-1) + (-0.09766*-1) + (0.58574*0) + (0.58594*2) + (-0.09766*3) + (0.01172*3) = 1.0000. Because of the symmetry of the nodal values this is the same value as calculated with the four-point system in Table 1-1.

1.3 TWO-DIMENSIONAL SLIDING POLYNOMIALS

The three levels of complexity of sliding polynomials presented in Section 1.2 are all designed to produce lines or curves. The dependent variable, $\hat{y}$, is a free-form function of a single independent variable, x, or its unitized form, z. In many problems, we must be concerned with fields of information that require a structure for surfaces instead of simple curves. The dependent variable, $\hat{y}$, becomes a function of two independent variables.

The line-function sliding polynomials of Section 1.1 can be used to develop form-free but mathematically continuous sliding polynomial surfaces. This development is based on a simple concept of "function-of-functions". We will illustrate such development with the two-point system for simplicity.

Consider the grid of nodes in Fig. 1-3. At each of the points visualize a nodal value, $Y_{j,i}$, perpendicular to the grid, i.e., projecting upward from the paper on which the grid is drawn. These nodal values give the height of the surface over the grid. Subscript j is the row number, and subscript i is the column number of the lines forming the grid. In the lower left unit surface, define $\hat{y}_1$ on nodes $Y_{1,1}$ and $Y_{1,2}$ using Eq. 1-4:

$$\hat{y}_1 = C_1(z_1)\ Y_{1,1} + C_2(z_1)\ Y_{1,2} \tag{1-36}$$

Define $\hat{y}_2$ on nodes $Y_{2,1}$ and $Y_{2,2}$:

$$\hat{y}_2 = C_1(z_1)\ Y_{2,1} + C_2(z_1)\ Y_{2,2} \tag{1-37}$$

$\hat{y}_1$ and $\hat{y}_2$ are straight lines along the bottom and top edges of the unit grid. Now make line $\hat{y}_1$ change to line $\hat{y}_2$ as a two-point sliding polynomial. We write directly

$$\hat{y} = C_1(z_2)\ \hat{y}_1 + C_2(z_2)\ \hat{y}_2 \tag{1-38}$$

Expanding with Eqs. 1-36 and 1-37 gives:

$$\hat{y} = C_1(z_2)\ C_1(z_1)\ Y_{1,1} + C_1(z_2)\ C_2(z_1)\ Y_{1,2} + C_2(z_2)\ C_1(z_1)\ Y_{2,1} + C_2(z_2)\ C_2(z_1)\ Y_{2,2} \tag{1-39}$$

Equation 1-39 will give the height of the surface in the first unit span where z_1 and z_2 are given as unitized values in the x_1 and x_2 scales, respectively.

The values of z_1 and z_2 are zero at the center point of the unit surface. From Eq. 1−3, we note that all C(z) in Equation 1-39 are 1/2 for $z = 0$. Therefore, all the products $C_j(z)\ C_i(z)$ are 1/4 and

$$\hat{y} = (Y_{1,1} + Y_{1,2} + Y_{2,1} + Y_{2,2})/4$$

which is logical.

At a point on the edge of the unit surface midway between $Y_{1,1}$ and $Y_{1,2}$, $z_1 = 0$ and $z_2 = -1/2$, which gives $C_1(z_1) = C_2(z_1) = 1/2$, $C_1(z_2) = 1$, and $C_2(z_2) = 0$. Therefore, for this point, $\hat{y} = (Y_{1,1} + Y_{1,2})/2$. At a point $z_1 = -1/4$ and $z_2 = -1/4$, $C_1(z_1) = 3/4$, $C_2(z_1) = 1/4$, $C_1(z_2) = 3/4$, and $C_2(z_2) = 1/4$. Then from Eq. 1-39:

$$\hat{y} = \frac{3}{4} \cdot \frac{3}{4}\ Y_{1,1} + \frac{3}{4} \cdot \frac{1}{4}\ Y_{1,2} + \frac{1}{4} \cdot \frac{3}{4}\ Y_{2,1} + \frac{1}{4} \cdot \frac{1}{4}\ Y_{2,2}$$

The coefficient products are 9/16, 3/16, 3/16, and 1/16 and add to unity showing that the weighing is unbiased.

Equations similar to Eq. 1-36 can be written for any unit surface provided proper nodal values are specified. Any two adjacent units will have the same line function at their common boundary. The ensemble of all unit surfaces will be continuous, but the surfaces will kink at the common boundaries since the first derivatives are not continuous.

Equation 1-36 can be written in the more concise form

$$y = \sum_{i=1}^{2} C_i(z)\ Y_i \tag{1-40}$$

and the two-dimensional form of Eq. 1-39 can be written concisely

$$\hat{y} = \sum_{j=1}^{2} \sum_{i=1}^{2} C_j(z_2)\ C_i(z_1) Y_{j,i} \tag{1-41}$$

Double summations, such as in Equation 1-41, are evaluated from "outside to inside". This means that the outside index j is set to 1 and then i is set to 1 and 2, in order. Next, the outside index is set to 2 and again i is set to 1 and 2. The concept is the same as the nested DO-loop in computer programming, and the "function-of-functions" construction of a two-dimensional sliding polynomial surface is computer efficient.

1.3.1 Four-Point Two-Dimensional Sliding Polynomial Surfaces

A four-point sliding polynomial surface can be constructed directly following the "function-of-functions" concept. The development will be illustrated using the unit surface A in Fig. 1-4. First, four lines of four-point sliding polynomials on the rows below, at the edges, and above the unit surface are defined:

$$\hat{y}_1 = C_1(z_1)\ Y_{1,1} + C_2(z_1)\ Y_{1,2} + C_3(z_1) Y_{1,3} + C_4(z_1)\ Y_{1,4} \tag{1-42a}$$

$$\hat{y}_2 = C_1(z_1)\ Y_{2,1} + C_2(z_1)\ Y_{2,2} + C_3(z_1)Y_{2,3} + C_4(z_1)\ Y_{2,4} \qquad \text{(1-42b)}$$

$$\hat{y}_3 = C_1(z_1)\ Y_{3,1} + C_2(z_1)\ Y_{3,2} + C_3(z_1)Y_{3,3} + C_4(z_1)\ Y_{3,4} \qquad \text{(1-42c)}$$

$$\hat{y}_4 = C_1(z_1)\ Y_{4,1} + C_2(z_1)\ Y_{4,2} + C_3(z_1)Y_{4,3} + C_4(z_1)\ Y_{4,4} \qquad \text{(1-42d)}$$

Operating on these four functions with a four-point sliding polynomial function at right angles, or parallel to the x_2 dimension, gives:

$$\hat{y} = C_1(z_2)\ \hat{y}_1 + C_2(z_2)\ \hat{y}_2 + C_3(z_2)\ \hat{y}_3 + C_4(z_2)\ \hat{y}_4 \qquad \text{(1-43)}$$

Expanding Eq. 1-43 yields:

$$\begin{aligned}\hat{y} = {} & C_1(z_2)\ C_1(z_1)\ Y_{1,1} + C_1(z_2)\ C_2(z_1)\ Y_{1,2} + C_1(z_2)\ C_3(z_1)\ Y_{1,3} + C_1(z_2)\ C_4(z_1)\ Y_{1,4} \\ & + C_2(z_2)\ C_1(z_1)\ Y_{2,1} + C_2(z_2)\ C_2(z_1)\ Y_{2,2} + C_2(z_2)\ C_3(z_1)\ Y_{2,3} + C_2(z_2)\ C_4(z_1)\ Y_{2,4} \\ & + C_3(z_2)\ C_1(z_1)\ Y_{3,1} + C_3(z_2)\ C_2(z_1)\ Y_{3,2} + C_3(z_2)\ C_3(z_1)\ Y_{3,3} + C_3(z_2)\ C_4(z_1)\ Y_{3,4} \\ & + C_4(z_2)\ C_1(z_1)\ Y_{4,1} + C_4(z_2)\ C_2(z_1)\ Y_{4,2} + C_4(z_2)\ C_3(z_1)\ Y_{4,3} + C_4(z_2)\ C_4(z_1)\ Y_{4,4}\end{aligned} \qquad \text{(1-44)}$$

In a concise double summation form, we can write

$$\hat{y} = \sum_{j=i}^{4} \sum_{i-1}^{4} C_j(z_2)\ C_i(z_1) Y_{j,i} \qquad \text{(1-45)}$$

From Table A-1.1 for $z_1 = 0$, we get the four-point coefficients: -0.0625, 0.5625, 0.5625 and -0.0625. If z_2 is also set to 0, we get the same coefficients in this dimension. The 16 products of coefficients indicated in Eq. 1-45 are shown on a 4x4 grid in Fig. 1-5. The locations, of course, are controlled by the indices j and i. Now if we had 16 nodal values made up, say, of the values 10 along the edges and 20 at the four interior nodes, the approximate value of the midpoint of the unit surface (where z_1 and z_2 are zero) is calculated as in Table 1-2. The unit surface is thus a dome, rising symmetrically from 20 at the four corners to 22.66 at the center.

Referring again to Fig. 1-4, the unit surfaces B, C, and D can also be approximated by an equation such as Eq. 1-45, provided the appropriate nodes $Y_{j,i}$ are used. The unit surfaces will be common and have common first derivatives along their common boundaries since at the boundaries, they reduce to common four-point line functions. The ensemble of all such unit surfaces forms a mathematically continuous, but free-form, surface for the entire problem space.

Snyder et al. (1984) developed a two-dimensional, four-point sliding polynomial surface based on four sub-function surfaces centered on each of the four interior nodes. A surface one order higher, common and tangent to these subfunctions at the interior nodes, was solved algebraically to yield expressions for the 16 coefficients. These are identical to the products of coefficients indicated in Eq. 1-45. Such a solution is tedious and error-prone since a set of 16 simultaneous equations must be solved. However, this work does confirm the validity of the function-of-functions concept. The advantages of the function-of-functions concept are that it is simpler in structure, more easily programmed, and extendable to more complex forms.

1.3.2 Six-Point Two-Dimensional Sliding Polynomial Surfaces

It is possible to proceed through Eqs. 1-42 to 1-49 using the coefficients for six-point sliding polynomials instead of those for the four-point system. The following equation, which is comparable to Eq. 1-45, results:

$$\hat{y} = \sum_{j=1}^{6} \sum_{i=1}^{6} C_j(z_2)\, C_i(z_1)\, Y_{j,i} \qquad (1\text{-}46)$$

Interpolation now requires a 6 x 6 grid as in Fig. 1-6. The coefficients shown in this figure are used for evaluating the height of the surface at the mid-point of the unit surface. The $C_i(z)$ values for $z = 0$ can be found in Table A-1.2 and are as follows: 0.01171875, -0.09765625, 0.58593750, 0.58593750, -0.09765625, and 0.01171875. The 36 products of coefficients, which are indicated in Eq. 1-46, are the plotted values in Fig. 1-6.

Now consider the problem of approximating a value at the mid-point of 36 nodes. This set is made up of nodal values of 5 around the outer edge, values of 10 along the next square inward, and values of 20 for the four interior nodes. Calculations are shown in Table 1-3. The value of 23.5 is slightly higher than the value in Table 1-2. Such differences should be expected since the grids are not identical and the approximating functions differ.

Calculations to get approximate values manually as in Table 1-3 are cumbersome. However, efficient computer programs can be written to compute both the coefficients and the products of these coefficients times the nodal values. Such programs will be introduced in Chapter 2.

1.4 MULTI-DIMENSION SLIDING POLYNOMIALS

In Section 1.3, two-dimensional sliding polynomials were constructed under a concept of function-of-functions. Equations 1-41, 1-45, and 1-46, covering two-, four-, and six-point surfaces were developed.

The function-of-functions concept can be extended to additional dimensions. Consider a series of polynomial surfaces. The change from one surface to the next in the series can be made to change as a sliding polynomial function. A three-dimensional two-point sliding polynomial function is given by:

$$\hat{y} = \sum_{k=1}^{2} \sum_{j=1}^{2} \sum_{i=1}^{2} C_k(z_3)\ C_j(z_2)\ C_i(z_1)\ Y_{k,j,i} \tag{1-47}$$

The expansion of Eq. 1-47, following the outside-to-inside principle of procedure, is given in Table 1-4. The C(z) values are all obtained from Equations 1-3. The arrangement of the nodes in Eq. 1-47 is shown as a cube in Figure 1-7. When $z = 0$ in the three dimensions, all of the coefficients have values of 1/2, and all of the triple products in Eq. 1-47 become 1/8. Therefore, at the center of the cube in Fig. 1-7, the approximate value of a property is the sum of the eight nodes expressing specific values of the property divided by eight, which is the mean value. This is only true at the mid-point of a two-point, or linear, system. At other points, the appropriate coefficients must be evaluated following Table 1-4. Again, this is a tedious manual computation but can be readily programmed as nested DO-loops for a highly efficient computer procedure.

Equation 1-47 can be generalized as:

$$\hat{y} = \sum_{k=1}^{n_3} \sum_{j=1}^{n_2} \sum_{i=1}^{n_1} C_k(z_3)\ C_j(z_2)\ C_i(z_1)\ Y_{k,j,i} \tag{1-48}$$

It is not necessary to have $n_1 = n_2 = n_3$. For example, for $n_3 = 2$ and $n_2 = n_1 = 4$, we would have a four-point sliding polynomial surface in the bottom plane of Fig. 1-7; this surface would change linearly to a different four-point sliding polynomial surface in the top plane. Such three-dimensional structures are necessary for some problems. For example, estimation of the temperature of the air at any point in a room when we only know the temperature at a few points in the room requires a three-dimensional approximating system. More abstractly, if we have a surface of a relationship based on variables x_1 and x_2 and we wish to see how the pattern of the relationship changes as a third variable x_3 changes, we must use a three-dimensional structure. It is possible to go beyond three dimensions but such problems demand large amounts of data to establish the super-grids of nodal values and they are difficult to visualize and display.

1.5 SUBJECTIVE COMPARISON OF SLIDING POLYNOMIALS AND SPLINES

Splines and sliding polynomials are both piece-wise polynomial structures and have some superficial resemblances. However, the mathematical form of the two differs significantly.

Sliding polynomials were shown to be piece-wise arcs that are common to and have common derivatives with a sub-function at their left and right edges. Thus, two adjacent arcs are mathematically continuous because they have these common properties with common sub-functions. Splines are not based on such sub-functions. Adjacent arcs of splines are only made continuous with each other. Therefore, sliding polynomials are functionally continuous but splines are only point continuous. Because of the sub-function structure, sliding polynomials are a restrained system. In general, they have must less tendency to polynomial overswing than splines do.

Sliding polynomials are symmetrical in their weighting of the nodal values through which they pass. Thus, they are reversible. Splines are not symmetrical, and if data are reversed, approximations at equivalent points will not be the same. A false hysteresis is introduced.

Sliding polynomials are an unbiased system. The coefficients always total unity. This may not be true with splines.

Sliding polynomials are readily expandable to multi-dimensions. No additional algebraic manipulation is necessary for such expansion from one dimension. This solution for the sliding polynomial coefficients was a one-time effort because of the unitized local variables used in the development.

EXERCISES

1-1. In a given problem values of x range from 1. to 5. At what x-values will uniformly spaced nodes be placed to break the range into four spans? What are the z-values for x-values of 2.2, 3.4, and 4.9?

1-2. Given five nodal values of 0.5, 1.0, 2.0, 4.0, and 8.0 that correspond to the x values of 1 to 5, calculate $\hat{y}$ for the three fractional x-values of Exercise 1-1.

1-3. Label the five nodal values of Exercise 1-1 as Y_1, Y_2, Y_3, Y_4, and Y_5. If we write Eq. 1-4 in generalized form, what are the values of index i for the three fractional x-values?

1-4. Show that the composite of index i and unit variable z constitute a linear transform for z to x.

1-5. Given that x ranges from 0 to 10. Calculate the x-position of nodes if the range is broken into four spans. Include two outlier nodes for the four-point system.

1-6. Given nodal values -0.1, 0.0, 0.2, 0.5, -0.1, -0.3, and -0.8 at the positions calculated in Exercise 1-5, calculate values of $\hat{y}$ for x-values of 4.5, 4.75, 5.0, 5.25, and 5.5. Plot these with the nodal values on an x-y graph. Discuss continuity at x = 5.

1-7. If we write Equation 1-16 in the generalized form $\hat{y} = \sum_{i=1}^{4} c_i(z)Y_{n+1}$, what are the values of n for values of x of 1.0, 6.0, and 9.0 of Exercise 1-5.

1-8. Compute the eight coefficient-products in the equation $\hat{y} = \sum_{j=1}^{2} \sum_{i=1}^{4} c_j(z_2)c_i(z_1)Y_{j,i}$, for $z_1 = -0.1$, and $z_2 = 0.2$. Sketch the grid, show the point of interpolation, and plot the coefficients on the grid. Use Table A-1.1 for $C_i(z_1)$.

1-9. Given the nodal temperatures:

$T_{1,1,1}$	$T_{1,1,2}$	$T_{1,2,1}$	$T_{1,2,2}$	$T_{2,1,1}$	$T_{2,1,2}$	$T_{2,2,1}$	$T_{2,2,2}$
10	15	12	18	13	19	16	25

calculate the approximate temperature for the point $z_1 = 0.2$, $z_2 = 0.4$, and $z_3 = 0.1$ using Equation 1-47.

1-10. Given nodes

x:	0.2	0.3	0.4	0.5	0.6	0.7
y:	0.15	0.3	0.45	0.6	0.75	0.9

Compute $\hat{y}$ for x = 0.44 and 0.46. Use coefficients from Table A-1.2.

1-11. Show the calculations to obtain the coefficient values of Eqs. 1-10.

1-12. Show the derivation of Eq. 1-13.

1-13. Derive Eq. 1-46 in the same way that Eq. 1-45 was developed in section 1.3.1.

Table 1-1. Calculations for Four-Point Approximation

x	z for span 1	z for span 2	Y	ŷ(z)
xL	-1.5		-1	
xL + dx	-0.5	-1.5	0	
xL + 5/4 dx	-0.25			(-.0703*-1)+(.8672*0)+(.2266*2)+(-.0234*3) = 0.4533
xL + 6/4 dx	0.0			(-.0625*-1)+(.5625*0)+(.5625*2)+(-.0625*3) = 1.0000
xL + 7/4 dx	0.25			(-.0234*-1)+(.2266*0)+(.8672*2)+(-.0703*3) = 1.5469
xL + 2 dx	0.5	-0.5	2	
xL + 9/4 dx		-0.25		(-.0703* 0)+(.8672*2)+(.2266*3)+(-.0234*3) = 2.3440
xL + 10/4 dx		0.0		(-.0625* 0)+(.5625*2)+(.5625*3)+(-.0625*3) = 2.6250
xL + 11/4 dx		0.25		(-.0234* 0)+(.2266*2)+(.8672*3)+(-.0703*3) = 2.8439
xL + 3 dx	1.5	0.5	3	
xL + 4 dx	1.5		3	

Table 1-2. Two-Dimensional Approximation for Figure 1-5.

Nodal Location	Nodal Value	Nodal Weights
4 corners	10	4 (10) (.0039062) = 0.1562
8 edges	10	8 (10) (-.0351562) = -2.8125
4 interior	20	4 (20) (.3164062) = 25.3125
		$\hat{y}$ = 22.6562

Table 1-3. Two-Dimensional Approximation for Figure 1-6.

Nodal Location	Nodal Value	Nodal Weights		
4 outer corners	5	4 (5) (.0001373)	=	.00275
8 outer edges	5	8 (5) (-.0011444)	=	-.04578
8 outer edges	5	8 (5) (.0068664)	=	.27466
4 inner corners	10	4 (10) (.0095367)	=	.38147
8 inner edges	10	8 (10) (-.0572204)	=	-4.57763
4 interior	20	4 (20) (.3433227)	=	27.46582
		$\hat{y}$	=	23.50129

Table 1-4. Expansion of Equation 1-47

Coefficient	Node
$C_1(z_3) \cdot C_1(z_2) \cdot C_1(z_1)$	$\cdot Y_{1,1,1}$
$C_1(z_3) \cdot C_1(z_2) \cdot C_2(z_1)$	$\cdot Y_{1,1,2}$
$C_1(z_3) \cdot C_2(z_2) \cdot C_1(z_1)$	$\cdot Y_{1,2,1}$
$C_1(z_3) \cdot C_2(z_2) \cdot C_2(z_1)$	$\cdot Y_{1,2,2}$
$C_2(z_3) \cdot C_1(z_2) \cdot C_1(z_1)$	$\cdot Y_{2,1,1}$
$C_2(z_3) \cdot C_1(z_2) \cdot C_2(z_1)$	$\cdot Y_{2,1,2}$
$C_2(z_3) \cdot C_2(z_2) \cdot C_1(z_1)$	$\cdot Y_{2,2,1}$
$C_2(z_3) \cdot C_2(z_2) \cdot C_2(z_1)$	$\cdot Y_{2,2,2}$

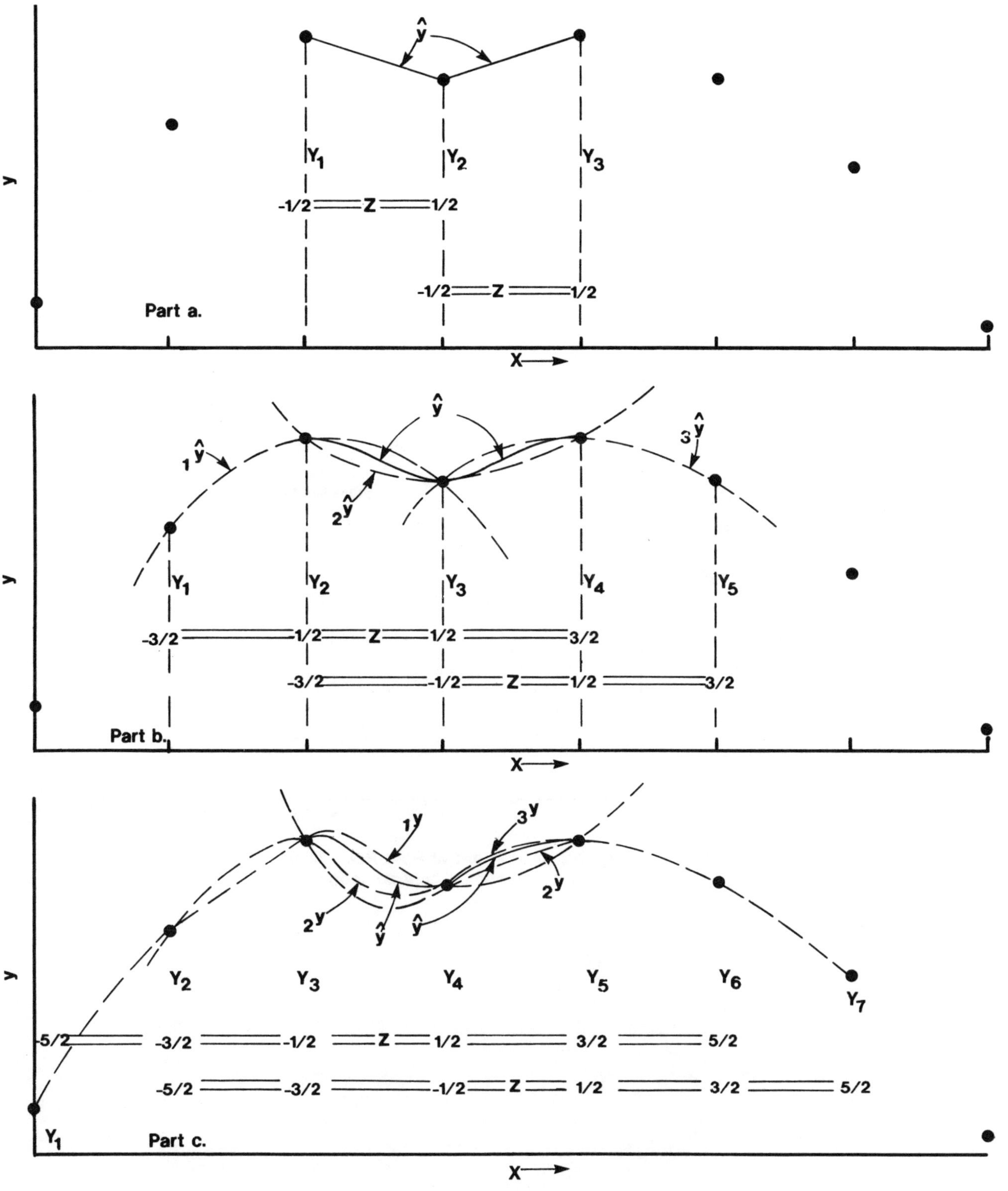

Fig. 1-1. Graphical Definition of Sliding Polynomials

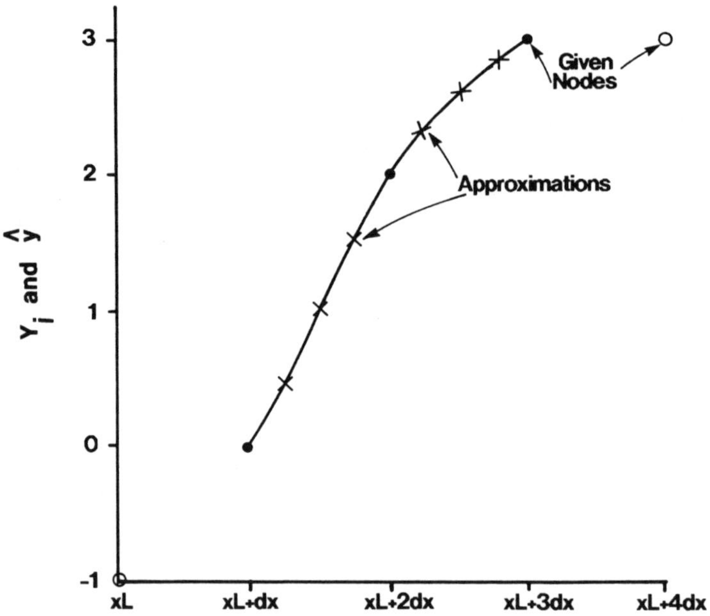

Fig. 1-2. Alignment of Four-Point Sliding Polynomial Approximations.

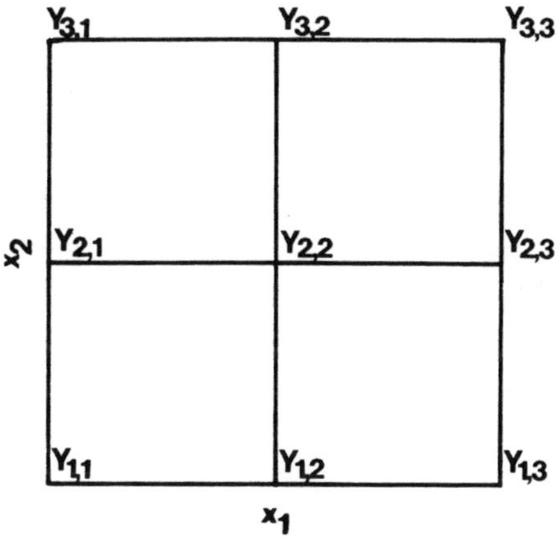

Fig. 1-3. Grid for Two-Dimensional Sliding Polynomials

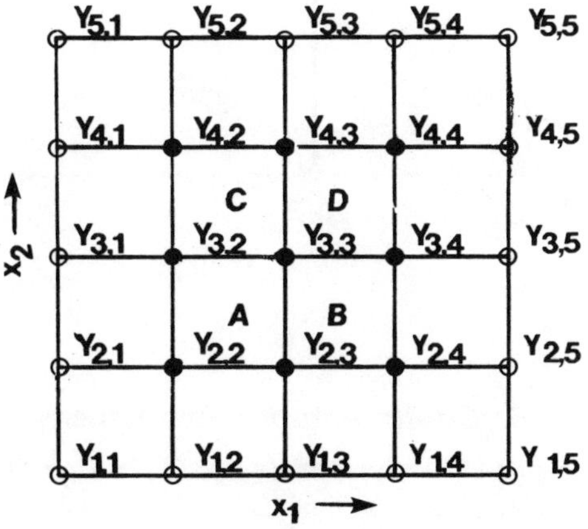

Fig. 1-4. Grid of Nodes for Two-Dimensional
Four-Point Sliding Polynomial Surface

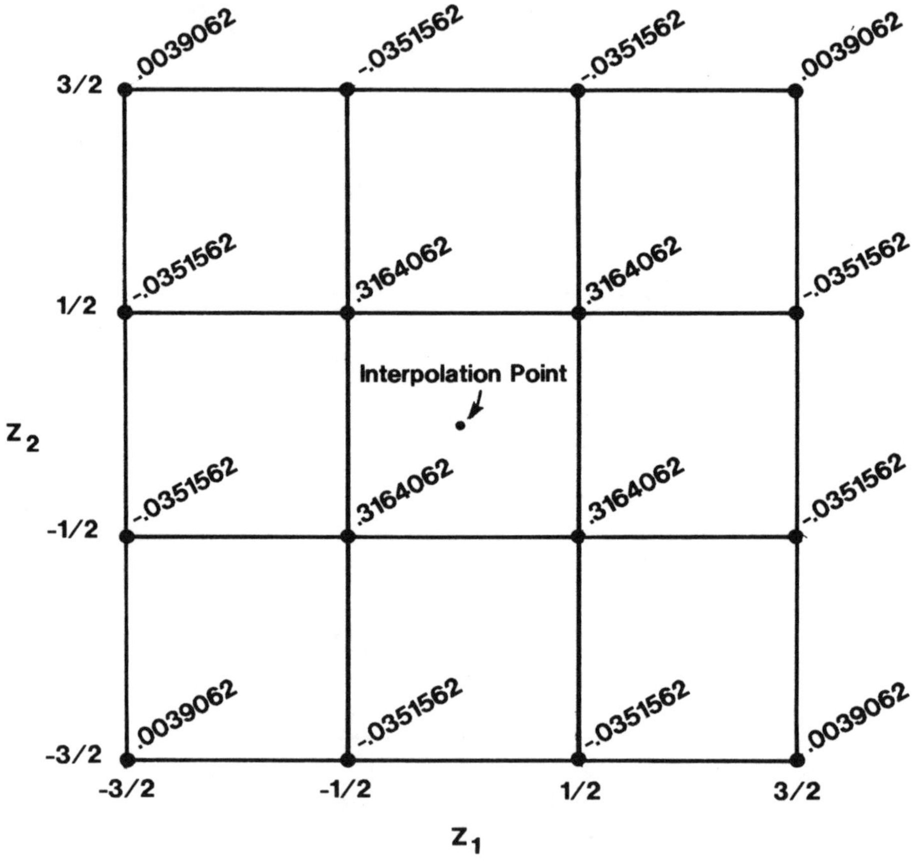

Fig. 1-5. Coefficients for Two-Dimensional Four-Point Sliding Polynomials for $Z_1 = Z_2 = 0$

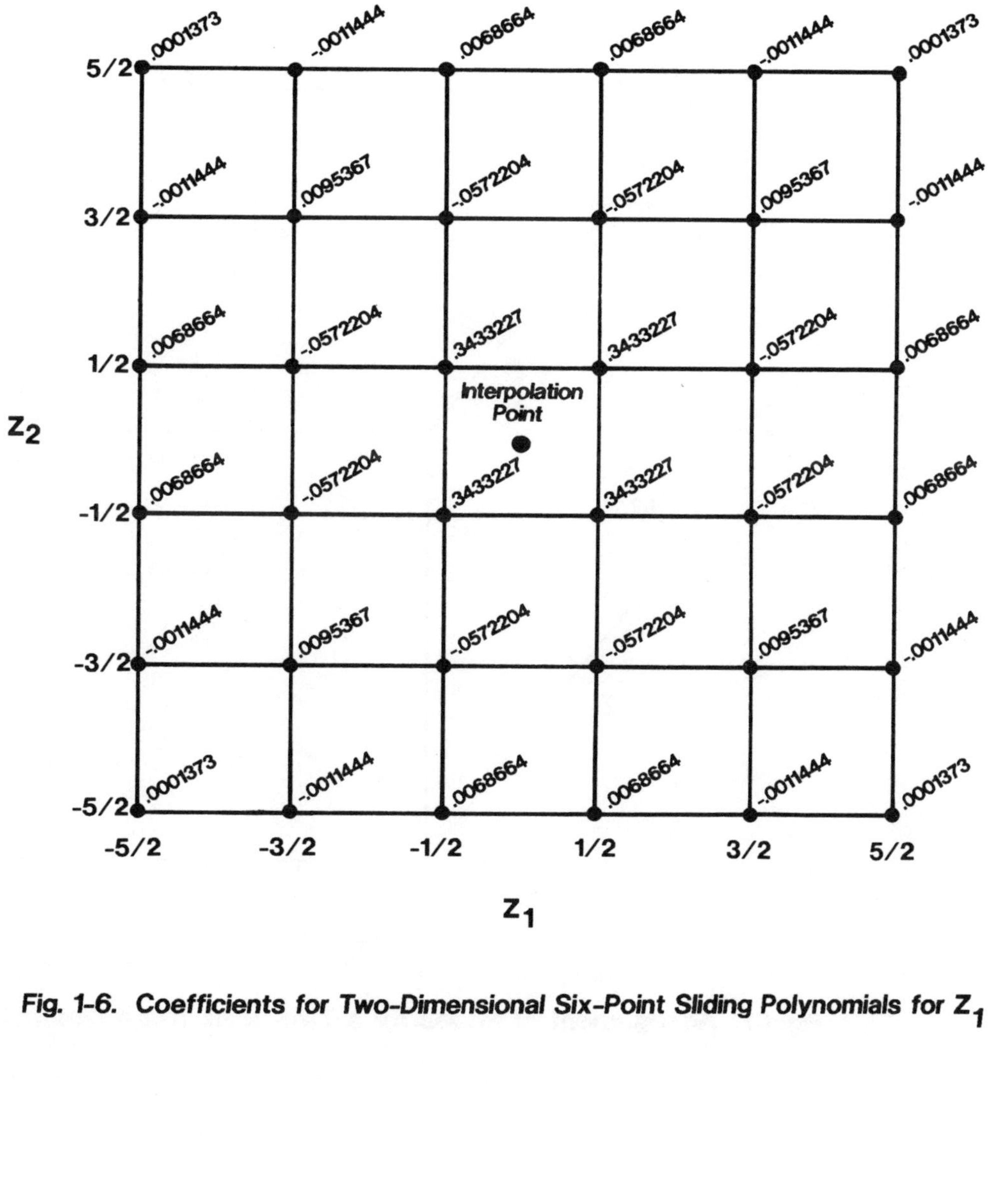

Fig. 1-6. Coefficients for Two-Dimensional Six-Point Sliding Polynomials for $Z_1 = Z_2 = 0$

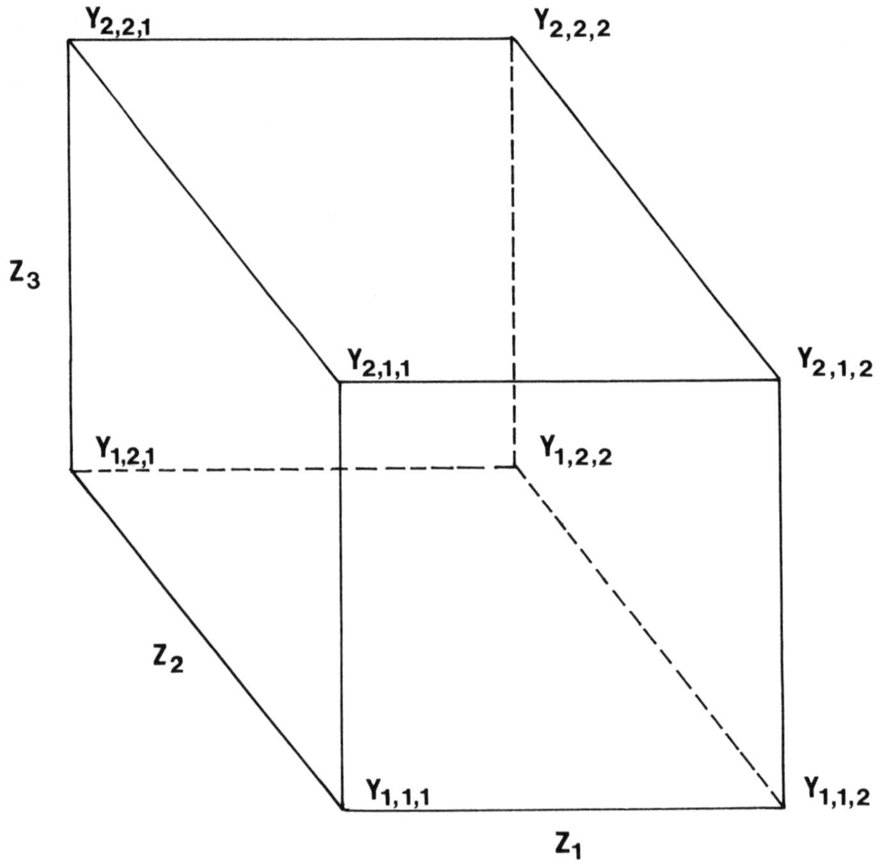

Fig. 1-7. Arrangement of Nodes for a Two-point Three-Dimensional System

CHAPTER 2

Interpolation and Contouring

2.1 INTRODUCTION

In Chapter 1 the development and structure of sliding polynomials were appraised. The objective in the development of these piecewise forms was for approximation when actual functions were unknown. Such calculation of approximate values is usually called interpolation.

Interpolation is an important task in solving many engineering problems and will serve as a helpful starting point for the more advanced numerical methods to be discussed in later chapters. Historically, interpolation was the starting point in the development of sliding polynomials, but it is also a fundamental concept in numerical analysis. While linear interpolation is the most frequently used method of interpolation, other common methods include the method of finite differences with Newton's equation, the Gregory-Newton method for equal intervals, and Lagrangian interpolation. The task of extrapolation is closely tied to the concept of interpolation.

Everybody interpolates sometimes. For example, suppose an automobile gets 18 miles per gallon at 60 miles per hour and 25 miles per gallon at 40 miles per hour. Then we should expect to get about 21.5 miles per gallon at 50 miles per hour:

$$18 + \left(\frac{60 - 50}{60 - 40}\right)(25 - 18) = 18 + \frac{1}{2}(7) = 21.5 \tag{2-1}$$

Also, we would expect to get 18.35 miles per gallon at 59 mph:

$$18 + \left(\frac{60 - 59}{60 - 40}\right)(25 - 18) = 18 + \frac{1}{20}(7) = 18.35 \tag{2-2}$$

In Eqs. 2-1 and 2-2, the values of 1/2 and 1/20 serve as weights that are a function of the location of the point for which an interpolated value is desired. For a value of 60 mph the weight would be zero, while for 40 mph the weight would equal 1.0. Of course, when we interpolate in this way, we assume that miles per gallon changes in a straight line relationship with the speed of the automobile. We know that the assumption of linearity is not correct, but we expect that the use of linear interpolation will provide a more accurate estimate than the use of either of the end-point values (i.e., 18 mpg or 25 mpg.).

Since in our simple linear example we had only two points, we could have used the two-point sliding polynomial Eq. 1-4 in the form:

$$M = C_1(z)\ M_1 + C_2(z)\ M_2 \tag{2-3}$$

where M_1 is 25 mpg (at 40 mph); M_2 is 18 mpg (at 60 mph); and M is miles per gallon at some other speed S between 40 and 60 mph. The $C_i(z)$ are defined in Eq. 1-3. To compute $C_1(z)$ and $C_2(z)$ a transform from S to z is needed. Such transforms are usually linear:

$$z = a + bS \tag{2-4}$$

Simple algebra gives a = $-SM/\Delta S$ and b = $1/\Delta S$ where ΔS is the speed difference between M_1 and M_2 (60 - 40 = 20 mph) and SM is the speed midway between M_1 and M_2 (50 mph). Then,

$$z = (S - SM)/\Delta S \tag{2-5a}$$

and

$$S = \Delta S\ z + SM \tag{2-5b}$$

At S = SM, z = 0 and both $C_1(z)$ and $C_2(z)$ are 1/2. M would be interpolated as 21.5 mpg, which agrees with Eq. 2-1. At S equal to 59 mph, z equals 9/20. Then $C_1(z)$ equals 1/20 and $C_2(z)$ equals 19/20. M is now interpolated as 18.35 mpg, which agrees with Eq. 2-2.

If we had a third point in our miles per gallon example, say 16 mpg at 70 mph, then we could also interpolate on a straight line between 60 and 70 mph. We would, in our total problem, have two straight-line pieces that join at 60 mph. We can use Eq. 2-3 to find values of M for the interval between 60 and 70 mph. At 61 mph, z equals (61-65)/10 = -2/5, $C_1(z)$ equals 9/10, and $C_2(z)$ equals 1/10. Then M = (9/10) 18 + (1/10) 16 = 17.8 mpg.

Between 40 and 60 mph linear interpolation indicates a rate of change in gas mileage of 0.35 mpg per mph. Between 60 and 70 mph the indicated change is 0.20 mpg per mph. There is no reason to believe that gas mileage should change abruptly at 60 mph. This is just a result of using linear interpolation. Remember that interpolation gives approximate values, and the approximations are the values that lie along the straight lines between 40 and 60 mph and between 60 and 70 mph. Linear interpolation is quick and easy. If the lack of precision in approximation is unsatisfactory for a particular problem, then the four-point or the six-point sliding polynomial system could be used.

2.2 COMPUTER PROGRAMS FOR INTERPOLATION

Some simple interpolation can be performed manually, but computerization of the method is necessary for any practical level of work. Also, single interpolations can be performed using the coefficient tables A-1.1 and A-1.2, which are especially useful when using hand calculators. But the manual calculation of many points is too tedious and error-prone; computers are required.

There are many problems where the approximate answer is not a single interpolated value between a set of nodes, but rather consists of a systematic set of interpolations closely spaced. When these closely spaced values are connected, they are a visual presentation of an approximating function. The total approximating functions may need to cover several piecewise spans between nodes. Particularly in multi-dimensional approximations, millions of calculations may be required.

Program B-2.1 can be used to construct four-point sliding polynomial functions through a set of nodes. Program B-2.2 can be used to construct six-point sliding polynomials. These programs differ only slightly. The student should compare them closely to see how they express repetitious calculations of Eqs. 1-18 and 1-35.

2.2.1 Example 2-1: Use of Program B-2.1

Program B-2.1 can be used to calculate and plot a continuous four-point sliding polynomial through the nodal values 0, 1, 3, 5, 5, 1, 0, 1. The first and last values are outlier nodes. The left-hand value of x at the nodal value of 1 is 0.6. The right-hand value of x at the nodal value of 0 is 1.0.

Program B-2.1 interpolates values at the eighth-points of each span. Values of x are also calculated at each of these points. The given nodes and the interpolated values are plotted in Fig. 2-1. and a smooth curve is drawn through them. Note that the total ensemble is a smooth approximate function. While each span between nodes is a cubic arc, the collection of all arcs has no simple algebraic form. However, since it is mathematically continuous in the first derivative, we can use it in most ways that we normally use formal functions. Later, we will see that we can integrate and differentiate this approximating ensemble of arcs.

2.2.2 Example 2-2: Use of Program B-2.2

Program B-2.2 can be used to calculate and plot a continuous six-point sliding polynomial through the nodal values of Example 2-1. The additional outlier nodes of -1 to the left of the original set and 5 to the right of the original set are added to the data set.

Program B-2.2 also calculates values at the eighth-points of each span. These values are also plotted in Fig. 2-1. The four-point and six-point interpolations differ noticeably at only two places. These are places of sharp curvature at the apex of the function and at x equal to 9.5. However, as we will see in Chapter 3, the derivatives of the two approximate functions would differ more noticeably.

2.3 OUTLIERS AND BOUNDARY CONDITIONS

Both the four-point and the six-point systems of sliding polynomials require nodes to the left and right beyond the problem boundary. When such nodes are unknown, they can be estimated from boundary conditions. If the outlier node is Y_1, Y_2 is on the boundary, and Y_3 is the first interior node, then the simplest condition to assume is that Y_1 lies on the straight line projected through Y_2 and Y_3. Simple algebra yields the relationship $Y_1 = 2Y_2 - Y_3$. If the approximating function is curving rapidly near the boundary, then it is better to assume that Y_1 lies on the parabola projected through Y_2, Y_3, and Y_4. For this condition it can be shown that $Y_1 = 3Y_2 - 3Y_3 + Y_4$. At the right boundary such a parabolic projection would give $Y_n = 3Y_{n-1} - 3Y_{n-2} + Y_{n-3}$.

An outlier node can also be calculated if the slope of the approximating function is known at the boundary. The slope of the four-point sliding polynomial at the boundary is $(Y_3 - Y_1)/2$. Therefore, $dy/dx = dy/bdz = (Y_3 - Y_1)/2b$ from which Y_1 can be calculated.

The effect of these boundary conditions is illustrated in Fig. 2-2. A series of nodes 5, 10, 7, 8, and 9 are assumed. The last value is the outlier node on the right. The left-hand outlier was computed three ways. In linear projection $Y_1 = 2(5)-10 = 0$. In parabolic projection $Y_1 = 3(5) - 3(10) + 7 = -8$. If $dx = 2dz$ and the slope is set to one, then $Y_1 = Y_3 - 4 = 6$.

Program B-2.1 was run three times to calculate the approximating function under the three different boundary conditions on the left. The three runs are plotted in Fig. 2-2. The parabolic projection with a left outlier node of -8 gives the steepest curve. The linear projection with the outlier node of zero rises less steeply. A set slope of 1.0 at x equal to 2 produces even less steepness. All three approximations are the same after the first span.

If there is no specific information on which to base a choice, then a parabolic projection is probably the best to use. We should be careful not to consider the requirement for outlier nodes as a problem or disadvantage. They should be regarded as positive attributes that can be used to control approximations as we approach our problem boundaries. We often have knowledge of how a problem solution must behave at the boundary, and the outliers are used to impose this extra knowledge. Boundary conditions will be discussed many times in later chapters.

2.4 CYCLIC INTERPOLATION

Cyclic functions may be thought of as curves plotted on the sides of cylinders. When joined smoothly, such functions have no beginning nor ending. Perhaps the best known cyclic functions are sines and cosines of angles. The seasonal march of climate also generates annual cyclic values of climatic variables. For example, monthly average temperature for a city is a cyclic function.

Sliding polynomials are excellent devices for constructing approximate cyclic functions because they can be made to overlap and thus eliminate boundaries. Also, because of their symmetrical structure of coefficients, they will cyclically repeat identical forms, though the forms themselves need not be symmetrical.

2.4.1 Example 2-3: Constructing a Six-point Cyclic Function

Program B-2.2 can be used to generate a cyclic function through the repeating nodes 70, 65, 32, 40, 70, 65, 32, 40, 70. Set an independent variable t to a value of 1 at the third node from the left and a value of 13 at the third node from the right.

The computer output is shown in Fig. 2-3. Two waves were plotted to show the smooth transition from the first to the second cycle. Note that, while the two waves are identical, the waves themselves are not symmetric about their vertical center lines.

2.5 REVERSE INTERPOLATION

In most problems, it is required that a value of y be estimated when a value of x is given. The sliding polynomials are constructed to calculate y on a curve passing through a given set of nodes Y_i when the unit variable z is known.

Problem situations may arise where a value of y is known and we need to know what value of x corresponds to this value of y. This problem cannot be solved directly. Some type of iterative procedure is necessary. The "interval-halving" method is frequently used. This method works as follows: Given that the value of x corresponding to y knowingly lies between a left value x_l and a right value x_r, the midpoint $x_m = (x_l + x_r)/2$ can be calculated, and then interpolate for y_l, y_m, and y_r for the three values of x. Now calculate $(y_l - y)$, $(y_m - y)$, and $(y_r - y)$ for the known value y. If $(y_l - y)$ and $(y_m - y)$ have the same algebraic sign, then the x corresponding to y lies between x_m and x_r. Alternatively, if the

$(y_m - y)$ and $(y_r - y)$ pair have the same sign, then x lies between x_l and x_m. The half-interval containing x now becomes the search interval and this interval is divided in the same manner. The two halves are examined to see which half contains x. The process continues until the interval $x_l < x < x_r$ becomes very small. Since the process is iterative, it is readily programmed for computer application.

Program B-2.3 is an elemental program to perform reverse interpolation. It can be used for specific simple problems but is probably of more value when adapted for use in larger software packages.

2.5.1 Example 2-4: Example of Reverse Interpolation

Given the four nodes (0,6.0), (2,4.0), (4,2.5) and (6,1.0), program B-2.3 can be used to calculate x corresponding to a y of 3.2. The successive trials are shown in Table 2-1. In order to illustrate the method, this example was started with a large initial range. Usually it is more efficient to narrow the initial range by linear interpolation and then use the iterative procedure for increased precision.

2.6 NON-UNIFORM SPACING OF NODES

The development of sliding polynomials was based on a uniform spacing of nodes in the scale of the local unitized variable z. A linear transform from z to the global variable x means that the nodes are also uniformly spaced in x. The uniform spacing in z was used because it produces a unique structure that requires only a one-time algebraic solution, and also it makes the solution algebraically simpler. However, there are situations where non-uniform spacing of nodes in the x-scale is necessary to enhance the solution to a problem.

Perhaps the most frequently encountered problem in interpolation is polynomial overswing, sometimes called polynomial bounce. Overswing occurs at places of rapid change of curvature in the approximating function. Often times, the function descends rapidly, then bends abruptly to a near-horizontal orientation. Polynomials tend to oscillate above and below the near-horizontal portion. Because of the restrained structure that is achieved by using subfunctions to enforce functional continuity at the nodes, sliding polynomials tend to have little overswing. When necessary, this small amount of overswing can be suppressed by non-uniform spacing of the nodes. Briefly, the nodes must be closely spaced through the zone of rapid change of curvature.

A typical example of polynomial overswing is shown in Fig. 2-4. A four-point sliding polynomial was passed through the nodes (12, 9.0), (14, 2.9), (16, 1.2), and (18, 1.1). A left outlier of 15.0, and a right outlier of 1.0 were used. Values for the curve were calculated using program B-2.1. In Fig. 2.4 it may be noted that the approximating curve reaches a minimum near x equal to 16.7, then rises slowly to the boundary node (18, 1.1). Such an approximation is unsatisfactory, particularly if we are interested in the slope of the function as it approaches the right boundary. In order to correct the problem in Fig. 2-4, the nodes must be more closely spaced around the maximum curvature at x equal to 15. Non-uniform spacing can be accomplished in a number of ways, but in order to preserve the continuity in the derivatives of the sliding polynomials, the method should be mathematically rigorous. A simple device is to insert a third variable v between global variable x and local variable z. Nodes will be uniformly spaced in z and v but not in x. The variable v is global.

A mathematically continuous transform from x to v is given by:

$$v = V - V \exp \frac{b(x-C)}{2} \qquad x \le C \tag{2-7a}$$

$$v = V \exp \frac{b(C-x)}{2} \qquad x > C \tag{2-7b}$$

Equations 2-7 form two exponentials joined at a common point C. Parameter V sets the scale between x and v, and parameter b shapes the exponentials.

An example is shown in Fig. 2-5 with C equal to 15, V equal to 10, and b equal to 0.5. The nodes are spaced uniformly in the v-scale at values of 8.876, 6.936, 4.996, 3.056, and 1.116. The mid-value of 4.996 transforms to an x of 15. By happenstance, the other values transform very closely to the x-values of the original four nodes. The placement of nodes in the x-scale, which correspond to the uniform spaced v-values above, are shown at the bottom of Fig. 2-5. Note that the nodal spacing in the center is about one-half the outer spacing.

A nodal value for x equal to 15 was read from the curve in Fig. 2-4. This was added to the original four-node set. Program B-2.1 was used to interpolate values between these nodes, which are uniform in v-scale. The x-values of the points of interpolation are then calculated from the uniform v-values using Eqs. 2-7. These (x, y) interpolations were then plotted in Fig. 2-4. There is no difference in the two curves from x equal to 12 to x equal to 15, but they do differ beyond x equal to 15. With the close spacing of nodes, the polynomial overswing is suppressed and the approximating curve drops smoothly to the right boundary value of (18, 1.1).

2.7 PROFILE DISCONTINUITIES

Sliding polynomials were developed based on the mathematical continuity of the sub-functions to which they are common and tangent. This continuity is usually the controlling criterion in problem solution, since we expect cause-and-effect mechanisms to function in a continuous manner. A different circumstance may arise, however, in working with geophysical or topographic data. Landforms are often rounded and smoothly continuous for normal working scales for, say, small watersheds. But a topographic map will show slope discontinuities where the elevational contour lines intersect major stream channels.

Discontinuities can be generated with sliding polynomials by using double sets of nodes. Each set is used to approximate a land profile, and the two profiles join but have a discontinuous derivative. The technique is illustrated in Fig. 2-6. On the left, a curve is interpolated on nodes (4, 10), (5, 10), (6, 9), (7, 7), (8, 3), and (9, -1). The first and last of these nodes are outliers. A right-hand curve is interpolated on nodes (7, -3), (8, 3), (9, 9), (10, 11), and (11, 11). Again the first and last nodes are outliers. Two runs of program B-2.1 were made using these two sets of nodes. The interpolated curves are common at node (8, 3). The composite of the two curves appears as a cross-section through a deeply incised valley.

One should note that this method of creating profile discontinuities, when necessary, depends very simply upon the placement of two "spurious outlier" points, one to either side of the discontinuity. While the example in Figure 2-6 was constructed using the four-point sliding polynomial, the six-point system could also be used if double "spurious outliers" are specified. Such outliers can usually be estimated by a linear or parabolic projection of the valley side slopes.

2.8 EXAMPLE 2-5: ESTIMATING A SNOW WATER DEPLETION CURVE

The data of Table 2-2 (Rango and Martinec, 1982) can be used to further illustrate the concepts of the previous sections. The zonal snow coverage, in percent, which is the dependent variable, is the percent of a watershed covered by snow at a particular point in time. The relationship between the zonal snow coverage and the accumulated degree-days is the snow water depletion curve. The accumulated degree-days is the summation from some point in time of the number of degrees above some threshold. For example, if the threshold is 2°C and the mean temperatures on five consecutive days are -1°C, 4°C, 2°C, 5°C, and 1°C, then the accumulated degree-days for the five days would be 0, 2, 2, 5, and 5.

The values of Table 2-2 were first plotted in Fig. 2-7. These are the solid dots in the figure. In order to construct an approximation curve through these points, the left and right boundary conditions must be set. A simple outlier point was added on the right. This is not a critical value since snow coverage is asymptotically approaching zero. At the left the point (50, 100) was considered the actual problem boundary. A discontinuity exists here because at zero accumulated degree-days the snow cover is also 100 per cent. Therefore, the first data point, (0, 100), was used as the left outlier and the interpolation begins at the second point (50, 100).

In order to interpolate snow cover at any degree-day value, a transformation from global variate degree-days to the local variate z is necessary. If we keep this transform linear (the data are not highly curved), then

$$z = a + bD \tag{2-8}$$

where D is the accumulated degree-days and a and b are the transform coefficients. For the first span, $50 < D < 100$, we can write the following for Equation 2-8:

$$-\tfrac{1}{2} = a + 50\,b$$

$$\tfrac{1}{2} = a + 100\,b$$

from which a = -3/2 and b = 1/50. For the second span, $100 \leq D \leq 150$, a = -5/2, b = 1/50, and for the third span, $150 \leq D \leq 200$, a = -7/2, b = 1/50. Now it is obvious that "a" of Eq. 2-8 changes by the span number and b is a constant 1/50. We can generalize this to

$$z = -\left(S + \frac{1}{2}\right) + \frac{D}{50} \tag{2-9}$$

where S is now the span number. As an example, for D = 225 accumulated degree-days, which is in the fourth span, z = -(4 + 1/2) + 225/50 = 0. From Table A-1 for z = 0, we find $C_1 = C_4 = -0.0625$, and $C_2 = C_3 = 0.5625$. Then the interpolated snow cover at this number of accumulated degree-days is

$$-0.625\ (63) + 0.5625\ (44) + 0.5625\ (33) - 0.0625\ (20) = 38.125\%$$

This single interpolation is plotted in Fig. 2-7.

Program B-2.1 was used to calculate the interpolated curve through the data nodes in Fig. 2-7. This curve agrees exactly with the single value interpolated at 225 accumulated degree-days. The approach to zero snow cover appears reasonable. The approach at the left boundary of 100 percent cover does not appear unreasonable, but there is really no way to judge the precision of the interpolated values without additional information on the boundary conditions set by the initial melting of snow.

2.9 TWO-DIMENSIONAL INTERPOLATION

Two-dimensional interpolation is necessary whenever we have a "double-entry" problem. This is illustrated in Table 2-3, where values of some property are given for four values of an argument x_1 and four values of an argument x_2. Suppose we want an estimate of the property at x_1 = 3.66 and x_2 = 2.18. We must interpolate on both arguments simultaneously. The unit variable for z_1 is 0.1 at 3.66 and the unit variable for z_2 is 0.3 at 2.18. For the two-point system Equation 1-3 gives the following coefficients:

$$C_1(z_1) = \frac{1}{2} - 0.1 = 0.4 \tag{2-10a}$$

$$C_2(z_1) = \frac{1}{2} + 0.1 = 0.6 \tag{2-10b}$$

$$C_1(z_2) = \frac{1}{2} - 0.3 = 0.2 \tag{2-10c}$$

$$C_2(z_2) = \frac{1}{2} + 0.3 = 0.8 \tag{2-10d}$$

Then using Eq. 1-41:

$$\hat{y} = (0.2\ (0.4)\ (2.23)) + (0.2\ (0.6)\ (4.17)) + (0.8\ (0.4)\ (3.37))$$
$$+ (0.8\ (0.6)\ (5.68)) = 4.473$$

The four-point sliding polynomial can also be used to estimate the value of the property at some interior point. If we estimate the value at the same point as above, then z_1 and z_2 remain the same. From Table A-1 we read directly the coefficients as follows:

	C_i			
	1	2	3	4
z_1:	-0.048	0.424	0.696	-0.072
z_2:	-0.016	0.168	0.912	-0.064

Using Eq. 1-45

$$y = (-0.016\ (-0.048)\ (0.96)) + (-0.016\ (\ 0.424)\ (1.76))$$
$$+ (-0.016\ (\ 0.696)\ (3.50)) + (-0.016\ (-0.072)\ (6.18))$$
$$+ (\ 0.168\ (-0.048)\ (1.63)) + (\ 0.168\ (\ 0.424)\ (2.23))$$
$$+ (\ 0.168\ (\ 0.696)\ (4.17)) + (\ 0.168\ (-0.072)\ (7.15))$$
$$+ (\ 0.912\ (-0.048)\ (2.40)) + (\ 0.912\ (\ 0.424)\ (3.37))$$
$$+ (\ 0.912\ (\ 0.696)\ (5.68)) + (\ 0.912\ (-0.072)\ (8.70))$$

+ (-0.064 (-0.048) (2.95)) + (-0.064 (0.424) (4.23))

+ (-0.064 (0.696) (6.42)) + (-0.064 (-0.072) (9.44))

= 4.388

This value is a better estimate of the property than the 4.473 obtained with the two-point system, because there is curvilinearity in the data in Table 2-3. However, this increased precision is gained only with added effort. Only four arithmetic products were calculated with the two-point system. If data were available, 36 products would have to be calculated to use the six-point system. In general, one would use the simplest system that would yield the precision necessary for the particular numerical problem.

These simple examples are intended only to illustrate the arithmetic effort required in any problem of multi-dimensional interpolation. Because of this effort, little manual, or single point, double interpolation is performed compared to one-dimensional interpolation. Multi-dimensional interpolation is usually programmed as a part of surface contouring, which will be discussed next.

As noted earlier, one-dimensional sliding polynomials yield interpolation curves. Two-dimensional sliding polynomials yield surfaces of the property under study. Curves or graphs are easy to display, as we can note in Figs. 2-6 and 2-7, for example. Surfaces can be displayed also, but the amount of work is much greater. Surfaces are usually displayed as contours that are "iso-lines" of the property. A temperature map shows iso-therms, or lines of equal temperature. A topographic map shows lines of equal elevation above some datum. In order to draw such contours a large number of double interpolations must be made and plotted.

Suppose we consider contouring a given grid of nodes using the six-point sliding polynomials. Now Eq. 1-46 shows that each interpolation requires 36 triple multiplications. Suppose that we will plan to draw the contours through interpolations at each tenth-point at the edges and interior of each unit surface. This requires 11 * 11, or 121, interpolated values. It also requires 36 * 121, or 4356, triple multiplications. If our problem space were a 3x3 ensemble of unit surfaces, then the total number of multiplications would be 9*4356, or 39,204. We see immediately that programmed computer calculation is the only feasible approach and that even a single manual double interpolation is cumbersome.

It is an advantage of the function-of-function construction of multi-dimensional sliding polynomials that computer programming of the concept is compact and efficient.

Some effort can be saved by using the four-point or two-point systems of interpolation. The penalty paid is in the loss of precision and discontinuities in the derivatives. Since a considerable portion of applied work must deal with slopes and gradients of properties, as well as the properties, the extra effort in using the six-point polynomial system is justified. Program B-2.4 was prepared to perform the multiple interpolations for six-point, two-dimensional contouring of a given grid of nodes. This program is useful as presented, but it can also be considered a model for simpler four-point or two-point sliding polynomial programs. Also, with modification it could serve as a sub-program for inclusion with larger software packages.

When hardware and software are available, the outputs from Program B-2.4 should be stored for mechanical plotting, either contouring or the more elaborate perspective drawings of the surfaces. Without this capability, the interpolated values must be hand-plotted and free-hand contoured.

2.9.1 Example 2-6: Simple Contouring Example

Program B-2.4 was used to perform the two-dimensional interpolation for contouring the surface defined by the grid of nodes in Table 2-4. The problem space consists of four unit areas outlined by the dashed lines in the center of the table. The surrounding values are outlier nodes that are necessary for the six-point system. All of the data in Table 2-4 are synthetic, selected only to illustrate contouring. In a real problem the outliers would be set by using boundary controls. The outliers are always only two deep, a 10 x 10 ensemble of unit surfaces would still have only two rows and columns of outliers.

Program B-2.4 performs double interpolation at the tenth-points of the unit surfaces. These values were entered on a 21x21 point grid in Figure 2-8. Then iso-lines, or contours, were drawn through these values. The corner values for the four unit surfaces should be checked against the nodal values in Table 2-4. Scales of the independent variable x_2, which is indexed as "j", were added to Fig. 2-8. The values of x_1 and x_2 for each interpolated value are calculated and given as output by Program B-2.4.

Note in Figure 2-8 that all contours run smoothly across all common boundaries of the four unit surfaces. Remember that each unit surface in the six-point sliding polynomials is a two-dimensional, fully interactive fifth-order polynomial (Equations 1-20 and 1-46). The sliding polynomials are essentially a means to assemble these complex unit surfaces into larger groupings with mathematical continuity across the boundaries. The ensemble of four unit surfaces in Figure 2-8 is essentially free-form. This ensemble could re-shape itself by changing one or more nodes in Table 2-4. Yet the free-form new ensemble would always be a continuous two-dimensional interactive expression of the property under study.

Figure 2-8 shows a shallow valley, the lowest point located just right of the center at the bottom of the figure. The valley floor rises as we move upward in the figure. The valley flanks rise more steeply to high points in the upper left and right corners.

2.9.2 Example 2-7: Contouring with a Discontinuity

In Section 2.7 the construction of discontinuities in curves was discussed. Figure 2-6 presented a profile that could be the discontinuity of land-slope in a valley cross-section at the stream channel. Discontinuities in two-dimensional surfaces can also be constructed. Such discontinuities occur by natural processes in land forms or they may result from cultural practices. Table 2-5 presents a grid of nodal values used with Program B-2.4 to construct a contour map with a significant discontinuity at one point, which gradually transforms into a surface with continuous form.

In Table 2-5 the grid of nodes covers a 3 x 2 ensemble of unit surfaces. Variable x_2, indexed by "j", is the dimension with three unit surfaces. For index values of 1, 2, and 3, two separate rows of nodes are shown. The rows of nodes with index 1a, 2a, and 3a are used for the left hand column of three unit surfaces. The rows of nodes with index 1b, 2b, and 3b are used for the right-hand column of three unit surfaces. Program B-2.4 was run twice, once for the left-hand 3 x 1 ensemble of surfaces, and once for the right-hand 3 x 1 ensemble. The two runs will have no discontinuity of the surface values since nodal values for index "i" equal to 4 have equal values in the runs. The derivatives will be discontinuous until nodal values of index "j" equal to 3 drop from the calculations.

However, the differential discontinuity will diminish with the diminishing weights of the sliding polynomial coefficients.

The 3 x 2 ensemble of unit surfaces with contours is shown in Fig. 2-9. The discontinuity of differentials is clearly evidenced by the kinking of the contours on the boundary between the two columns of unit surfaces. The kinking is strong for low values of x_2, but fades and disappears in the top set of two surfaces. The overall pattern in Fig. 2-9 is similar to that in Fig. 2-8, an upward sloping valley. The difference is the deliberately constructed discontinuity at the lower end of the valley, which might represent the head of an incised stream.

2.10 THREE-DIMENSIONAL INTERPOLATION

The function-of-functions expansion of sliding polynomials to three dimensions was presented in Eqs. 1-47 and 1-48. These equations are compact and lead to compact and efficient computer programs. The required number of multiplications to estimate one value of a property by triple interpolation is too great for most manual effort.

Triple interpolation is necessary if we need to estimate the value of a property inside a cube, as sketched in Fig. 1-7. As mentioned, estimation of air temperature in a room serves as an example. One great difficulty is how to display the results of such approximating calculations. It cannot be done by single curves or contours, as in one- or two-dimensional problems. One method of display is to present a series of two-dimensional contour maps. Each such map shows a contour pattern in say x_1 and x_2 for a selected value of third independent variable x_3. We must visualize the total series as a display of the effect of x_3 on the contour patterns in x_1 and x_2.

Program B-2.5 was prepared for triple interpolation. This program uses the four-point sliding polynomial. Values of the property are approximated at the eighth-points in x_1 and x_2, and at the third-points in x_3. Outputs are organized for plotting a series of four contour maps, one for each value of x_3 by thirds.

2.10.1 Example 2-8: Illustration of Three Dimensional Interpolation

Program B-2.5 can be used to illustrate the display of three-dimensional contouring. Data used in the illustration are given as four 4 x 4 sets of nodal values in Table 2-6.

The interpolation will take place in the unit cube, defined at the bottom, where k is equal to 2, by the four nodes enclosed by the dashed square, and at the top, where k is equal to 3, also by four nodes enclosed by a dash square.

The results of interpolation using program B-2.5 are shown in Fig. 2-10. Parts 1 to 4 of the figure show contour maps at the bottom, thirds, and top of the cube. At the bottom, Part 1, the contours are concave toward the bottom of the figure. At the one-third level, Part 2, the contours are still shallowly concave toward the bottom of the figure. At the two-thirds level, Part 3, the contours are convex downward, and at the top, Part 4, they have changed to a shallow circular pattern.

This three-dimensional interpolation is thus displayed as contours in the two-dimensional space of variables x and w, with the patterning of the contours changing with the variable, v. This simple example covers only one unit cube. However, Program B-2.5 allows for up to three unit cubes in each dimension and this number could be increased.

EXERCISES

2-1. Assume that runoff (Q) equals 0.0833 inch for a rainfall (P) of 1 inch and that Q equals 0.5625 inch for P equal to 2 inches. Develop the equation to transform rainfall to the standardized variable z. Also develop the two-point sliding polynomial for estimating Q. Determine the coefficients $C_1(z)$ and $C_2(z)$ for values of P for 1 to 2 at an increment of 0.20.

2-2. The following table gives the runoff (Q, inches) and rainfall (P, inches):

Q | 0	0.083	0.5625	1.25	2.042	2.893
P | 0	1.0	2.0	3.0	4.0	5.0

Using the first and last points as the outlier nodes, use program B-2.1 to estimate the runoff over the range $1 \leq P \leq 4$. Plot the Q versus P relationship.

2-3. The following table gives the discharge rate (q, cfs) for storm times (t, hours):

q	334	647	1010	628	217	147	123	104	86
t	11.9	12.0	12.1	12.2	12.3	12.4	12.5	12.6	12.7

Using the first and last points as the outlier nodes, use program B-2.1 to estimate the discharge rate with the four-point polynomial over the range $12.1 \leq t \leq 12.5$. Plot the q versus t relationship.

2-4. Using the data from problem 2-3 and program B-2.2, estimate the discharge rate with the six-point polynomial over the range $12.1 \leq t \leq 12.5$. Plot the q versus t relationship (if problem 2-3 was solved, compare the two plots).

2-5. Using program B-2.1, estimate the value of sin θ for values of θ in the range $30° \leq \theta \leq 35°$, with an increment of 1°. Use values of θ of 29° and 36° as the outlier nodes. Estimate the maximum error from the true value.

2-6. The following table gives the station number (s, feet) and the corresponding elevation of the channel cross-section (y, feet)

y	25	22	21	22	19	18	15	14	17	19	21	23	25
x	0	10	20	30	40	50	60	70	80	90	100	110	120

Using program B-2.1, estimate the cross-section elevations in the range $0 \leq x \leq 120$. Obtain values for the outlier nodes for each of the following methods: (a) horizontal extension from the end-point node; (b) linear extension of the two end points; (c) vertical extension form the end-point node. Evaluate the effect of the three outlier methods on the estimated cross-section elevations in the intervals $0 \leq x \leq 10$ and $110 \leq x \leq 120$.

2-7. The following table gives a three-year record of the mean monthly temperature (°F) for Jan., Mar., May, July, Sept., and Nov. at the Wildcat Ck. near Lawrenceville, GA:

year	Jan	Mar	May	July	Sept	Nov
1961	38.9	54.4	64.7	76.0	72.3	54.7
1962	40.6	47.5	74.0	78.7	70.4	49.2
1963	37.3	54.7	66.8	76.2	72.0	52.5

Use program B-2.2 to generate a cyclic function through the nodes. Plot the three waves to compare the cycles across a 12-month period. Estimate values for the outlying nodes using the mean values for the appropriate months.

2-8. The following table gives one-year records of precipitation P(inches) and runoff R(inches) for the Chattooga River near Clayton, GA:

date	11/54	12/54	1/55	2/55	3/55	4/55	5/55	6/55
P	0.54	8.41	2.68	8.97	4.63	6.70	8.52	5.26
R	0.84	2.05	1.70	3.54	3.08	4.48	5.07	2.88

date	7/55	8/55	9/55	10/55	11/55	12/55	1/56	2/56
P	9.53	3.39	2.90	1.44	2.80	1.95	2.23	12.77
R	2.96	2.38	1.26	1.19	1.04	1.02	0.86	4.20

Use program B-2.2 to generate cyclic function for the 1955 precipitation and runoff. Plot the two waves to compare the cycles across the 12-month period.

2-9. The following data give the standard normal deviate z and the corresponding probability P for the cumulative normal distribution:

z	1.00	1.02	1.04	1.06	1.08	1.10
P(z)	0.8413	0.8461	0.8508	0.8554	0.8599	0.8643

The cumulative probability is usually estimated for a given value of z. Using the reverse interpolation program B-2.3, estimate the value of z for a probability of 0.8525.

2-10. The following table gives temperature values over a 4x5 grid of nodes. Use Program B-2.4 to contour this grid. Draw isotherms for each 5 degrees.

N-S	E-W	Temp.	N-S	E-W	Temp.
3	2	82	6	2	37
3	3	82	6	3	40
3	4	81	6	4	48
3	5	77	6	5	51
4	2	59	7	2	30
4	3	61	7	3	35
4	4	70	7	4	42
4	5	69	7	5	45
5	2	42			
5	3	49			
5	4	60			
5	5	61			

Table 2-1

Successive Values in Reverse Interpolation

Round	xl	xr	$\hat{y}m$
0	2.0	4.0	3.21875
1	3.0	4.0	2.863281
2	3.0	3.5	3.040527
3	3.0	3.25	3.129333
4	3.0	3.125	3.173943
5	3.0	3.0625	3.196319
6	3.0	3.03125	3.207527
7	3.015625	3.03125	3.201921
8	3.023438	3.03125	3.199119
9	3.023438	3.027344	3.200520
10	3.025391	3.027344	3.199820

Table 2-2

Depletion Curves of Snow Coverage vs. Accumulated Degree Days in Elevation Zone C of the South Fork Basin in 1979

Zonal Snow Coverage (%)	Accumulated Degree-Days (°C)
100	0
100	50
91	100
63	150
44	200
33	250
20	300
14	350
9	400
6	450
4	500

Table 2-3

Data for Two-Dimensional Interpolation

		x_1			
		3.5	3.6	3.7	3.8
	2.0	0.96	1.76	3.50	6.18
	2.1	1.63	2.23	4.17	7.15
x_2	2.2	2.40	3.37	5.68	8.70
	2.3	2.95	4.23	6.42	9.44

Table 2-4

The Grid of Nodal Values $Y_{j,i}$ for Example 2-6

INDEX i

INDEX j	1	2	3	4	5	6	7
7	11	11	11	11	11	11	11
6	13	13	13	13	13	13	13
5	14	15	15	12	14	13	12
4	15	13	11.5	10	11	11	10
3	14	12	10	9	9.5	11	12
2	6	6	7	8	7	6	6
1	5	4	3	2	2	3	4

Table 2-5

The Grid of Nodal Values $Y_{j,i}$ for Example 2-7

INDEX j \ INDEX i	1	2	3	4	5	6	7
8	65	60	55	50	55	60	65
7	60	55	50	45	50	55	60
6	50	45	40	35	37	38	39
5	37	36	34	30	32	32.5	33
4	32	31	30	27	28	28.5	29
3a	29	28	27	25	21	15	
3b		20	23	25	26	26.5	27
2a	25	26	26	24	22	18	
2b		18	22	24	25	25	24
1a	24	25	24	23	22	21	
1b		18	21	23	24	25	24

Table 2-6

Nodal Values $Y_{k,j,i}$ for Example 2-8

INDEX K = 1

INDEX j \ INDEX i	1	2	3	4
4	1	2	2	1
3	2	3	3	2
2	2	3	3	2
1	1	2	2	1

INDEX K = 2

INDEX j \ INDEX i	1	2	3	4
4	1	2	2	1
3	2	3	3	2
2	3	4	4	3
1	2	3	3	2

INDEX K = 3

INDEX j \ INDEX i	1	2	3	4
4	4	3	3	4
3	3	2	2	3
2	3	2	2	3
1	4	3	3	4

INDEX K = 3

INDEX j \ INDEX i	1	2	3	4
4	3	2	2	3
3	2	1	1	2
2	2	1	1	2
1	3	2	2	3

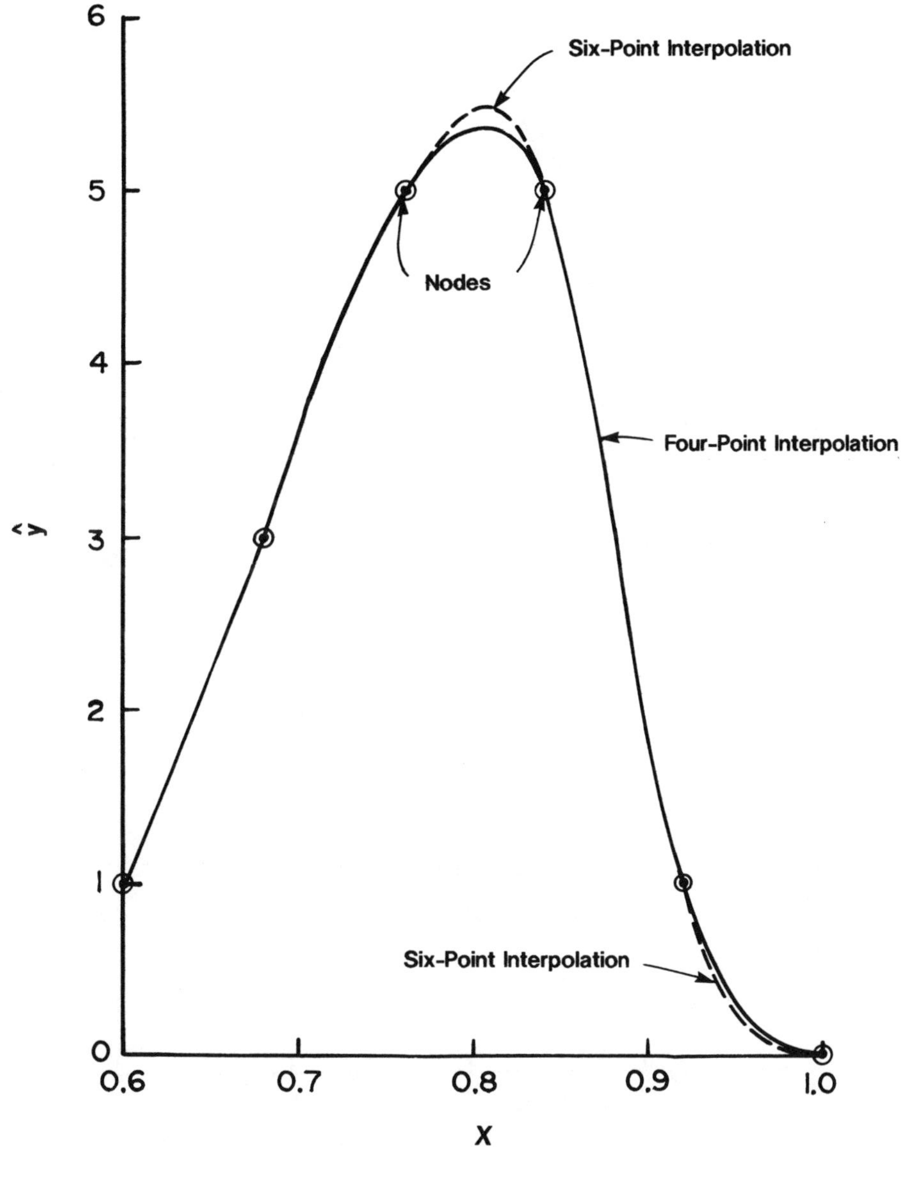

Fig. 2-1. Comparison of Four-Point and Six-Point Interpolations

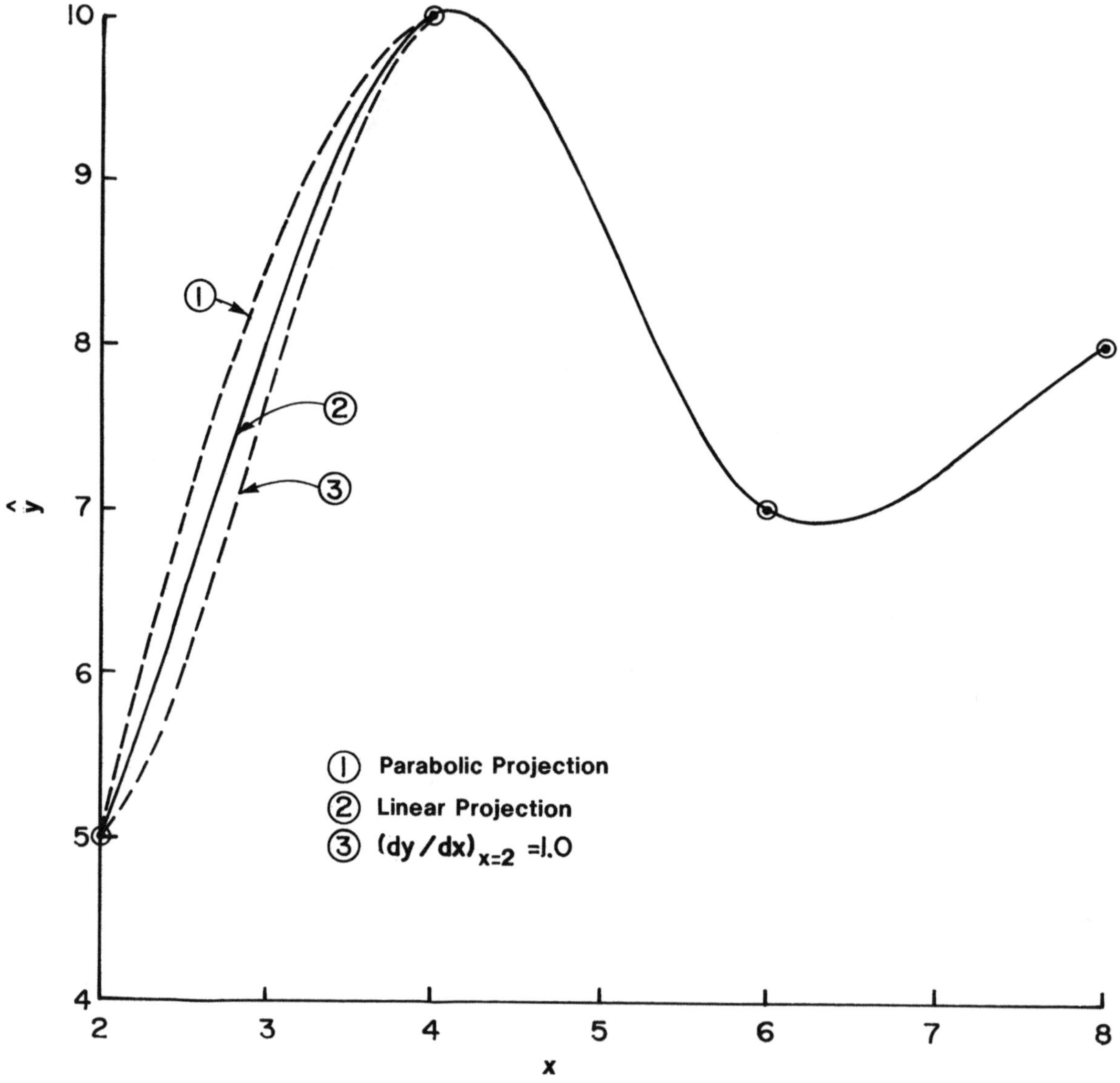

Fig. 2-2. Effect of Boundary Condition

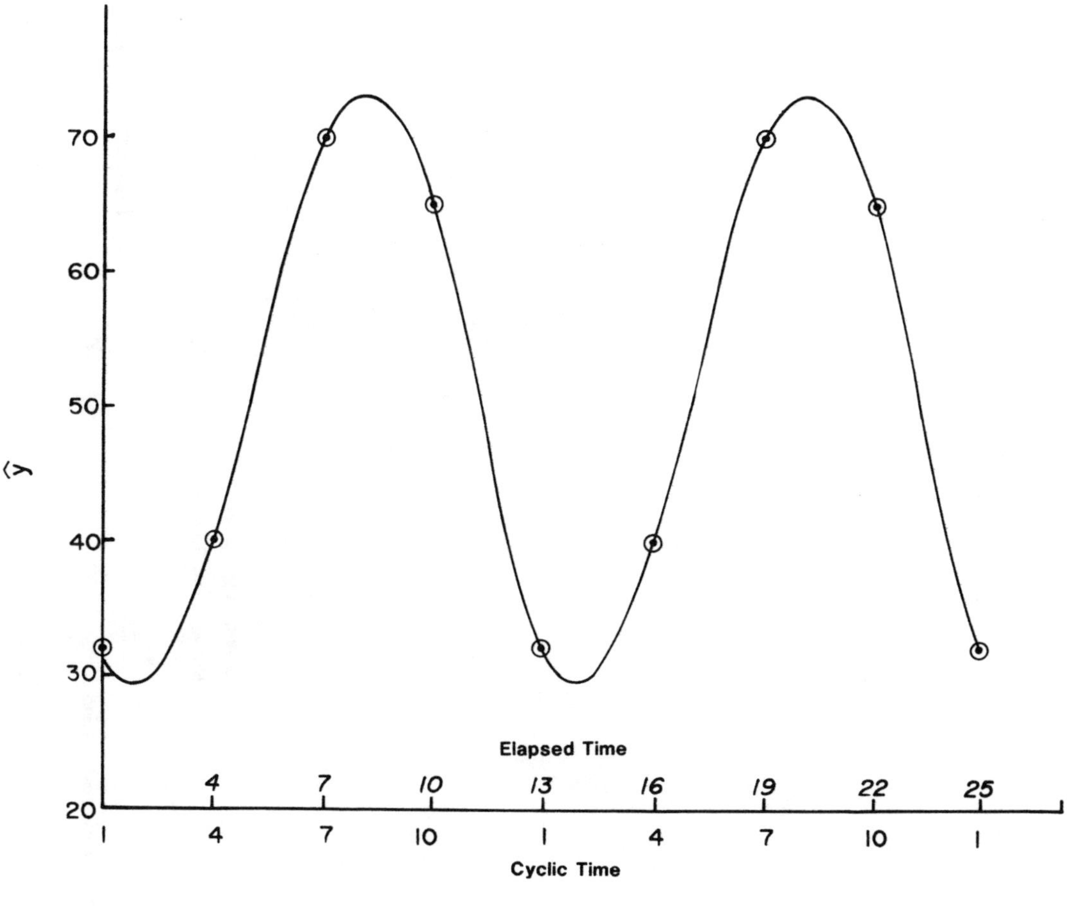

Fig. 2-3. A Cyclic Sliding Polynomial

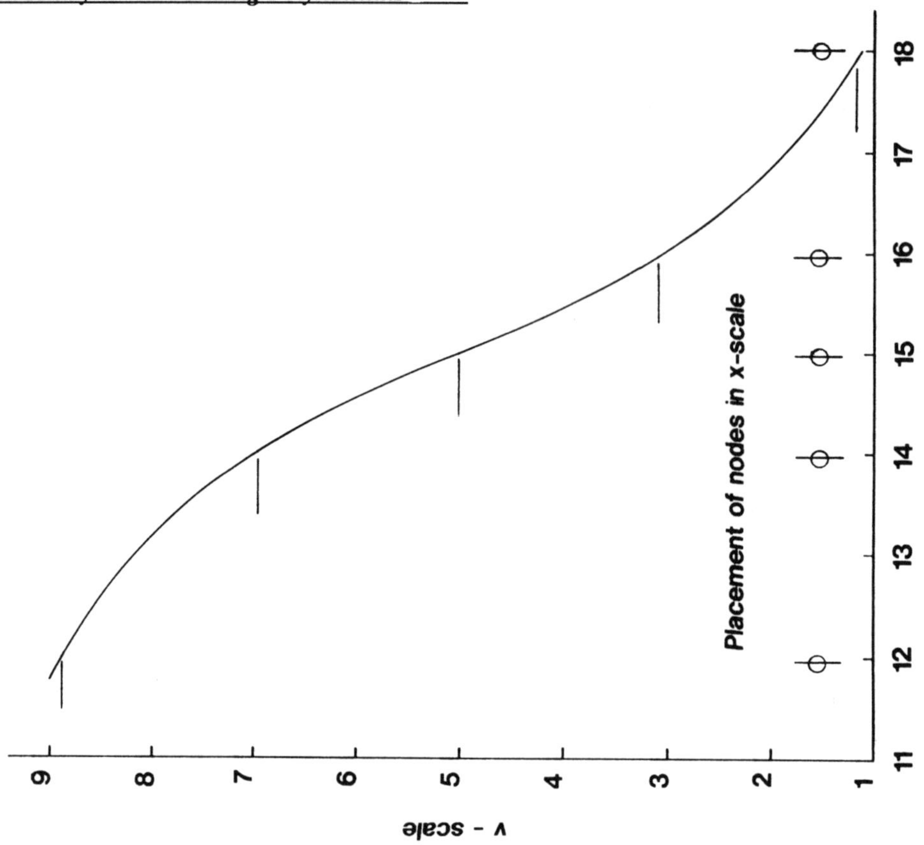

Fig. 2-5. Spacing of Nodes With X–V Transform

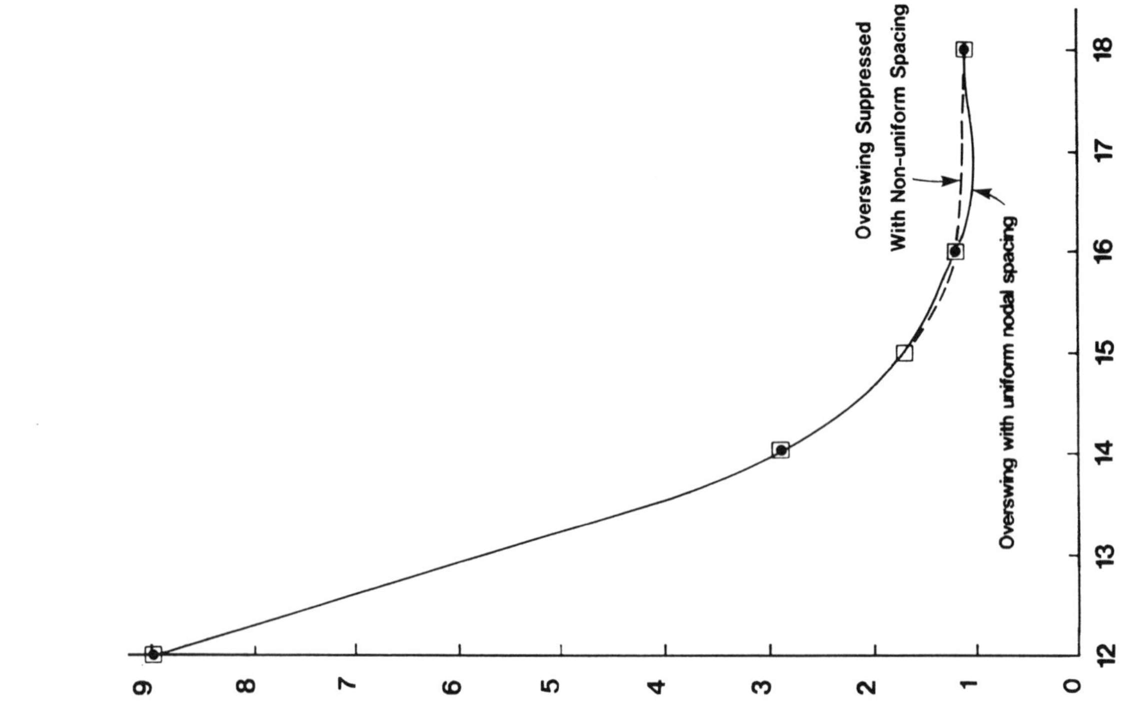

Fig. 2-4. Correction of Polynomial Overswing

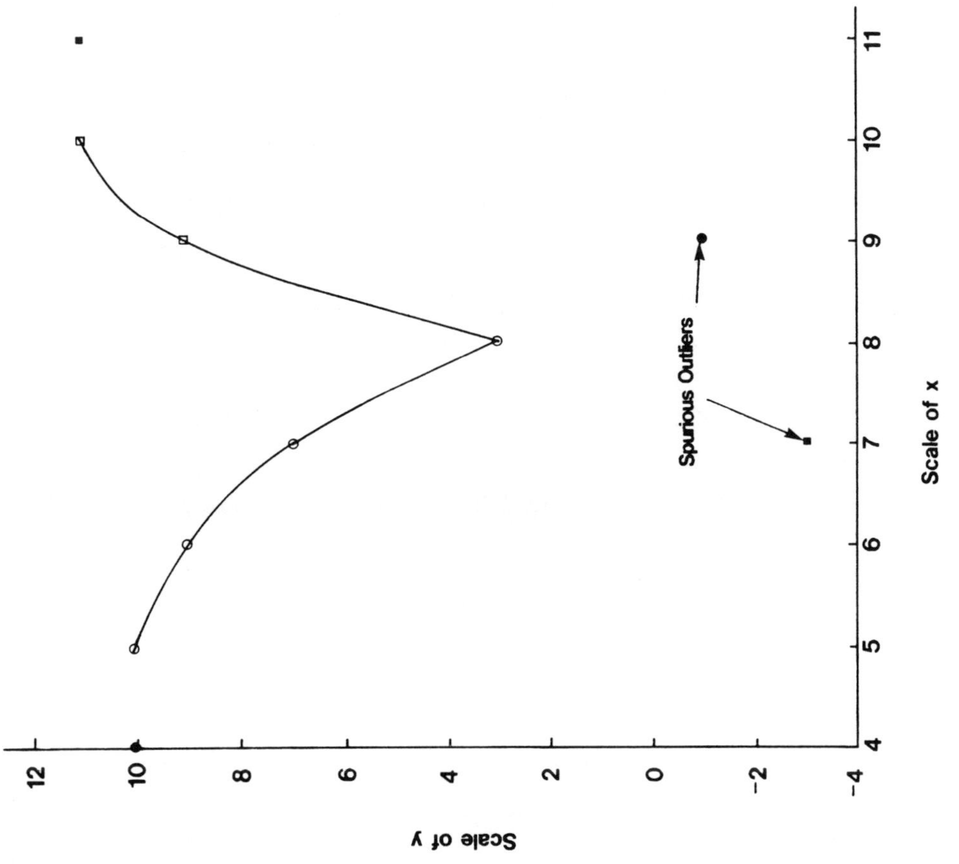

Fig. 2-6. Example of Profile Discontinuity

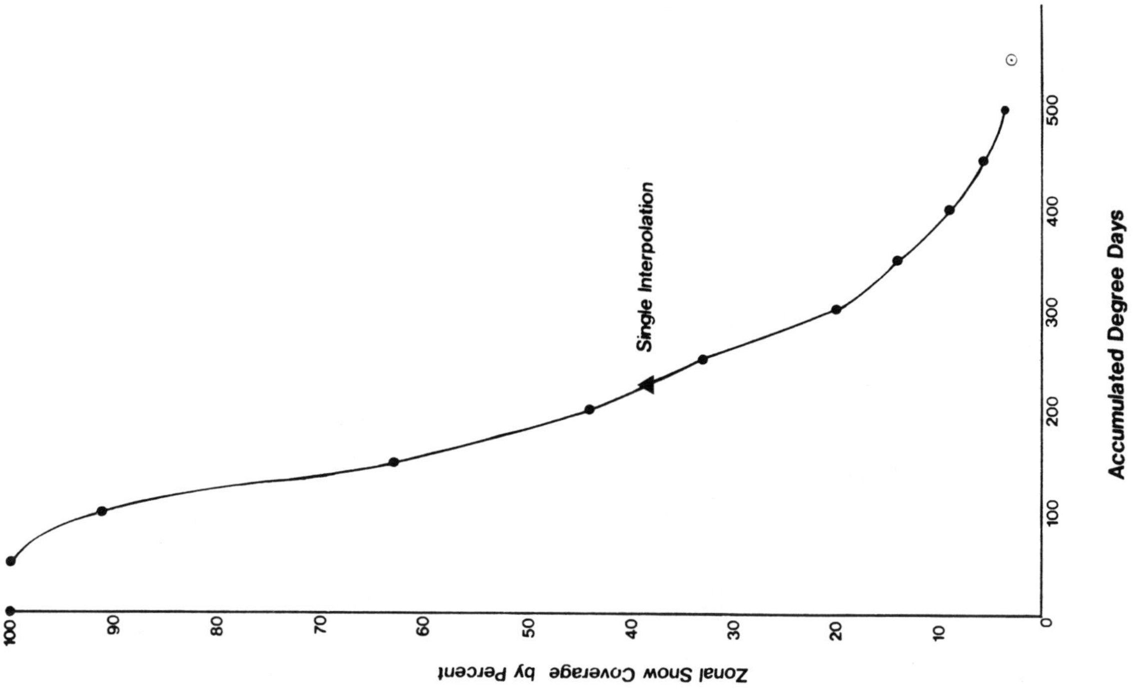

Fig. 2-7. Zonal Cover vs. Accumulated Degree Days

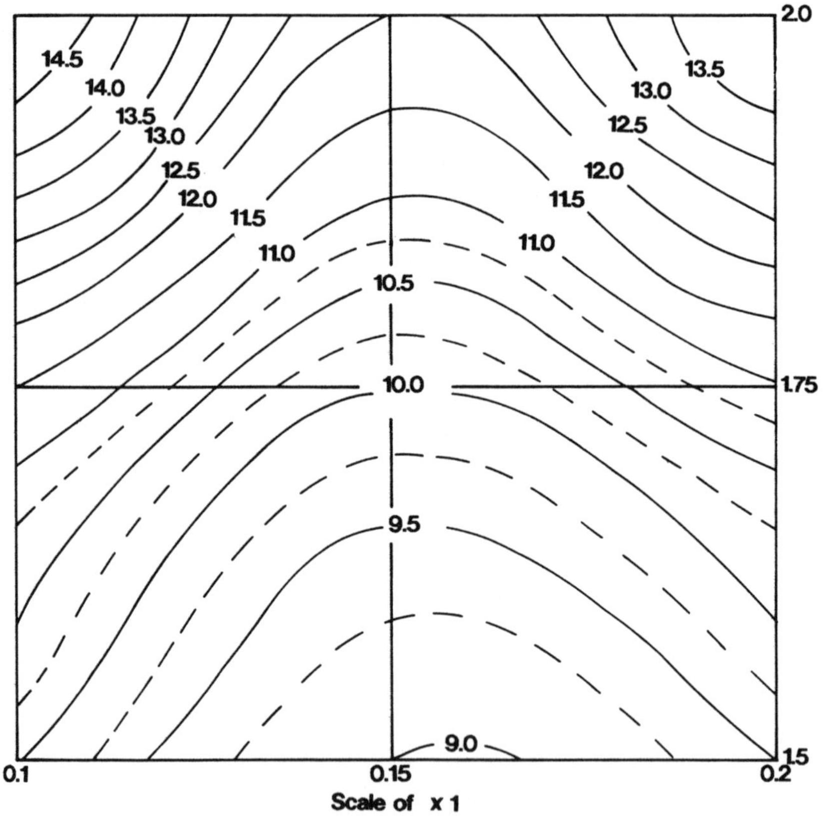

Fig. 2-8. Contours for Example 2-6

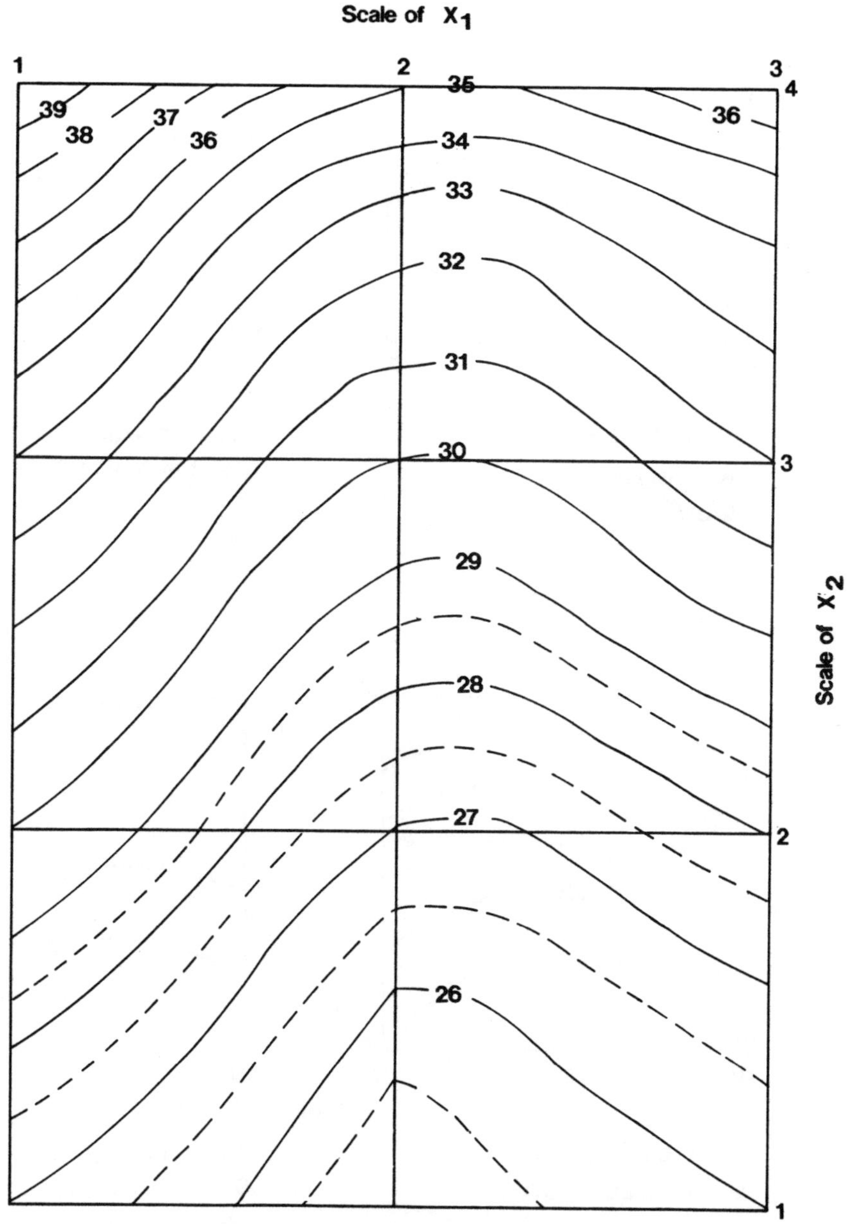

Fig. 2-9. Contours for Example 2-7.

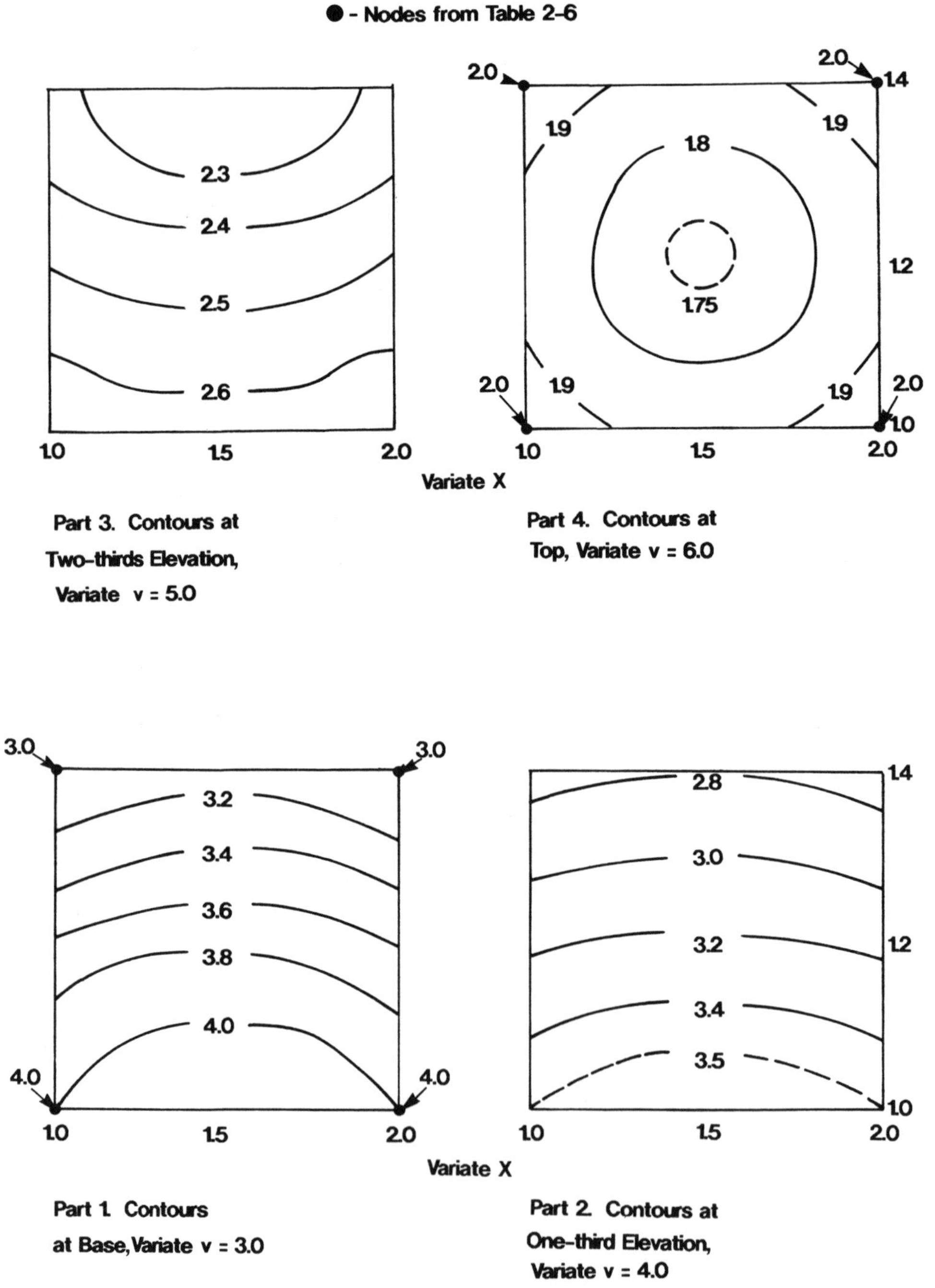

Fig. 2-10. Contours for Example 2-8

CHAPTER 3
Numerical Differentiation

3.1 INTRODUCTION

The material in Chapter 2 should not suggest that the sliding polynomial method only applies to the linear interpolating polynomial form. In many cases, we need to derive gradients or accumulations of variables when we know the function that describes the relationship between variables. These processes require differentiation or integration, which involves changing the mathematical order of our information. That is, calculation of either gradients or accumulations indicates a need to change the mathematical order of our information away from the basic relationship by either differentiation or integration. Obviously, if a problem solution is dependent on a simple readily integrated or differentiated function, then one should proceed with the direct analytical integration or differentiation. But if only ordinates, or observations of some unknown process function, are available, then it is necessary to use some approximating function.

The approximation of the functional y(x) by ŷ(x), which was the subject of Chapter 2, should be regarded as only one of several possible levels in the mathematical hierarchy. More completely, we should visualize the following forms:

$$\int_a^b \hat{y}(x) \approx \int_a^b y(x) \tag{3-1a}$$

$$\hat{y}(x) \approx y(x) \tag{3-1b}$$

$$d\hat{y}(x) \approx dy(x) \tag{3-1c}$$

$$d^2\hat{y}(x) \approx d^2y(x) \tag{3-1d}$$

We usually consider Eq. 3-1b to be the desired level in problem solution. However, if we need rates or gradients, we must move from Eq. 3-1b to Eq. 3-1c by differentiation. If we need rates of change of rates, or accelerations, we must move from Eq. 3-1b to Eq. 3-1d by performing a second differentiation. Such differentiation is the subject of this chapter. Integration, or accumulation, requires a movement in the hierarchy from Eq. 3-1b up to Eq. 3-1a. Integration will be the subject of Chapter 4.

There are problem situations where we may have information at the levels of Eqs. 3-1a or 3-1c. Our problem then consists of ways to move to the desired level of Eq. 3-1b. This requires solution of integral or differential equations. Discussion of such changes in the mathematical hierarchy is reserved mainly to Chapters 7 and 8, but will have some treatment in Chapters 5 and 6.

3.2 DIFFERENTIAL SLIDING POLYNOMIALS

In Chapters 1 and 2 the approximation $\hat{y}(x)$ of Eq. 3-1b was expanded as:

$$\hat{y}(x) = \hat{y}(z) = \sum_{i=1}^{n} C_i(z)\, Y_i \tag{3-2}$$

The $C_i(z)$ are the sliding polynomial coefficients stated in terms of the local variate z, and the Y_i are the nodal values, with n being 2, 4, or 6 depending upon which system of sliding polynomials is specified. Differentiating Eq. 3-2 yields:

$$\frac{d\hat{y}(x)}{dx} = \left(\frac{1}{b}\right)\frac{d\hat{y}(z)}{dz} = \frac{1}{b}\sum_{i=1}^{n}\frac{dC_i(z)}{dz}\, Y_i + \frac{1}{b}\sum_{i=1}^{n} C_i(z)\,\frac{dY_i}{dz} \tag{3-3}$$

Now the nodes are not functions of z; therefore, $dYi/dz = 0$ and,

$$\frac{d\hat{y}(x)}{dx} = \left(\frac{1}{b}\right)\frac{d\hat{y}(z)}{dz} = \frac{1}{b}\sum_{i=1}^{n}\frac{dC_i(z)}{dz}\, Y_i \tag{3-4}$$

The parameter b results from the linear transformation from global variate x to unit local variate z by x = a + bz. The coefficient b can be evaluated as dx/dz = $\Delta x/\Delta z = \Delta x$ since Δz is unity, with Δx being the increment of the independent variable between nodes. The second derivative of Eq. 3-2 is:

$$\frac{d^2\hat{y}(x)}{dx^2} = \left(\frac{1}{b^2}\right)\frac{d^2\hat{y}(z)}{dz^2} = \frac{1}{b^2}\sum_{i=1}^{n}\frac{d^2C_i(z)}{dz^2}\,Y_i \qquad (3\text{-}5)$$

In summary, note that the Y_i of Eqs. 3-2 through 3-5 are identical. In other words, given a set of nodes Y_i, we can interpolate on this set using the $C_i(z)$ of Eq. 3-2. We can construct a first derivative, or slope, by operating on the Y_i with the $dC_i(z)/dz$ of Eq. 3-4. We can construct a second derivative by operating on the same Y_i by the coefficients $d^2C_i(z)/dz^2$ of Eq. 3-5.

3.2.1 Differential Sliding Polynomial Coefficients

For the two-point sliding polynomial, the equations for $C_i(z)$ are given in Eqs. 1-3. Then

$$\frac{dC_1(z)}{dz} = -1 \qquad (3\text{-}6a)$$

$$\frac{dC_2(z)}{dz} = 1 \qquad (3\text{-}6b)$$

For any pair of nodes

$$\left(\frac{1}{b}\right)\frac{d\hat{y}(z)}{dz} = \frac{1}{b}\sum_{i=1}^{2}\frac{dC_i(z)}{dz}\,Y_i = \frac{1}{b}\,(Y_{i+1} - Y_i) \qquad (3\text{-}7)$$

This simply says that the slope of the straight-line segment is a constant. For the next segment, we apply

$$\left(\frac{1}{b}\right)\frac{d\hat{y}(z)}{dz} = \frac{1}{b}\left(Y_{i+2} - Y_{i+1}\right) \tag{3-8}$$

This slope is also a constant and is different from the first segment, except for the special case of the three nodes lying on a straight line. Since the first derivatives are constants, the second derivatives are always zero.

The equations for the $C_i(z)$ of the four-point sliding polynomials are given in Eqs. 1-18. The first derivatives of these coefficients are:

$$\frac{dC_1(z)}{dz} = \frac{(-12z+4)z+1}{8} \tag{3-9a}$$

$$\frac{dC_2(z)}{dz} = \frac{(36z-4)z-11}{8} \tag{3-9b}$$

$$\frac{dC_3(z)}{dz} = \frac{(-36z-4)z+11}{8} \tag{3-9c}$$

$$\frac{dC_4(z)}{dz} = \frac{(12z+4)z-1}{8} \tag{3-9d}$$

The second derivatives of the coefficients are:

$$\frac{d^2C_4(z)}{dz^2} = \frac{1-6z}{2} \tag{3-10a}$$

$$\frac{d^2C_3(z)}{dz^2} = \frac{-1+18z}{2} \tag{3-10b}$$

$$\frac{d^2C_2(z)}{dz^2} = \frac{-1-18z}{2} \tag{3-10c}$$

$$\frac{d^2C_1(z)}{dz^2} = \frac{1+6z}{2} \tag{3-10d}$$

These values can be used directly in Eqs. 3-4 or 3-5 to compute first or second derivatives of the approximating function through a set of nodes. It should be noted that an ensemble of first derivatives across multiple spans will "kink" at the nodes since the second derivatives are not continuous at the nodes.

Equations for the six-point sliding polynomial are given in Equations 1-35. The first derivatives of the coefficients are:

$$\frac{dC_1(z)}{dz} = \frac{(((-400z+32)z+216)z-40)z-13}{384} \tag{3-11a}$$

$$\frac{dC_2(z)}{dz} = \frac{(((2000z-96)z-1272)z+312)z+81}{384} \tag{3-11b}$$

$$\frac{dC_3(z)}{dz} = \frac{(((-4000z+64)z+2736)z-272)z-562}{384} \tag{3-11c}$$

$$\frac{dC_4(z)}{dz} = \frac{(((4000z+64)z-2736)z-272)z+562}{384} \tag{3-11d}$$

$$\frac{dC_5(z)}{dz} = \frac{(((-2000z-96)z+1272)z+312)z-81}{384} \tag{3-11e}$$

$$\frac{dC_6(z)}{dz} = \frac{(((400z+32)z-216)z-40)z+13}{384} \tag{3-11f}$$

The second derivatives of the six-point coefficients are:

$$\frac{d^2C_1(z)}{dz^2} = \frac{((-200z+12)z+54)z-5}{48} \tag{3-12a}$$

$$\frac{d^2C_2(z)}{dz^2} = \frac{((1000z-36)z-318)z+39}{48} \tag{3-12b}$$

$$\frac{d^2C_3(z)}{dz^2} = \frac{((-2000z+24)z+684)z-34}{48} \tag{3-12c}$$

$$\frac{d^2C_4(z)}{dz^2} = \frac{((2000z+24)z-684)z-34}{48} \tag{3-12d}$$

$$\frac{d^2C_5(z)}{dz^2} = \frac{((-1000z-36)z+318)z+39}{48} \tag{3-12e}$$

$$\frac{d^2C_6(z)}{dz^2} = \frac{((200z+12)z-54)z-5}{48} \tag{3-12f}$$

These values can also be used directly in Eqs. 3-4 or 3-5 to evaluate first or second derivatives of the approximating function through a set of nodes. The six-point sliding polynomials have a smooth first derivative through the nodes because the second derivative is continuous. In this system it is the second derivatives that "kink" at the nodes.

Since the six-point system has a smoothly continuous first derivative, it is recommended for most problems where calculation of slopes or gradients is necessary. Table A-3.1 in the appendix gives values of the derivatives of Eqs. 3-11 for each one-hundredth of the interval $-0.5 \leq z \leq 0.5$. These values can be used for manual calculation where only a few values of slopes are needed. For any extensive work a computer program is required. Program B-3.1 was prepared to approximate the function and first and second derivatives through a set of nodes.

3.2.2 Example 3-1. Calculation of Four-Point Derivatives

Given the set of values

x:	-4	-2	0	2	4	6	8	10	12	14	16
Y:	15	11	7.5	4.7	3.0	1.8	1.3	1.1	1.05	1.03	1.01

the four - point coefficients can be used to calculate the first and second derivatives at x = 2.2. For x equal to 2.2, the nodes for x at 0, 2, 4, and 6 are used, with the local variate $z = -0.5 + (x - x_2)/(x_3 - x_2) = -0.5 + (2.2 - 2)/(4 - 2) = -0.4$. Then Eqs. 3-9 give the values $C_1' = -0.315$, $C_2' = -0.455$. $C_3' = 0.855$, and $C_4' = -0.085$. Then using Eq.3-4, we get the first derivative at $x = 2.2$:

$$\frac{d\hat{y}}{dx} = \frac{1}{2}\,[(-0.315)\cdot(7.5) + (-0.455)\cdot(4.7) + (0.855)\cdot(3.0) + (-0.085)\cdot(1.8)]$$

$$= -1.0445$$

Then following Eq. 3-5:

$$\frac{d^2\hat{y}}{dx^2} = \frac{1}{4}\,[(1.7)\cdot(7.5) + (-4.1)\cdot(4.7) + (3.1)\cdot(3.0) + (-0.7)\cdot(1.8)] = 0.38$$

3.2.3 Example 3-2. Calculation of Six-Point Derivatives

The data of Example 3-1 can be used to calculate the first and second derivatives at $x = 2.2$ with the six-point sliding polynomial coefficients. The value of the unit variate z remains -0.4. For the first derivative the values of the coefficients can be read directly from Table A-1.3. Then we calculate directly

$$\frac{d\hat{y}}{dx} = \frac{1}{2}\,[0.06581\ (11) + (-0.49473)\cdot(7.5) + (-0.31754)\cdot(4.7) + (0.86288)\cdot(3.0)$$

$$+ (-0.12327)\cdot(1.8) + (\ 0.00685)\cdot(1.3)] = -1.0517$$

For the second derivative the coefficients can be evaluated using Eqs. 3-12. These values are then used in the calculation:

$$\frac{d^2\hat{y}}{dx^2} = \frac{1}{4}\,[(-0.2475\cdot 11) + (\ 2.0092\cdot 7.5) + (-3.6617\cdot 4.7) + (2.4050\cdot 3.0)$$

$$+ (-0.6242\cdot 1.8) + (\ 0.1192\cdot 1.3)] = 0.346$$

Note that values of $d\hat{y}/dx$ and $d^2\hat{y}/dx^2$ are in reasonable agreement with the values found using the four-point coefficients in Example 3-1.

3.2.4 Example 3-3. Comparison of Two-, Four-, and Six-Point Approximations

In this example, $\hat{y}$, $d\hat{y}/dx$, and $d^2\hat{y}/dx^2$ are calculated for the range $0 \leq x \leq 12$ of the data in Example 3-1 using two-, four-, and six-point systems.

Values of $\hat{y}$ are shown in Figure 3-1a. Approximations of $\hat{y}$ for the four-point and six-point systems cannot be distinguished. The straight-line segments of the two-point system are clearly seen.

Values of $d\hat{y}/dx$ for the three systems are shown in Fig. 3-1b. The two-point system appears as a series of constants, which are plotted as bars for each span. The four-point system is a series of parabolas computed with Eqs. 3-9 and 3-4. The parabolas are joined at the nodes, but "kinking" occurs at most nodes. The six-point system is a series of quartic arcs given by Eqs. 3-11 and 3-4. In contrast to the four-point system, these arcs are smoothly continuous across the nodes. We should note that values from all three systems are in general agreement in magnitude. The differences lie in the smoothness caused by the degree of mathematical continuity.

The values of $d^2\hat{y}/dx^2$ are plotted in Fig. 3-1c. Since there is no second derivative for the two-point system, only the four-point and six-point second derivatives are shown. The four-point second derivatives are discontinuous straight lines computed with Eqs. 3-10 and 3-5. The six-point second derivatives are connected cubic arcs given by Eqs. 3-12 and 3-5. Note again that the magnitudes computed with the four-point and six-point systems are in general agreement. It is the continuity of the second derivative in the six-point system that causes the smooth continuity of its first derivative.

If values of the second derivative are critical in the analysis of a particular problem, the "kinking" at the nodes should be removed by smoothing. This can be done "by eye" and can be done more rigorously by methods that will be presented in Chapter 5. We should keep a sense of proportion concerning the second derivatives in Fig. 3-1c. If these are differentials of the differentials in Fig. 3-1b, then it follows that Fig. 3-1b represents integrals of Fig. 3-1c. Even though the differentials in Fig. 3-1c are quite different, their integrals in Fig. 3-1b are not very different. This is characteristic of numerical analysis. Differentiation exaggerates differences, integration smooths over differences. Numerical integration will be discussed in depth in Chapter 4.

3.2.5 Example 3-4. Further Comparison of Two-, Four-, and Six-Point Approximation

The problem in this example is to calculate $\hat{y}$, $d\hat{y}/dx$, and $d^2\hat{y}/dx^2$ for the range $1.0 \leq x \leq 3.0$ for the following data:

x:	0	0.5	1.0	1.5	2.0	2.5	3.0	3.5	4.0
Y:	3.0	4.0	6.0	9.0	9.7	9.8	9.2	8.2	7.0

Values of the Y_i and the approximating function are plotted in Figure 3-2a. Again, there is very little difference between the four- and six-point approximations, but the two-point approximations are very different, particularly across the sharp bend in the range $1.5 \leq x \leq 2.0$. The difference between the function in Fig. 3-1a and Fig. 3-2a should be noted. The former is a smoothly transforming function with no abrupt changes in curvature. The latter shows a sharp change of curvature at about $x = 1.7$ and again near $x = 2.4$.

The slopes of the approximating functions for the three systems are shown in Fig. 3-2b. In general magnitude, all slopes are in agreement, but the two-point system cannot show any detail. The four-point and the six-point systems both place the maximum slope near $x = 1.2$ and both show an abrupt change of slope near $x = 1.8$. However, the four-point system has a strong cusp at $x = 2$, while the six-point system closely approximates the near-constant slope from $x = 1.9$ to $x = 2.2$ in Fig. 3-2a.

The second derivatives for the four-point and six-point systems in Fig. 3-2c are also in good agreement in general magnitude. The four-point system shows significant discontinuity at both $x = 1.5$ and $x = 2.0$. The six-point second derivatives are continuous but have pronounced cusps at these positions in x.

Lastly, we should again note that the six-point system with its continuous second derivative is the better system to use when numeric differentiation is necessary to get a slope or gradient.

3.2.6 Example 3-5: The Slope of the Saturation Vapor Pressure Curve

The slope of the saturation vapor pressure (P_v) curve with respect to temperature (T) is an input to some methods of estimating evaporation rates. Table 3-1 gives values of the P_v for selected values of T, with an increment of 10°F. To interpolate values in the range from 70°F to 80°F with the six-point sliding polynomial, the following interpolation formula can be used:

$$\hat{P}_v = \sum_{i=1}^{6} C_i(z)\ P_v \tag{3-13}$$

in which $\hat{P}_v$ is the estimated value of P_v; and the C_i can be obtained from Table A-2.2. For T = 75°F, z = 0 and the C_i values are: 0.01172, -0.09766, 0.58594, 0.58594, -0.09766 and 0.01172. Substituting these values into Eq. 3-13 yields $\hat{P}_v$ = 29.3245 mb. The true value is 29.32, which indicates an error of 0.00 mb for the six-point sliding polynomial. Linear interpolation gives a value of 29.70 mb, which is in error by 0.38 mb.

The slope of the P_v versus T curve at any point in the interval from 70°F to 80°F is:

$$\frac{d\hat{P}_v}{dT} = \frac{dP_v}{dz}\left(\frac{dz}{dt}\right) = \frac{1}{b}\sum_{i=1}^{6} dC_i(z)\ P_v \tag{3-14}$$

For T = 75°F the values of dC_i are obtained from Table A-3.1 as -0.03385, 0.21094, -1.46354, 1.46354, -0.21094, and 0.03385. Substituting these into Eq. 3-14 yields $(d\hat{P}_v/dT)_{T=75}$ = 0.9770 mb/°F. Linear interpolation would yield a value of 0.982 mb/°F.

3.3 PARTIAL DIFFERENTIAL SLIDING POLYNOMIALS

Equation 1-46 expressed a two-dimensional six-point sliding polynomial based on the function-of-functions concept. The equation can be differentiated with respect to either independent variable x_1 or x_2. Such derivatives are partial derivatives:

$$\frac{\partial \hat{y}}{\partial x_1} = \frac{1}{b_1}\frac{\partial \hat{y}}{\partial z_1} = \frac{1}{b_1}\sum_{j=1}^{6}\sum_{i=1}^{6} C_j(z_2)\ \frac{dC_i(z_1)}{dz_1}\ Y_{ji} \tag{3-15}$$

$$\frac{\partial \hat{y}}{\partial x_2} = \frac{1}{b_2}\frac{\partial \hat{y}}{\partial z_2} = \frac{1}{b_2}\sum_{j=1}^{6}\sum_{i=1}^{6} C_i(z_1)\ \frac{dC_j(z_2)}{dz_2}\ Y_{ji} \tag{3-16}$$

$$\frac{\partial^2 \hat{y}}{\partial x_1^2} = \frac{1}{b_1^2}\frac{\partial^2 \hat{y}}{\partial z_1^2} = \frac{1}{b_1^2}\sum_{j=1}^{6}\sum_{i=1}^{6} C_j(z_2)\,\frac{d^2C_i(z_1)}{dz_1^2}\,Y_{ji} \tag{3-17}$$

$$\frac{\partial^2 \hat{y}}{\partial x_2^2} = \frac{1}{b_2^2}\frac{\partial^2 \hat{y}}{\partial z_2^2} = \frac{1}{b_2^2}\sum_{j=1}^{6}\sum_{i=1}^{6} C_i(z_1)\,\frac{d^2C_j(z_2)}{dz_2^2}\,Y_{ji} \tag{3-18}$$

Program B-3.2 can be used to calculate the partial derivatives given in Equations 3-15 to 3-18. This program additionally contours the surface through the given set of nodes. Values are computed at the tenth-points of the unit surface for both surface and derivatives. The program uses the six-point sliding polynomial to produce smoothly continuous first derivatives.

3.4 DIRECTIONAL SLOPES

The partial derivatives of Eqs. 3-15 and 3-16 measure the slope of a surface parallel to the x_1 and x_2 axes, respectively. In many problems we need the slope across the contours. This slope measures the maximum gradient. We can think of this cross-contour slope as the slope that we would see in the natural landscape. This slope perpendicular to the contours is the vector sum of the two partial derivatives as shown in Fig. 3-3. If we assume the convention that the arrows point from high to low across the contours, then s_d in Fig. 3-3 is the "downhill" slope. To repeat, it is the steepest gradient we would see if we were standing at some point in a sloping field. The value of s_d can be computed by:

$$s_d = \sqrt{(\partial\hat{y}/x_1)^2 + (\partial\hat{y}/\partial x_2)^2} \tag{3-19}$$

While Equation 3-19 gives the magnitude of the downhill slope, it does not give the direction of that slope. If we maintain the convention of downhill direction, we need to know the direction, measured from north, of s_d. The calculation of this direction is shown in Figure 3-4. First, an angle can be defined:

$$\tan \theta = \text{abs}\left[\left(\frac{\partial \hat{y}}{\partial x_1}\right) \Big/ \left(\frac{\partial \hat{y}}{\partial x_2}\right)\right] \tag{3-20}$$

Then θ is the arctan ($\tan \theta$), with θ in radians. Now, depending upon whether $\partial\hat{y}/\partial x_1$ and $\partial\hat{y}/\partial x_2$ are positive or negative, θ can be converted to an aspect angle measured clockwise from north. In Fig. 3-4 the partial derivatives have been shown as s_1 and s_2 for brevity. If s_1 and s_2 are both positive, then the aspect angle is $\pi + \theta$. If s_1 is negative and s_2 is positive, then the aspect angle is $(\pi - \theta)$. For s_1 positive and s_2 negative, the angle is $(2\pi - \theta)$. For both s_1 and s_2 negative, the angle is θ. The calculation of s_d and the aspect angle of s_d are included in Program B-3.2.

3.4.1 Example 3-6: Illustration of Program B-3.2

The set of nodes in Table 3-2 can be used to calculate the surface, the directional slope, and the slope aspect. The problem space is the interior set of four unit surfaces.

The surface contours across the ensemble of four unit surfaces are shown in Fig. 3-5. The surface has a high knob in the upper center. This transforms to a shallow swale in the lower center. A faint valley extends up the right side, and convex curvature from the knob extends downward along the left side.

Contours of the magnitude of the downhill slope are shown in Figure 3-6. The slope must be zero at the top of the knob in the upper center. The slope increases to 0.7 at ($x_1 = 5$, $x_2 = 2.5$). There are very shallow slopes in the lower left of the surface. The slopes are given as dimensionless values. If we thought of natural land slopes, we would be measuring the drop in elevation in, say, feet, for each foot of distance along the downhill slope. In this case the value of 0.7 is very high. A more normal value for natural land slopes might be about 0.007, or a drop of seven feet in a thousand feet. If Figure 3-5 were a mathematical surface representing some physical process, then the slopes might be the rate of change of the process and the numerical values would be appropriate for that process. In a decay process, the dimensions would be the loss of mass per unit time, or gm/sec.

Figure 3-7 displays the direction downhill of the slopes in Fig. 3-6. These directions must also reflect the shape of the surface in Fig. 3-5. The arrows are the direction of the aspect angle at that point. The computer output is digital in radians but sufficient arrows were constructed to portray the pattern. The arrows point outward in all directions from the knob at top center. On the left they point down and out across the contours in Fig. 3-5. In the center they point mainly downward. At the right they point down and inward into the shallow valley. If Fig. 3-5 were a smooth natural surface and we sprinkled it with water, the arrows in Fig. 3-7 show the directions of the downhill flow as the water runs off. The slopes in Fig. 3-6 could be used to compute the velocity of the flow.

3.5 MULTI-DIMENSIONAL DERIVATIVES

Equation 1-48 was given as the generalized form for a sliding polynomial with three independent variables. This equation can be differentiated partially with respect to any or all of the three independent variables. For example, the following equation is the partial derivative with respect to x_3:

$$\frac{\partial \hat{y}}{\partial x_3} = \frac{\partial \hat{y}}{b_3 \partial z_3} = \frac{1}{b_3} \sum_{k=1}^{n_3} \sum_{j=1}^{n_2} \sum_{i=1}^{n_1} \frac{dC_k(z_3)}{\partial z_3} \cdot C_j(z_2)\ C_i(z_i)\ Y_{k,j,i} \qquad (3\text{-}21)$$

A mixed partial derivative with respect to x_3 and x_1 is:

$$\frac{\partial^2 \hat{y}}{\partial x_3 \partial x_1} = \frac{\partial^2 \hat{y}}{b_3 b_1 \partial z_3 \partial z_1} = \frac{1}{b_3 b_1} \sum_{k=1}^{n_3} \sum_{j=1}^{n_2} \sum_{i=1}^{n_1} \frac{dC_k(z_3)}{dz_3}\ C_j(z_2) \cdot \frac{dC_i(z_1)}{dz_1}\ Y_{k,j,i} \qquad (3\text{-}22)$$

Many other forms are possible. These multi-dimensional forms have not yet had much use in problem application, but if required, they can be readily and efficiently programmed.

EXERCISES

3-1. Derive the first and second derivatives of Eqs. 3-9 and 3-10 for the four-point sliding polynomial.

3-2. Derive the first and second derivatives of Eqs. 3-11 and 3-12 for the six-point sliding polynomial.

3-3 Evaluate values of the interpolating coefficients for the four-point system at x = 6.6 with the data of Example 3-1 (section 3.2.2) for both the first and second derivatives. Then evaluate the derivatives at x = 6.6.

3-4 Evaluate values of the interpolating coefficients for the six-point system at x = 6.6 with the data of Example 3-1 (section 3.2.2) for both the first and second derivatives. Then evaluate the derivatives at x = 6.6.

3-5. Using the data of Example 3-1 (section 3.2.2), compare the first and second derivatives at x = 9 with the four-point and six-point systems.

3-6 Using the data of Example 3-1 (section 3.2.2) show that the first derivative computed with the four-point system kinks at the node for x = 2. Plot the value of the second derivative across the node.

3-7 Using the four-point system, evaluate the slope of the saturation vapor pressure curve in the range from 80°F to 90°F. Compare the results with linear interpolation at 82°, 84°, 86°, and 88°. The data are given inn Table 3-1.

3-8 The data from the following tabulation are vlaues of a statisitcal objective function F for a model having two unknown coefficients, b_o and b_1:

	b_o					
b_1	1.0	1.2	1.4	1.6	1.8	2.0
5.3	3.72	3.48	3.40	3.38	3.48	3.63
5.4	3.65	3.42	3.29	3.28	3.37	3.46
5.5	3.61	3.37	3.24	3.23	3.31	3.54
5.6	3.57	3.34	3.23	2.25	3.32	3.57
5.7	3.60	3.36	3.26	2.28	3.41	3.68
5.8	3.77	3.47	3.34	2.35	3.49	3.82

Using program B-2.3 compute and plot the surface and its first derivative. Evaluate the derivatives in the region bounded by $1.4 \leq b_o \leq 1.6$ and $5.5 \leq b_1 \leq 5.6$. Estimate the values of b_o and b_1 where F is a minimum.

3-9 Using Program B-3.1 estimate the minimum of Y in the following data set.

x:	3.2	3.4	3.6	3.8	4.0	4.2
y:	5.3	3.9	1.1	1.0	2.8	5.1

Assume two reasonable outliers right and left. Change the outliers slightly and find the new minimum. Is there a difference? Use Program B-3.1.

3-10 Use Program B-3.2 to find the maximum "downhill" slope in the data of Exercise 2-10.

Table 3-1

Saturation Vapor Pressure (P_v) Versus Air Temperature

T (°F)	P_v (mb)
50	12.19
60	17.51
70	24.79
80	34.61
90	47.68
100	64.88

Table 3-2

Nodes for Example 3-6

x_1

		0			5			10
	10	5	5	4	4	9	9	9
		5	6	5	6	8	8	9
		5	6	7	8	8	9	10
x_2	5	4	5	6	7	7	9	11
		3	4	5	4	5	7	9
		4	5	5	3	3	5	7
	0	4	4	4	2	2	3	5

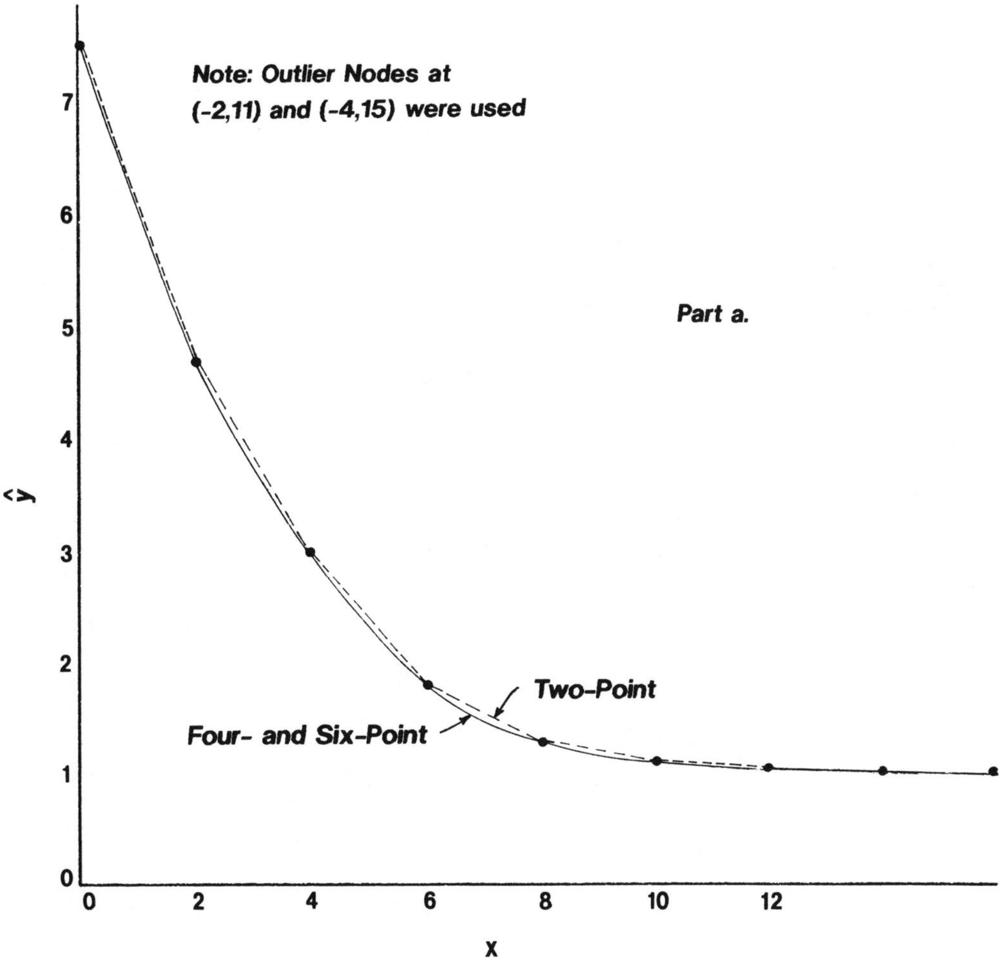
Note: Outlier Nodes at
(-2,11) and (-4,15) were used
Part a.
Two-Point
Four- and Six-Point
ŷ
7
6
5
4
3
2
1
0
0
2
4
6
8
10
12
x

Fig. 3.1.

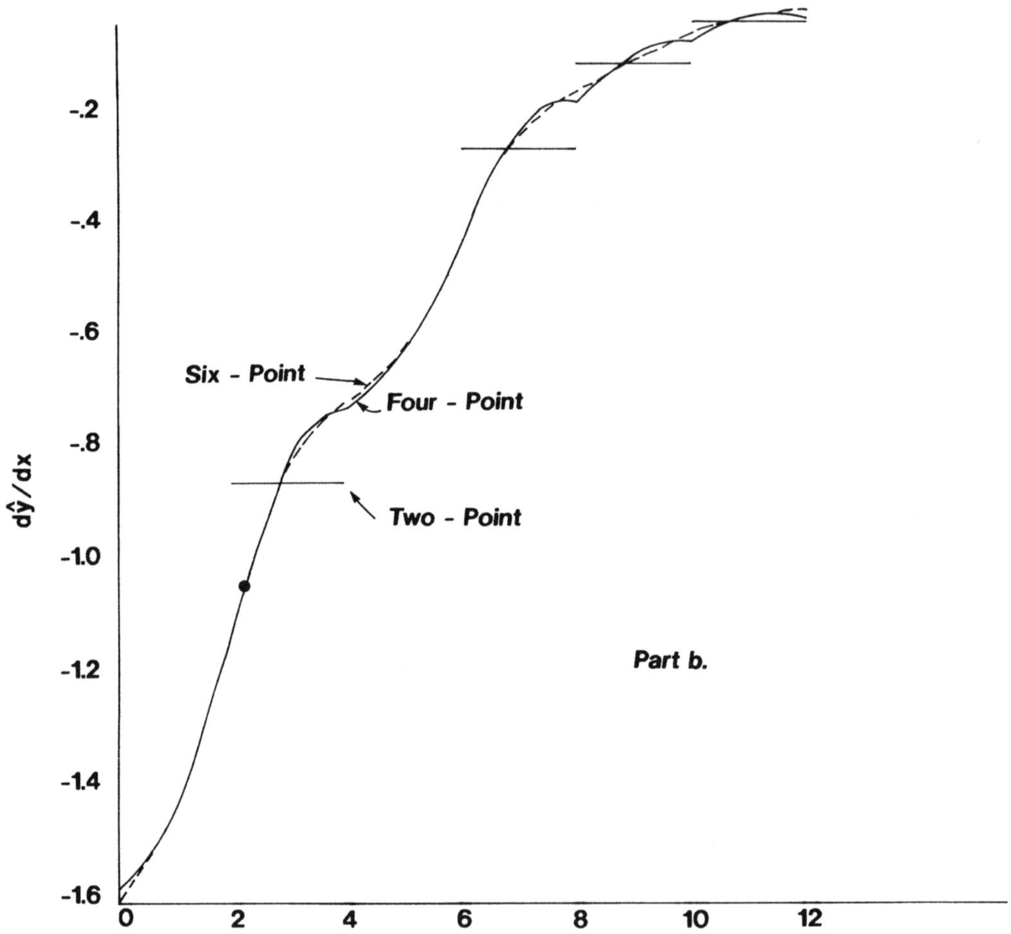
-.2
-.4
-.6
-.8
-1.0
-1.2
-1.4
-1.6
dŷ/dx
0
2
4
6
8
10
12
Six - Point
Four - Point
Two - Point
Part b.

Fig. 3.1.

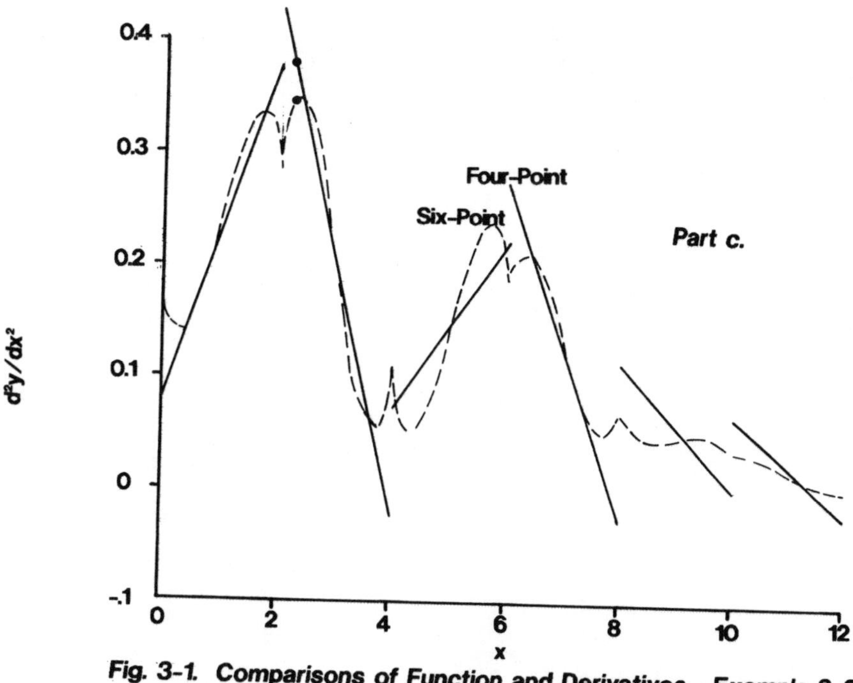

Fig. 3-1. Comparisons of Function and Derivatives - Example 3-3

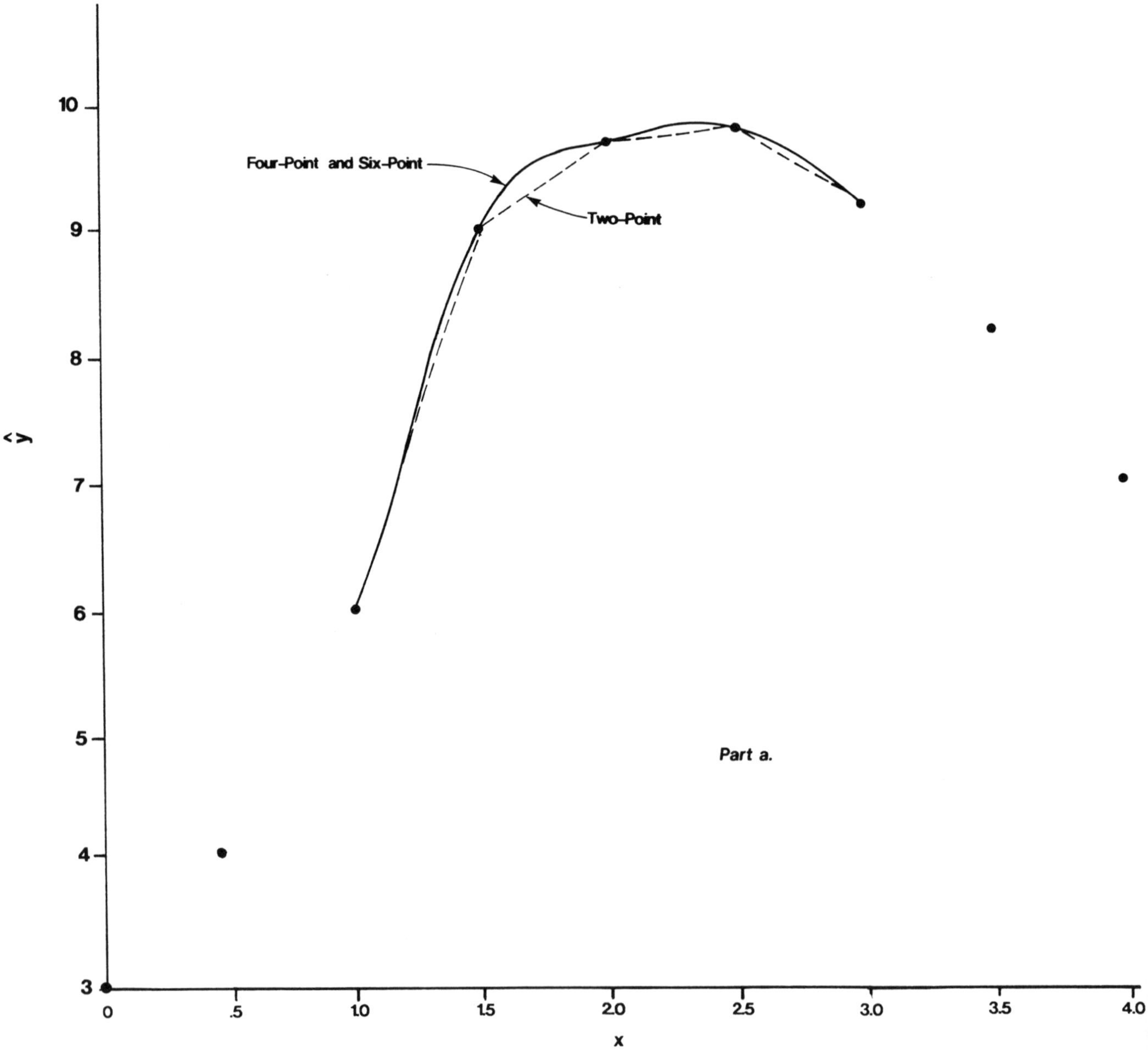

Fig. 3-2. Comparison of Function and Derivatives- Example 3-4.

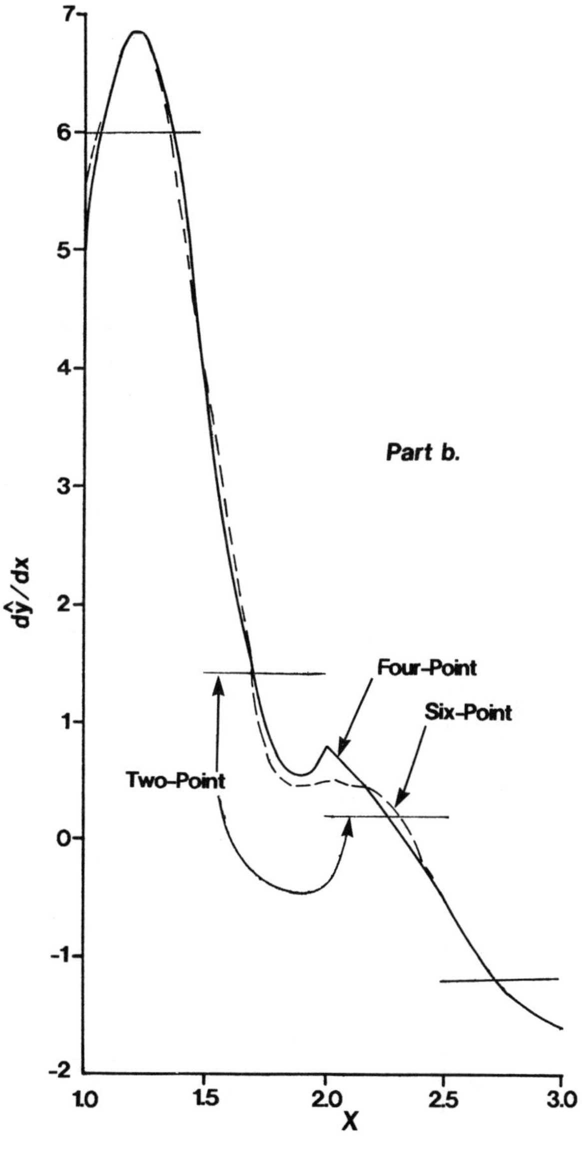

7
6
5
4
3
2
1
0
-1
-2
dŷ/dx
Part b.
Four-Point
Six-Point
Two-Point
1.0
1.5
2.0
2.5
3.0
X

Fig. 3-2.

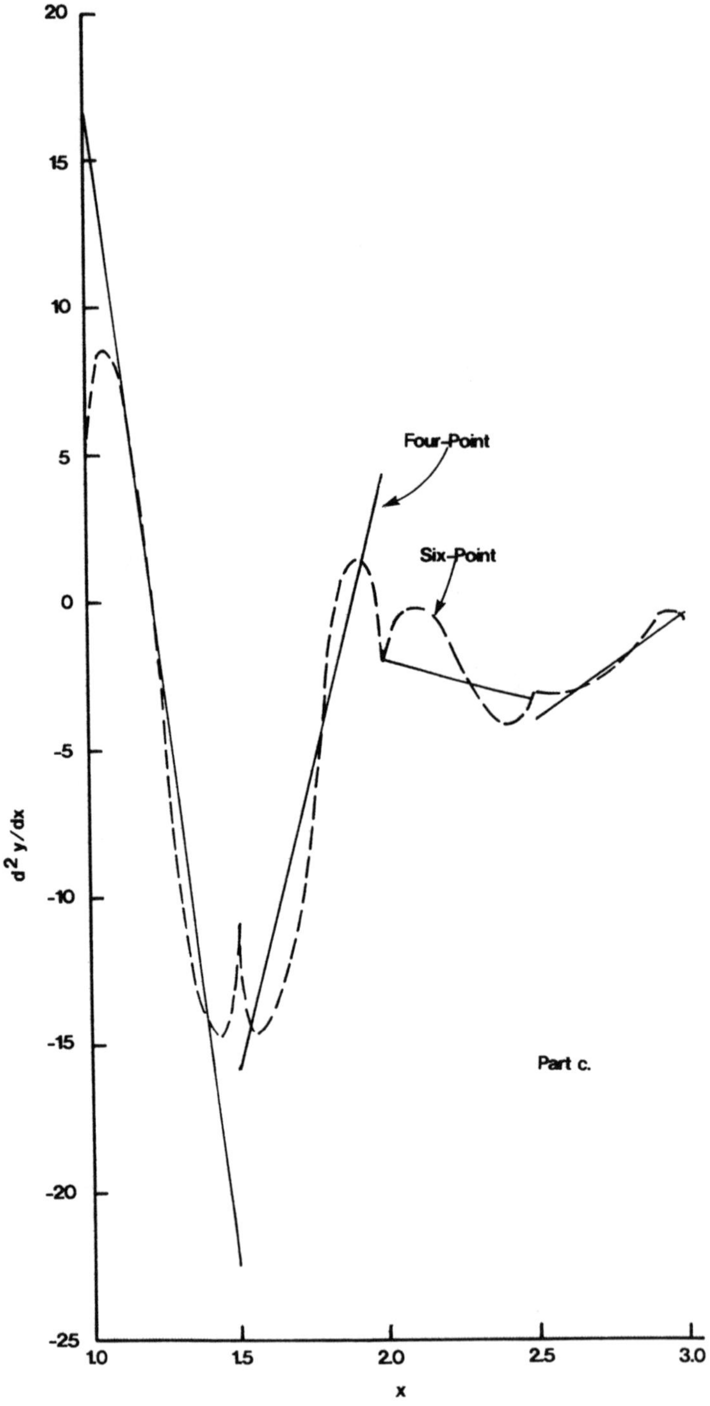
20
15
10
5
0
-5
-10
-15
-20
-25
d2 y/dx
1.0
1.5
2.0
2.5
3.0
x
Four-Point
Six-Point
Part c.

Fig. 3-2.

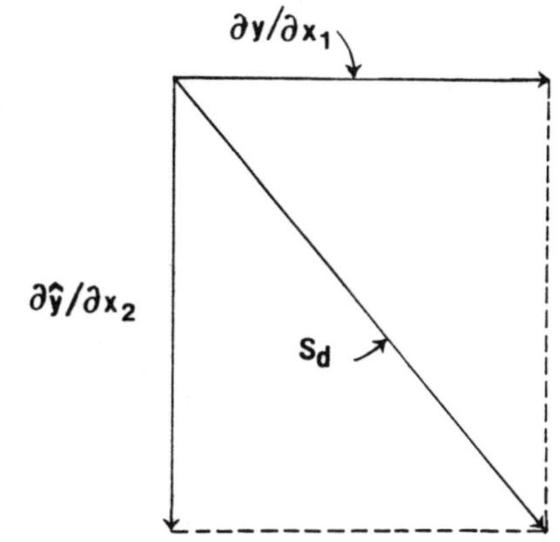

Fig. 3-3. Definition of Downhill Slope

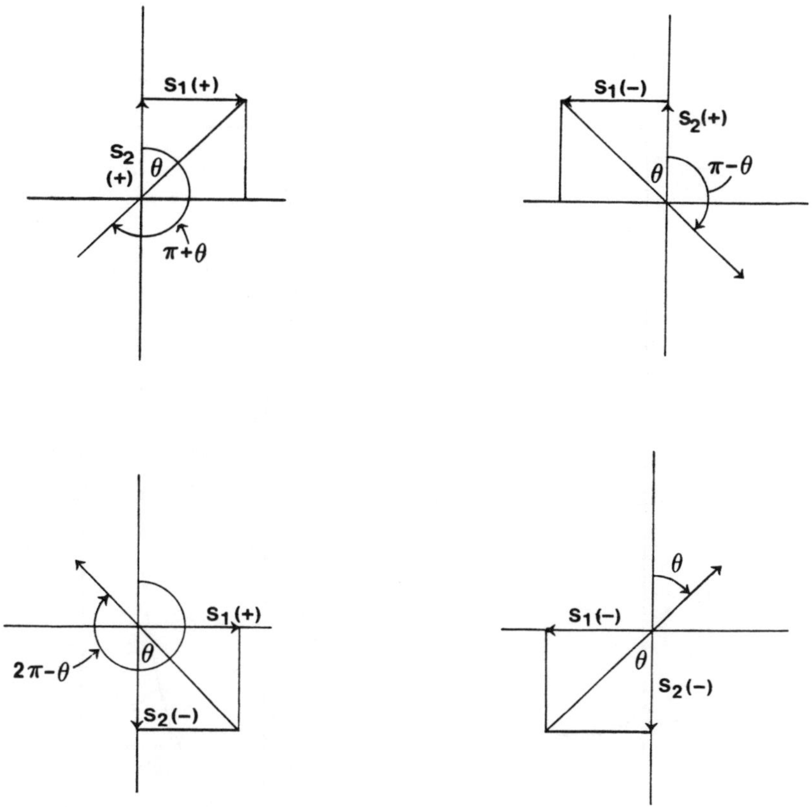

Fig. 3-4. Determination of Slope Direction

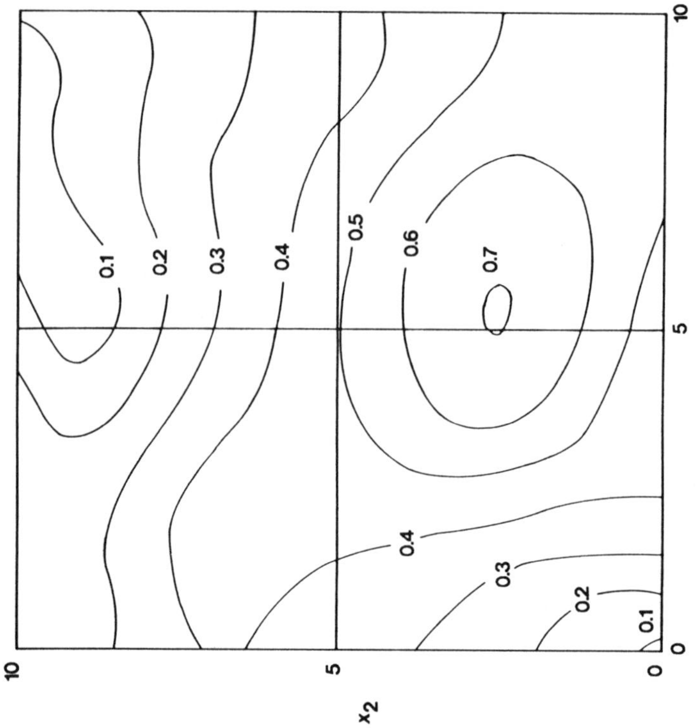

Fig. 3-6. Values of Downhill Slope for Example 3-6

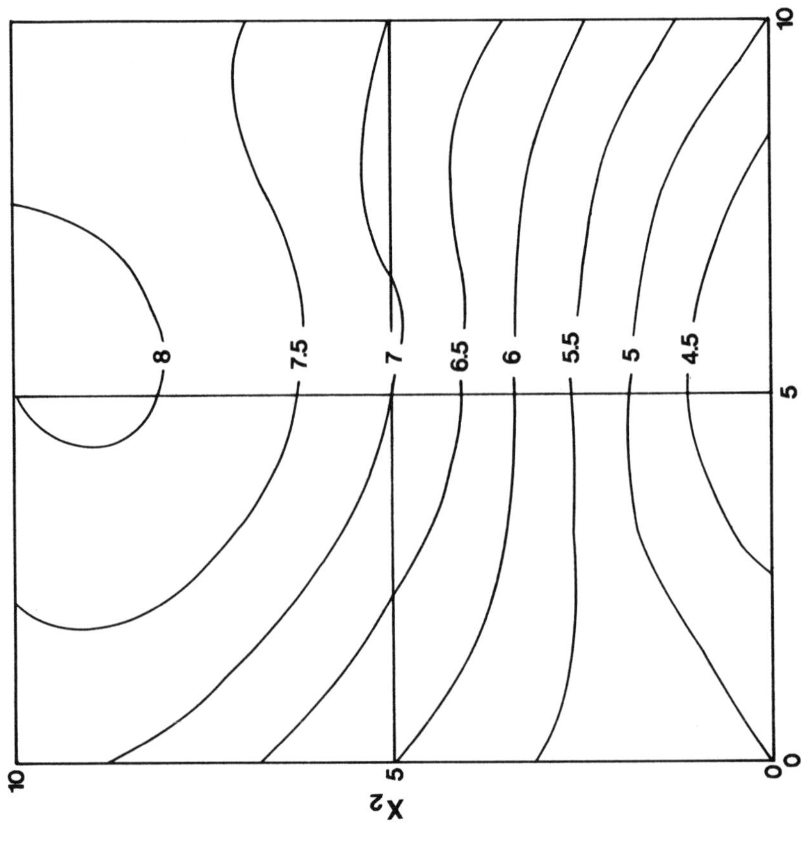

Fig. 3-5. Contours of the Surface of Example 3-6

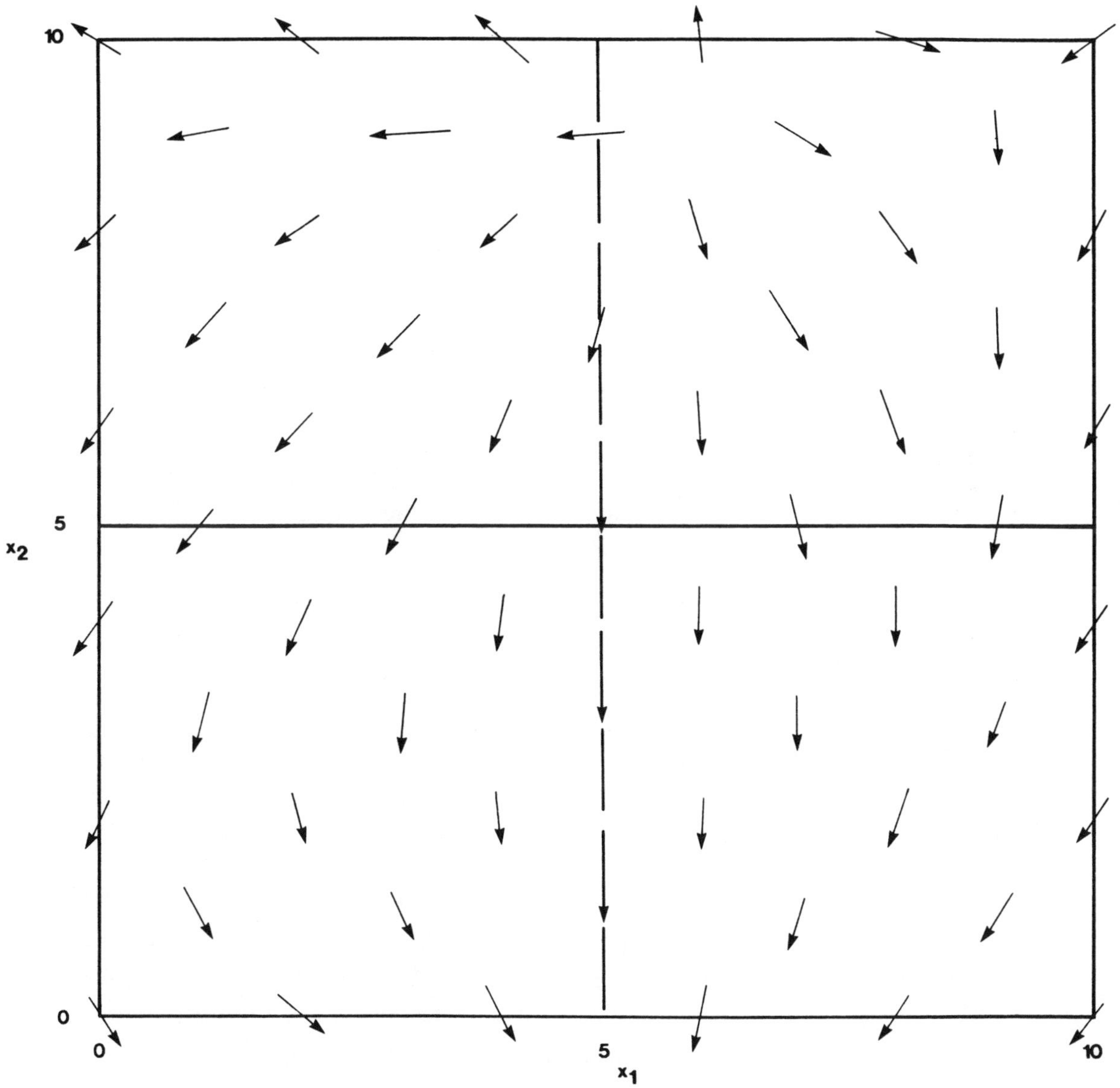

Fig. 3.7. Direction of Downhill Slope for Example 3-6.

CHAPTER 4
Numerical Integration

4.1 INTRODUCTION

Integration was mentioned in Chapter 3 as one of the alternative levels in a hierarchy built around a basic function. Differentiation moved us downward from the basic function of Eq. 3-1b to calculate rates of change, or gradients of the basic function, which are in the form of Eqs. 3-1c and 3-1d. Integration moves us upward in the hierarchy to give us summations, or accumulations, of the basic function, i.e., to go from the function of Eq. 3-1b to that of Eq. 3-1a.

Geometrically, integration produces the area under a graph of the basic function between a left and a right limit. This definition is sketched in Figure 4-1 as it applies to a numerical integral. $\hat{y}(x)$ is an approximating function through a set of nodes. The cross-hatched area is expressed as $\int_{x_1}^{x_2} \hat{y}(x)dx$ and is an approximation of the true integral $\int_{x_1}^{x_2} y(x)dx$. If y(x) is known, then usually it can be integrated and a numerical approximation is not needed. However, if we only know values of y(x) at a set of nodes, then a numeric approximation of the integral is necessary. Numerical integration is also used when y(x) is known but its direct mathematical integration is complicated and not feasible for the gain in precision.

4.2 SLIDING POLYNOMIAL INTEGRALS

Given the sliding polynomial functional approximation $y(x) \approx \hat{y}(x) = \hat{y}(z) = \sum_{i=1}^{n} C_i(z)Y_i$ then the approximate integral of y(x) is given by:

$$\int_{x_1}^{x_2} y(x)dx \approx \int_{x_1}^{x_2} \hat{y}(x)dx = \int_{z(x_1)}^{z(x_2)} \hat{y}(z)(bdz) = b \sum_{i=1}^{n} \left[\int_{z(x_1)}^{z(x_2)} C_i(z)\ dz \right] Y_i \quad (4\text{-}1)$$

For the two-point system the integral of the coefficients in Eqs. 1-3 can be written as:

$$\int_{z_1}^{z} C_1(z)dz = \frac{1}{2}\ z - \frac{1}{2}\ z^2 + k_1 \quad (4\text{-}2a)$$

$$\int_{z_1}^{z} C_2(z)dz = \frac{1}{2} z + \frac{1}{2} z^2 + k_2 \qquad \text{(4-2b)}$$

in which k_1 and k_2 are constants of integration. If we start the integration at the left of each span of the sliding polynomials, then the integrals are zero when the local variate z equals -0.5. Substituting these values into Eqs. 4-2 yields the following values for k_1 and k_2:

$$-\frac{1}{4} - \frac{1}{8} + k_1 = 0; \qquad k_1 = \frac{3}{8} \qquad \text{(4-3a)}$$

$$-\frac{1}{4} + \frac{1}{8} + k_2 = 0; \qquad k_2 = \frac{1}{8} \qquad \text{(4-3b)}$$

The full expression of the integrals of the coefficients are:

$$\int_{-.5}^{z} C_1(z)\ dz = \frac{3}{8} + \frac{1}{2} z - \frac{1}{2} z^2 \qquad \text{(4-4a)}$$

$$\int_{-.5}^{z} C_2(z)\ dz = \frac{1}{8} + \frac{1}{2} z + \frac{1}{2} z^2 \qquad \text{(4-4b)}$$

It is important to note fully that Eqs. 4-4 are the two-point coefficients for integration when the integration starts with zero at the left edge of the span. Geometrically, they express the area under a straight-line segment between z = -0.5 and any value of z up to 0.5.

The analogous integral coefficients for the four-point sliding polynomial are given by:

$$\int_{-0.5}^{z} C_1(z)\ dz = \frac{(((-48z + 32)z + 24)z - 24)z - 11}{384} \qquad \text{(4-5a)}$$

$$\int_{-0.5}^{z} C_2(z)\ dz = \frac{(((144z - 32)z - 264)z + 216)z + 161}{384} \qquad \text{(4-5b)}$$

$$\int_{-0.5}^{z} C_3(z)\ dz = \frac{(((-144z - 32)z + 264)z + 216)z + 47}{384} \qquad \text{(4-5c)}$$

$$\int_{-0.5}^{z} C_4(z)\ dz = \frac{(((48z + 32)z - 24)z - 24)z - 5}{384} \qquad \text{(4-5d)}$$

Note again that the integration starts at the left edge of the span $-0.5 \leq z \leq 0.5$.

Table A-4.1 gives values of the integral coefficients for each one-hundredth of the interval $-0.5 \leq z \leq 0.5$. These can be used for manual calculations. Program B-4.1 can be used for more extensive calculations.

The six-point integral coefficients are given by:

$$\int_{-0.5}^{z} C_1(z)dz = \frac{(((((-1600z+192)z+2160)z-800)z-780)z+540)z+261}{46080} \quad (4\text{-}6a)$$

$$\int_{-0.5}^{z} C_2(z)dz = \frac{(((((8000z-576)z-12720)z+6240)z+4860)z-4500)z-2033}{46080} \quad (4\text{-}6b)$$

$$\int_{-0.5}^{z} C_3(z)dz = \frac{(((((-16000z+384)z+27360)z-5440)z-33720)z+27000)z+1980}{46080} \quad (4\text{-}6c)$$

$$\int_{-0.5}^{z} C_4(z)dz = \frac{(((((16000z+384)z-27360)z-5440)z+33720)z+27000)z+5862}{46080} \quad (4\text{-}6d)$$

$$\int_{-0.5}^{z} C_5(z)dz = \frac{(((((-8000z-576)z+12720)z+6240)z-4860)z-4500)z-943}{46080} \quad (4\text{-}6e)$$

$$\int_{-0.5}^{z} C_6(z)dz = \frac{(((((1600z+192)z-2160)z-800)z+780)z+540)z+91}{46080} \quad (4\text{-}6f)$$

Equations 4-6 are included here to complete the possible methods of piece-wise integration. However, the four-point system is recommended for most work.

4.2.1 Example 4-1: Two-point Integral of an Entire Span

Given the nodes Y_1 and Y_2 at the left and right boundaries of a span of width Δx, the area under the line connecting the nodes can be computed. At the right boundary z equals 0.5 and Eqs. 4-4 give $\int C_1 = \int C_2 = 1/2$. Equation 4-1 then leads directly to the following integral:

$$\int_{-0.5}^{0.5} \hat{y}(z)(bdz) = \Delta x\left[\frac{1}{2} Y_1 + \frac{1}{2} Y_2\right] = \frac{\Delta x}{2} [Y_1 + Y_2] \qquad (4\text{-}7)$$

This agrees with the conventional geometrical expression for the area of a trapezoid.

4.2.2 Example 4-2: Two-Point Integration of the Left and Right Halves of a Span

Given Y_1 and Y_2 at the left and right boundaries of a span of width Δx, the areas under the line can be computed for the intervals $-0.5 \le z \le 0$ and $0 \le z \le 0.5$.

At the midpoint z equals 0; at this point, $\int C_1 = 3/8$ and $\int C_2 = 1/8$. Equation 4-1 yields the integral:

$$\int_{-0.5}^{0} \hat{y}(z)(bdz) = \Delta x\left[\frac{3}{8} Y_1 + \frac{1}{8} Y_2\right] = \frac{\Delta x}{8} [3\ Y_1 + Y_2] \qquad (4\text{-}8)$$

The area under the right half of the line equals the area under the whole interval (Eq. 4-7) minus the area under the left half of the interval (Eq. 4-8):

$$\int_{0}^{0.5} \hat{y}(z)(bdz) = \frac{\Delta x}{2} (Y_1 + Y_2) - \frac{\Delta x}{8} (3\ Y_1 + Y_2) = \frac{\Delta x}{8} (Y_1 + 3\ Y_2) \qquad (4\text{-}9)$$

4.2.3 Example 4-3: Two-Point Integration Across Two Spans

Given the data

x:	0	5	10
Y:	10	20	25

The integral $\int \hat{y}dx$ for the range $0 \le x \le 10$ can be computed with an increment of 1.25 to show the continuity of the integration process.

The values of x, the local variate z, and the coefficients $\int C_1$ and $\int C_2$ are given in Table 4-1. The piecewise integration across the two spans is given in Table 4-2. For example, the integral at x = 1.25 is computed as [0.21875 (10) + 0.03125 (20)] 5 = 14.0625. At x = 6.25 the integral for the interval $5.0 \le x \le 6.25$ is [0.21875 (20) + 0.03125 (25)] 5 = 25.78125. However, for the total integral from x = 0, we must add to this the entire first-span integral of 75.0. Then the integral for the span $0 \le x \le 6.25$ is 100.78125.

4.2.4 Example 4-4: Four-Point Integration

Given the following data the integral for the interval $0 \leq x \leq 10$ can be computed at an increment of 1.25 in x:

x:	-5	0	5	10	15
Y:	-5	10	20	25	25

The integral coefficients in Table A-4.1 can be used.

The values of x, the local variate z, and the integral coefficients are given in Table 4-3. The piecewise integrals are shown in Table 4-4. Using the integral at x = 1.25 as an example, again, the calculation is [(-0.010905)(-5)+(0.238444)(10)+(0.024577)(20)+(-0.002116)(25)] = 14.3880. The values in the second span are increased by the total integral of the first span to give the total accumulation from 0 to the indicated value of x.

The values of the integrals are plotted in Fig. 4-2, along with the values of the two-point integrals of Example 4-3. The values of the two-point system form parabolas as indicated by Eqs. 4-4. The values of the four-point system form quartic arcs as indicated by Eqs. 4-5. The integrals for the four-point system are slightly larger because the straight lines of the two-point system cut across the curvature of the approximating function through the nodes. This was noted in Figs. 3-1 and 3-2.

4.2.5 Example 4-5: Rate to Volume Integral

Consider the case of a tank that contains a liquid which leaks from the tank at an irregular rate. The following hourly measurements of the volume rate of leakage were made for a period of five hours:

Time (hours):	0	1	2	3	4	5
Leakage (cm^3/sec):	10	10	7	5	8	10

The problem is to estimate the total volume of liquid that leaked from the tank over the five-hour period. Program B-4.1, with its four-point integrals, can be used. Also an outlier node of 7 cm^3/sec at the left and 11 cm^3/sec at the right are used. Note that volume = $\int_{t_1}^{t_2}$ rate*dt.

The results of the computer run are plotted in Fig. 4-3. The leakage rate and accumulating volume are both shown. The total volume leaked at the end of hour 5 is about 4800 cm^3. Note that the volume curve is relatively smooth with little of the curvature of the rate curve. Such smoothing has been mentioned previously as typical of the process of integration.

4.2.6 Example 4-6: Velocity to Distance Integral

A log floating in the sea drifts with the currents. The east-west component of the drift velocity is as follows:

Time (hours):	0	12	24	36	48	60	72	84	96
Velocity (km/hr):	0.8	1.0	0.78	-0.2	-0.5	-0.5	-0.4	-0.2	0.0

Assuming that negative velocities are to the west, it is of interest to evaluate the east-west position of the log over the 96 hour period. A left outlier of 0.6 km/hr and a right outlier of 0.2 km/hr are used. Note that distance = $\int_{t_1}^{t_2}$ velocity*dt.

The output from Program B-4.1 is plotted in Fig. 4-4. The log stayed east of its initial point for the entire 96 hours. At the end of the period, it was 3.6 km to the east of its initial position. The maximum eastward drift was 25.2 km.

4.2.7 Example 4-7: Pressure to Total Force Integral

The variation of pressure along a strip of material 1 cm by 300 cm was measured as follows:

Distance (cm)	0	100	200	300
Pressure (Newtons/cm^2)	15.0	9.3	13.2	7.8

The problem is to find the total force on both the left 150 cm and the right 150 cm. A left outlier of 30.3 Newton/cm^2 and a right outlier of -12.6 Newtons/cm^2 were assumed. Force = $\int_{x_1}^{x_2}$ pressure*dx.

The pressure and force curves are plotted in Fig. 4-5. The force at 150 cm is 1630 Newtons, and this is obviously the total force on the left half. The total force at 300 cm is 3420 Newtons. Therefore, the total force on the right half is 3420-1630 = 1790 Newtons.

4.3 CENTER OF GRAVITY

The center of gravity of an area is the point in x-scale where the portion of the area to the left of the point in x exactly balances the portion to the right. The center of gravity may also be called the first moment of the area. The center of gravity for continuous functions can be computed by:

$$x_{CG} = \frac{\int_{x_1}^{x_2} x\, f(x)dx}{\int_{x_1}^{x_2} f(x)dx} \qquad (4\text{-}10)$$

In order to approximate the center of gravity with the four-point sliding polynomial, we can use the following:

$$x_{CG} = \frac{\sum_{i=1}^{4} \left(\int_{z(x_1)}^{z(x_2)} (a + bz)\ C_i(z)bdz \right) Y_i}{\sum_{i=1}^{4} \left(\int_{z(x_1)}^{z(x_2)} C_i(z)bdz \right) Y_i} \qquad (4\text{-}11)$$

The denominator of Eq. 4-11 is the integral of the function approximated by the set of nodes as already discussed in previous examples. The numerator requires the evaluation of new integral coefficients. If the new integral coefficients are designated $MC_i(z)$ for moment coefficient, then we need to integrate the four equations:

$$MC_i(z) = b \int_{-0.5}^{z} (a + bz)\ C_i(z)dz$$

as i takes on values of 1 through 4. The $C_i(z)$ are, of course, the same coefficients given in Eqs. 1-17. These integrals are:

$MC_1(z)=$

$$\left[\frac{\frac{-110a+23b}{240} - az + \frac{(2a-b)z^2}{2} + \frac{(4a+2b)z^3}{3} - \frac{(8a-4b)z^4}{4} - \frac{8bz^5}{5}}{16} \right] \qquad (4\text{-}12a)$$

$MC_2(z)=$

$$\left[\frac{\frac{1610a-439b}{240} + 9az - \frac{(22a-9b)z^2}{2} - \frac{(4a+22b)z^3}{3} + \frac{(24a-4b)z^4}{4} + \frac{24bz^5}{5}}{16} \right] \qquad (4\text{-}12b)$$

$MC_3(z)=$

$$\left[\frac{\frac{470a-71b}{240} + 9az + \frac{(22a+9b)z^2}{2} - \frac{(4a-22b)z^3}{3} - \frac{(24a+4b)z^4}{4} - \frac{24bz^5}{5}}{16} \right] \qquad (4\text{-}12c)$$

$MC_4(z)=$

$$\left[\frac{\frac{(-50a+7b)}{240} - az - \frac{(2a+b)z^2}{2} + \frac{(4a-2b)z^3}{3} + \frac{(8a+4b)z^4}{4} + \frac{8bz^5}{5}}{16}\right] \quad (4\text{-}12d)$$

Program B-4.2 was prepared to compute the center of gravity of an area under the approximating curve $\hat{y}(x)$ that passes through a set of nodes Y_i. The program computes the center of gravity on a continuous basis as the area accumulates from the beginning value of zero at the left boundary.

4.3.1 Example 4-8: First Center of Gravity Approximation

Program B-4.2 can be used to compute the center of gravity of the area under the function through the following nodes:

x:	0.5	1.0	1.5	2.0	2.5	3.0	3.5	4	4.5
Y_i:	26	10	6	4	3	2	1	0	-1

The first and last nodes are the required outliers for the four-point sliding polynomial.

The results of the computer run are plotted in Fig. 4-6. The function diminishes rapidly from 10 to zero and, therefore, the area accumulates rapidly near the left boundary and reaches a maximum of about 10.25 units at x equal to 4. The center of gravity starts at a value of 1.0, the left boundary of x, and shifts to the right as the area accumulates to the right. It reaches a value of 1.91 units of x at x equal to 4. This final value is shown in x-scale on the approximating function $\hat{y}(x)$. If a fulcrum were placed at this point, the function would be in balance.

4.3.2 Example 4-9: Second Center of Gravity Approximation

Program B-4.2 can be used to compute the center of gravity of the area under the approximating function through the following nodes:

x:	0.5	1.0	1.5	2.0	2.5	3.0	3.5	4.0	4.5
Y:	20	10	6	4	4	8	14	22	32

As in Example 4-8, the first and last nodes are outliers.

The results of this analysis are plotted in Figure 4-7. In this example the function first falls, then rises fairly rapidly to the right. The area accumulation curve goes through an inflection near x equal to 2.4 as the function passes through its minimum. The center of gravity continues to shift to the right as more area is added to the right. It goes through a slight inflection near x equal to 2.5. The final value is 2.86 units of x-scale. This final value is plotted on the function $\hat{y}(x)$ at x equal to 2.86. Again, this is the balance of y(x) for the range $1 \leq x \leq 4$.

4.4 TWO-DIMENSIONAL INTEGRATION

Following the function-of-functions form in Eq. 1-45, we can write the following three equations directly:

$$A_1\hat{y} = \sum_{j=1}^{4} \sum_{i=1}^{4} C_j(z_2) \int C_i(z_1) b_1 \, dz_1 \; Y_{j,i} \tag{4-13}$$

$$A_2\hat{y} = \sum_{j=1}^{4} \sum_{i=1}^{4} \int C_j(z_2) b_2 \, dz_2 \; C_i(z_1) \; Y_{j,i} \tag{4-14}$$

$$V\hat{y} = \sum_{j=1}^{4} \sum_{i=1}^{4} \int C_j(z_2) b_2 \, dz_2 \int C_i(z_1) b_1 \, dz_1 \; Y_{j,i} \tag{4-15}$$

Equations 4-13 and 4-14 could be called partial integrals of $\hat{y}(x)$. Equation 4-13 expresses how the area under the curve $\hat{y}(x)$ produced by integration in the x_1 dimension, $A_1\hat{y}$, varies in the x_2 dimension. Equation 4-14 expresses the way that the area produced by integration in the x_2 dimension, $A_2\hat{y}$, varies in the x_1 dimension. Equation 4-15 is a double integral that yields the volume under the surface $\hat{y}(x_1,x_2)$. To get the total volume under an ensemble of unit sliding polynomial surfaces, it is necessary to sum the individual volumes.

Equations 4-13 and 4-14 appear to have had little use in practical problem solution but are included here for completeness of the integral structures. The volume from Eq. 4-15 has more practical value. Program B-4.3 can be used to compute four-point sliding polynomial volume approximations.

4.4.1 Example 4-10: Two-Point Volumes

Equations 4-4 can be used to develop an expression for the volume under the two-point surface that passes through the nodes $Y_{1,1}$, $Y_{1,2}$, $Y_{2,1}$, and $Y_{2,2}$. The nodes are a distance Δx_1 apart in x_1 scale and a distance Δx_2 apart in the x_2 scale.

At z = 0.5, $\int C_1 = \int C_2 = 1/2$. Then the volume is

$$\frac{1}{2}\Delta x_2\left(\frac{1}{2}\Delta x_1\right)Y_{1,1} + \frac{1}{2}\Delta x_2\left(\frac{1}{2}\Delta x_1\right)Y_{1,2} + \frac{1}{2}\Delta x_2\left(\frac{1}{2}\Delta x_1\right)Y_{2,1} + \frac{1}{2}\Delta x_2\left(\frac{1}{2}\Delta x_1\right)Y_{2,2} = \frac{\Delta x_2\,\Delta x_1}{4}\left(Y_{1,1} + Y_{1,2} + Y_{2,1} + Y_{2,2}\right)$$

This agrees with the conventional geometric equation. At z = 0.0, $\int C_1 = 3/8$ and $\int C_2 = 1/8$. Then the volume is

$$\frac{3}{8}\Delta x_2\left(\frac{3}{8}\Delta x_1\right)Y_{1,1} + \frac{3}{8}\Delta x_2\left(\frac{1}{8}\Delta x_1\right)Y_{1,2} + \frac{1}{8}\Delta x_2\left(\frac{3}{8}\Delta x_1\right)Y_{2,1} + \frac{1}{8}\Delta x_2\left(\frac{1}{8}\Delta x_1\right)Y_{2,2} = \frac{\Delta x_2\,\Delta x_1}{64}\left(9Y_{1,1} + 3Y_{1,2} + 3Y_{2,1} + Y_{2,2}\right)$$

4.4.2 Example 4-11: Four-point Sliding Polynomial Volumes

Program B-4.3 can be used to compute the volume under the four-point sliding polynomial surface passed through the nodes in Fig. 4-8a. The outer edges are outlier nodes.

The results are plotted in Fig. 4-8b. The program accumulates the volumes under the individual unit surfaces from left to right and from bottom to top. These sub-volumes are all plotted and contours of the accumulated volumes are drawn. The lower left-hand unit surface has a volume of 0.52 units. The unit surface between 1.5 and 1.75 for both x_1 and x_2 has a unit volume (5.21-3.43)-(3.38-2.24)=0.64. However, such differences are difficult to detect in the contours. Most practical problems require only the total volume of 23.53 units.

EXERCISES

4-1. Using Eqs. 1-17, derive Eqs. 4-5 for the four-point system.

4-2. Using Eqs. 1-34, derive Eqs. 4-6 for the six-point system.

4-3. Derive an expression for the two-point interpolating polynomial for computing the integral of the left-hand one-fourth of the interval (i.e., for the localized unit variate z interval $-0.5 \leq z \leq -0.25$). Express the integral in terms of the nodal values Y_1 and Y_2 and the interval width x.

4-4. For the data of Example 4-3 (section 4.2.3) compute the value of the integral using an increment of 2.5. Compare the result with the value of the integral computed in Example 4-3.

4-5. Given the ordinates of a hydrograph q(t)

t (min)	0	20	40	60	80	100
q(t)(cfs)	0	25	45	30	20	0

compute the volume of runoff using a 10-minute time increment and the two-point sliding polynomial.

4-6. Using the two-point system with the data for Example 4-5, compute the total volume of liquid that leaks from the tank in the five-hour period. Use a 0.5-hour time increment. Compare the computed volume with that obtained in Example 4-5.

4-7. Using the data of Problem 4-5, compute the volume of runoff with the four-point system. Use a 10-minute increment and the integral coefficients of Table A-4.1.

4-8. The daily average temperature is often used in snowmelt modeling in computing the number of degree days. The following are point measurements of temperature at six-hour intervals:

hour	18	24	6	12	18	24	6
temp (^oF)	38	31	29	34	37	33	30

Using a 3-hour time increment compute the average temperature for the 24-hour period (midnight to midnight) using four-point interpolation. Use the integral coefficients of Table A-4.1.

4-9. Given the following nodes

x	0	2	4	6	8	10	12
Y_i	6	5	8	13	10	7	8

find the center of mass using a four-point system for the range $2 < x < 10$. Program B-4.3 can be used.

4-10. In computing a time of concentration from a measured storm hydrograph, the time to the center of mass of runoff is used. Find the center of mass of the hydrograph of Problem 5-4.

4-11. The shear stress along the side of a river bank is computed at various depths, as follows:

depth (ft)	0	0.5	1.0	1.5	2.0	2.5	3.0
stress	0	1.9	5.2	7.8	8.3	5.4	0

Compute the depth above which one-half of the shear stress occurs. Use a four-point system. Assume the water depth is 3 feet.

Table 4-1

Coefficients for Example 4-3

x	z	$\int C_1$	$\int C_2$
0	-0.5	0	0
1.25	-0.25	0.21875	0.03125
2.5	0	0.375	0.125
3.75	0.25	0.46875	0.28125
5.0	0.5	0.5	0.5
6.25	-0.25	0.21875	0.03125
7.5	0	0.375	0.125
8.75	0.25	0.46875	0.28125
10.0	0.5	0.5	0.5

Table 4-2

Integration for Example 4-3

x	Y	$\int \hat{y} dx$
0	10	0
1.25		14.0625
2.5		31.25
3.75		51.5625
5.0	20	75.0
6.25		25.78125 + 75.0 = 100.78125
7.5		53.125 + 75.0 = 128.125
8.75		82.03125 + 75.0 = 157.03125
10.0	25	112.5 + 75.0 = 187.5

Table 4-3

Coefficients for Example 4-4

x	z	$\int C_1$	$\int C_2$	$\int C_3$	$\int C_4$
0	-0.5	0	0	0	0
1.25 - 6.25	-0.25	-0.010905	0.23844	0.024577	-0.002116
2.50 - 7.50	0	-0.028646	0.419271	0.122396	-0.013021
3.75 - 8.75	0.25	-0.039551	0.517090	0.303223	-0.030762
5.00 - 10.00	0.5	-0.041667	0.541667	0.541667	-0.041667

Table 4-4

Integration for Example 4-4

x	Y	$\int \hat{y}dx$
-5	-5	- - -
0	10	0
1.25		14.3880
2.50		32.2917
3.75		53.3202
5.00	20	77.0835
6.25		26.1068 + 77.0835 = 103.1903
7.50		54.1667 + 77.0835 = 131.2502
8.75		83.7891 + 77.0835 = 160.8726
10.0	25	114.5834 + 77.0835 = 191.6669
15.0	25	- - -

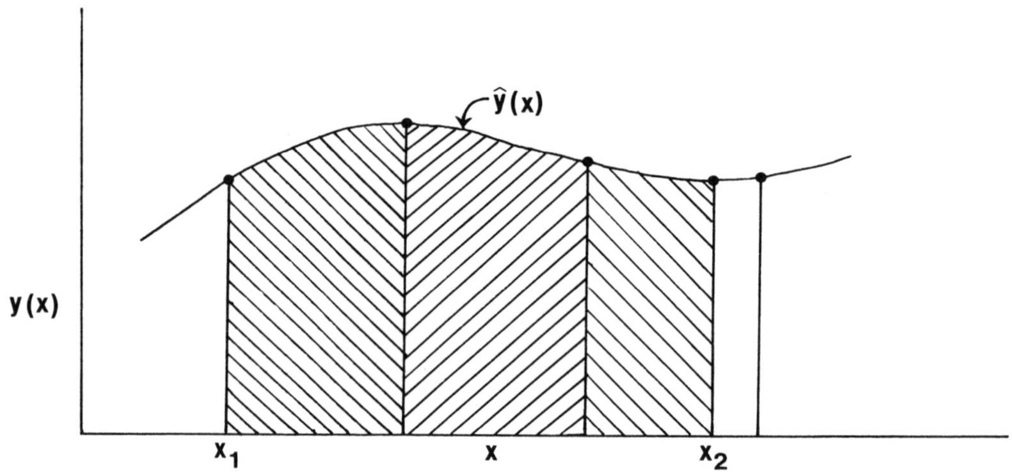

Fig. 4-1. Approximate Numeric Integral

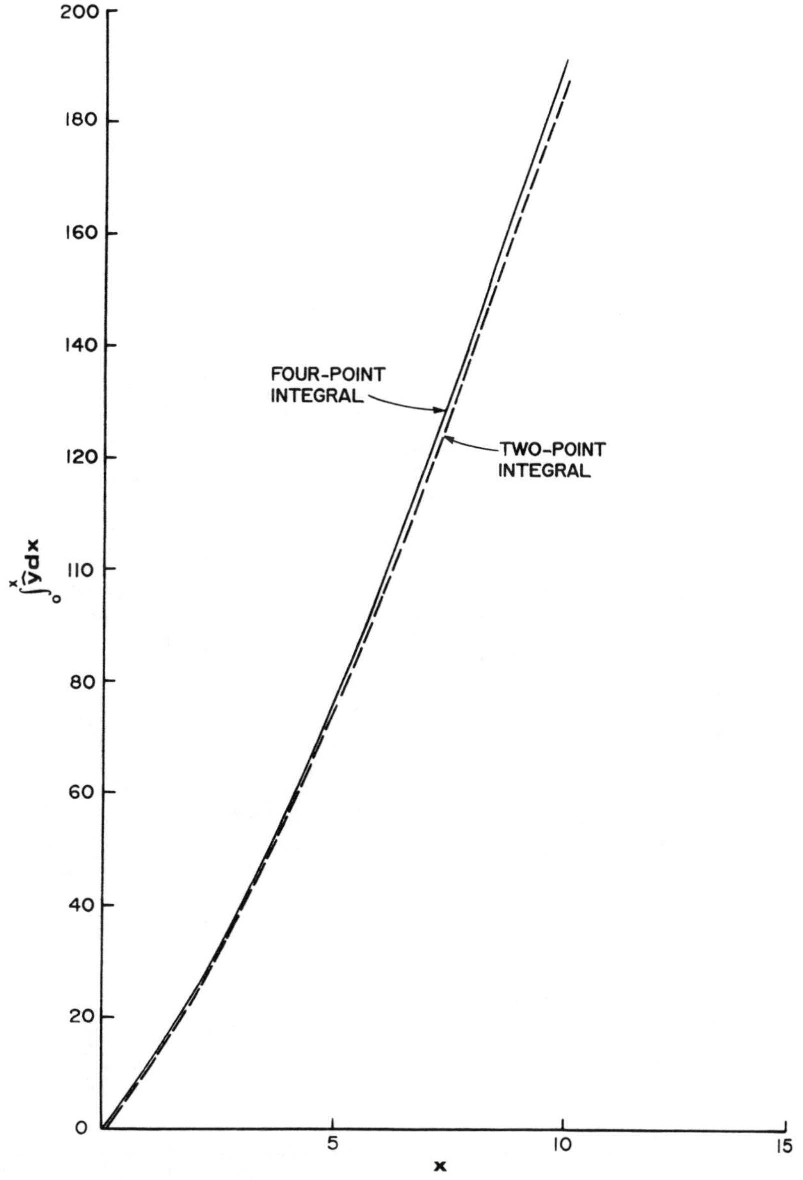

Fig. 4-2. Comparison of Two- and Four-Point Integrals

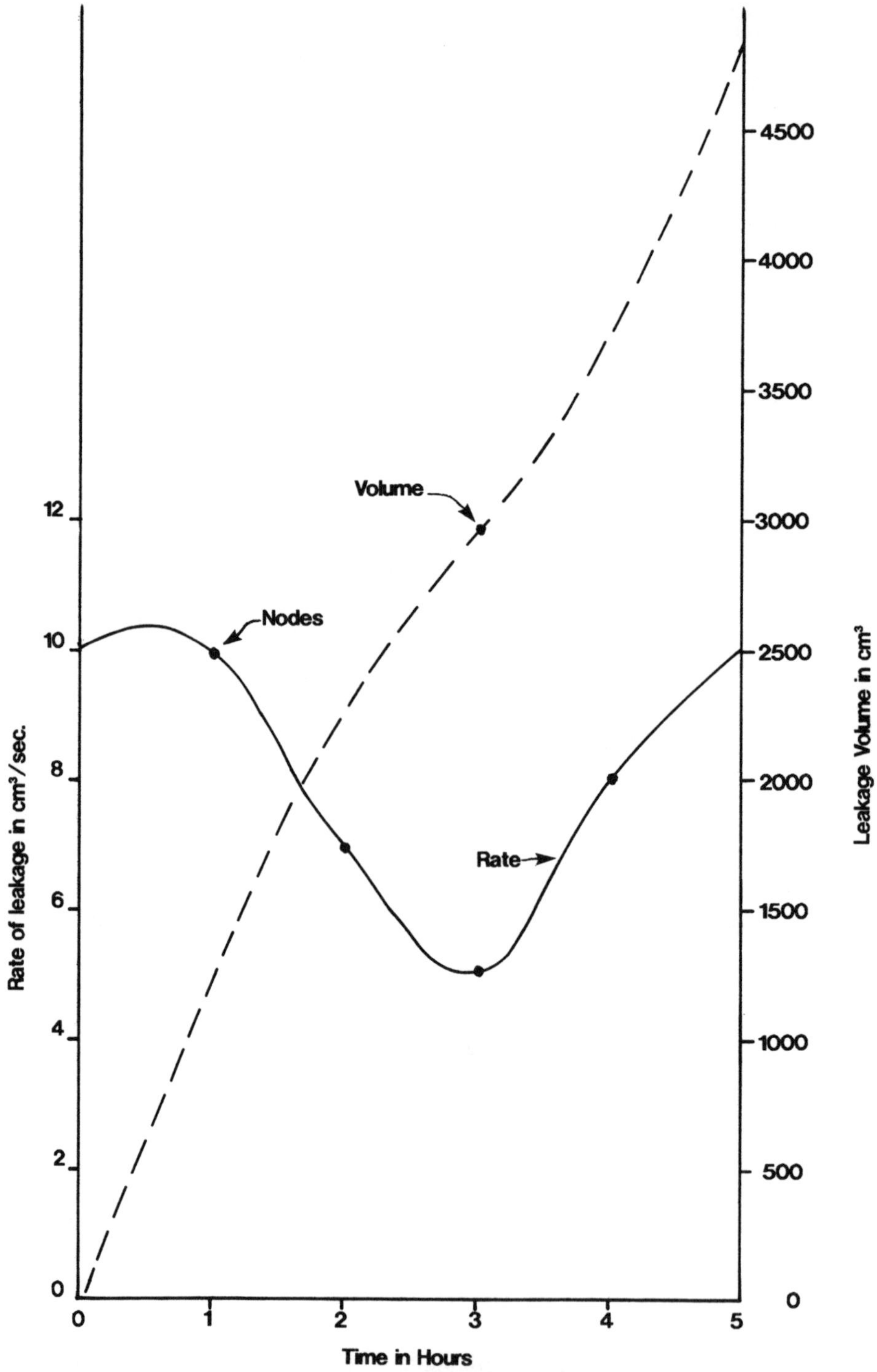

Fig. 4-3. Example 4-5, Rates and Volumes

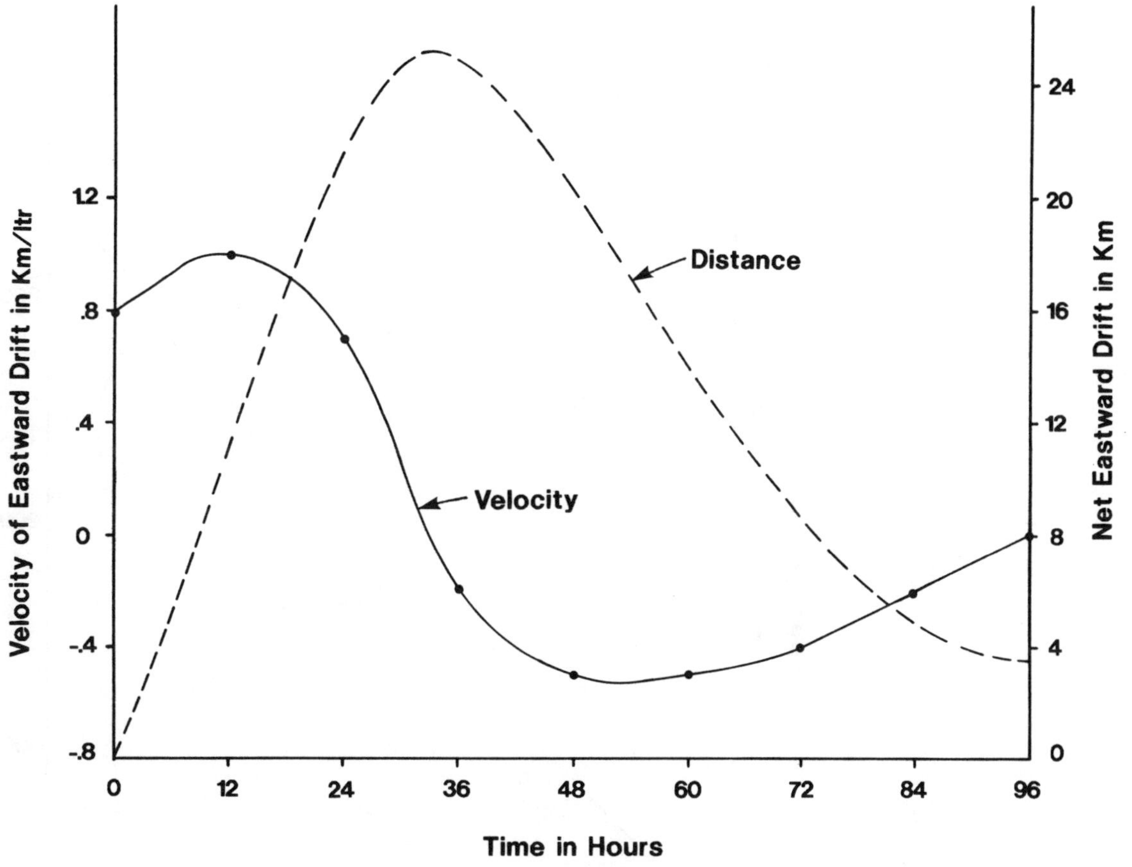

Fig. 4-4. Example 4-6, Distance of Drift

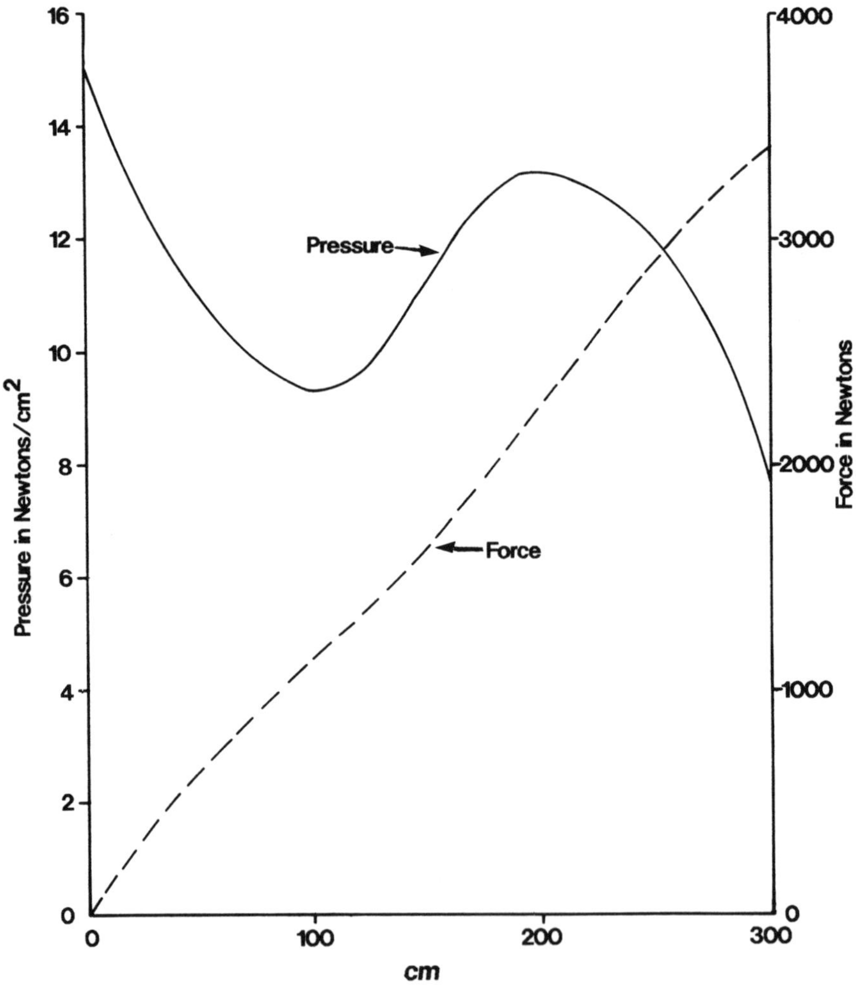

Fig. 4-5. Example 4-7. Pressure-Force Relationship

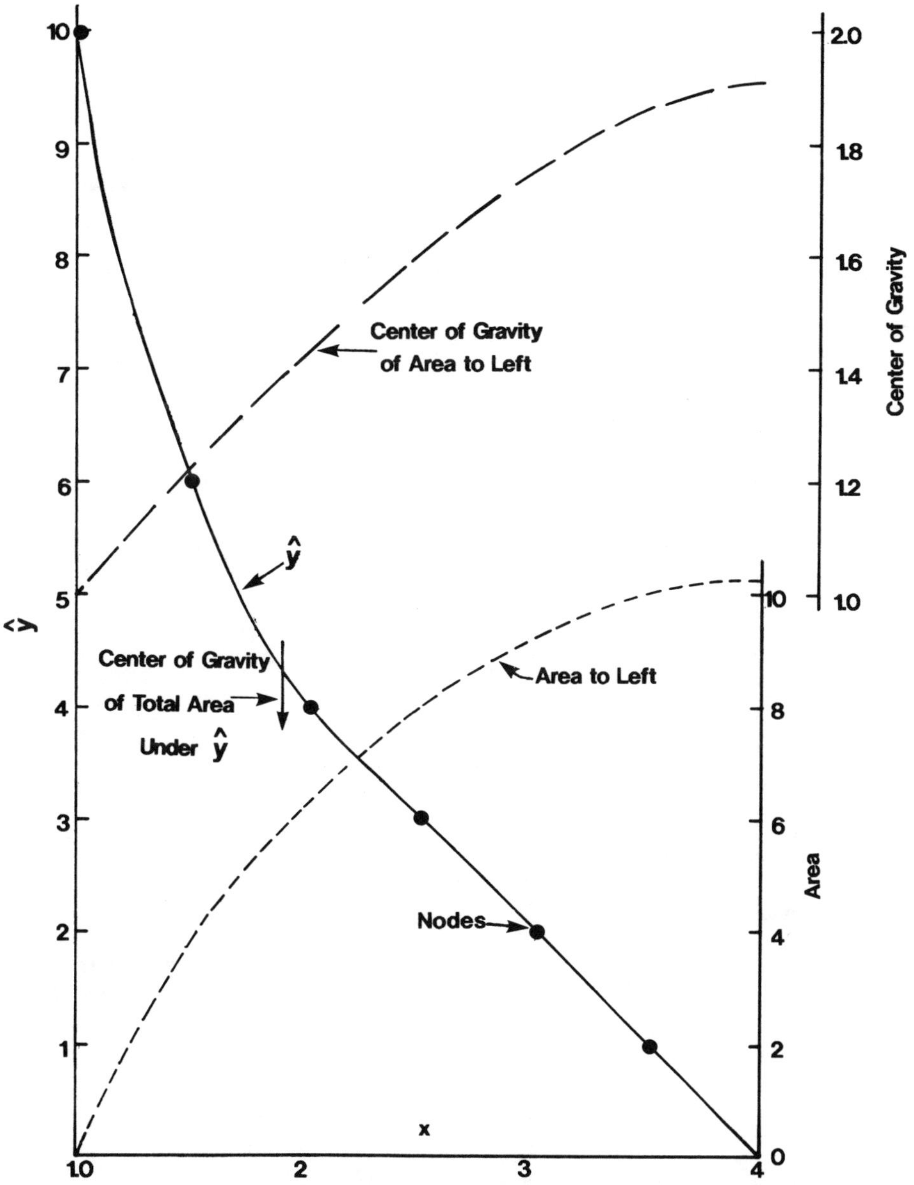

Fig. 4-6. Example 4-8, Center of Gravity

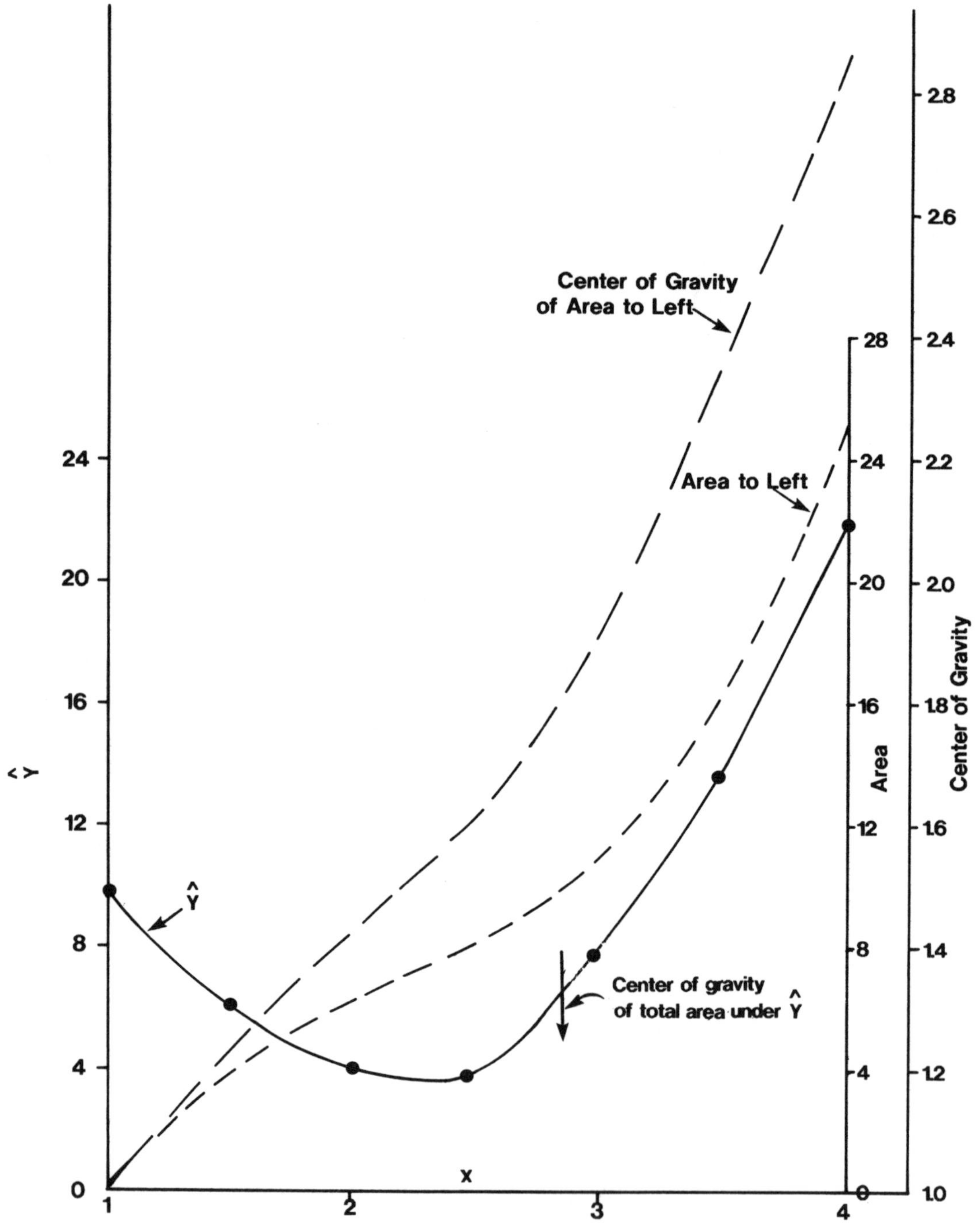

Fig. 4-7. Example 4-9, Center of Gravity

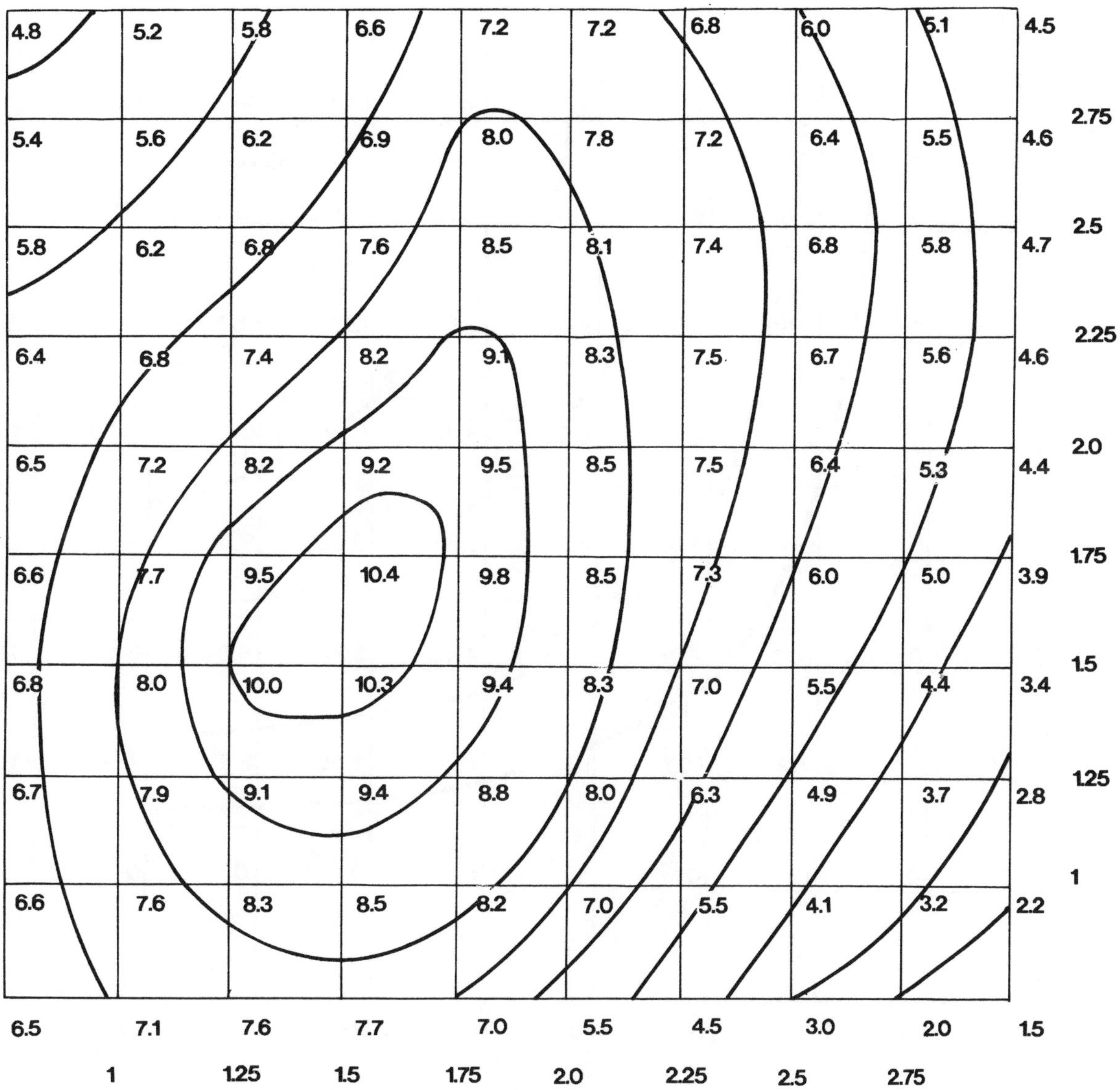

Part a. The Given Grid of Nodes

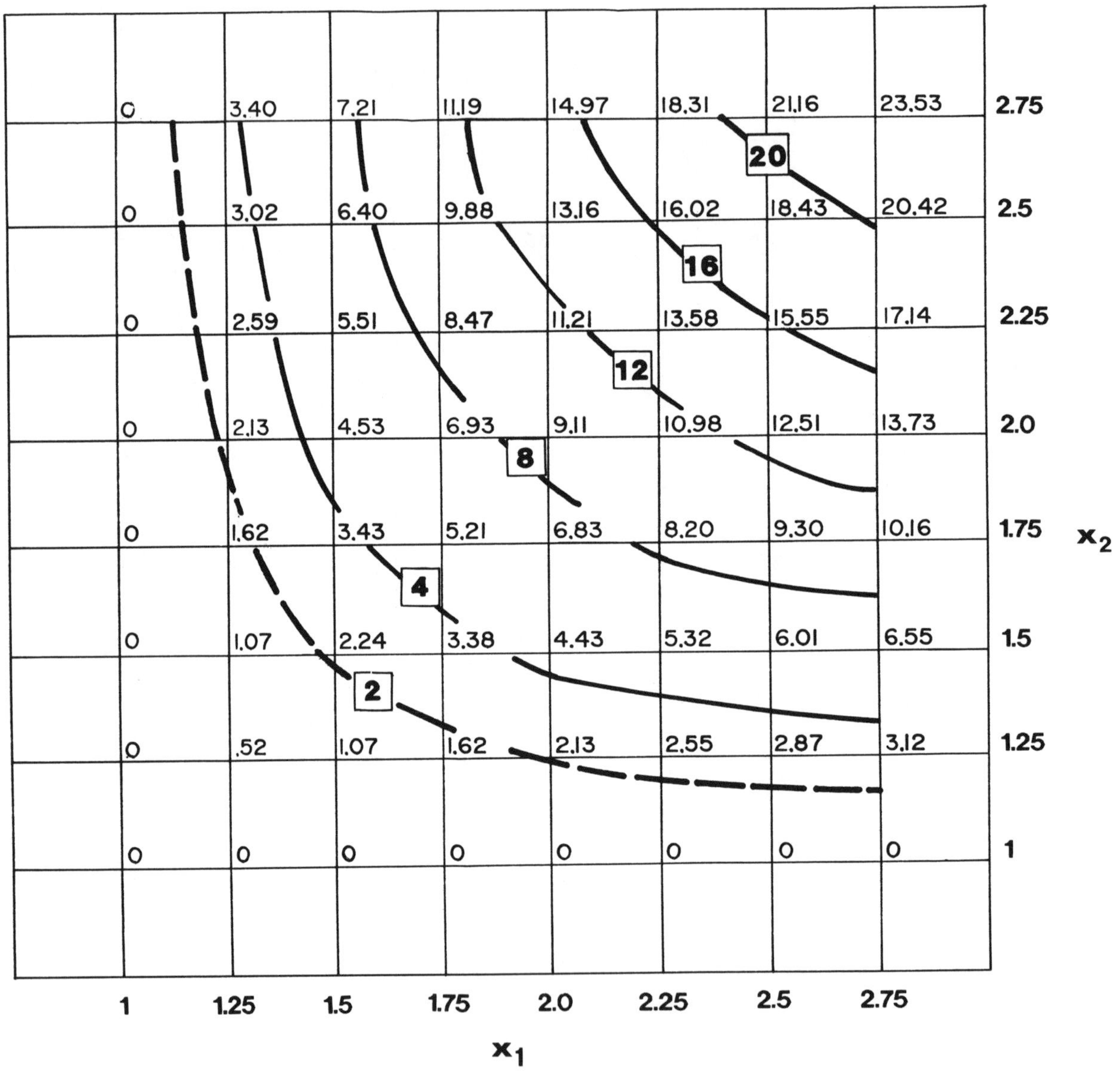

Part b. Accumulated Volumes by Grids

Fig. 4-8. Example 4.11. Double Integration

CHAPTER 5

Quantifying Sliding Polynomials With Experimental Or Historical Data

5.1 INTRODUCTION

With this chapter we begin to study ways of constructing the functions that we used in the previous chapters. Methods of interpolation, differentiation, and integration were discussed. In each of these methods we assumed that the function, expressed as a series of nodal values, was available as a starting point. Now we will explore methods to derive these functions when we are given information other than the functions as our starting point.

Two different kinds of information are generally available as a starting point. The first of these is numerical information. Quantifying functions with such data will be taken up first, in Chapters 5 and 6. The second type of information is mathematical, usually a statement using algebraic symbols that expresses our understanding of physical, chemical, or other processes that control the problem. We usually call them differential or integral equations. Quantifying functions with such statements amounts to solving these equations numerically and will be taken up in Chapters 7 and 8.

The numerical information we explore in this chapter is generally considered either historical data or experimental data. Historical data accumulate from record keeping. We have regular information on rainfall and temperature, on daily stock market activities, on crop production, or on the gross national product. Such historical data are usually recorded without direct consideration of cause-and-effect processes. The purpose is to establish inventories of resources or levels of activities. However, we can analyze such data to see how inventories or activities change, perhaps explore why and how they change in relation to other bits of information in our record-keeping. We can attempt to answer questions such as the following: What is the average temperature for a particular month? In what month does the greatest seasonal shift of temperature take place? Do high temperatures and high rainfalls occur together? Which economic indicators appear to swing with the stock market? Can we correct changes in market indexes for the usual seasonal cycles?

Experimental data are yielded by observation and measurement for a specific purpose. Such measuring and recording form one step in scientific methodology. We are attempting to confirm or reject a specifically stated hypothesis. Here we are very close to, and sometimes overlap, the thinking that we do in analyzing information that is a statement of process, or a differential or integral equation. In fact, we wish to confirm or reject our ideas on the processes at work. When working with experimental data, however, our methods must be conditioned by the fact that measurements contain errors; such errors may be minor "mistakes" or they may be due to a lack of precision in our measurements. We assume that the best estimate of the function we are searching for is given by smoothing across the errors in the experimental data. We also smooth across the minor and unexplained irregularities in historical data.

5.2 THE METHOD OF LEAST SQUARES WITH SLIDING POLYNOMIALS

The method of least squares is the most commonly used method for smoothing across minor fluctuations and errors. This simply means that we locate the smoothing function in such a way that the sum of the squared distances parallel to the axis of the dependent variable, from the data points to the function, is a minimum. Data smoothing by a least-squares operation using piecewise functions is similar to regression analysis in statistics, but we will later show some important differences.

The mechanics of a least-squares smoothing operation on numerical data will now be developed. We will use one span of the four-point sliding polynomials to develop the principle and illustrate it with detailed numerics. As developed in earlier chapters, any point on the curve, in the effective range of the four-point polynomial, can be predicted, or interpolated, by the equation:

$$\hat{y} = C_1Y_1 + C_2Y_2 + C_3Y_3 + C_4Y_4 \tag{5-1}$$

In Eq. 5-1, $\hat{y}$ means a calculated value of the variable y, the C_i are the coefficients, given as functions of z in Chapter 1, and Y_i are the four nodal values defining the sliding polynomial span. When we are smoothing, we imply the calculation of $\hat{y}$ for every point where we have a measurement y. The squared error between the smoothed value, $\hat{y}$, and the measured value, y, is:

$$\epsilon^2 = (\hat{y} - y)^2 \tag{5-2}$$

Suppose we have n measurements and we wish to smooth across these measurements. The total squared error is:

$$\sum_{j=1}^{n} \epsilon^2 = \sum_{j=1}^{n} (y - \hat{y})^2 = \sum_{j=1}^{n} y^2 - 2 \sum_{j=1}^{n} y\hat{y} + \sum_{j=1}^{n} \hat{y}^2 \tag{5-3}$$

By least-squares smoothing we mean that we want those values of Y_1, Y_2, Y_3, and Y_4 for which ϵ^2 is a minimum. It is a basic principle of calculus that to find a minimum we must differentiate $\Sigma\epsilon^2$ with respect to each Y_i in Eq. 5-1 after substituting Eq. 5-1 into Eq. 5-3:

$$\frac{\partial(\Sigma\epsilon^2)}{\partial Y_1} = -\sum 2y \frac{\partial \hat{y}}{\partial Y_1} + \sum 2\hat{y} \frac{\partial \hat{y}}{\partial Y_1} = -\sum C_1 y + \sum C_1 \hat{y} = 0 \tag{5-4}$$

Expanding $\hat{y}$ in Eq. 5-4 yields:

$$\frac{\partial(\Sigma\epsilon^2)}{\partial Y_1} = -\sum C_1 y + \sum (C_1C_1Y_1 + C_1C_2Y_2 + C_1C_3Y_3 + C_1C_4Y_4) = 0 \tag{5-5}$$

Similar equations can be written for $\partial(\Sigma\epsilon^2)/\partial Y_2$, $\partial(\Sigma\epsilon^2)/Y_3$, and $\partial(\Sigma\epsilon^2)/\partial Y_4$. Rewriting Eq. 5-5, plus the additional three equations for Y_2, Y_3, and Y_4, gives the set of four simultaneous equations:

$$(\Sigma C_1C_1)Y_1 + (\Sigma C_1C_2)Y_2 + (\Sigma C_1C_3)Y_3 + (\Sigma C_1C_4)Y_4 = \Sigma C_1 y \tag{5-6a}$$

$$(\Sigma C_2C_1)Y_1 + (\Sigma C_2C_2)Y_2 + (\Sigma C_2C_3)Y_3 + (\Sigma C_2C_4)Y_4 = \Sigma C_2 y \tag{5-6b}$$

$$(\Sigma C_3C_1)Y_1 + (\Sigma C_3C_2)Y_2 + (\Sigma C_3C_3)Y_3 + (\Sigma C_3C_4)Y_4 = \Sigma C_3 y \qquad (5\text{-}6c)$$

$$(\Sigma C_4C_1)Y_1 + (\Sigma C_4C_2)Y_2 + (\Sigma C_4C_3)Y_3 + (\Sigma C_4C_4)Y_4 = \Sigma C_4 y \qquad (5\text{-}6d)$$

Now for every data point of the set of n, we know the z-value, and therefore, we can compute the values of the C_i for each point. From these values of C_i we next calculate both the products in parentheses on the left-hand side of Eqs. 5-6 and also the right-hand side vector. We sum the coefficient products as indicated in Eqs. 5-6. Only the four nodes Y_i are now unknown. If we solve the four simultaneous equations, we get the values of these four Y_i, and it follows that this set of Y_i produces the minimum values of $\Sigma\epsilon^2$. We can call this the "best-fit" smoothing operation as defined by least squares.

While we used only one span with four nodes to develop this principle of least squares for a sliding polynomial approximation, any number of spans can be used. Two adjacent spans can be smoothed by solving a set of five simultaneous equations that have a structure similar to Eq. 5-6. Three spans can be smoothed by solving a set of six simultaneous equations. We can also smooth data using the two-point and six-point systems of sliding polynomials. We need only redefine the coefficients C_i to these systems. In the following examples we will also illustrate differential and integral smoothing.

The discussion of smoothing experimental or historical data by least squares has tended toward line or curve smoothing. This is the simplest and easiest introduction to smoothing using the sliding polynomial structures. We will first develop this methodology in the following sections of this chapter. However, smoothing of surfaces will also be discussed. This progression is a natural extension of analysis from one dimension to two dimensions.

5.3 ONE-DIMENSIONAL ANALYSIS

By one-dimensional analysis we mean a type of problem controlled by one independent variable that is functionally related to a dependent variable. Smoothing in one dimension produces lines or curves that average the data and are readily plotted on graph paper. Two-dimensional smoothing, to be discussed later, uses two independent variables related to a dependent variable. Mathematical surfaces, rather than curves, are produced by smoothing. Plotting of two-dimensional problems is not simple and requires perspective drawings or contours of the surface. Smoothing with sliding polynomials can be extended to greater than two dimensions. However, the computer programs for such multi-dimensional analysis require large amounts of memory for storage of data matrices. For this reason we will limit our discussion to one- and two-dimensional analyses that are readily performed on micro-computers.

5.3.1 Example 5-1: Basic Principles of Smoothing

The basic steps in least squares smoothing will be illustrated using the small data set at the top of Table 5-1 and displayed in Figure 5-1. Five data points are uniformly spaced in the x-scale. Such uniform spacing of data is not necessary but is used for simplicity in this example. If we had only data points 1, 3, and 5, we could draw the two straight-line segments as shown by the dashed lines in Figure 5-1. If we add points 2 and 4, however, then where do we place the best two joined straight-line segments? We specify three nodes for solution: Y_1 at x = 1, Y_2 at x = 1.5, and Y_3 at x = 2. We seek the best simultaneous values of these nodes.

At the top left in Table 5-1 are shown the values of the unit local variate z for the two spans. Note that any interior node such as Y_2 can be considered to belong to either the left or the right span. Given the values of z, the two-point sliding polynomial coefficients can be evaluated with Equations 1-3. These values are shown tabulated under their respective nodes. The y-values of the data are shown to the right. The sums of squares and sums of products of the C-values, shown in Eq. 5-6, are given in the center of Table 5-1. The solution for the best nodal values for the 5-point data set is shown at the bottom of Table 5-1. This solution is also displayed as the solid line in Fig. 5-1. Note that the effect of data point number 2 is to raise the left portion slightly, and the effect of data point number 4 is to lower the right portion slightly.

5.3.2 Example 5-2: Elementary Four-point Smoothing

The second data-smoothing example will use the four-point sliding polynomial. Eleven data points are plotted in Fig. 5-2. This synthetic data set was designed to have pronounced curvature to illustrate the suitability of the four-point system. The data points are tabulated at the top of Table 5-2. The z-values for two spans of the polynomial are shown, and the coefficients for five nodes are listed. Node Y_1 is the outlier node to the left of the data and node Y_5 is the outlier node to the right. These coefficients can be found in Appendix A. Alternatively, they can be calculated with Eq. 1-18. Program B−5.1 was used to calculate the sums of squares matrix in the middle of Table 5-2 and to yield the solution of the simultaneous equations at the bottom.

The four-point sliding polynomial interpolation program, Program B-2.1, was used to construct the curve through the nodes Y_1, Y_2, and Y_3. This smoothing curve is plotted in Figure 5-2. We should note the approach of the curve to node Y_3. The second data point from the right is relatively low, with a y-value of 23. At $x = 3$, the data point has a y-value of 25. The smoothing curve swings low and then re-curves to approach the data boundary at $x = 3$.

This synthetic data set was constructed to illustrate the re-curvature characteristic that was noted in the previous paragraph. There is no way to tell from the given data whether the re-curvature near x equal to 2.8 is real, or simply a happenstance of errors in the last two data points. The analyst has three options. The smoothing line may be accepted as derived, if there is no valid reason to reject it. If this re-curvature is believed to be suspicious, then additional data may be collected to try to confirm it. Lastly, if some additional knowledge of the controlling processes can be obtained, then the slope of the curve at the right boundary might be controlled by specifying appropriate boundary conditions. This procedure will be developed in future examples.

5.4 DIFFERENTIAL AND INTEGRAL SMOOTHING

In Examples 5-1 and 5-2 we saw how to smooth data and from this smoothing derive an approximate functional representation of the data. We are not limited to this use of sliding polynomials to get functions from data. It is also possible to smooth the data in such a way as to yield approximate integrals or approximate differentials of the data. This is one of the advantages of using the piecewise structures instead of simple regression analysis. The piecewise structures can produce functions, their integrals, or their differentials that are concentrated on local portions of the data. Contrarily, regressions, which assume the same functional form for the entire span of the data, may produce distortions in the analysis.

In functional smoothing we write the smoothing operation for one span as:

$$\sum_{i=1}^{n} C_i(z)Y_i = \hat{y} \approx y = f_1(x) \tag{5-7a}$$

$$z = f_2(x) \tag{5-7b}$$

This equation says we have a true but unknown function $y = f_1(x)$. We approximate this y with $\hat{y}$, which we derive from the sliding polynomial operation and which is a simple weighted summation on a local set of ordinates Y_i. In practice we write the direct operation:

$$\sum_{i=1}^{n} C_i(z)Y_i \approx f_i(x) \tag{5-8a}$$

$$z = f_2(x) \tag{5-8b}$$

If $f_1(x)$ is given as a set of observational data, as in Examples 5-1 and 5-2, and the $C_i(z)$ are calculated as functional or interpolation coefficients from the equations in Chapter 1, then smoothing yields the set of nodal values Y_i which represent the function. But suppose we differentiate Eq. 5-8 and get the following:

$$\sum_{i=1}^{n} C_i'(z)Y_i \approx df_i(x)/dz \tag{5-9a}$$

$$z = f_2(x) \tag{5-9b}$$

$C'_i(z)$ means the first derivative of $C_i(z)$. These derivative equations are given in Chapter 3. Now dz = $f'_2(x)dx$ and we can clear Eq. 5-9 to read as:

$$f'_2(x) \sum_{i=1}^{n} C'_i(z)Y_i \approx df_i(x)/dx \tag{5-10}$$

If a set of nodes Y_i in Eq. 5-10 are operated on with $C'_i(z)$, we get the differential of the approximating function $f_i(x)$. Y_i is an order higher than $df_1(x)/dx$. Now it follows that, if we consider a set of data points to be measures of the rate of change, such as slopes or gradients, and do least-squares smoothing using the $C'_i(z)$, then the resulting Y_i are integrals of the data.

We may also write the integral of Eq. 5-8 as:

$$\sum_{i=1}^{n} \int_{-0.5}^{z} C_i(z)dzY_i \approx \int_{x_1}^{x} f_1(x)dz \tag{5-11a}$$

$$z = f_2(x) \tag{5-11b}$$

Clearing Eq. 5-11 yields:

$$(1/f'_2(x)) \sum_{i=1}^{n} \int_{-0.5}^{z} C_i(z)dzY_i \approx \int_{x_1}^{x} f_1(x)dx \tag{5-12}$$

According to Eq. 5-12, if the data are treated as measures of the integration process $\int_{x_1}^{x} f_1(x)dx$ and these are smoothed by least squares with the $\int^{z} C_1(z)dz$, we will derive Y_i that are one order lower, and are, therefore, the smoothed differentials of the data. Integral and differential smoothing will be illustrated in the next two examples.

5.4.1 Example 5-3: Smoothing Differentials

We will consider the data in Fig. 5-2 to be measures of a differential dy/dx and we will use the structure of Eq. 5-10 to produce a smooth integral of the data. For $C'_i(z)$ will use the six-point sliding polynomial. The differential coefficients are given in Appendix A or can be computed using the equations in Chapter 3. The data are organized for least squares smoothing at the top of Table 5-3. The sums of squares matrix is shown at the center of this table, and the derived integral nodes are shown at the bottom.

One boundary condition was imposed on this solution. Since we are approximating an integral with our derived Y_i, we require the integral to be zero at the left edge of the data where x equals 1. Node Y_3 in Table 5-3 is at this left edge, with Y_2 and Y_1 lying to the left and outside the data. We set Y_3 equal to zero and drop this column of coefficients from the least squares solution. Note additionally that $f_2'(x) = 1$. If $f_2'(x)$ were some value other than unity, we would simply multiply the $C_i'(z)$ by that value as shown in Equation 5-10.

The solution shown at the bottom of Table 5-3 is plotted as the integral curve in Figure 5-3. The curve through the nodes was calculated using the four-point sliding polynomial interpolation Program B-2.1 in the appendix. Note that this curve is in agreement with the functional smoothing curve in Fig. 5-2. Y_4 in Table 5-3 has a value of 19.325. This is the integral of the first span in Fig. 5-2 and can be seen to be the average value for the unit spacing of the nodes. Y_5 in Table 5-3 is the total integral from x = 1 to x = 3. Y_5 minus Y_4 is 25.775 and is the integral of the second span in Figure 5-2.

5.4.2 Example 5-4: Smoothing Integrals

In this example an approximate differential of the data in Fig. 5-2 will be developed using the structure of Eq. 5-12. Remember that smoothing with integral coefficients yields a differential.

Smoothing with Eq. 5-12 requires advanced preparation of coefficients. This advanced preparation is not required in functional smoothing or when using differential coefficients to get integrals. Consider the term on the right side of Eq. 5-12, $\int_{x_1}^{x} f_1(x)dx$. Our measurements, or data, are now considered integrals from a left boundary, x_1, to the point x where the integral is measured. For one span x_1 is the left edge of the span and the integral can start at zero for that point and run to the right edge of the span. However, for the second span the total integral must be the value of the integral for the first span plus the within-span integration of the second span.

In further exploration of this integration concept, two special sets of the integral coefficients of the four-point sliding polynomial need to be considered. The integral of an entire span from z = -0.5 to z = 0.5 is given by:

$$\int_{x_1}^{x_2} y dx = \frac{1}{f_2'(x)} \int_{-0.5}^{0.5} \hat{y} dz = \frac{1}{f_2'(x)} \left(\begin{matrix} -0.041667Y_1 + 0.541667Y_2 \\ + 0.541667Y_3 - 0.041667Y_4 \end{matrix} \right) \quad (5\text{-}13)$$

Node Y_2 is located at z = -0.5, and node Y_3 is located at z = 0.5. The second set of integral coefficients used for illustration here is given by:

$$\int_{x_2}^{x} y dx = \frac{1}{f_2'(x)} \int_{-0.5}^{-0.3} y dz = \frac{1}{f_2'(x)} \left(\begin{matrix} -0.007533Y_2 + 0.193933Y_3 \\ + 0.014733Y_4 - 0.001133Y_5 \end{matrix} \right) \quad (5\text{-}14)$$

Equation 5-14 expresses the integral at the two-tenths point of the span to the right of the span in Eq. 5-13. This is a within-span integral. The total integral from the left-hand edge of the first span to the two-tenths point of the second span is the sum of Eq. 5-13 and 5-14, shown by:

$$\int_{x_1}^{x} y dx = \frac{1}{f_2'(x)} \left\{ \begin{matrix} -0.041667Y_1 + (0.541667 - 0.007533)Y_2 + (0.541667 \\ + 0.193933)Y_3 - (0.041667 - 0.014733)Y_4 - 0.001133Y_5 \end{matrix} \right\} \quad (5\text{-}15)$$

This concept of adding coefficients needs to be broadened to a general statement of procedure. For any point where measurement provides an integral we must add the coefficients of the within-span portion for one span to the coefficients giving the total integral at the left of the span.

Table 5-4 shows the within-span coefficients at the top. The sum of the within span and the left-edge coefficients is shown at the bottom. One particular feature must be pointed out. At the left edge of the first span, the first data point has a value of six. It is thus doubtful that we should start the integration while assuming that the smoothed integral is zero at x = 1. Therefore, we must use an extra span to the left of x = 1. This span is defined on nodes Y_0, Y_1, Y_2, and Y_3, and the integral at the right of this extra span is identically equal to the integral at the left of the first real span. For these reasons the extra node Y_0 was placed at the extreme left. This node only enters the data matrix at the top of Table 5-4 for the first data point. The value of Y_0 will thus not be well defined by the smoothing operation.

Since node Y_0 is not well defined, it will be algebraically eliminated by making it a function of other nodes. We assume that Y_0 will lie on the extension of a straight line through Y_1 and Y_2. This imposes a boundary condition that is well to the left of the two real data spans, and the effect can be expected to be minimal. If Y_0 lies on a line through Y_1 and Y_2, then simple algebra shows that:

$$Y_0 = 2Y_1 - Y_2 \tag{5-16}$$

Any row of a data matrix where Y_0, Y_1, and Y_2 occur, such as the following, can be considered an interpolation point for the four-point sliding polynomial:

$$C_0Y_0 + C_1Y_1 + C_2Y_2 + C_3Y_3 = \hat{y} \tag{5-17}$$

Substituting Eq. 5-16 for Y_0 yields:

$$C_0(2Y_1 - Y_2) + C_1Y_1 + C_2Y_2 + C_3Y_3 = \hat{y} \tag{5-18}$$

Combining terms yields:

$$(2C_0 + C_1)Y_1 + (C_2 - C_0)Y_2 + C_3Y_3 = \hat{y} \tag{5-19}$$

Equation 5-19 implies that Y_0 exists, but is expressed by Y_1 and Y_2, and is not expressed independently. We do not assign a value to Y_0; we only state that, whatever the values given by least squares for Y_1 and Y_2, Y_0 will lie on the straight line through them. Of course, the influence of such a Y_0 is exerted on the solution by the summed coefficients in Eq. 5-19.

The algebraic elimination of Y_0 was performed after the calculation of the total integral coefficients in the center of Table 5-4. Least squares smoothing was then performed, with the solution given at the bottom of the table. This solution is plotted in Fig. 5-3 as the differential, with four-point interpolation used to fill in the curve between nodes. It is important to observe the correspondence between this differential and the structure of the smoothed function in Fig. 5-2. The differential shows the curve has the greatest positive slope near the left boundary. The slope diminishes to zero near x = 2.2 where the function passes through a maximum. The greatest negative slope is at x = 2.55 and rises from this point to zero as the function passes through a minimum near x = 2.85 and continues to increase to about 16 at the right boundary.

5.4.3 Example 5-5: An Integral Smoothing Problem

For this problem data that represent observations of an accumulating process will be analyzed. The problem is a common one in hydraulics. Specifically, the data are measurements of the increasing volume of water in a storage tank. In our analysis we wish to smooth the data in such a way that we estimate the rate of inflow with time.

The simple integral covering this problem is:

$$\int_0^t Q\,dt = V \tag{5-20}$$

The equation simply says that the integral of the flow rate, Q (cubic meters per hour), integrated over a period t hours, is the volume, V (cubic meters).

For the solution of this problem Eq. 5-12 can be applied. Data are given at the top left of Table 5-5. These data consist of measurements taken systematically in time. Four spans of the sliding polynomial will be used to cover the time scale from 0 to 4 hours. Any number of spans that is consistent with the number of data points could be used. The within-span coefficients are given at the top of Table 5-5. The total integral coefficients calculated from these within-span values are shown at the center of the table.

Two boundary conditions were imposed before applying the least squares solution. First, the ordinate Q_0 at time zero was set to zero to mark the start of inflow. The column under Q_0 thus drops from the matrix of total integral coefficients prior to the sum-of-squares calculations. Second, Q_5, which lies beyond the time of measurements, was algebraically eliminated. Elimination was established by arbitrarily setting a small negative slope at t = 4 hours. For z = 0.5 the coefficients $C_i'(z)$ have the values 0, −0.5,

0, and 0.5. Therefore, for span number 4 in Table 5-5, and with unit spacing of the nodes, we can write:

$$dV/dt = dV/dz = 0.0\ Q_2 - 0.5\ Q_3 + 0.0\ Q_4 + 0.5\ Q_5 \tag{5-21}$$

For the boundary condition, dV/dt was set to the small value of -0.1 cubic meters per hour. Using this value and Eq. 5-21 gives:

$$Q_5 = Q_3 - 0.2 \tag{5-22}$$

This value for Q_5 was used to eliminate it from the last two data equations in the total integral coefficient matrix of Table 5-5. The coefficients 0.919271 and 1.041667 for node Q_3 become 0.906250 and 1.0, respectively, prior to sums-of-squares calculation. The column under Q_5 drops from the matrix of total integral coefficients. The volumes 4.3 and 4.45 become 4.297396 and 4.441667, respectively, as shown in the right column of Table 5-5.

The solution by least squares is shown at the bottom of Table 5-5. Q_0 and Q_5 have been filled in from their imposed boundary values. The original data and the derived inflow curve are plotted in Fig. 5-4. Direct interpolation was used to fill in the curve through the nodes of inflow rate. Integral interpolation was used to construct the smooth curve through the original measurements of volume.

5.4.4 Example 5-6: A Differential Smoothing Problem

This example of functional analysis of data will illustrate the reverse of the order change in Example 5-5. A set of differential data will be smoothed and integrated.

Imagine a small sailing vessel without sophisticated navigation equipment. An on-board compass is used to maintain a constant heading. Also, equipment is on board to log the speed of the vessel, and a clock is on board so that the time of the velocity measurements can be taken systematically. The vessel departs from some known reference point at time zero. The velocities logged during the voyage are given at the top left of Table 5-6. The problem is to compute the distance traveled during five hours sailing time.

The simple differential equation covering this example is given by:

$$\frac{dD}{dt} = V \tag{5-23}$$

in which D is the distance traveled and V is the velocity. Equation 5-10 with the four-point sliding polynomial will be used for this smoothing example. We can write directly the set of observation equations in the upper right of Table 5-6. For boundary conditions, D_0 was set to zero, and D_6 was projected as lying on the parabola that passes through D_3, D_4, and D_5. D_6 can be eliminated algebraically based on this boundary condition:

$$D_6 = D_3 - 3D_4 + 3D_5 \tag{5-24}$$

While not used in this example, the following equation could be used to eliminate D_{-1} on a left boundary condition of a parabolic projection:

$$D_{-1} = 3D_0 - 3D_1 + D_2 \tag{5-25}$$

Equations 5-24 and 5-25 are derived algebraically for uniform spacing of nodes in the x-scale. The solution for this problem by least squares, with columns under D_0 and D_6 dropped, is shown at the bottom of Table 5-6 and is plotted, with the velocity measurements, in Fig. 5-5. Values for D_0 and D_6 are given by the boundary conditions. The solution is validated by the construction of the differential of the distance. This construction of the velocities was obtained by differentiating the derived distance curve with the six-point sliding polynomial differential coefficients. The extra boundary nodes were estimated by parabolic projection. This constructed differential agrees closely with the measured velocities in Fig. 5-5.

5.4.5 Example 5-7: Integral Approximation with Scale Change

In this example the advantages of a change of scale for the independent variable for some problems will be illustrated. With a change of scale that is nonlinear, the solution nodes can be made to cluster, instead of being positioned uniformly in the scale of independent variable.

Figure 5-6 shows an accumulation process similar to that in Fig. 5-4. The difference is that in Fig. 5-6 the accumulation is initially rapid, and after some interval, tends toward zero change. It is well known that polynomials tend to "bounce" when attempting to approximate such processes. Typically, they will fit the initial rapid change, but cannot bend around the abrupt change of curvature. They will oscillate above and below the correct low rate of accumulation. As an alternative, sliding polynomials with close spacing of the nodes at the left and into the bend will more closely (i.e., more accurately) approximate such functions.

In order to cluster the solution nodes to the left, an intermediate variate, v, will be used. This variable is defined in Eq. 5-26. It should be understood that this equation for scale change only applies to this example:

$$v = 10 - 9 * \exp(-0.3t) \tag{5-26}$$

At t = 0, v = 1, and at t = 10, v = 9.5519. We will use five spans of the four-point sliding polynomial to cover the range from t = 0 to t = 10. The placement of nodes for these spans will be uniform in the v-scale with nodes separated by (9.5519 - 1)/5 = 1.7104 units of v. Other logarithmic and exponential functions could be used but must be designed for particular problem requirements.

Values of v at a uniform interval, with corresponding values of t computed from Eq. 5-26, are given in Table 5-7. Values at the mid-points in the v-scale are also included. In conducting an experiment, we would measure v, the accumulating volume, at the indicated times t. This is not essential, but it is convenient to specify measurements at uniform values of v to keep this example simple.

Table 5.7 shows the organization of the data that was used to derive a solution for the inflow function. The solution takes place in the v-scale where the nodes are uniformly distributed. The change form global variate v to local variate z is linear, and a simple calculation gives dz = 1.7104dv. The governing equation is the following integral:

$$\int_{1}^{9.5519} Q dv = V \qquad (5\text{-}27)$$

This equation can be approximated with span equations such as:

$$Q_1 \int_{-0.5}^{z} C_1(z)dz + Q_2 \int_{-0.5}^{z} C_2(z)dz + Q_3 \int_{-0.5}^{z} C_3(z)dz$$
$$+ Q_4 \int_{-0.5}^{z} C_4(z)dz = V/1.7104 \qquad (5\text{-}28)$$

The integral coefficients in Equation 5-28 are the same as those in Table 5-5. Total integral coefficients are calculated as in the center of Table 5-5.

Boundary conditions were set by eliminating Q_{-1} and Q_6 from free solution. Both of these nodes were eliminated by parabolic extension using Eq. 5-24 and 5-25. The reduced set of total integrals was then used to smooth the dependent variable V/1.7104. The solution is plotted in both the v and t scales in Figure 5-6. As stated earlier, the solution takes place in the v-scale and yields values of dV/dv. This rate is changed to dV/dt by (dV/dv)(dv/dt), where dv/dt from Eq. 5-26 is [2.7 exp(-0.3t)]. The solution was validated by constructing the integral of the inflow function for comparison with the original accumulation data in the v-scale. Each fifth point of this integral was then changed to its location in the t-scale and also plotted in Fig. 5-6.

Several features of the information in Fig. 5-6 are noteworthy. First, the rate curve dV/dt is a smoothly descending monotonic curve through nodes clustered to the left and more loosely spaced to the right. Second, both the V-curve in the v-scale and the dV/dv curve are rather shallowly curved and, therefore, easy to smooth. Both the V-curve in the t-scale and the dV/dt curve have sharper changes in curvature. Third, a curve of rate, or mathematical slope, changes in magnitude with a change in variate from t to v. An integral curve is also scale dependent. However, basic functions such as the V−curves in Figure 5-6 are not changed in magnitude by a change in scale. As a further explanation of this requirement, Eq. 5-8, 5-9, and 5-11, should be reviewed. Wherever the term dz occurs in an equation, the change of variate to dv and then to dt must be included in the calculations.

5.5 FUNCTIONAL UNCERTAINTY

A scientist smoothes an experimental data set when it contains errors. Errors should be taken herein to mean the agglomeration of all elements that produce an observation or measurement that cannot be regarded as totally precise or accurate. When the scientist smoothes data, an average, either a simple average of a single property or the average way the property relates to one or more variables, is sought. However, the scientist must also be concerned with the reliability of the average.

Piecewise functions are used in smoothing when a scientist feels that averages or average relationships should be localized, or when the piecewise functions are taken, in the ensemble, to represent some unknown or intractably complex function. Piecewise functions can also be used to localize the uncertainty of the averages that the scientist extracts from the data (Snyder, 1980). The technique is given a geometrical basis in Fig. 5-7.

Two methods of uncertainty analysis following piecewise smoothing are illustrated using Fig. 5-7. Smoothing was accomplished by deriving the values for four nodes by least squares. These nodes establish three linear segments. The method of evaluation is given in Table 5-8. The two-point coefficients are tabulated as a data matrix in the left portion of Table 5-8. Following the ordinate form of the equations, the dependent variable, y, is regressed on the C-values, and the nodal ordinates, Y_i, are obtained as regression coefficients. This is simply another way of describing least-squares smoothing.

The conventional calculation of the variance of a regression coefficient is explained in any standard statistical text. The sums-of-squares matrix of the C-values is inverted. The elements on the main diagonal of the inverse matrix are multiplied by the error variance residual to fitting to yield the variances of the respective regression coefficients. Results of this calculation are shown at the bottom left of Table 5-8. Note that this calculation assumes homoscedasticity (i.e., uniform variance of y for all values of x) and that the residual error can be averaged along the span of the function.

The consequence of these assumptions is evident in Fig. 5-7 and Table 5-8. This synthetic sample was deliberately constructed with large variability associated with the node at x = 0, yet conventional analysis of nodal variability yields the largest variance, 0.0334, for node 2.

An alternative method of calculation of the variability of the nodes yields a piecewise uncertainty that compliments the piecewise structure of the three segments in Fig. 5-7. This method weights the residual errors to the nodes using the C-values. The weighting function is simply a plot of the two-point sliding polynomial coefficients centered on Node 1 for illustration. The C-values for Node 1 in Table 5-8 are values of this weighting function at the data points. For example, the data points 5, 6, and 7 in Fig. 5-7 are weighted only to Node 1 since the C-values are unity. Data point number 3, however, weights by only 0.5 to Node 1, with the other one-half being weighted to Node 0. If the weighted data points are summed and if the sum is divided by the sum of the weights, obviously a weighted average results. In the same way, using the variance of a linear function, the squared errors can be weighted to the nodes. The form of the calculations is shown to the right of Table 5-8. The lower right-hand part of Table 5-8 shows two further adjustments to the weighted squared errors. First, they must be adjusted by the residual degrees of freedom, and second, they must be converted to the variance of a mean value by dividing by the sum of the weights.

Note the difference in nodal variability using the piecewise, localized uncertainty. The variance of Node 0 has increased significantly and is now roughly an order of magnitude greater than the variance of the other three nodes. This corresponds to the wide scattering of data points 1, 2, and 3 in Figure 5-7. The variances of Nodes 1, 2, and 3 are all reduced under localization and reflect the narrow scattering of residual errors in this portion of the piecewise function. The implication is that predictions in this span can be used for design or system management with greater assurance than could be assumed under the variance of regression coefficients.

Localized uncertainty based on smoothing with any sliding polynomial system is presented more formally as follows. Define an error, $e_i = y_i - \hat{y}_i$, where y_i is the measured value for data point i, and $\hat{y}_i$ is the interpolated value of the function at that point based on the nodal values derived by smoothing. The residual variance weighted to the node is given by:

$$E(e_i)^2 = \frac{\sum_{i=1}^{n} C_k^2 e_i^2}{\sum_{i=1}^{n} C_k^2} \tag{5-29}$$

The expected residual variance, $E(e_i)^2$, around the kth node is the weighted square error summed over the n points in the function span of that node. The C_k are the sliding polynomial coefficients weighting the data points to the nodes. They are exactly the same coefficients that are used in smoothing. The variance of the nodal value, s_k^2, is then computed as the variance of an expected value adjusted for the residual degrees of freedom as:

$$s_k^2 = E(e_i)^2 \text{ x } ND/(ND - NN)/\sum_{i=1}^{n} C_k \tag{5-30}$$

The value $\sum_{i=1}^{n} C_k$ is the portion of ND, the sample size, weighted to the kth node. The terms ΣC_k totalled for all nodes, NN, is identically equal to the number of data points, ND.

This section on functional uncertainty demonstrates the strong difference between smoothing by regression and smoothing by the piecewise sliding polynomials. We saw earlier that piecewise smoothing allows easy adjustment of the smoothing curve to local averages of the data, and connects these averages to form a mathematically continuous curve. On the other hand, regression forces a fit of one function across the entire span of data.

Using the concept of localized variance in this section we have developed a method to match residual errors locally to the localized averages derived by piecewise smoothing. The uncertainty of our functions is localized in conformance with the localized variance in the data. In regression analysis a principle of homoscedasticity is applied. This assumes a uniform distribution of variability across the entire span of data. The variance in many data sets does not conform to this assumption, and the functional uncertainty as measured by regression is then questionable.

5.6 AUTOMATION OF SLIDING POLYNOMIAL SMOOTHING

In the smoothing examples discussed to this point, we have been primarily concerned with illustrating and explaining the various steps and controls in smoothing with sliding polynomials. The coefficients were first constructed, boundary controls were applied where necessary, and then a least-squares computer program was used to generate the sums of squares matrix and to solve the simultaneous equations.

The entire operation of sliding polynomial smoothing can be incorporated into a computer program. Coefficients for interpolation, differentiation, or integration can readily be programmed for calculations using equations from Chapter 1. Boundary controls from equations such as Eqs. 5-16, 5-24, and 5-25 can be imposed. A change of scale can be included. Following the solution of the simultaneous equations, the residual errors and variances of the nodes can be computed.

Several small micro-computer programs are given in Appendix B. These are specific in utilization of two-, four-, or six-point systems and in imposing boundary conditions. They are general, however, in terms of the number and location of data points and in the number of nodes. These programs can serve in two ways. They can be used directly on problems for which they are suited and they can also serve as models or starting points for programs specific to other problems.

5.6.1 Example 5-8: Automation of Two-Point Smoothing

This example illustrates the use of Program B-5.2 in Appendix B, which finds an optimized set of nodes connected by linear segments and gives the statistical variance of these nodes using Eqs. 5-29 and 5-30. Data used in this example are given in Table 5-9. Seven nodes were chosen for the solution. The left-most node was placed at $x = 1$, and the right-most node was placed at $x = 5.5$. These nodes could have been placed beyond the data, say at $x = 0.9$ or at $x = 5.75$. However, they cannot be placed inside the range of data. The data and the smoothing curve are shown in Figure 5-8. Approximate confidence curves located one standard deviation above and below the smoothing curve are also shown. These are based on the computed standard deviations of the nodes. Values of the standard deviation for the confidence curve between nodes are calculated by:

$$\Delta y_z = \sqrt{\sum_1^2 C_i(z)^2 \, V_i} \tag{5-31}$$

in which Δy_2 is the standard deviation of the curve at point z between nodes i and i+1; the $C_i(z)$ are the sliding polynomial coefficients for point z; and V_i is the variance of the nodes.

The values of the nodes and the elements of their variance calculation from Eq. 5-29 and 5-30 are given in Table 5-10. The pattern for the standard deviation of the nodes should be particularly noted. Nodes 1 and 2 on the left of the data, where values of y are high and changing rapidly, have a high standard deviation. Nodes 3 through 7 have smaller and decreasing standard deviations. This synthetic data set was designed to illustrate experimental conditions where observations are subject to larger errors of measurement when values are high, but where errors are smaller for smaller values of y. Note the contrast with the regression smoothing approach where the assumption is made that the errors are uniform along the smoothing curve, i.e., homoscedasticity is assumed.

5.6.2 Example 5-9: Automation of Four-Point Smoothing

This example illustrates the use of Program B-5.3 in Appendix B. As in the previous example using two-point smoothing, this program performs all steps in the smoothing operation. The wave-form data in Table 5-11 and in Fig. 5-9 were smoothed. Five nodes were equally spaced from x = 10.2 to x = 11. Note that some data are negative and that the smoothing curve takes on negative values. Approximate confidence curves of one standard deviation plus and minus are also plotted to reflect the uncertainty. Program B-5.4 calculates the errors that are residual to smoothing, i.e., the values of the data points minus the interpolated values on the arcs of the sliding polynomials. The 30 residual errors are plotted in Fig. 5-10. The nodal standard deviations from Table 5-12 are also shown as vertical bars at the nodal positions. We should note that the errors are not uniform across the span of the data. The nodal standard deviations reflect this non-uniformity of error, but additionally, reflect the concentration of data around the nodes. In both Tables 5-10 and 5-12 we may note that the sum of the $\Sigma\ C_k$ values across all nodes is equal to the number of data points. Thus, the values $\Sigma\ C_k$ express the way data are concentrated around each node.

5.7 CYCLIC FUNCTIONS

Many cultural and natural processes are cyclic in form. Climatological data exhibit annual patterns of temperature, rainfall, and streamflow. Meteorological data have daily patterns superimposed on longer time trends. Power consumption shows the daily and weekly patterns of our everyday activities of work and recreation. Business indices show annual patterns reflecting holiday buying, vacations, tax times, and many other events. The study of time trends in activities that are normally cyclic requires that an average cyclic value be factored from the data. The residual values can then be studies for trends. The average values become base values of the cyclic process, and data at a specific time can be compared to the cyclic base to determine if events are running high or low compared to normal.

The sliding polynomial functions are readily adapted to smoothing cyclic data. The primary requirement is to set the boundary conditions so that the beginning left-most nodes overlap the ending right-most nodes to form a repeating cycle of all nodes. Conceptually, the end result is to produce a function wrapped around a cylinder. The process is shown schematically in Fig. 5-11. Part a of the figure shows nodes in the usual linear arrangement. Nodes are placed every third month with a total span that is sufficient to cover an entire year. For the four-point system we must first visualize some extra nodes. For example, an interpolation for December would be based on Node 3, and Node 4, Temporary Node 5, and Temporary Node 6. Now visualize wrapping the linear arrangement of nodes around a cylinder, which makes January and Temporary Node 5 coincide. Then Temporary Node 0 will coincide with Node 4, and Temporary Node 6 will coincide with Node 2. Thus, the cyclic function reduces to the curve running through the set of Nodes 1 through 4, and this set repeats for each year. Obviously, other cycles and patterns of nodes can be used so long as the beginning and end of cycles overlap.

The wrap-around of coefficients to produce a cyclic function is illustrated in Table 5-13 for a four-point system. In the top part of the table, nodes are arranged to cover an annual cycle with one node every third month, at January, April, July, and October. The four-point system requires the advance node N_0 and the two following nodes N_5 and N_6, after the scheme in Figure 5-11. To wrap-around Nodes 1 and 5, Nodes 2 and 6, and Nodes 4 and 0 are added. After summing, the reduced matrix of nodes is shown at the bottom of Table 5-13. The annual cycle is covered by four spans defined on four nodes with the coefficients cycling within the year for each of the four nodes. If a two-year span of data were to be analyzed, the pattern of nodal coefficients at the bottom of Table 5-13 would be repeated for the second year. Each additional year would require another repetition of the pattern. In actual problem analysis, say observed temperature, the dependent variable must be added as the right-hand vector.

5.7.1 Example 5-10: The Annual Cycle of Monthly Temperatures

Data on monthly average temperatures are widely available. Smoothing such data to develop the continuous trend of temperature through the seasons forms an interesting application of cyclic functions. Figure 5-12 displays the average monthly temperature for the years 1968 to 1970 at Lawrenceville, GA., which is close to Atlanta but removed from the effects of the metropolitan environment and the major airport. These data were smoothed by four spans of the four-point sliding polynomial with coefficients arranged as in Table 5-13. The smoothed curve is superimposed on the data in Figure 5-12. We should note the symmetry of the data and the curve, with the coldest temperature occurring in January just after the shortest day in December. The hottest temperatures occur in July, just after the longest day in June. Note additionally that there is month-to-month averaging by the smoothing curve, as well as within-month averaging. The September high value tends to balance the cluster of low values in November.

5.8 COMPOSITE SMOOTHING

All of the discussion and examples of smoothing to this point have dealt with data from a single function. Only one process was represented by the set of nodal values derived by smoothing. Such an assumption of a single process is not always correct. For example, we could visualize a container holding a certain amount of material. Material can be either added to or removed from the container. Any function to represent the amount of material in the container must be based on a composite of input and output. As another example, a cyclic process, such as annual climatic values, may change due to a long-term change in climate. An average value for any month must then be based on a composite function of cycle plus time trend.

Sliding polynomials can be used to smooth data supplied from composite functions. Certain restrictions must be observed. The data set must be sufficient in length and in detail to define each of the separate functions. We must know enough about the processes generating the data to enable us to construct good sliding polynomial approximations. The separate functions must be distinguishable. A descending exponential is difficult to separate from a linear time trend, especially when the measured data are over a limited range. However, cycles and linear time trends are readily separable.

The variances of the separate parts of the composite functions can be calculated using Eq. 5-29 and 5-30 if the coefficients, C, are re-defined. In single-function smoothing, the sum of the coefficients for each data point is unity, and therefore, the sum of all coefficients for all nodes is equal to the number of data points. If two sliding polynomial approximations composed a complex function, then the sum of all of the coefficients for every data point would be two. Now the coefficients in Eqs. 5-29 and 5-30 act as weights to apportion the residual error for each data point to the nodes directly affected by that point. Such partitioning of the errors requires a unit sum of weights to prevent bias. Therefore, before using Eqs. 5-29 and 5-30 for the nodal variance in composite functions, the sum of the coefficients must be unitized for each observation. This is easily accomplished by first summing the coefficients for each observation, and then dividing each of the coefficients by this sum.

Computer Program B-5.4 was prepared to smooth monthly data that are a composite of an annual cycle and a time trend. The cycle is defined by nodes at times 1, 3, 5, 7, 9, and 11 on a cyclic scale of 12, and the four-point sliding polynomial is used for the representation of the cyclic function. The time trend is defined on the two-point system, or a straight line, running from zero at time 1 to a single node at time one plus 12 times the number of cycles. While written primarily for trend analysis in monthly data, the program can be used for other data provided the time scale is set for a cycle length of 12.

Program B-5.4 was used to derive the average annual cycle of monthly temperature and the time trend for the years 1966 to 1970 at Lawrenceville, GA. The data are shown plotted in the top portion of Fig. 5-13. The derived smoothing curve plus time trend is superimposed on the data. The residual errors, observed value minus computed value, are plotted in the bottom portion of Fig. 5-13. We should note that there is no particular time trend nor cyclicity in the residuals.

Table 5-14 shows the derived nodal values, weighted sums for Eqs. 5-29 and 5-30, and the nodal variances and standard deviations. The nodal values are very stable; only the values for the winter months of January and March show standard deviations as much as one degree Fahrenheit. The node quantifying the time trend has a standard deviation of only one-half of one degree. We are now faced with a statistical decision. Is the warming trend of nearly two degrees in five years significant, or can it be disregarded as being a likely amount of variability due to random temperature fluctuations? To answer this question we use the following simple statistical equation:

$$T_c = T \pm t_{\alpha,\nu} s_T \tag{5-32}$$

This equation states that a test temperature change, T_c, will differ from a base value of change, T, by the amount $t_{\alpha,\nu}\ s_T$, where s_T is the standard deviation of T and $t_{\alpha,\nu}$ is a value of the t-distribution for level of significance α and degrees of freedom ν. In our example, the degrees of freedom for the time trend node can be estimated at 18.160 x (60-7)/60 = 16.04 (use 16). This is the sample size to degrees of freedom correction used in Eq. 5-30. If we ask whether or not the time trend could be zero, then T_c is zero and $t_{\alpha,16}$ = 1.924/0.545 = 3.53. Any standard statistical textbook will have tables of the t-distribution. Such a table will show that $t_{0.995, 16}$ is 2.92. Therefore, we must reject the null hypothesis of randomness and conclude that the probability that the computed t-value could occur at random is extremely small. The time trend of 1.924 degrees in the period 1966 to 1970 is statistically significant.

5.9 MOVING AVERAGES

Moving averages are frequently used to produce a simplistic smoothing of data that are ordered in time or space. The averages are the weighted totals across a selected number of data points, and the weighting system is moved through the data to produce a locus of averages. Simple moving averages should be considered limited in utility, but they are introduced here to contrast with the more elegant averaging of sliding polynomials.

The method of calculation of moving averages is illustrated using the data in Table 5-15 and Fig. 5-14. A synthetic set of cyclic data is to be averaged. Some set of weighting coefficients must be chosen. For the illustration here we will use the weights 0.25, 0.5, and 0.25. Note that these weights total unity, so no bias is introduced. As shown in Table 5-15, these weights are centered on the time for which a moving average value is desired. For time t = 0.5 the calculation is (0.25 x 0) + (0.5 x 1) + (0.25 x 3) = 1.25. For the next calculation the weights would be moved one data interval ahead, which gives (0.25 x 1) + (0.5 x 3) + (0.25 x 6) = 3.25.

The data and the calculated moving averages are shown in Fig. 5-14. Several features should be noted. During the nearly linear rising and falling portions of the cycle the moving averages are close to the data. At the shallow trough the average smoothes across the bottom with a moderate difference from the data point. At the sharp peak the averaging reduces the peak dramatically. Moving averages are often used as filters. The averaging process filters out all sharp irregularities with wave lengths that are shorter than the span of the set of weighting coefficients. Longer wave lengths pass through the filter.

Simple linear weighting systems such as in Table 5-15 are used frequently in moving average calculations; however, there are few rules that can be used to guide the choice of weighting scheme. One difficulty with a linear system is that the data points enter and leave the weighting abruptly as the calculations move through the data. Sliding polynomial coefficients can be used as weights in moving average calculations and offer the advantage of gradually phasing in and phasing out the data points as the averaging moves. The coefficients for the four-point sliding polynomials are plotted in Figure 5-15. The distance on the abscissa is measured on the z-scale and covers a range from 0 to 2. From 0 to 1 the curve represents values of coefficient $C_2(z)$ as z takes on values from -0.5 to 0.5. From 1 to 2 the curve represents values of coefficient $C_1(z)$ as z takes on values from -0.5 to 0.5. The mathematical structure of the sliding polynomials requires the curve to be continuous across the point where the distance is one and where the two coefficients join. Since sliding polynomials are symmetrical the coefficients $C_3(z)$ and $C_4(z)$ would form an identical but reversed plot. Figure 5-16 is a plot of the six-point coefficients: $C_3(z)$ for the distance from 0 to 1, $C_2(z)$ for the distance from 1 to 2, and $C_1(z)$ for the distance from 2 to 3. We can imagine that if we were to move these curves from left to right across the data points, the weights on the points would develop slowly from 0, going first slightly positive or negative, then swinging smoothly upward to 1 at a distance of 0. Phasing out would be simply the reverse as the distance increases. In contrast to this smoothly transitional phasing in and out, think of a triangle of linear weights moving across a data set.

In order to implement the use of sliding polynomial coefficients as weights in moving average calculations, we must develop the relationship between three scales as illustrated in Fig. 5-17 for both the four-point and six-point systems. The first scale is the distance scale used in Figs. 5-15 and 5-16. This distance is measured in units of the local variate z. Therefore, we need a relationship between the unit variate z, running from -0.5 to 0.5, and the d-scale. These are simple linear relationships of the form $z(c_i) = a + d$ where d is the distance and a takes on values -0.5, -1.5, and -2.5 for the ranges $0 \leq d \leq 1.$, $1.0 \leq d \leq 2.0$, and $2.0 \leq d \leq 3.0$, respectively. These simple relationships serve mainly to define rigorously the d-scale in Figs. 5-15 and 5-16. The more important relationship to be developed is that between the z-scale and the positioning of the data points within the averaging span. The number of data points to be used in the averaging span is an option of the user, and the relationship between the z-scale and the data-point positions depends on this choice. In Part a of Fig. 5-17, three spans of data are shown equivalent to a distance of two z-units. These were chosen for graphical illustration only. Utilizing symmetry, Part a implies weighting five data points at z-distances of -1.333 -0.667, 0, 0.667, and 1.333 from the moving average point.

A more general relationship between position and z-scale is needed. This generality is expressed for the four-point system as:

$$z(c_2) = -0.5 + (2/sdx)p \qquad (-0.5 \le z(c_2) \le 0.5) \qquad (5\text{-}33a)$$

$$z(c_1) = -1.5 + (2/sdx)p \qquad (-0.5 \le z(c_1) \le 0.5) \qquad (5\text{-}33b)$$

The following equations are the general relationship for the six-point system:

$$z(c_3) = -0.5 + (3/sdx)p \qquad (-0.5 \le z(c_3) \le 0.5) \qquad (5\text{-}34a)$$

$$z(c_2) = -1.5 + (3/sdx)p \qquad (-0.5 \le z(c_2) \le 0.5) \qquad (5\text{-}34b)$$

$$z(c_1) = -2.5 + (3/sdx)p \qquad (-0.5 \le z(c_1) \le 0.5) \qquad (5\text{-}34c)$$

In Eqs. 5-33 and 5-34, dx is the interval between data points, s is the number of such intervals to be used to either side of the averaging point, and p is an integral number of intervals of dx that gives the positions of the data points. For example, the positioning in Part a of Fig. 5-17 shows that p can take on values of $0 \cdot dx$, $1 \cdot dx$, or $2 \cdot dx$. For a chosen $s = 3$, we get $z(c_2) = 0$; for $p = 0 \cdot dx$, $z(c_2) = 0.167$ for $p = 1 \cdot dx$, and $z(c_1) = -0.167$ for $p = 2 \cdot dx$.

In summary, a value, s, must be chosen, where s is the number of data intervals to cover two unit spans of the four-point system or to cover three unit spans of the six-point system. Equations 5-33 and 5-34 then yield the values of the unit variate z for the data positions, p. The coefficients to weight the data points can be calculated from these z-values using the equations in Chapter 1. Alternatively, one can read the weights directly from either Fig. 5-15 or 5-16. In order to prevent bias, the weights for any system must be unitized. This is easily accomplished by dividing each weight by the sum of the weights.

The calculation of moving averages is readily programmed for micro-computers. Program B-5.5 in the Appendix is an elemental program based on the schematic in Part a, Figure 5-17. This program was used to calculate the moving average of monthly temperatures for Lawrenceville, Ga., for the period from 1966 to 1970, which is the same data used in Example 5-11. The data and calculated moving averages are plotted in Fig. 5-18. This figure may be compared with the sliding polynomial smoothing of the same data in Fig. 5-13. In general, moving averages tend to be irregular, with less smoothing across individual irregular data points.

A moving average calculation is simple and may suffice for some problems. However, it has serious limitations. We should note that moving averages do not evaluate a mathematically continuous function. The moving average locus of points, therefore, cannot be readily integrated or differentiated, or otherwise mathematically operated on. The moving averages are optimal only in a limited local sense; they are not an optimal expression of functions. Moving averages are not directly suited to calculation of mathematically continuous but localized variances.

5.10 AN ADVANCED SMOOTHING PROGRAM

The smoothing programs presented in this chapter have been comparatively simple and specific to particular problems. General programs applicable to a variety of problems are longer and more complex in order to deal with the many control options. Program B–5.6 is such a general program. This program smoothes a data set and calculates the first derivative of the smoothing curve using the six-point sliding polynomial. Standard deviations of the nodes are also calculated (Snyder, 1980).

The utility of Program B-5.6 is enhanced by the incorporation of two degrees of optional control on the smoothing. First, the nodes can be given non-uniform spacing and made to cluster to either the left or right edge, or to the center, of the data. This allows use of the program to smooth data from a wide variety of problems with a rapidly varying functional form. Second, a number of different boundary conditions can be set independently at either the left, right, or both, boundaries. These control options can be used to make the smoothing curve follow known processes for initial or boundary conditions. Additionally, the boundary control options can be set to produce smoothing of cyclic functions.

Detailed instructions for using Program B-5.6 are given with the program in the Appendix. Two examples that illustrate the use of the program to smooth cyclic climatic data are given in the following sections.

5.10.1 Example 5-12: Smoothing of Average Monthly Temperatures

Figure 5-19 illustrates the smoothing of average monthly temperatures for Athens, Georgia, for the years from 1972 to 1976, using a cyclic spacing of five nodes (NOAA, 1972-1976). The temperature is highest in early August and, of course, the temperature change is zero at that time. The temperature is minimum in early January. The striking seasonal difference between heating and cooling is best shown by the temperature change curve. From March through June temperatures increase steadily at about 6°F/month. As the days shorten after June, the temperature change curve decreases abruptly to a low of -11°F/month near the autumnal equinox. Then the temperature change curve rises steadily until near the vernal equinox.

The envelope curves in Fig. 5-8 were constructed by incrementing and decrementing the derived nodal values by one standard deviation. Nodal variability for these data is shown in Table 5-16. Wintertime temperatures are more variable than summertime temperatures, and this is reflected in the higher variance of the winter nodes. Note that the envelope curves represent uncertainty of the average temperatures given by the nodes and do not represent the residual scatter of the data.

5.10.2 Example 5-13: Smoothing of Monthly Rainfall Volumes

Figure 5-20 shows a second application of sliding polynomials to cyclical data, where ten years of monthly rainfall data at the Elliot gage (Staff, 1977) are smoothed. The volume of rainfall peaks twice annually, in January and in July to August, with low points in April and November. The summer months show greater rainfall variability than the remainder of the year. The varying width of the envelope curves reflects this seasonal variability, and it is also shown by the variances of the six nodes in Table 5-16.

5.11 TWO-DIMENSIONAL SMOOTHING

In Chapter 1 we saw that the sliding polynomial structure for one dimension could be written concisely as:

$$\hat{y} = \sum_{i=1}^{n} C_i(z)Y_i \tag{5-35}$$

where n takes values of 2, 4, or 6 depending upon which system is being used. Similarly, using function-of-functions, the two-dimensional structure could be written concisely as:

$$\hat{y} = \sum_{j=1}^{n} C_j(u) \qquad \sum_{i=1}^{n} C_i(z)Y_{ji} \tag{5-36}$$

In this equation z is the local unit variate in a dimension x, and u is the local unit variable in a dimension w. If Eq. 5-36 is expanded specifically for the four-point system of sliding polynomials, we get:

$$\begin{aligned}\hat{y} = {} & C_1(u)\ C_1(z)\ Y_{11} + C_1(u)\ C_2(z)\ Y_{12} + C_1(u)C_3(z)\ Y_{13} + C_1(u)\ C_4(z)\ Y_{14} \\ & + C_2(u)\ C_1(z)\ Y_{21} + C_2(u)\ C_2(z)\ Y_{22} + C_2(u)C_3(z)\ Y_{23} + C_2(u)\ C_4(z)\ Y_{24} \\ & + C_3(u)\ C_1(z)\ Y_{31} + C_3(u)\ C_2(z)\ Y_{32} + C_3(u)C_3(z)\ Y_{33} + C_3(u)\ C_4(z)\ Y_{34} \\ & + C_4(u)\ C_1(z)\ Y_{41} + C_4(u)\ C_2(z)\ Y_{42} + C_4(u)C_3(z)\ Y_{43} + C_4(u)\ C_4(z)\ Y_{44}\end{aligned} \tag{5-37}$$

Equation 5-37 defines a unit surface. For any data point (x, w, y) falling in this unit surface, we could calculate the unit local variates u and z from global variates w and x and then calculate all the coefficient-products $C_j(u)\ C_i(z)$. These coefficient-products are the weights for the data point to each of the 16 nodes Y_{ji}. Since y is known, least squares can be used to calculate the optimum values of Y_{ji} by minimizing the total squared error:

$$\sum_{K=1}^{ND} \epsilon^2 = \sum_{K=1}^{ND} (y - \hat{y})^2 \tag{5-38}$$

This is the same optimization objective that was used when smoothing in only one dimension. Smoothing in two dimensions yields a best-fit surface of y in the w-space versus x-space of the data, in contrast to smoothing in one dimension that yields a best-fit curve in just the x-space of the data.

We must be aware of one practical aspect of smoothing in two dimensions. Consider the grid of nodes in Fig. 5-21. This schematic is based on the four-point system, but the concept also applies to the two- and six-point systems. For data points with coordinates x, w, and y, we show a grid with $0 \leq x \leq 5$ and $0 \leq w \leq 6$. Now if w is varied across its full range for all values of x, then some data points would be expected to lie in all of the unit surfaces for reasonable sized samples. We would normally eliminate algebraically the boundary nodes, leaving a 6 x 7 mesh of nodes, or 42 unknowns to be determined by least squares. If all unit surfaces are occupied and we have, say, 84 data points, the solution should be straight-forward. Now suppose that in the data, w does not vary across its full range for all values of x. Then the points would plot within some bounded area as shown in Figure 5-21. No data points would fall on the surfaces marked "D". Therefore, some columns of the $C_j(u)\ C_i(z)$ matrix would contain only zeroes for the entire data set, and the sums of squares matrix would contain one or more rows or columns of zeroes. For this condition the simultaneous normal equations can not be solved. The obvious answer is to scan the $C_j(u)\ C_i(z)$ matrix and delete the nodes that have a zero total of weights summed across the full set of data. After using least-squares to obtain the solution of the reduced matrix, an irregular mesh of nodes is constructed. This mesh outlines the data space which can be contoured. Only those unit surfaces that have a positive sum of coefficients at their four corners can be contoured. All unit surfaces remaining in the least squares solution must be joined by at least one common boundary.

Program B-5.7 in the Appendix uses the two-point system to fit a surface to a data set. Unit surfaces that are not occupied by data are deleted. Program B-5.8 allows for non-uniform spacing of nodes. Program B-5.8 uses the four-point system for two-dimensional smoothing and also will delete unoccupied surfaces.

5.11.1 Example 5-11: Two-Point Surface Smoothing

Program B-5.7 can be used to smooth the data given in Table 5-15 and to generate the contours of the retained unit surfaces. The data and the solutions are shown in Fig. 5-22. Columns of nodes were specified at x-values of 2.0, 2.1, 2.3, and 2.6. The horizontal lines of nodes were specified at w-values of 0.0, 0.2, 0.4, and 0.6. No data points fall in either the top and middle left-hand or the middle and bottom right-hand unit surfaces. Twelve nodes that define the retained contiguous surfaces were determined by least squares. The nodal values and their standard deviations are shown in Fig. 5-22.

The unit surfaces are joined by only one common boundary along any one line of nodes. These single boundaries are sufficient to establish a continuous surface in an irregular pattern from the bottom left to the top right of the figure. The contours derived using the two-point system do show kinking where they cross the common boundaries.

5.11.2 Example 5-12: Four-Point Surface Smoothing

Program B-5.8 can be used to smooth the data given in Table 5-17, and the results using four-point smoothing can be compared with the results of Example 5-11.

Program B-5.8, which uses four-point smoothing, is of necessity more complex than Program B-5.7, which uses two-point smoothing. Some unit surfaces in a four-point smoothing system will be unoccupied just as in the two-point smoothing. Some of these surrounding nodes in the four-point system will serve as boundary nodes. Consequently, Program B-5.8 drops nodes with a zero sum of coefficients, and then additionally, eliminates all unoccupied unit surfaces from contouring. In Program B-5.8 the nodes are uniformly spaced in both the x- and w-scales.

Figure 5-23 shows the results of smoothing the data of Table 5-17, using two-dimensional, four-point smoothing. Five nodes were specified in the range $2.0 \leq x \leq 2.6$, and four nodes were specified in the range $0 \leq w \leq 0.6$. The middle and lower right-hand unit surfaces are unoccupied by data and, therefore, were deleted in a manner similar to Fig. 5-22. However the middle and upper left-hand surfaces were not deleted due to the different position of the x-nodes. With two exceptions, the two figures show similarity in the contours of y. In Fig. 5-23 a peak in the contours is found at x = 2.0, w = 0.4. This is in the zone of no data as shown in Fig. 5-22, and the contours result form extrapolation from the right. Obviously, the contours should not be used to predict y from both w and x in this region of no data. The second difference between Figs. 5-22 and 5-23 is in the upper right. The contours of Fig. 5-23 are much higher at (x = 2.6, w = 0.6) than those shown in Fig. 5-22. A shallow trough appears just below this peak. We have no way of knowing, just from the smoothing, which is the better set of contours. In an actual problem, we would answer the question based on the general processes at work. Do we expect a highly non-linear structure to occur at the right-hand side of our data, or do we expect a continuance of the quasi-linearity to the left? If time, money, and circumstances permit, then more measurements in the questionable zones of the data could be considered.

Smoothed functions based on the two-point and four-point systems can also be compared using their residual errors. The differences between the smoothing surface and the measured values of y are shown for the 34 data points in Table 5-18. The errors for the four-point system are considerably smaller, and the root-mean-square (RMS) error, $\sqrt{\Sigma\epsilon^2/34}$ for four point smoothing is only 60 percent of the RMS error for the two-point smoothing. The four largest errors from the two-point analysis, which have magnitudes of 0.12 or larger, are greatly reduced when the four-point analysis is used.

5.11.3 Example 5-13: Smoothing Rainfall-Runoff Data

Program B-5.8 can be used to smooth measured rainfall and measured runoff from a medium-sized watershed. Monthly rainfall and runoff data for the period from 1953 to 1971 for the Chattooga River near Clayton, Georgia, will be used. This is a watershed with 207 square miles of drainage area.

Runoff during any month is caused by rainfall during that month, but it is also caused by a continuance of streamflow from rain that fell in antecedent months. We will, therefore, use rain of the current month as our independent variable x. Runoff of the previous month will be used as an index of the effect of past rainfall and will be our independent variable w. Runoff for the current month will be our dependent variable.

Examination of the record shows that the largest measured monthly rainfall was 16.01 inches. The largest measured monthly runoff was 8.80 inches. We will use five nodes in the x-scale to cover the range of rainfall from 0 to 16.4 inches. Four nodes in the w-scale will cover the range for the antecedent runoff from 0 to 9.0 inches.

The results of smoothing the 216 monthly data points are shown in Fig. 5-24. Note first that no nodes were dropped from the 5 x 4 node mesh, but that the upper right-hand unit surface was deleted. No data points fell within this portion of the nodal mesh. Figure 5-24 shows two general features. In the lower central portion of the figure, the contours are nearly linear, uniformly spaced, and slope downward to the right. A hatched boundary has been drawn around this portion of the contouring. In the remainder of the figure, an irregular pattern of highs and lows is evident. We know how a physical system such as a watershed works, with increasing amounts of rain producing increasing amounts of runoff. Also, high runoff in the antecedent month should increase runoff in the current month. These two conditions are only true within the drawn boundary of contours sloping down to the right. The irregular patterns outside this boundary are due primarily to a scarcity of data in these zones. There are insufficient data points for the smoothing operation to establish an average rainfall/runoff relationship.

We should not regard the irregularity in the contours in a portion of Fig. 5-24 as a failure of sliding polynomial smoothing. Rather, it should teach us to be cautious in attempting to predict runoff in data zones with few measurements. Actually, this is true of any data analysis method, including regression analysis. Also, the sliding polynomial smoothing has served to factor the information in our data-space. We now know where in the space we can comfortably predict, based on our historical data, and where we cannot.

5.12 TRANSFORMATION OF DATA IN TWO-DIMENSIONAL SMOOTHING

In Example 5-7 the method of scale change prior to smoothing was illustrated for one-dimensional analysis. A change of scale was used to make the placement of nodes non-uniform across the space of the dependent variable. This technique can also be used in two-dimensional smoothing to make the nodes cluster within the empirical range of their independent variable, and thus tend to suppress the occurrence of unoccupied classes.

A particular problem occurs frequently when cyclic data are measured at fixed intervals, such as measurements of monthly mean temperature, monthly total rainfall, or monthly total streamflow. These data go through annual ranges. In two-dimensional analysis we could use time by months as the independent variable x, and the magnitude of the measurement as the independent variable w. If we set our nodes uniformly in the w-scale to cover the range for the months with the highest values, we will find unoccupied classes for the months with low values. We need to bring the data for all months into some standardized alignment prior to smoothing.

A simple and useful data transformation to bring monthly data into alignment is linear rescaling. Monthly total rainfall will be used as an example. Suppose the largest January total rainfall is P_j'. We will set the position of the highest node slightly above this, say at $P_j' + \delta$. Now rescale this value to some standardized highest nodal placement, which we will designate SP_j'. Then any January rain of P_j will standardize to a value SP_j by the simple proportion:

$$\frac{P_j' + \delta}{P_j} = \frac{SP_j'}{SP_j} \tag{5-39}$$

We can use any value for SP_j'. Suppose we use unity. Then

$$SP_j = \frac{P_j}{P_j' + \delta} \tag{5-40}$$

Similarly, for February we can write:

$$SP_F = \frac{P_F}{P_F' + \delta} \tag{5-41}$$

Such a proportion is used for each calendar month, simply using the maximum of record, P_m', for each month. The transformed data for every month will lie in the range $0 \leq SP_m \leq 1.0$.

Smoothing is carried out in the fixed time scale of months for the independent variable x, and in the standardized scale for the independent variable w. We should note that any value SP_c in the standardized scale, such as the location of a node, or a point of interpolation for contouring the dependent variable y, can be transformed to the appropriate natural scale for a specific month, m, by the simple proportion:

$$P_c = \frac{SP_c}{P'_m + \delta} \tag{5-42}$$

Program B-5.9 in the Appendix can be used to smooth monthly data, or within limits, any data with measurements at twelve equally spaced points within a cycle. The month of the year is a fixed variable in the x-scale; a causative independent variable, say monthly rainfall, is used in the w-scale, and a dependent variable, say monthly streamflow, is smoothed and then contoured. The contours are transformed from the standardized w-scale back to the natural scale of rainfall for each month. The upper boundary, $P'_m + \delta$, is calculated from a sine-cosine curve quantified by input values of $P'_m + \delta$ for January, May, and September.

It should be clearly noted that the dependent variable, which is smoothed and contoured, is never transformed. A transformation of the dependent variable would cause a distortion of the residual errors and would invalidate the "least squares" definition inherent in Eqs. 5-2 to 5-6.

Program B-5.9 was initially written to smooth monthly rainfall and runoff data. However, it can be used for any situation where a dependent variable, y, is related to an independent variable, w, and this relationship varies by months. The smoothing produces a cyclic surface. The surface is a two-dimensional, four-point sliding polynomial based initially on a 6 x 7 mesh of nodes as in Fig. 5-25. By using boundary and cyclic conditions this mesh is reduced to a 4 x 4 for least squares smoothing.

5.12.1 Example 5-14: Seasonal Smoothing of Rainfall and Runoff

Program B-5.9 can be used to smooth a record of rainfall and runoff from two watersheds. The results can then be compared.

For the first watershed a record from a 117 square mile drainage near Oxford, Mississippi, will be used. Records of rainfall and runoff are available from January 1957 to December 1968, a total of 144 months. For the second watershed a 99 square mile drainage area of Taylor Creek in central Florida will be used. A great difference between these two areas is in the type of climate. The Oxford watershed lies in a the temperate zone with low streamflows in late summer and autumn. Taylor Creek lies in the subtropical zone with low streamflows during the winter.

Figures 5-26 and 5-27 show the results of seasonal smoothing of the records. In Fig. 5-26, for the Oxford watershed, we see that the contours show high runoff in March and April. The runoff decreases and reaches a minimum in September. After October the runoff increases again throughout the winter to the next seasonal maximum in March. The rainfall is not highly variable throughout the year. In Fig. 5-27, for the Taylor Creek watershed, the high variability of rainfall is immediately evident, from a low in December of about 8 inches to a high in August and September of nearly 17 inches. Low streamflows occur in February and March for low rainfalls, and in May and June for higher rainfalls. Streamflow is highest in November and December.

We see from these analyses that two-dimensional sliding polynomial smoothing can be used with a diversity of seasonal patterns. The scale transform of the independent variable w first standardizes the patterns and allows an efficient smoothed surface to be determined.

EXERCISES

5-1. Derive the normal equations for the least squares solution of a single span using a two-point system of sliding polynomials.

5-2. Derive the normal equations for the least squares solution of a two-span problem using a four-point system of sliding polynomials.

5-3. Given the following data, derive least squares estimators for the nodes with a four-point sliding polynomial:

time	0	10	20	30	40
discharge	0	5	16	11	7

Assume discharge is the dependent variable.

5-4. Using the data from Problem 5-3 and the normal equations from Problem 5-1, derive least squares estimators for the nodes.

5-5. Using program 5-1 for the four-point system, evaluate the nodes for the following data:

day	0	15	30	45	60	75	90	105	120
SCA	100	100	98	92	74	48	34	22	16

The snow covered area (SCA in percent) is the dependent variable.

5-6. Given the data:

x	0.90	1.00	1.05	1.10	1.20	1.30	1.35	1.40	1.48
y	1.00	1.90	4.00	6.10	6.25	6.15	4.63	2.30	61.26.

Use Program B-5.2 to smooth the data with the two-point sliding polynomial. Place nodes at 0.9, 1.1, 1.3 and 1.5. Plot data and smoothing lines.

5-7. Use Program B-5.3 to smooth the data of Problem 5-6, while keeping the same positioning of the nodes. Plot the smoothing curve over the lines of Problem 5-6. Which solution would you prefer? Why?

5-8. Obtain a 5-year record of monthly average temperature at some climatological station. Smooth the data with Program B-5.4. Is there a significant temperature trend? During which calendar month does the temperature rise most rapidly? Which month has the greatest rate of cooling?

5-9. Given the data as follows

x	0.02	0.03	0.09	0.11	0.17	0.15	0.13	0.19
w	1.2	1.8	1.4	1.1	1.3	1.9	2.2	2.4
y	4.0	3.8	2.5	1.8	1.3	1.5	1.8	1.0

x	0.15	0.15	0.18	0.14	0.09	0.05	0.02	0.01
w	2.8	2.5	3.8	3.3	3.5	3.6	3.9	3.1
y	1.4	1.5	1.0	1.6	2.6	4.1	5.0	5.3

Use Program B-5.7 to smooth this data field. Place nodes at 0, 0.1, and 0.2 in x-scale and at 1, 2, 3, and 4 in w-scale. Plot the data and the contours. What would be the result if the following points were added to the data set?

x	0.02	0.08	0.01
w	2.4	2.5	2.8
y	4.5	3.0	4.7

5-10. Find a record of monthly rainfall-runoff values from a watershed. The record should be 25 years long at a minimum. Use Program B-5.8 to smooth the data. Monthly rainfall is placed in x-scale, while runoff is the dependent variable y. For w-scale use the following cosine values by calendar months.

	J	F	M	A	M	J	J	A	S	O	N	D
Circular Degrees:	270	300	330	0	30	60	90	120	150	180	210	240
Cosine:	0	0.5	0.866	1.0	0.866	0.5	0	-0.5	-0.866	-1.0	-0.866	-0.5

5-11. Compare the approach to cyclic function definition in Problem 5-10 and in Example 5-14.

Table 5-1

Data and Analysis for Example 5-1

Data Point	x	z	Coefficients for Node Y_1	Y_2	Y_3	y
1	1	-1/2	1	0	0	4
2	1.25	0	0.5	0.5	0	2.9
3	1.5	1/2(-1/2)	0	1	0	1.0
4	1.75	0	0	0.5	0.5	0.4
5	2.0	1/2	0	0	1	0.5

Sums of Squares or Products

1.25	0.25	0	5.45
0.25	1.50	0.25	2.65
0	0.25	1.25	0.70

Least Squares Solution

$Y_1 = 4.157, \quad Y_2 = 1.014, \quad Y_3 = 0.357$

Table 5-2

Data and Analysis for Example 5-2

Data Point	x	z	Coefficients for Node Y_1	Y_2	Y_3	Y_4	Y_5	y
1	1.0	-.5	0	1.0	0	0		6
2	1.2	-.3	-.064	.912	.168	-.016		10
3	1.4	-.1	-.072	.696	.424	-.048		20
4	1.6	.1	-.048	.424	.696	-.072		24
5	1.8	.3	-.016	.168	.912	-.064		26
6	2.0	.5(-.5)	0	0	1	0		25
7	2.2	-.3		-.064	.912	.168	-.016	27
8	2.4	-.1		-.072	.696	.424	-.048	27
9	2.6	.1		-.048	.424	.696	-.072	26
10	2.8	.3		-.016	.168	.912	-.064	23
11	3.0	.5		0	0	1	0	25

Sums of Squares Matrix

.01184	-.13152	-.08928	.00896	0	-3.648
-.13152	2.53600	.76512	-.17856	.00896	38.296
-.08928	.76512	4.04832	.76512	-.08928	133.880
.00896	-.17856	.76512	2.53600	-.13152	75.544
0	.00896	-.08928	-.13152	.01184	-5.072

Solution

Y_1	Y_2	Y_3	Y_4	Y_5
-67.275	5.052	26.953	24.476	42.921

Table 5-3: Data for Example 5-3

Data Point	x	z	Differential Coefficient for Node Y_1	Y_2	Y_3	Y_4	Y_5	Y_6	Y_7	dy/dx
1	1.0	-.5	.083333	-.666667	.000000	.666667	-.083333	.000000		6
2	1.2	-.3	.037333	-.282000	-.698667	1.114667	-.192000	.020667		10
3	1.4	-.1	-.018000	.097333	-1.322667	1.464000	-.259333	.038667		20
4	1.6	.1	-.038667	.259333	-1.464000	1.322667	-.097333	.018000		24
5	1.8	.3	-.020667	.192000	-1.114667	.698667	.282000	-.037333		26
6	2.0	-.5	.000000	.083333	-.666667	.000000	.666667	-.083333		25
7	2.2	-.3		.037333	-.282000	-.698667	1.114667	-.192000	.020667	27
8	2.4	-.1		-.018000	.097333	-1.322667	1.464000	-.259333	.038667	27
9	2.6	.1		-.038667	.259333	-1.464000	1.322667	-.097333	.018000	26
10	2.8	.3		-.020667	.192000	-1.114667	.698661	.282000	-.037333	23
11	3.0	.5		.000000	.083333	-.666667	.000000	.666667	-.083333	25

Sums of Squares Matrix

.010584	-.081831	.005234	-.011509	.000151	.000000	.952022
-.081831	.648145	-.061760	.118594	-.016073	.000151	7.467283
.005234	-.061760	12.135614	-6.011281	-.061760	.005234	-40.608012
-.011509	.118594	-6.011281	6.267867	-.607226	.077370	134.138717
.000151	-.016073	-.061760	-.607226	-.648145	-.081831	6.794053
.000000	.000151	.005234	.077370	-.081831	.010584	-.871966

Solution

Y_1	Y_2	Y_3	Y_4	Y_5	Y_6	Y_7
12.542	8.268	(0.0)	19.325	45.100	64.178	74.450

Table 5-4

Data for Example 5-4

Data Point	x	z	Within Span Coefficients for Node Y_0	Y_1	Y_2	Y_3	Y_4	Y_5
1	1.0	-.5	-.041667	.541667	.541667	-.041667		
2	1.2	-.3		-.007533	.193933	.014733	-.001133	
3	1.4	-.1		.021867	.356267	.073067	-.007467	
4	1.6	.1		.034200	.468600	.185400	-.019800	
5	1.8	.3		.040533	.526933	.347733	-.034133	
6	2.0	.5		.041667	.541667	.541667	-.041667	
7	2.2	-.3			-.007533	.193933	.014733	-.001133
8	2.4	-.1			-.021867	.356267	.073067	-.007467
9	2.6	.1			-.034200	.468600	.185400	-.019800
10	2.8	.3			-.040533	.526933	.347733	-.034133
11	3.0	.5			-.041667	.541667	.541667	-.041667

Data Point	x	z	Total Integral Coefficients for Node Y_0	Y_1	Y_2	Y_3	Y_4	Y_5	$\int_{xl}^{x} f(x)dx$
1	1.0	-.5	-.041667	.541667	.541667	.041667			6
2	1.2	-.3	-.041667	.534134	.735600	-.026934	-.001133		10
3	1.4	-.1	-.041667	.519800	.897934	.031400	-.007467		20
4	1.6	.1	-.041667	.507467	1.010267	.143733	-.019800		24
5	1.8	.3	-.041667	.501134	1.068600	.306066	-.034133		26
6	2.0	.5	-.041667	.500000	1.083333	.500000	-.041667		25
7	2.2	-.3	-.041667	.500000	1.075800	.693933	-.026934	-.001133	27
8	2.4	-.1	-.041667	.500000	1.061466	.856267	.031400	-.007467	27
9	2.6	.1	-.041667	.500000	1.049133	.968600	.143733	-.019800	26
10	2.8	.3	-.041667	.500000	1.042800	1.026933	.306066	-.034133	23
11	3.0	.5	-.041667	.500000	1.041667	1.041667	.500000	-.041667	25

Solution

Y_1	Y_2	Y_3	Y_4	Y_5
-20.643	36.255	3.488	15.956	261.532

Table 5-5

Data for Example 5-5

Time Hours	Measured Volume	Z for Span 1	2	3	4	Within Span Coefficients for Node Q_{-1}	Q_0	Q_1	Q_2	Q_3	Q_4	Q_5
-1.0	-	-1.5										
0.0	0	-0.5	-1.5			0	0	0	0			
0.5	0.3	0				-.028646	.419271	.122396	-.013021			
1.0	1.3	0.5	-0.5	-1.5		-.041667	.541667	.541667	-.041667			
1.5	2.5		0			-.028646	.419271	.122396	-.013021			
2.0	3.3	1.5	0.5	-0.5	-1.5	-.041667	.541667	.541667	-.041667			
2.5	3.8			0				-.028646	.419271	.122396	-.013021	
3.0	4.1		1.5	0.5	-0.5			-.041667	.541667	.541667	-.041667	
3.5	4.3				0				-.028646	.419271	.122396	-.013021
4.0	4.45			1.5	0.5				-.041667	.541667	.541667	-.041667
5.0	-				1.5							

Time Hours	Volume	Total Integral Coefficients for Node Q_{-1}	Q_0	Q_1	Q_2	Q_3	Q_4	Q_5
0.0	0	0	0	0	0	0	0	0
0.5	0.3	-.028646	.419271	.122396	-.013021			
1.0	1.3	-.041667	.541667	.541667	-.041667			
1.5	2.5	-.041667	.513021	.960938	.080729	-.013021		
2.0	3.3	-.041667	.500000	1.083333	.500000	-.041667		
2.5	3.8	-.041667	.500000	1.054687	.919271	.080729	-.013021	
3.0	4.1	-.041667	.500000	1.041667	1.041667	.500000	-.041667	
3.5	4.297396	-.041667	.500000	1.041667	1.013021	.919271	.080729	-.013021
4.0	4.441667	-.041667	.500000	1.041667	1.000000	1.041667	.500000	-.041667

Solution

Q_1	Q_0	Q_1	Q_2	Q_3	Q_4	Q_5
-.199	(0.0)	2.492	1.226	.470	.286	(270)

Table 5-6

Data for Example 5-6

Time Hours	Velocity Km./Hr.	Z for Span 1	2	3	4	5	Differential Coefficient for Node D_{-1}	D_0	D_1	D_2	D_3	D_4	D_5	D_6
-1.0		-1.5												
0.0	0	-0.5	-1.5				-.5	0	.5	0				
0.5	1.0	0					.125	-1.375	1.375	-.125				
1.0	2.0	.5	-0.5	-1.5				-.5	0	.5	0			
1.5	3.5		0					-.125	-1.375	1.375	-.125			
2.0	5.0	1.5	.5	-.5	-1.5				-.5	0	.5	0		
2.5	6.0			0					.125	-1.375	1.375	-.125		
3.0	7.0		1.5	.5	-.5	-1.5				-.5	0	.5	0	
3.5	7.5				0					.125	-1.375	1.375	-.125	
4.0	8.0			1.5	.5	-.5					-.5	0	.5	0
4.5	8.25					0					.125	-1.375	1.375	-.125
5.0	8.5				1.5	.5						-.5	0	.5
6.0						1.5								

Solution

D_{-1}	D_0	D_1	D_2	D_3	D_4	D_5	D_6
1.038	(0.0)	1.016	4.559	10.650	18.245	26.485	35.36

Table 5-7

Transform for Example 5-7

v		t	V	V/1.7104
1	*	0	0	0
1.8552		.3328	0.96	0.5613
2.7104	*	.7026	1.86	1.0875
3.5656		1.1185	2.70	1.5786
4.4208	*	1.5939	3.47	2.0288
5.2760		2.1485	4.17	2.4380
6.1311	*	2.8142	4.78	2.7947
6.9863		3.6469	5.32	3.1104
7.8415	*	4.7594	5.75	3.3618
8.6967		6.4411	6.08	3.5547
9.5519	*	10.0000	6.25	3.6541

Table 5-8

Calculation of Nodal Variability for Fig. 5-7

Data Point Number	C-Values for Nodes 1	2	3	4	Measured y	Computed $\hat{y}$	Error, e, $y - \hat{y}$	$c_1^2e^2$	$c_2^2e^2$	$c_3^2e^2$	$c_4^2e^2$
1	1.00				0.30	0.6120	-0.3120	0.0973			
2	1.00				1.00	0.6120	0.3880	0.1505			
3	0.50	0.50			0.90	1.1091	-0.2091	0.0109	0.0109		
4	0.20	0.80			1.55	1.4073	0.1427	0.0008	0.0130		
5		1.00			1.45	1.6062	-0.1562		0.0244		
6		1.00			1.65	1.6062	0.0438		0.0019		
7		1.00			1.75	1.6062	0.1438		0.0207		
8		0.90	0.10		1.55	1.6282	-0.0782		0.0050	0.0001	
9		0.65	0.35		1.75	1.6834	0.0666		0.0019	0.0005	
10		0.15	0.85		1.70	1.7937	-0.0937		0.0002	0.0063	
11			0.65	0.35	2.35	2.2647	0.0853			0.0031	0.0009
12			0.20	0.80	2.90	2.8277	0.0723			0.0002	0.0033
13			0.05	0.95	2.90	3.0154	-0.1154			0.0000	0.0120
14				1.00	3.10	3.0780	0.0220				0.0005
Sum	2.70	6.00	2.20	3.10			-0.0001	0.2595	0.0780	0.0102	0.0167
Sum of Squares	2.29	5.14	1.32	2.66			0.4046				

Error Variance = 0.40460/(14 - 4) = 0.04046

Derived by Least Squares

Node Number	Node Value	Nodal Variability as Regression Coefficients: Main Diagonal	Variance	Standard Deviation	Nodal Variability as Localized Uncertainty	Variance	Stan. Dev.
1	0.6120	0.4432	0.0179	0.134	0.2595/2.29 x 14/(14 - 4)/2.70 =	0.0588	0.242
2	1.6062	0.2035	0.0082	0.091	0.0780/5.145 x 14/(14 - 4)/6.00 =	0.0035	0.059
3	1.8268	0.8265	0.0334	0.183	0.0102/1.32 x 14/(14 - 4)/2.20 =	0.0049	0.070
4	3.0780	0.3973	0.0161	0.127	0.0167/2.665 x 14/(14 - 4)/3.10 =	0.0028	0.053

Table 5-9

Data for Example 5-8

No	x	y	No	x	y	No	x	y
1	1.0	6.0	10	2.45	1.0	19	4.15	0.2
2	1.05	6.5	11	2.7	0.6	20	4.3	0.5
3	1.2	5.3	12	2.85	0.8	21	4.6	0.4
4	1.45	2.4	13	3.05	0.5	22	4.7	0.7
5	1.5	4.2	14	3.4	0.8	23	4.9	0.2
6	1.65	3.2	15	3.4	0.3	24	5.25	0.6
7	1.75	1.8	16	3.65	0.8	25	5.5	0.3
8	2.0	1.3	17	3.9	0.4			
9	2.1	0.8	18	4.05	0.8			

Table 5-10

Nodes and Nodal Variance for Example 5-8

x-location of Node	Nodal Value	$\Sigma c^2_k e^2_i$	Σc^2_k	Σc_k	S^2_k	S_k
1.0	6.3606	0.6512	2.6978	3.5333	0.09489	0.3080
1.75	1.8917	1.3658	3.3644	4.7333	0.11912	0.3451
2.5	0.6394	0.1364	2.0933	3.2667	0.02770	0.1664
3.25	0.6136	0.1092	2.3422	3.6667	0.01767	0.1329
4.0	0.5265	0.1652	3.0311	4.4000	0.01720	0.1312
4.75	0.4543	0.0988	2.4667	3.5333	0.01574	0.1255
5.5	0.3815	0.0257	1.4844	1.8667	0.01289	0.1135
				25.0000		

Table 5-11

Data for Example 5-9

No.	x	y	No.	x	y	No.	x	y
1	10.20	-1.2	11	10.42	3.2	21	10.77	3.1
2	10.20	-0.4	12	10.47	1.3	22	10.82	3.6
3	10.22	0.5	13	10.49	0.3	23	10.85	2.6
4	10.25	0.5	14	10.51	0.6	24	10.89	3.3
5	10.25	1.8	15	10.54	0.1	25	10.92	2.6
6	10.29	3.0	16	10.60	-0.5	26	10.97	1.9
7	10.29	2.3	17	10.65	0.3	27	11.00	0.7
8	10.33	3.3	18	10.73	2.0	28	10.98	-0.6
9	10.35	2.8	19	10.78	2.2	29	11.00	-1.4
10	10.39	2.9	20	10.71	-0.1	30	10.73	0.8

Table 5-12

Nodes and Nodal Variance for Example 5-9

x-location of Node	Nodal Value	$\Sigma c^2_k e^2_i$	Σc^2_k	Σc_k	S^2_k	S_k
10.2	-0.4463	1.1260	4.0525	5.1691	0.06451	0.2540
10.4	3.1458	1.3727	6.3000	8.2006	0.03188	0.1786
10.6	-1.2705	2.0560	3.8225	4.4162	0.14615	0.3823
10.8	3.0299	2.7642	6.4844	7.8894	0.06484	0.2546
11.0	-0.1119	4.1186	3.7342	4.3247	0.30604	0.5532
				30.000		

Table 5-13

Organization of Coefficients For Cyclic Functions

Data Point	*Node* N_0	N_1	N_2	N_3	N_4	N_5	N_6
				Basic Arrangement of Nodes			
1	0	1.0	0	0			
2	-.07407	.77778	.33333	-.03704			
3	-.03704	.33333	.77778	-.07407			
4		0	1.0	0	0		
5		-.07407	.77778	.33333	-.03704		
6		-.03704	.33333	.77778	-.07407		
7			0	1.0	0	0	
8			-.07407	.77778	.33333	-.03704	
9			-.03704	.33333	.77778	-.07407	
10				0	1.0	0	0
11				-.07407	.77778	.33333	-.03704
12				-.03704	.33333	.77778	-.07407
				Arrangement for Cyclic Function			
1		1.0	0	0	0		
2		.77778	.33333	-.03704	-.07407		
3		.33333	.77778	-.07407	-.03704		
4		0	1.0	0	0		
5		-.07407	.77778	.33333	-.03704		
6		-.03704	.33333	.77778	-.07407		
7		0	0	1.0	0		
8		-.03704	-.07407	.77778	.33333		
9		-.07407	-.03704	.33333	.77778		
10		0	0	0	1.0		
11		.33333	-.03704	-.07407	.77778		
12		.77778	-.07407	-.03704	.33333		

Table 5-14

Results of Composite Smoothing: Lawrenceville, Georgia Monthly Temperature, 1966-1970

Node	Nodal Value	$\Sigma c^z_k e^2_1$	Σc^z_k	Σc_k	S^2_k	S_k
Time trend	1.924	40.830	6.698	18.160	0.2965	0.545
January	39.407	43.653	4.515	7.202	1.1858	1.089
March	51.356	37.298	4.515	7.300	0.9997	1.000
May	68.011	12.640	4.290	7.089	0.3672	0.606
July	77.377	20.552	4.079	6.919	0.6433	0.802
September	70.589	17.085	3.884	6.757	0.5751	0.758
November	50.103	18.052	3.706	6.573	0.6547	0.809
				60.000		

Table 5-15

Simple Moving Average Calculation

t	y	Weights	Moving Average
0	0	.25:*	
0.5	1	.5 :	1.25
1.0	3	.25:	3.25
1.5	6		6.25
2.0	10		8.00
2.5	6		6.25
3.0	3		3.25
3.5	1		1.25
4.0	0		0.50
4.5	1		1.25
5.0	3	.25:**	3.25
5.5	6	.5 :	6.25
6.0	10	.25:	

*Positioning for average at t=0.5
**Positioning for average at t=5.5

Table 5-16

Variability of Nodes

Node Number	$\Sigma C^2_k e^2_i$	ΣC^2_k	ΣC_k	s^2_k	s_k
		Figure 5-19			
1	166.51	10.30	11.96	1.474	1.214
2	61.03	10.36	12.03	0.534	0.731
3	30.91	10.33	11.99	0.272	0.522
4	28.53	10.33	11.99	0.251	0.501
5	78.64	10.36	12.03	0.688	0.829
			60.00		
		Figure 5-20			
1	23.54	17.06	20.0	0.073	0.270
2	21.55	17.06	20.0	0.066	0.257
3	59.81	17.06	20.0	0.185	0.430
4	95.06	17.06	20.0	0.293	0.541
5	105.92	17.06	20.0	0.327	0.572
6	49.45	17.06	20.0	0.153	0.391
			120.0		

Table 5-17

Data for Example 5-11

Point No	x	w	y	Point No	x	w	y
1	2.52	0.55	2.67	18	2.16	0.39	2.01
2	2.58	0.53	2.51	19	2.13	0.31	2.37
3	2.42	0.54	2.09	20	2.30	0.25	1.98
4	2.45	0.59	2.45	21	2.20	0.24	2.36
5	2.55	0.42	2.11	22	2.11	0.26	2.51
6	2.41	0.41	1.81	23	2.26	0.19	2.39
7	2.47	0.46	2.09	24	2.29	0.11	2.57
8	2.31	0.41	1.63	25	2.20	0.14	2.63
9	2.39	0.49	1.88	26	2.19	0.18	2.58
10	2.25	0.43	1.60	27	2.10	0.16	2.73
11	2.24	0.47	1.59	28	2.01	0.17	2.86
12	2.19	0.59	1.99	29	2.05	0.14	2.95
13	2.13	0.46	2.01	30	2.29	0.01	2.82
14	2.30	0.56	1.89	31	2.20	0.00	2.89
15	2.21	0.51	1.65	32	2.15	0.05	2.99
16	2.22	0.60	1.96	33	2.09	0.02	3.13
17	2.28	0.37	1.77	34	2.06	0.07	3.06

Table 5-18

Comparison of Residual Errors,
Example 5-11 And 5-12

Point No	Residual Error 2 pt	4 pt	Point No	Residual Error 2 pt	4 pt
1	0.150	0.076	18	0.053	0.064
2	-1.21	-.048	19	.066	.036
3	-.078	-.047	20	-.085	-.062
4	.030	.000	21	.007	.008
5	-.001	-.002	22	-.006	-.040
6	.047	.003	23	.015	.009
7	.029	.059	24	.040	.030
8	.089	.020	25	-.003	-.016
9	-.057	-.096	26	.019	.012
10	-.089	-.027	27	-.074	-.053
11	-.171	-.056	28	-.012	-.024
12	.084	.049	29	.034	.067
13	.027	-.024	30	.009	-.017
14	.059	.059	31	-.076	.009
15	-.198	-.085	32	.060	.013
16	.049	-.001	33	.025	-.031
17	.100	.100	34	-.022	.017

	2 pt	4 pt
Sum errors, Σe	-.001	.002
Sum error^2, Σe^2	.1924	.0729
$\sqrt{\Sigma e^2/34}$	.0752	.0463

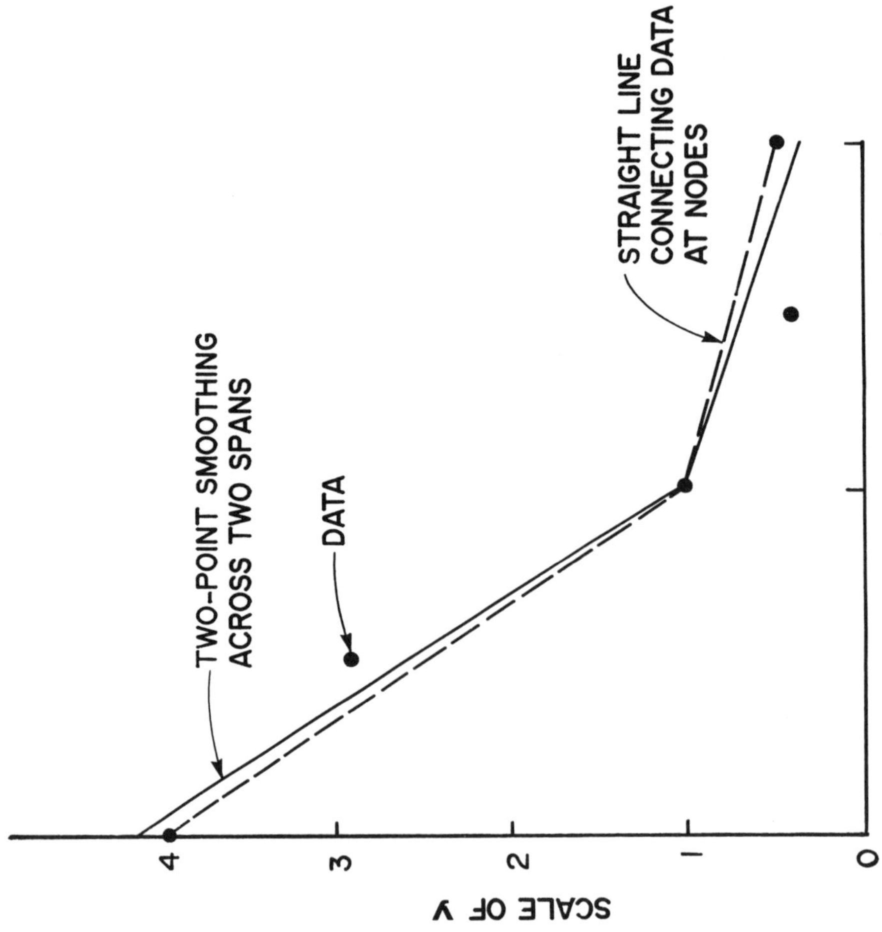

***Fig. 5-1.* Data and Piecewise Smoothing of Example 5-1**

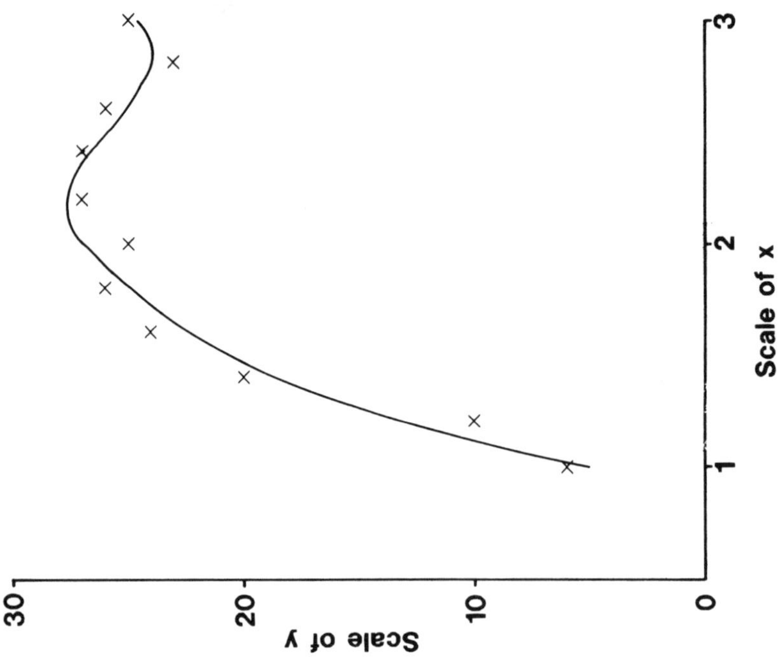

***Fig. 5-2.* Data and Piecewise Smoothing for Example 5-2**

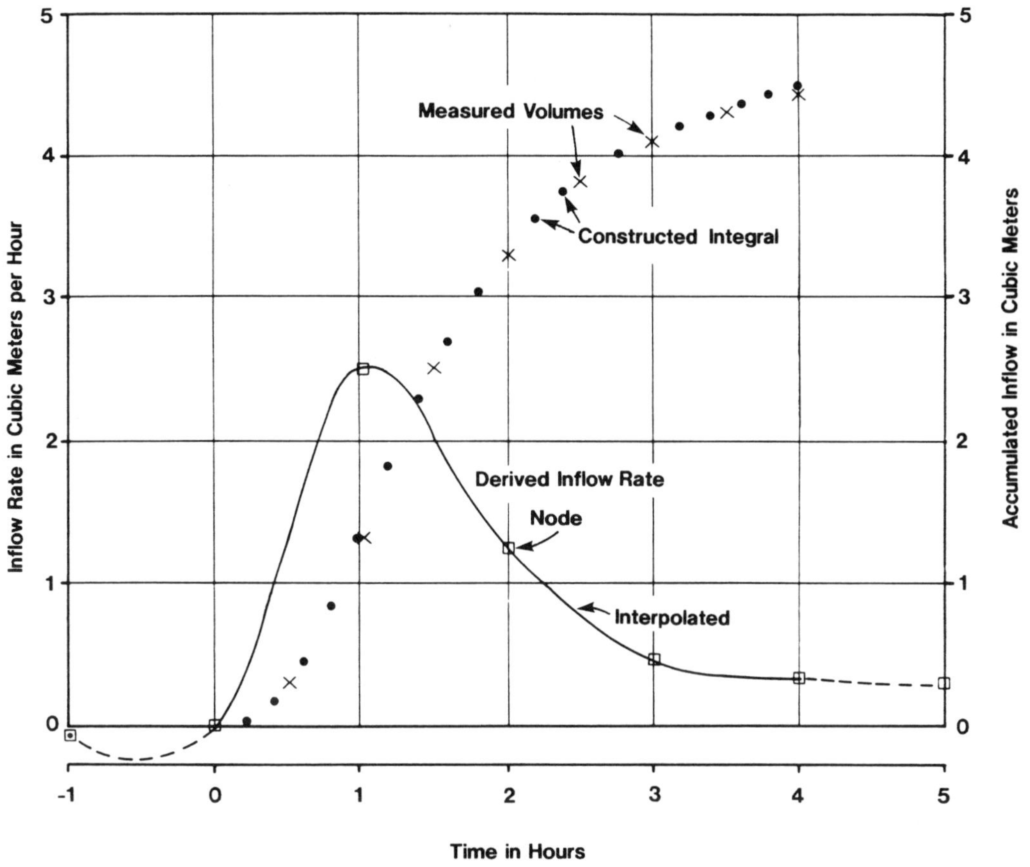

Fig. 5-4. Integral Approximation

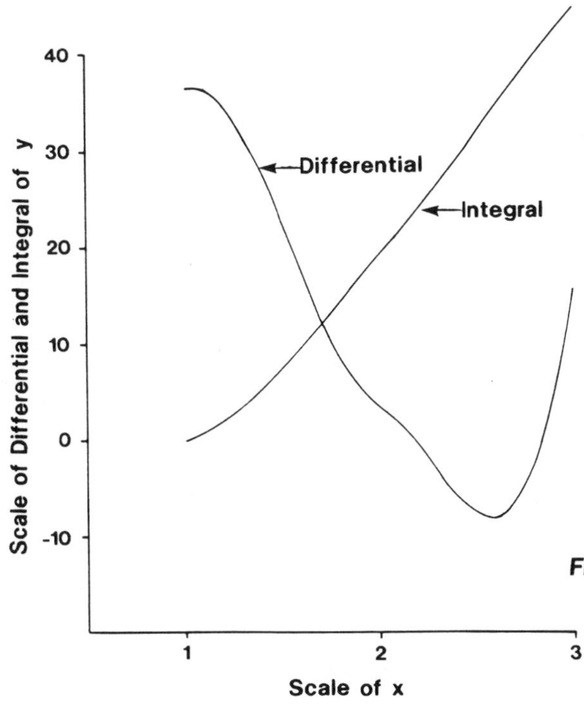

Fig. 5-3. Integral and Differential Smoothing

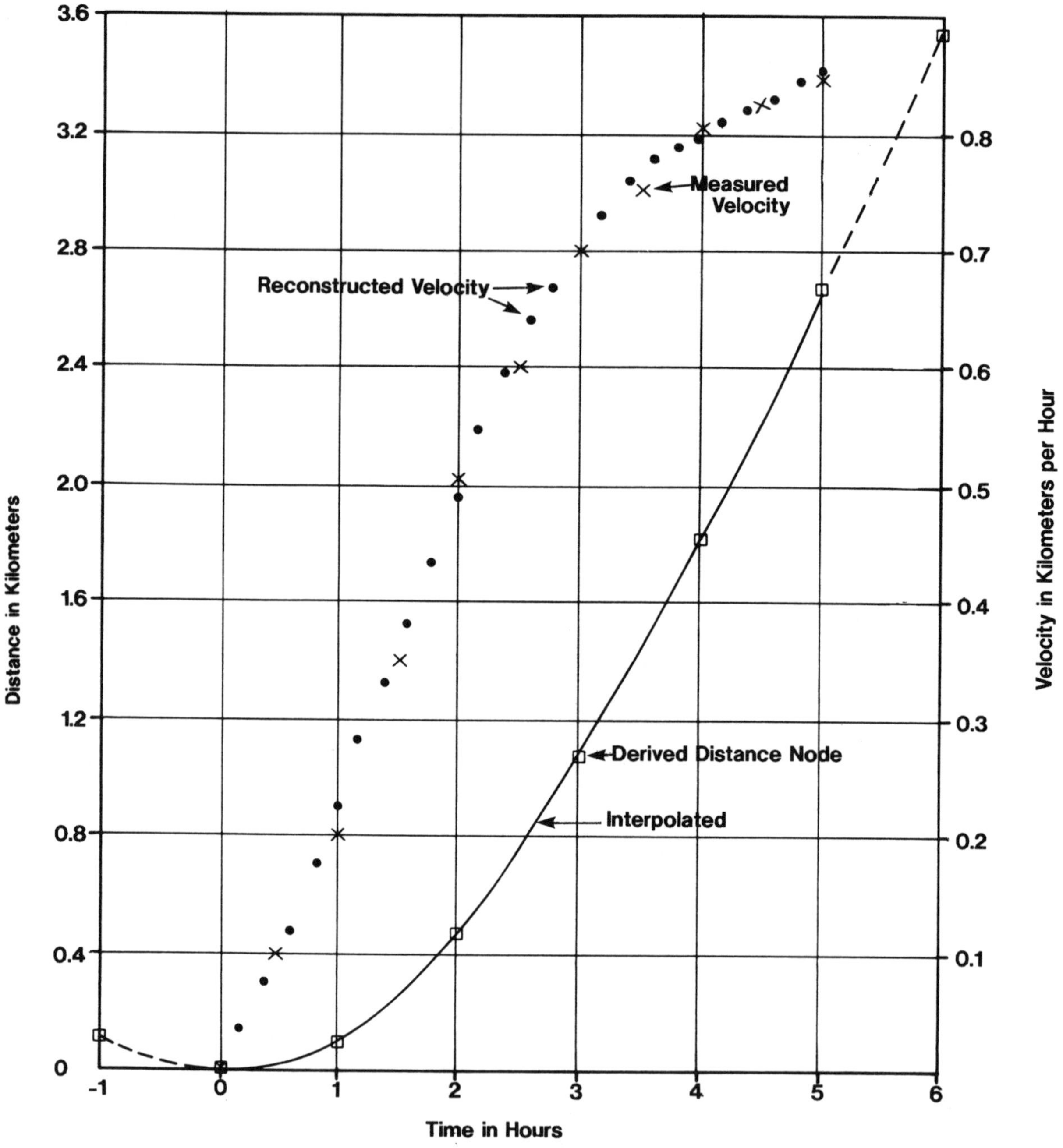

Fig. 5-5. Differential Approximation

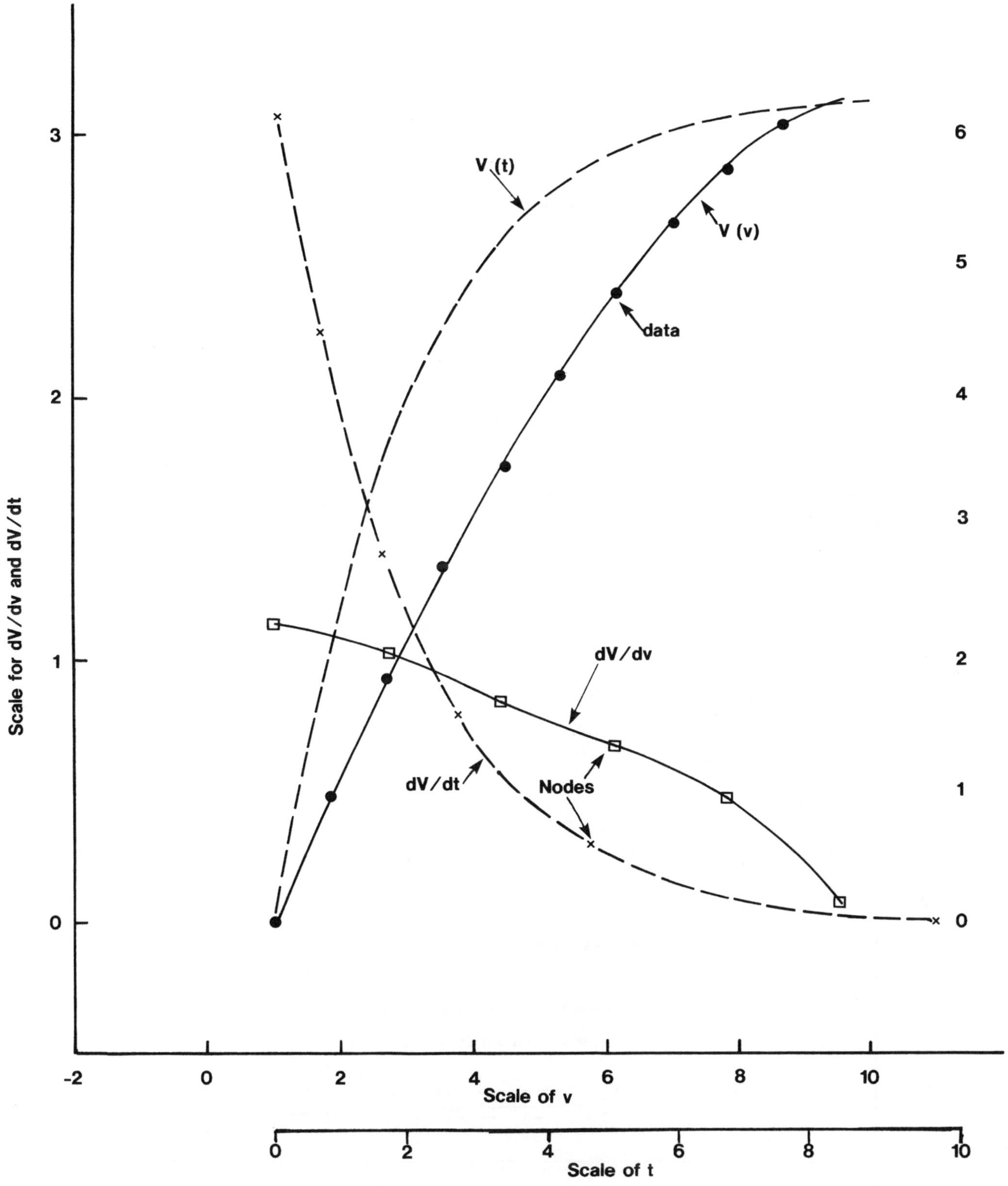

Fig. 5-6. Integral Approximation with Scale Change

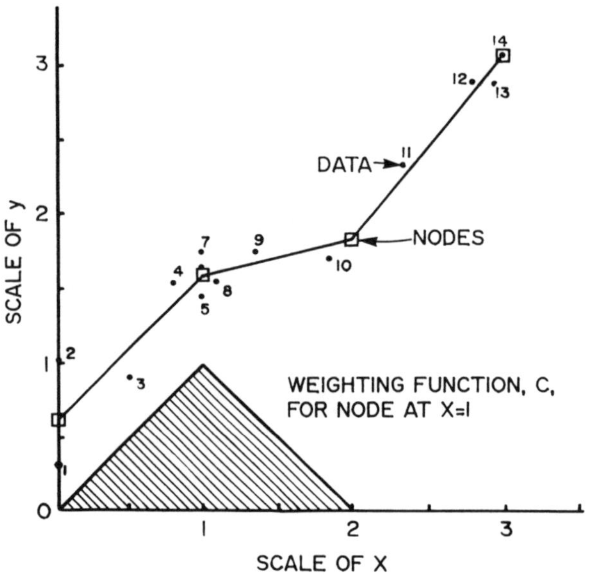

Fig. 5-7. The Geometry of Nodal Variability

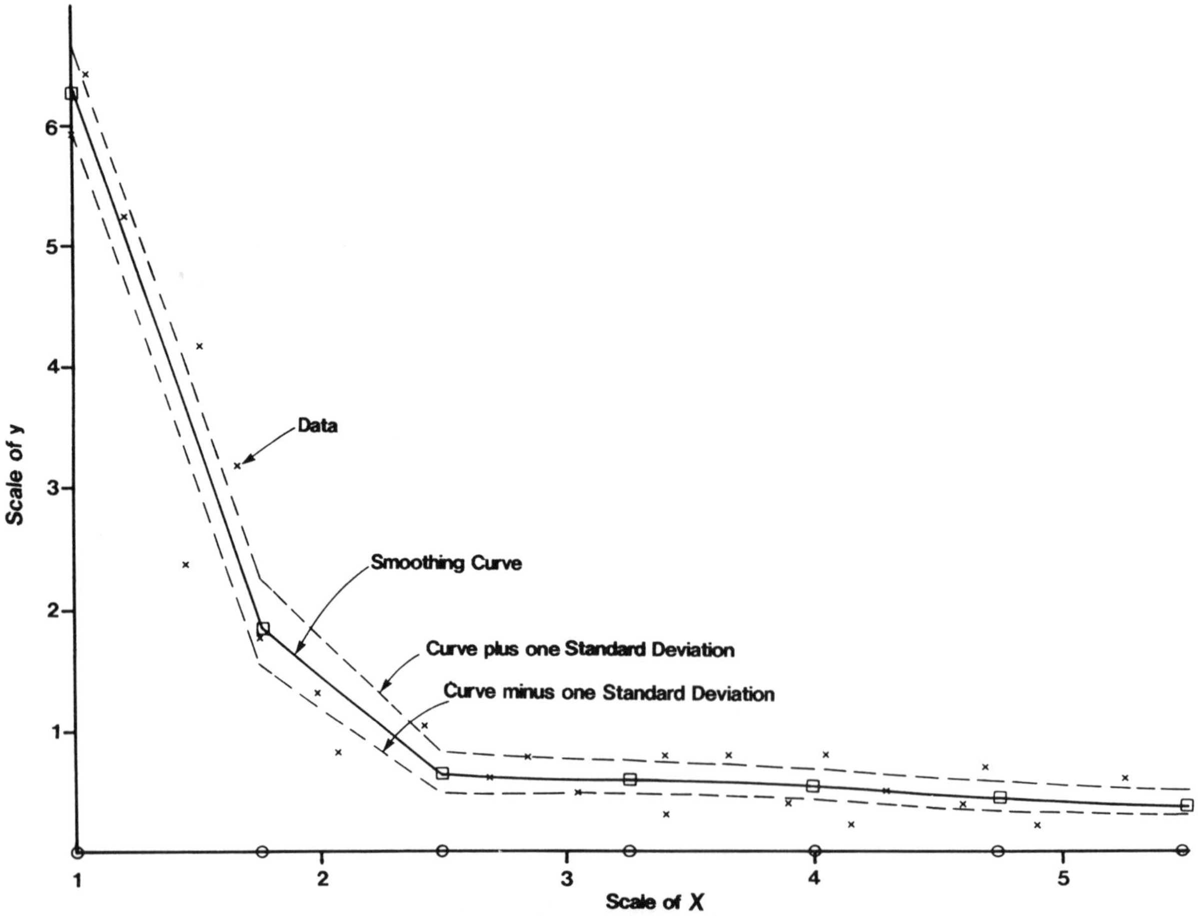

Fig. 5-8. Two-Point Smoothing with Localized Uncertainty

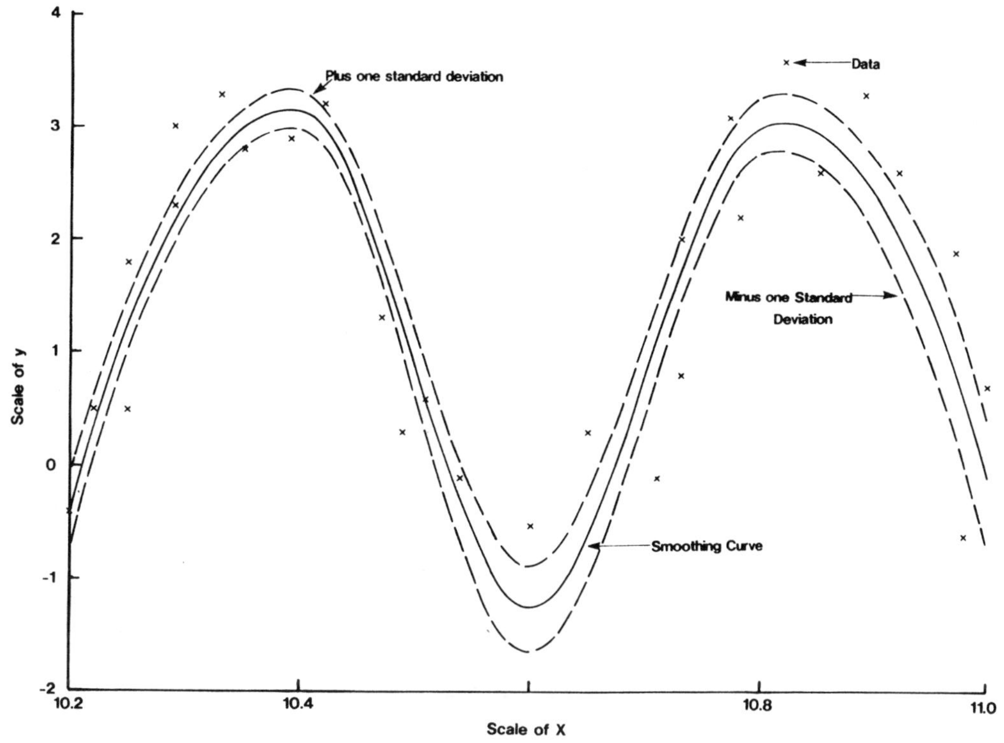

Fig. 5-9. Four-Point Smoothing with Localized Uncertainty

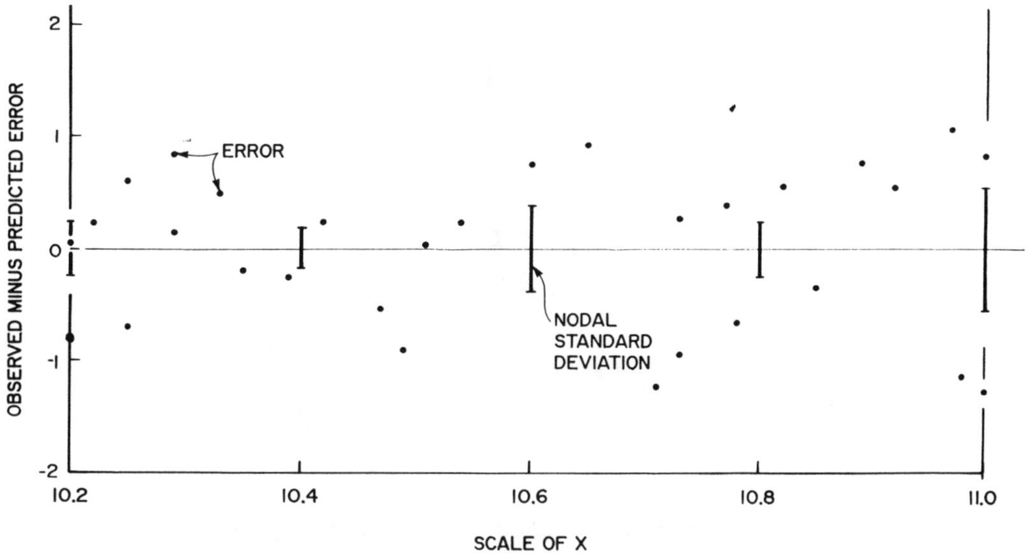

Fig. 5-10. Residual Error and Nodal Standard Deviation of Example 5.9.

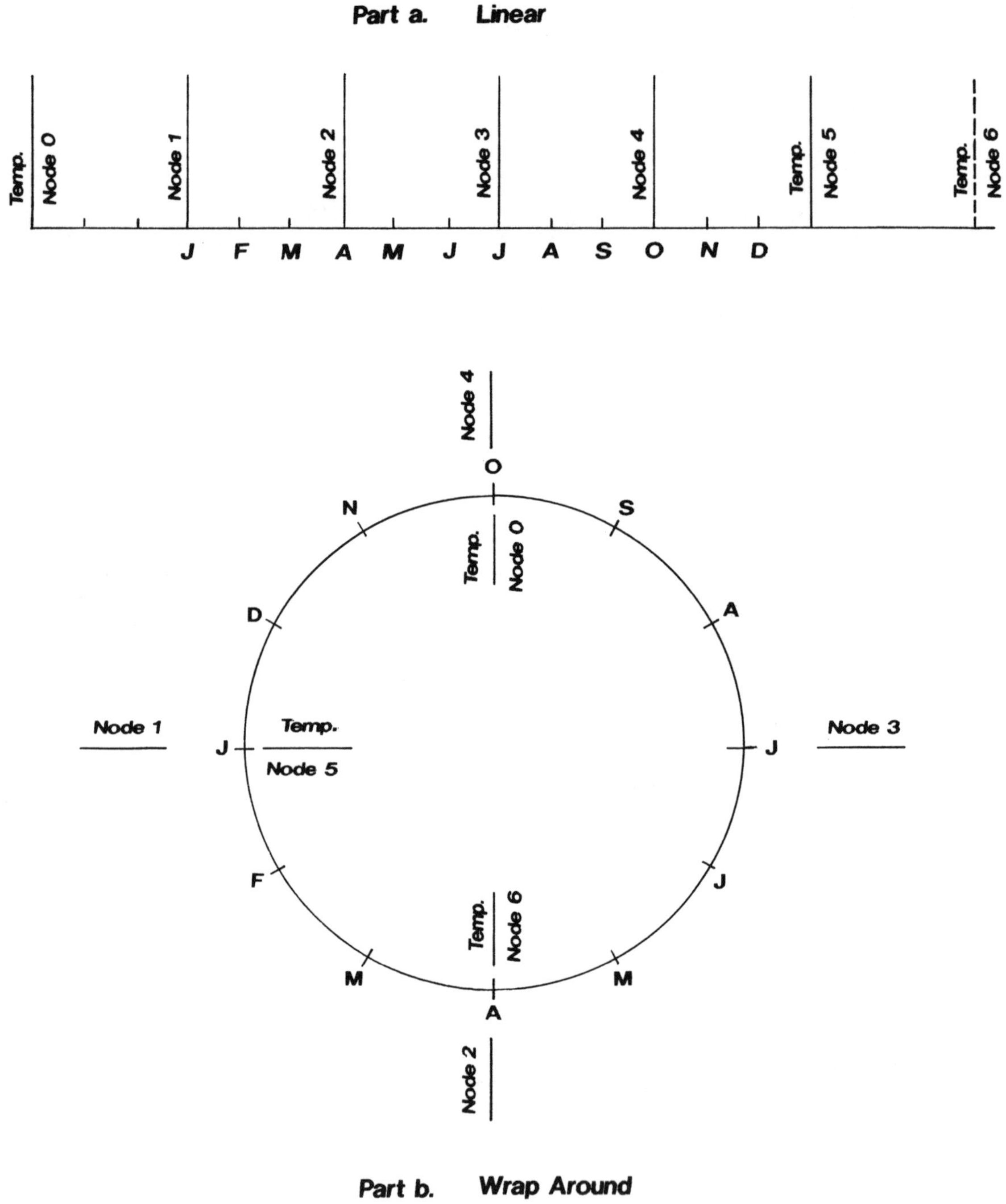

Fig. 5-11. **Schematic for Cyclic Nodes**

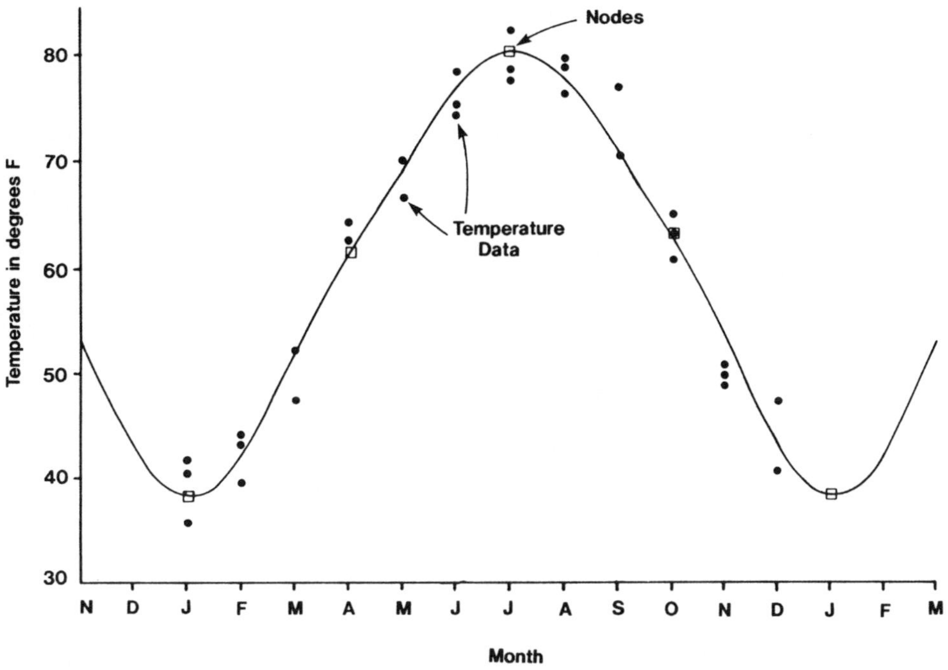

Fig. 5-12. Monthly Average Temperature 1968-1970, Lawrenceville, GA

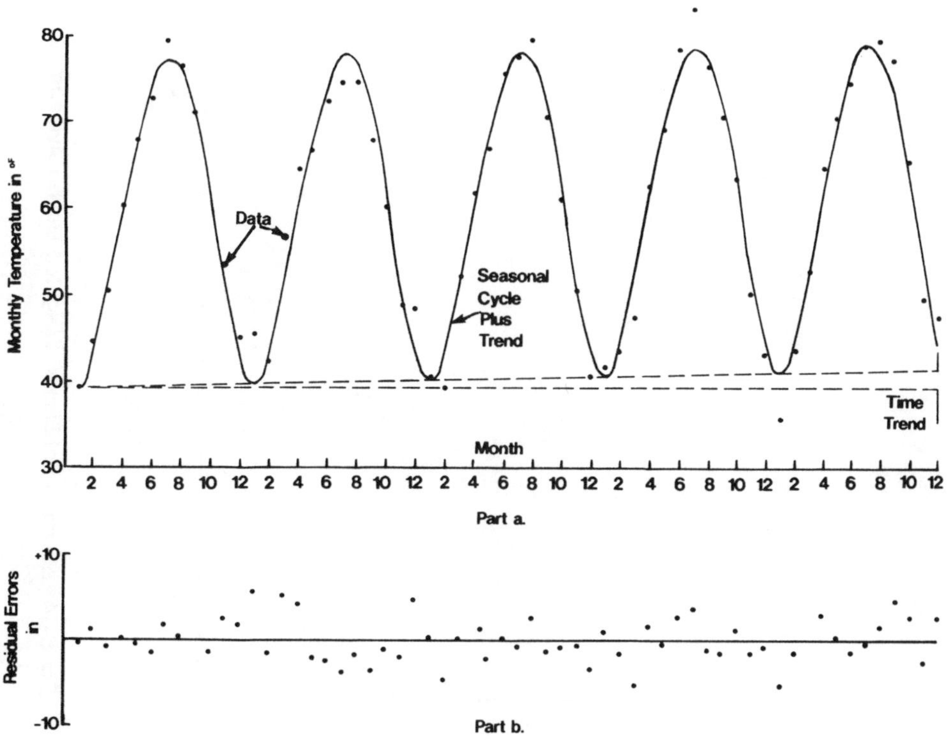

Fig. 5-13. Monthly Average Temperature 1966-1970 at Lawrenceville, GA Smoothing with Time Trend

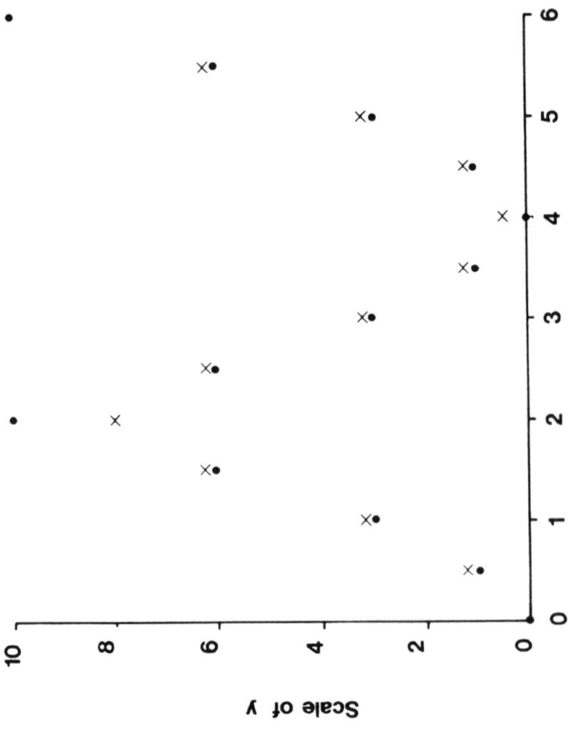

Fig. 5-14. **Illustration of Linear Weighted Moving Average**

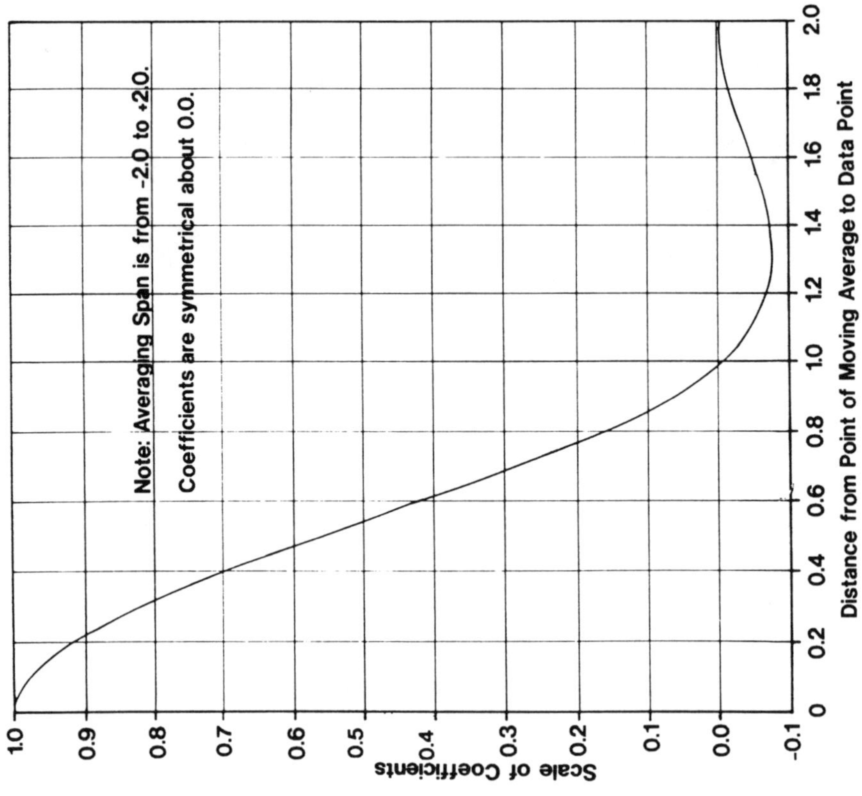

Fig. 5-15. Four-Point Weighting Coefficients

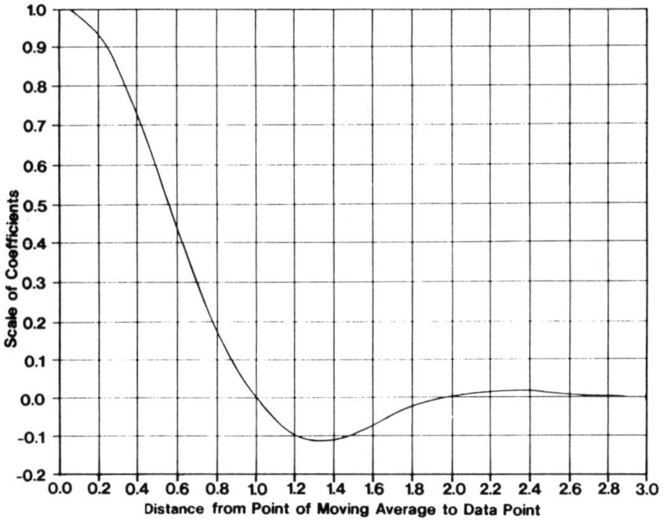

Fig. 5-16. Six-Point Weighting Coefficients

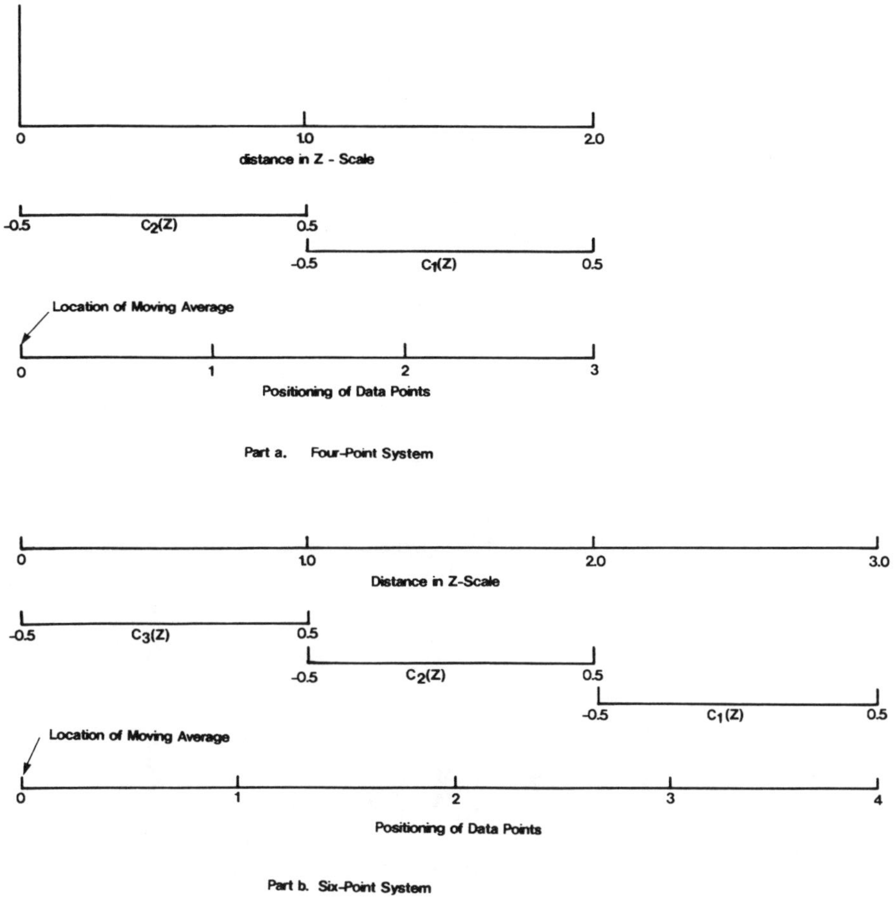

Fig. 5-17 Relationship of Scales

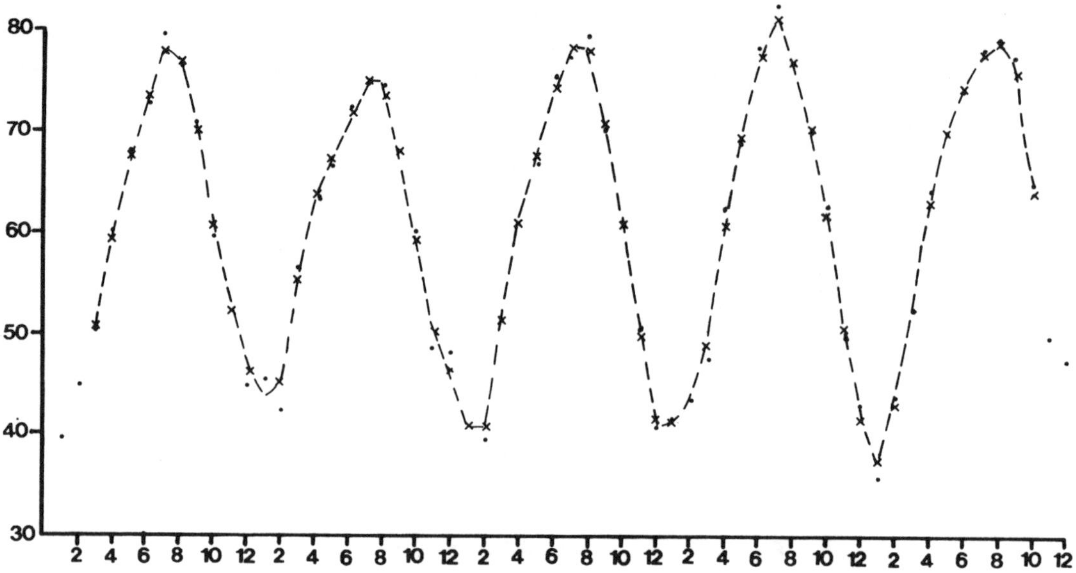

Fig. 5-18. Monthly Average Temperature 1966-1970 Lawrenceville, GA

Moving Average Smoothing

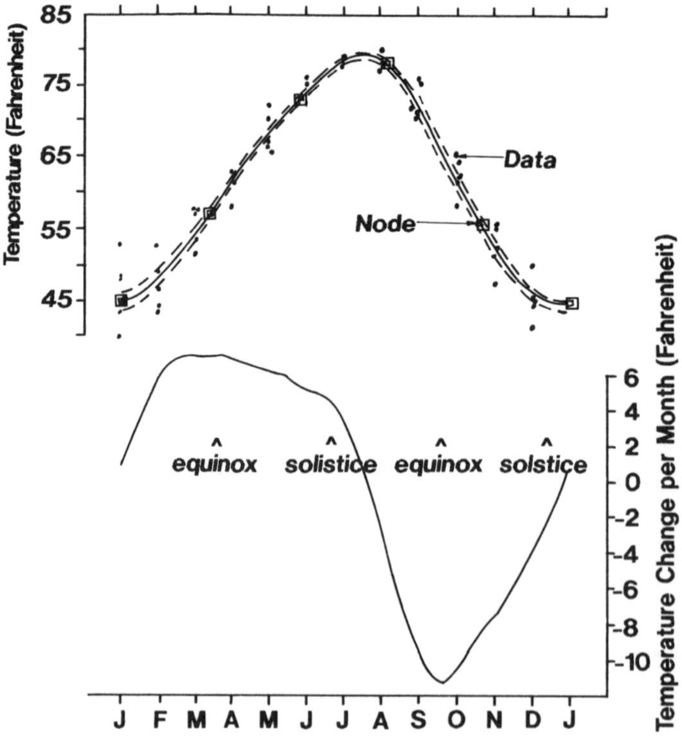

Fig. 5-19. Cyclic Smoothing of Average

Monthly Temperature at Athens, GA 1972-1976

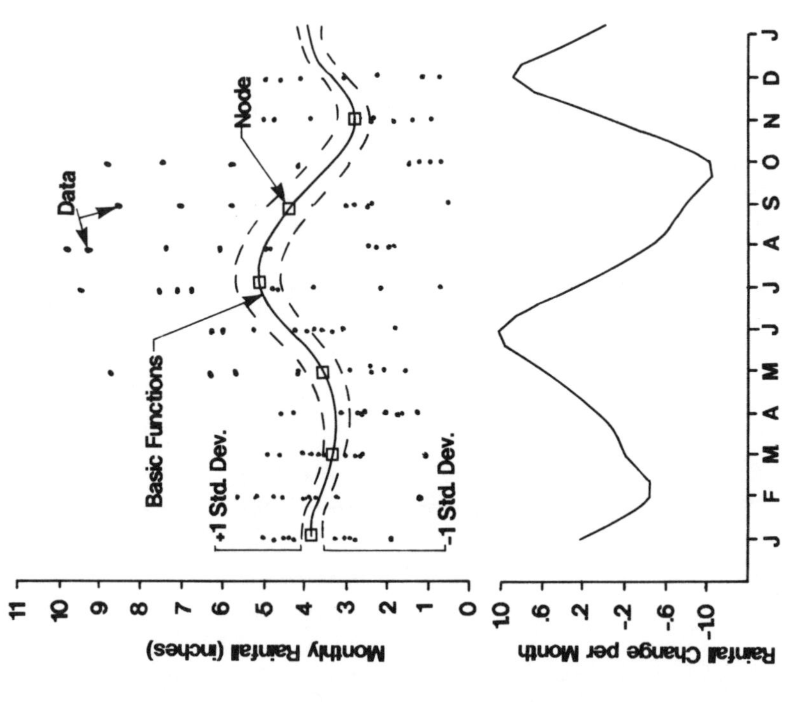

Fig. 5-20. Elliot Raingage Monthly Amounts 1963-1972 Inclusive

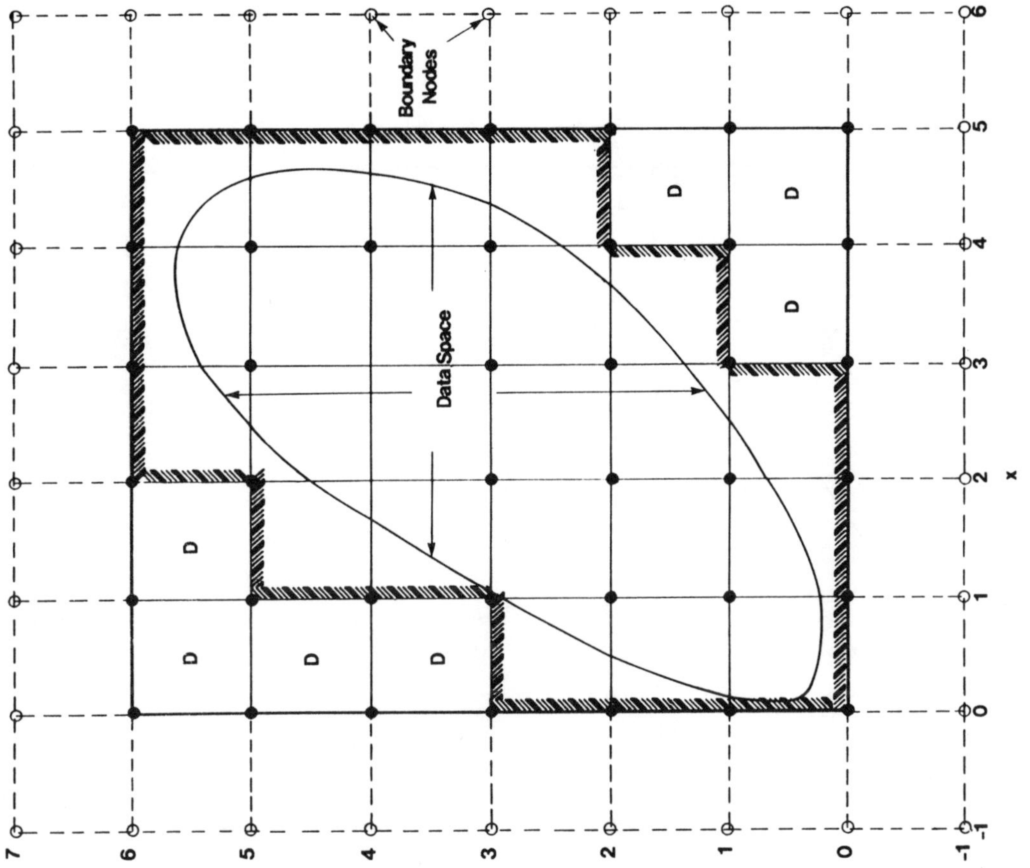

Fig. 5-21. Definition of Deletable Unit Surfaces

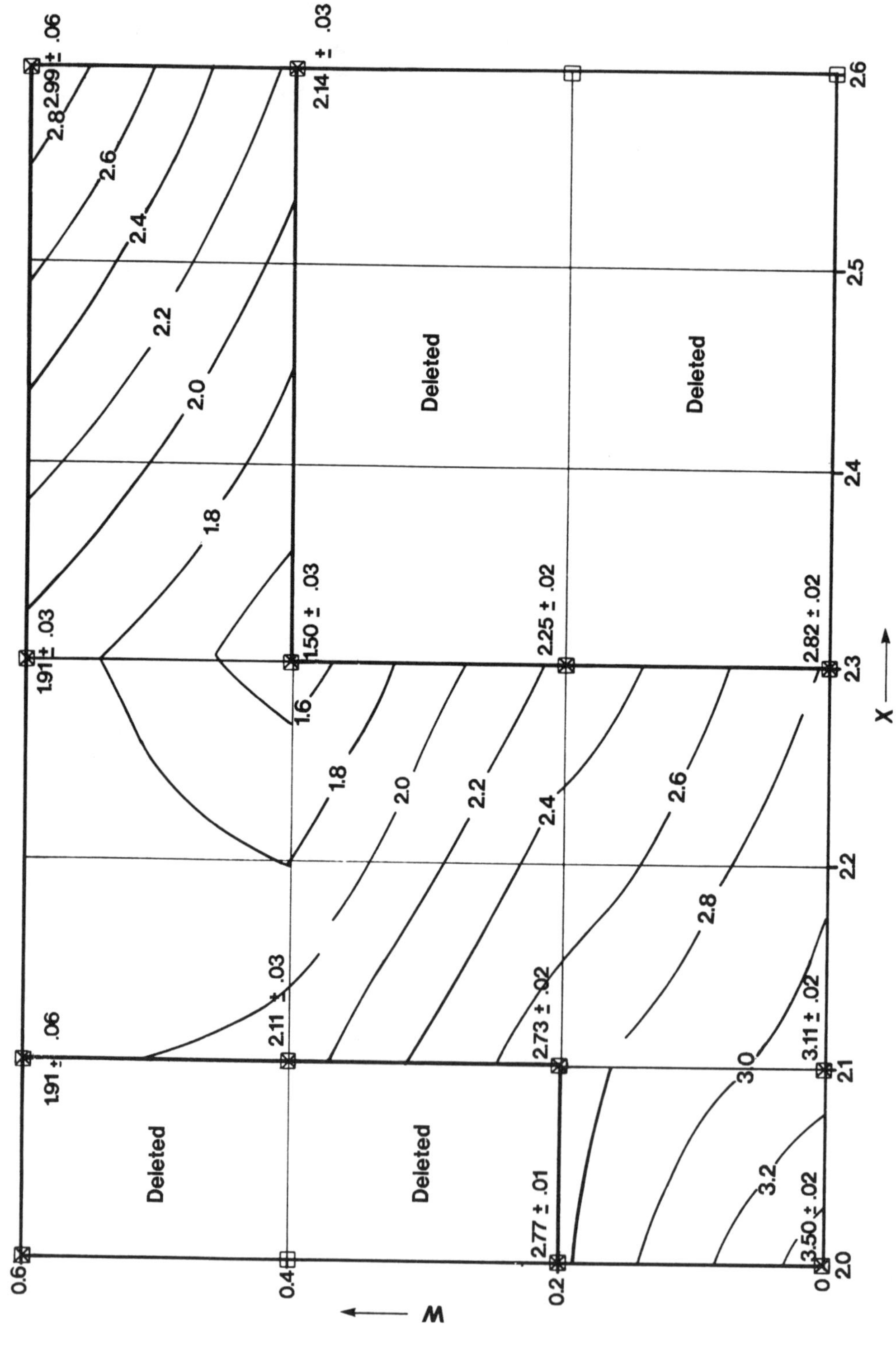

Fig. 5-22. Contours of the Smoothed Surface for Example 5-11.

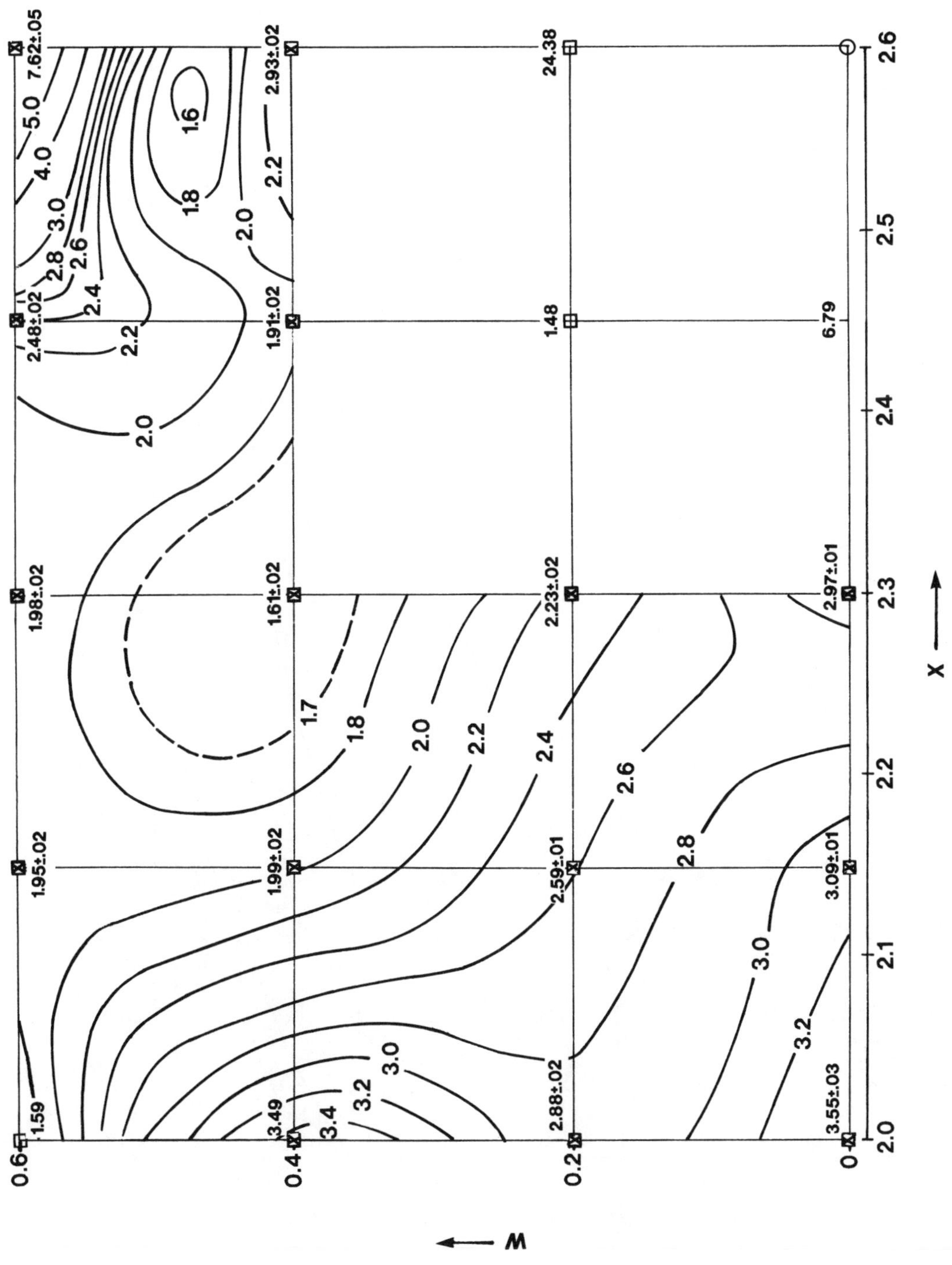

Fig. 5-23. **Contours of the Smoothed Surface for Example 5-12**

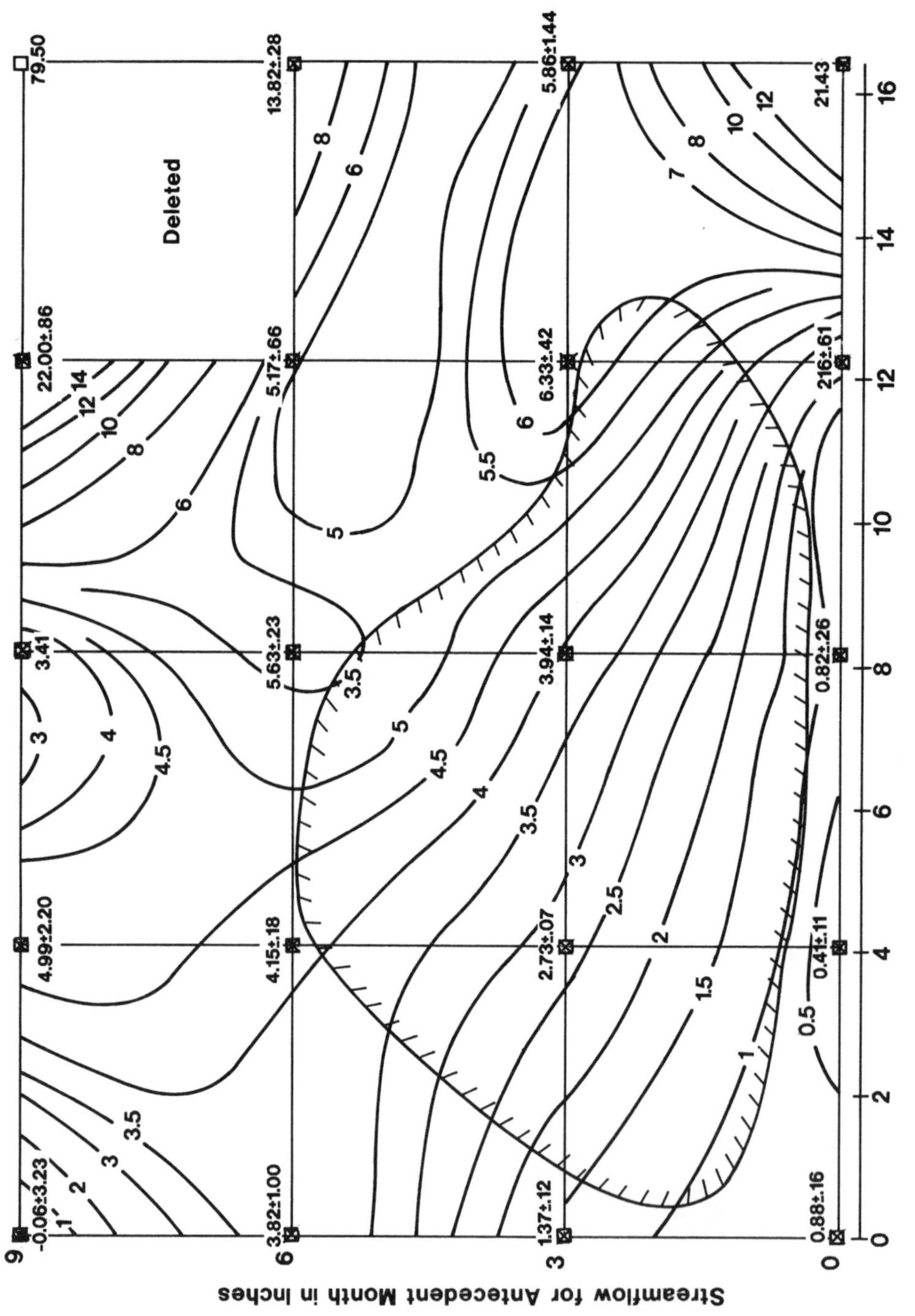

Fig. 5-24. Smoothed Monthly Rainfall-Runoff Data, Chattooga River near Clayton, GA 1953-1971

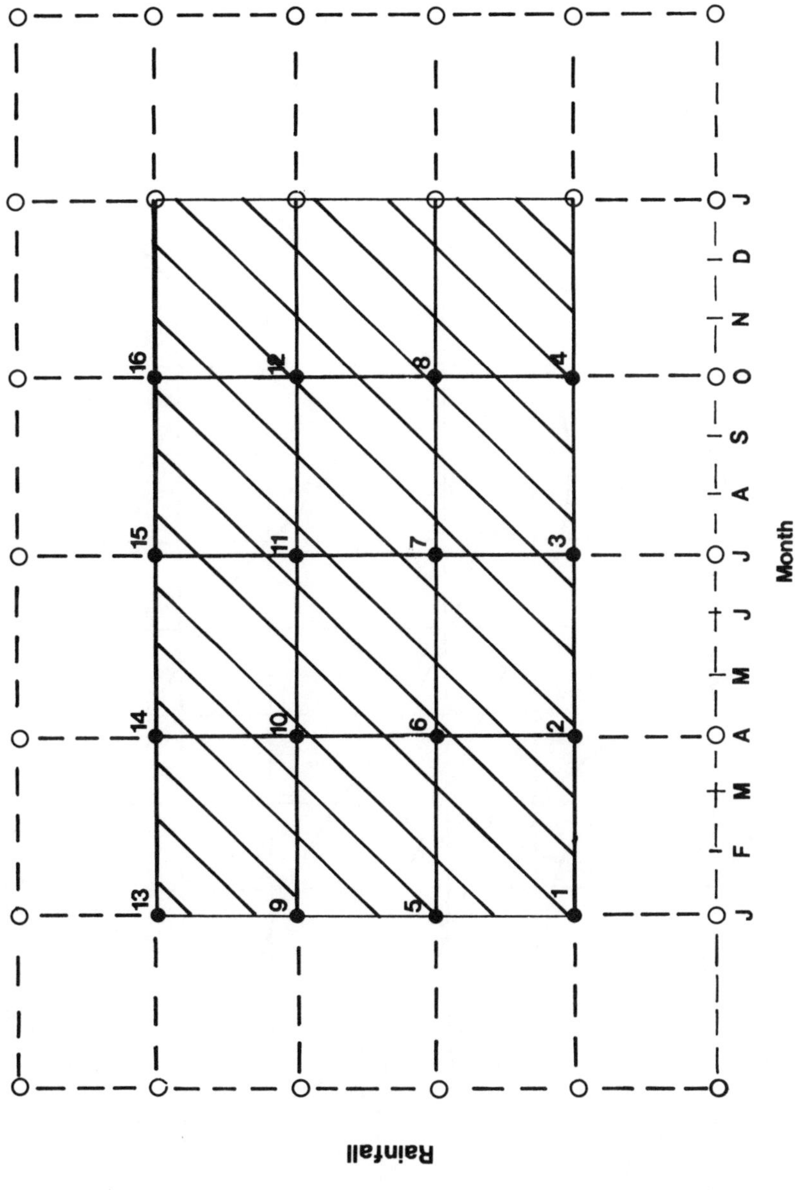

Fig. 5-25. **Arrangement of Nodes**

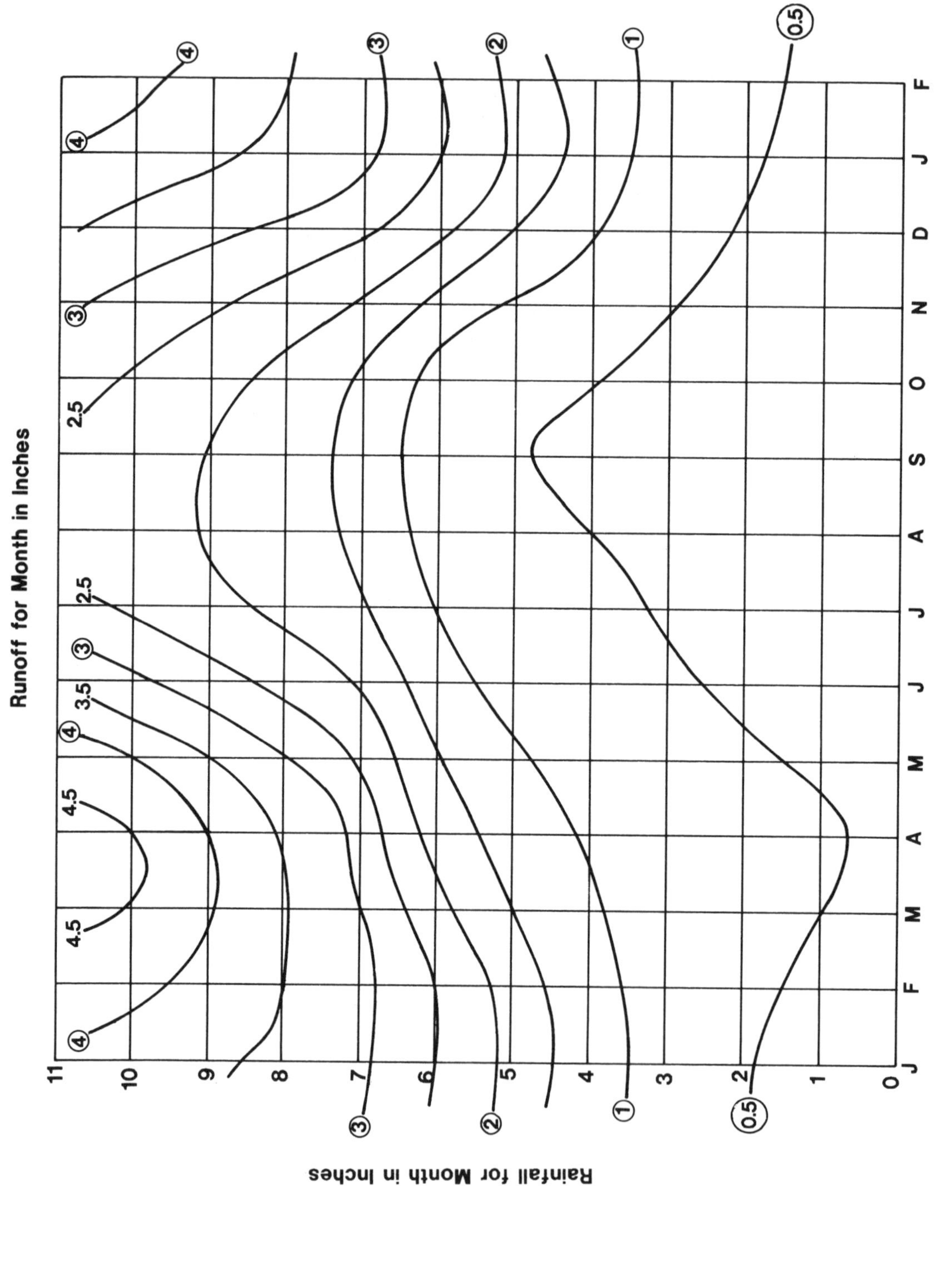

Fig. 5-26. Seasonal Rainfall-Runoff Patterns for Oxford, Mississippi

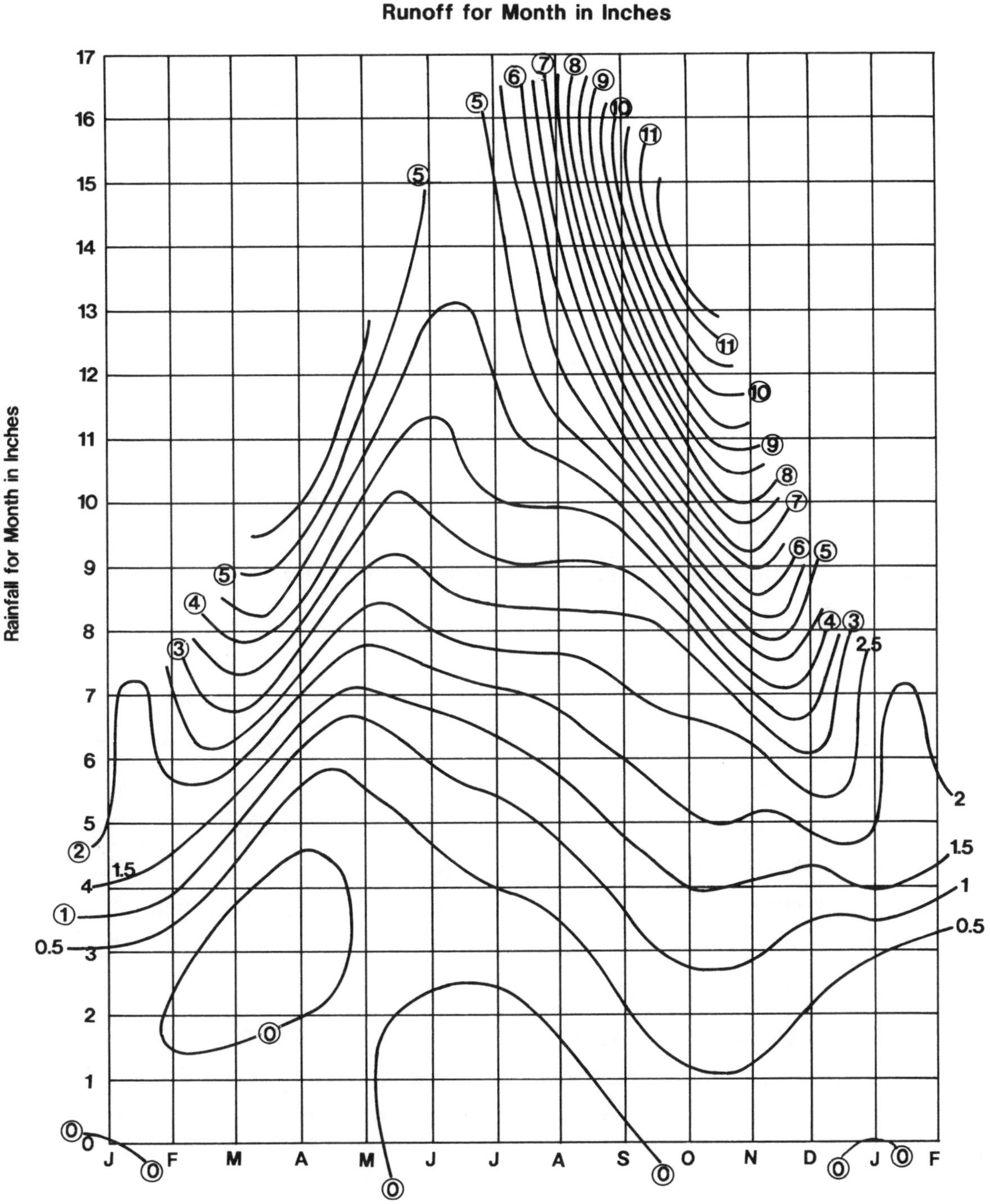

Fig. 5-27. Seasonal Rainfall-Runoff Patterns for Taylor Creek, Florida

CHAPTER 6

Probabilistic Analysis by Piecewise Functional Smoothing

6.1 INTRODUCTION

Chapter 6 is the second chapter dealing with analysis based on observations. In Chapter 5 we treated historical or experimental data to determine functional relationships between variables. We used piecewise sliding polynomials as a free-form representation of unknown functions. These functions could be differentials or integrals of higher or lower order functions. The guiding criterion was that we processed observations and measurements in such a way as to extract functional information. We searched for implied hidden relationships that, once revealed, would help us to understand the underlying system.

In this chapter, we also start with observations and measurements, but we search for different answers. For example, instead of getting an average temperature by months from an historical record, we ask how often the temperature either rose above or fell below a certain critical value in the past and how often we expect it to rise above or fall below such a value for some period in the future. We also ask with what confidence we can make such statements about the frequency of occurrences in the future.

When we ask such questions, we are, of course, entering the broad field of statistics and probability. We will not move deeply into this vast topic, here, because it is a highly specialized study with libraries of literature. We will be concerned with just one small facet, and this is in the overlap between mathematical statistics and numerical analysis. We find a common ground in prediction. In Chapter 5, we were mainly concerned with predicting the magnitude of an event. In Chapter 6, we will be mainly concerned with predicting the chance of that event occurring.

Probabilistic analysis is necessary whenever we deal with events beyond our control, or beyond our understanding of the processes at work. Consider the monthly average temperature again. Temperature can be critical in many human activities. A cold temperature may delay spring farming operations and have an impact on food prices. People may delay their usual spring shopping during a cold April, and business will be affected. Beyond temperature, we need to be concerned with floods, droughts, birth rates, rejects in industrial quality control, and queuing lines at the supermarket checkout; these other variables could be predicted in the same way that we predict temperatures of the future.

From a record of temperature we can compute the average of temperatures measured during all months of January of record, all the temperatures for the months of February of record, and so on. These monthly averages will form a seasonal cycle. We could in fact find these averages by smoothing with a cyclic form of the sliding polynomials. But now consider a particular month, say April, in some detail. We have the average April temperature, of course. However, very few months in the future would have a mean temperature that exactly equals the long-term average. In the past record, some were found to exceed the mean while some were below the mean, with a very few being extremely high or extremely low. Suppose we want to predict the temperature to expect for April for 10 years in the future. We do not presently understand how to predict world climate for 10 years in the future, so we cannot possibly predict actual April temperatures.

It is because uncontrollable events are so important in human activities that we predict the chances of things happening when we cannot predict with accuracy that a certain event will happen. We cannot predict an actual 10 years of April temperatures, but we can predict one or more sets of possible 10 year sequences of temperatures, and in each of these sets, we can see how many April temperatures fall above or below some values with which we may be concerned. We can then make decisions about our activities on the risks of these extreme temperatures in any one year. We cannot be certain, but we can decide to risk much or risk little.

6.2 BRIEFLY ABOUT SAMPLES AND PROBABILITIES

We need to learn a little about the stages of data processing to establish risks before we can discuss the use of sliding polynomials in certain of these stages. The first stage is always the acquisition of a record of the events in which we are interested. In Chapter 5, it was mentioned that records are generally considered to be historical data or experimental data. The same classification applies in probabilistic analysis, but most effort is directed toward historical data. Relatively long records are required for acceptable precision in calculating probabilities. Only well designed studies provide sufficient experimental data for probabilistic analysis.

The second stage in risk determination is organizing the record. This involves placing the events in the record in order of magnitude. Two general methods are used. The events can be placed individually in order of magnitude, largest first, next largest second, and so on. Or the individuals may be tallied into classes to form a histogram. We will be mainly concerned with the second method.

Before proceeding to the third stage, some more background development is necessary. In probabilistic analysis, we tend to call the records samples. An implied meaning is attached to this. We consider that the stochastic processes that produced our recorded events will continue to operate in the same way in the future. Records from the past have only sampled this implied infinite continuity of the process. But it is about the only information available on which to base some judgment about what the process will produce in the future. A public utility can serve as another practical example. The utility company knows the average number of degree-days for heating from past records, but it must be concerned with the amount of fuel that can reasonably be considered sufficient for the next season, which could be a cold one. The utility must make a projection about the heating oil to purchase ahead of time so that it will have a sufficient amount to meet the demand. If they don't purchase enough, then the heating oil deficit will have to be purchased later at a potentially higher price. This lack of certainty must be considered in the decision to purchase heating oil before the heating season begins. As another example, an engineer can look at the record of floods in the past, but some level of risk must be set to tell how high to place a bridge so that expected floods of the future will pass safely beneath it.

The third stage of risk determination helps us to make the transition from the sample observations of the past to probabilities of the future. To make this transition, we must smooth the sample histogram we constructed in stage two. We intend that the smoothed sample shall form an estimate of a population of all possible events. Smoothing can be done in several ways. Specific distribution functions, such as the normal, gamma, Poisson, or any other distribution, can be selected and then evaluated from certain properties of the sample. The chosen function, in effect, smoothes the sample and becomes a population estimate. We will follow a different approach herein. Instead of constraining the distribution to a preselected form, we will use sliding polynomials for smoothing the sample. Since the sliding polynomials are free-form, they should be capable of smoothing a wide variety of samples. Thus, we tend to diminish the need to choose a specific distribution with the attendant risk of choosing the wrong one. Also, we can smooth double-peaked samples, something most of the conventionally used specific distributions cannot do.

The fourth stage in our analysis of probabilities is the construction of a probability scale to associate with the various expected magnitudes of events. This could be considered part of the operation of smoothing the sample, but for clarity of presentation, we will treat it as a separate stage. The sample of events could be smoothed in many ways, but to use this smoothing operator as an estimate of the population from which the sample was drawn, a probability scale that runs from zero to one (or 0 to 100 percent) must be constructed. Another scale must cover all possible magnitudes of expected events for the dependent variable. For example, there is a value below which events cannot happen. This may be minus infinity or a finite value like zero. Then zero percent, probabilistically, expresses the fact that we cannot have events below that value. At the other end of the scale, there is a value above which events cannot happen. This value may be plus infinity or some other relatively large value. The probability of 1 (or 100 percent) associated with this large value expresses the fact that all events must be smaller. Any probability associated with a magnitude between these limits expresses the relative frequency with which we expect events larger or smaller to occur. Usually, the probability scale can be constructed by simple re-scaling of the sample frequencies.

The first four stages in analysis deal with the analysis of the sample. Stages five and six cover utilization of the analysis to establish risk and uncertainty. A value of risk is a selected value. A judgmental decision is to be made. A common risk level selected for engineering design is the 1-in-100 event. By this we mean the magnitude that is expected to be exceeded only once in 100 times. We can choose the flood we expect to be exceeded on the average only once in 100 years, and then design our dams and bridges to pass this 100-year flood safely. We can set farming operations so that temperature-sensitive crops are lost by freezing only one year in ten on the average, i.e., the 1-in-10 event. The 1-in-100 or 1-in-10 are chosen risks; they can be expressed probabilistically as one percent or ten percent risk, respectively. With the one percent or ten percent, we can go to our smoothing curve and find the size of the event that goes with that probability.

The sixth and last stage in probability analysis is the estimation of uncertainty. In stage 5 we chose a risk level, and from this we obtained the magnitude to expect. But we must remember that stage 1 began with obtaining our record, or one sample of the events in which we are interested. Suppose we had a second sample. If we worked through this sample, what magnitude could we expect for, say, a 1-in-100 event level? We most certainly cannot expect it to be the same as in the first sample. And if we had a third and fourth sample, how would the magnitude of the 1-in-100 event vary? This variation in the magnitude of our chosen risk level is what we will call uncertainty. As we will show later, we try to put a probability scale on the variation of our risk magnitude.

6.2.1 Example 6-1: Simple Probabilistic Analysis

The data in Table 6-1 will be used to illustrate the five stages of the probabilistic analysis of a sample. Stage 1, getting the sample, is already accomplished with the specification of the sample. The 30 events in Table 6-1 should be taken as a record of 30 years, or of 30 months, or at 30 points in the space of some stochastic process. It could indicate 30 random points in a field, or 30 trials of drawing numbers from a box. The magnitudes are the measurements that were made 30 times, perhaps the highest temperature for each month of July in the 30 years. We call these magnitudes the variate x. Stage 2, organizing the sample, is shown in a simple manner to the right in Table 6-1. The largest value of x is 5.974, and obviously has occurred once in 30 years or 30 trials. If the same recording is repeated in the future, we would expect an x of near this value in every 30 measurements. We would expect an x of around 5.5 or greater every 15 trials because the second largest measurement has a value near 5.5.

We do not wish, usually, to be limited to saying that the past record will repeat in future trials. We need to allow for sample-to-sample variation in the future. A better way to organize the sample for this more comprehensive analysis is to tally the sample into classes. In Table 6-2, classes were defined by setting class boundaries, or limits, of 0-1, 1-2, and so on up to 5-6. All of the x's in the sample will fit into these classes. The total number of x's in each class is shown next to the class limits. This set of class totals, or frequencies, is called a histogram. It is plotted in Part a, Figure 6-1. With the sample organized into histograms, we can usually make some rational decisions about smoothing operators. Remember that the sample is smoothed to get rid of happenstance irregularities in the sample. The smoothed histogram will become the estimate of the population of all events, a parent distribution. Therefore, if we have knowledge of this parent distribution, we should use the mathematical expression of it as the smoothing operator. Often we have no firm knowledge of the population, and we are reduced to trying to get the most information from the sample about possible distributions.

For the analysis of this sample, we will assume that we do not know the population from which the sample comes. The smoothing operator will be free-form, made up of four straight-line segments between nodes. The nodes will be placed at x-values of 0, 1.5, 3.0, 4.5, and 6. Further assumptions are made that the population is symmetrical, which makes the node at 1.5 equal to the node at 4.5, and that the population is bounded by $x = 0$ and $x = 6$, making the nodes at these locations equal to zero.

At the top of Table 6-2 are shown the values of x for the mid-point of each class of the histogram, the unitized variable z for sliding polynomial construction, and the values of the two-point sliding polynomial coefficients for the basic set of nodes. The reduced set of coefficients is shown in the center of Table 6-2. Nodes at $x = 0$ and $x = 6$ were set to zero, hence these coefficients are dropped. The nodes at 1.5 and 4.5 are equal so coefficients for these nodes were added. The remaining two columns of nodes, with the class frequencies from the histogram, form a data set for the ordinary least-squares solution shown at the bottom of Table 6.2. This solution is superimposed on the histogram in Part a of Fig. 6-1.

We should review briefly the meanings in Part a of Fig. 6-1. The histogram is the sample. The symmetrical four-segment figure is the estimate of the parent distribution from which the sample came. This parent distribution is obviously dependent on the assumptions we made concerning its structure. The main purpose of Part a of Fig. 6-1 is to confirm these assumptions. Could the sample have reasonably come from such a population? Answers to this question are based on techniques varying from simple visual confirmation to elaborate mathematical testing beyond our present purpose.

The analysis performed on the sample data becomes much more useful in understanding risks if the results are converted to a probability scale. This is stage 4 in our sequence of steps. Essentially, the integrals of the smoothing segments in Part a of Fig. 6-1 are required, and the total of these integrals must be rescaled from 0 to 100 percent. The equations for the four linear segments defined by the solution in Table 6-2 are given at the top of Table 6-3. The integrals of these four straight lines are also shown at the top of Table 6-3. Each of the segment integrals shows the mathematical area under the straight lines, each starting at zero at its left-hand side. The values of the integrals at selected x-values are given. At $x = 6$, the total integral is 28.686. Now in order to convert to a scale from zero to 100 percent, each value of the integrals is divided by 28.686 and multiplied by 100. The resulting values are shown at the lower right-hand side of Table 6-3. These values are also plotted as Part b of Figure 6-1. They are interpreted as follows. On the average, 13.7 percent of all values of x will be equal to or less than 1.5. Also, 23.9 percent of all values will be equal to or less than 2.0. Therefore, $23.9 - 13.7 = 10.2$ percent of all values of x will lie between $x = 1.5$ and $x = 2.0$. It is important to note that we derive these estimates from the smoothing function, which represents the estimated population, and not from the original sample of x-values.

The probability integrals may be used in another way. Suppose a management decision is made that we want to be sure that some system under design can handle 90 percent of all occurrences of x, but will be allowed to fail for the other 10 percent of x's. What value of x goes with this 90-10 decision on risk? From Part b of Fig. 6-1, we read immediately that at 90 percent probability, x is about 4.7. For 10 percent of the occurrences, x is expected to be equal to or greater than 4.7. If management wanted to risk less, they could allow for only five percent failures. This means safety for 95 percent of occurrences, and the associated value of x is about 5.0. Selecting the value to use is not part of the analysis. The decision on the appropriate risk is external to the analysis and represents policy. The determination of values of x for the various chosen levels of risk completes the fifth stage of the probability analysis.

The last stage in the analysis is concerned with uncertainty. In risk assessment, we determined what values of x are associated with the various levels of risk. But how sure are we of these values? Remember that the smoothing operator became the estimate of the population. However, only one sample was smoothed. Suppose a completely different sample from the same unknown population was available and was smoothed? How much would the estimate of the population change? How much change would there be in x at a selected level of risk? These are questions we attempt to answer with uncertainty analysis.

For certain well-studied probability distributions that are commonly used to smooth samples, a rigorous analysis of uncertainty is possible. The most noteworthy example is the normal distribution. Confidence intervals and tolerance limits are expressions of uncertainty that allow us to estimate how much the risk levels are apt to vary on a probabilistic basis. Also with known populations, order statistics can be used. In most instances of probability analysis, however, we are forced to slightly cruder, but extremely practical, procedures in uncertainty analysis. These procedures are only feasible when performed on a computer. This general approach has recently been named "computer intensive statistics" (Cover and Unny, 1986).

The basic procedure in computer-oriented uncertainty analysis is simulation. The computer is used to generate numerous synthetic but possible samples. By studying these simulated sets, one can see how any specific characteristics of a distribution vary from sample to sample. When we determine, in particular, how risk assessment varies from sample to sample, we can place rough probabilistic limits on the variation of risk. This procedure provides answers to the uncertainty question we asked above.

In the present example, a simulation was performed as follows. Thirty-five random numbers were generated by computer. Such random numbers are uniformly distributed, meaning they all have equally likely probabilities of occurrence. These 35 numbers were converted to values of x using the segment integral equations of Table 6-3. Alternatively, one could use the graph in Part b of Fig. 6-1. Now these 35 values of x are a simulated sample, different from our real sample of record in Table 6-1. This simulation was repeated 30 times. Now we have 30 simulated samples with 35 values of x in each.

As the next step, each of the 35 samples were rearranged. In each, the 35 values were placed in order of magnitude: largest first, next largest second, and so on down to the smallest in the 35th position. Consider the values with rank number 1. These occur once in samples of 35 values; therefore roughly, the risk of getting one this large in a sample is 1-in-35, or a probability of 2.8 percent. The values of rank number one for the 30 samples are plotted as a histogram at the top of Fig. 6-2. This histogram can be used as follows. Note that four values were greater than 5.8. We can now state that we are ((30-4)/30)100, or about 87 percent sure that our risk of 1-in-35 will not be greater than 5.8. Also, the median value of the simulations of rank 1-in-35 is 5.45, meaning 15 values are smaller than this, 15 are larger. Now note that this value lies near 100 - 2.8 = 97.2 percent probability in Part b of Fig. 6-1. This demonstration shows that, if we do not compute the uncertainty ranges by simulation and only use the risk values given in the figure, we are in effect assuming about 50 percent level of uncertainty. If we want to be more confident than this about the risk statements, we must do an uncertainty analysis.

Calculation of uncertainty is not limited to rank number one. As further examples, the distributions of rank numbers 2, 35, 34, 9, 18, and 27 are also shown in Fig. 6-2. The distributions of number 34 should be the reverse of number 2. This is difficult to see because such small samples of only 30 values are ranked. In practice, one would simulate say, 100 to 500 samples, thereby enhancing the precision of the uncertainty statements.

Rank number 9 marks the division between the upper one-fourth and the lower three-fourths of the samples. Rank number 27 marks the lower one-fourth, and rank number 18 the middle of the samples. These so-called quartiles are obviously risk levels of 25, 50, and 75 percent. Their distributions in Fig. 6-2 show the uncertainty attached to these risk levels. For example, we might note that in four samples in 30, the lower quartile would be marked at x-values of 1.8 or lower.

Program B-6.1 in the appendix was used to simulate the 30 samples of 35 values. This program was prepared specifically for this example, but it can also serve as a beginning point for more complex simulations. Typically, simulations are designed to the particular risk and uncertainty needs for specific planning problems.

6.3 SMOOTHING SAMPLE PROBABILITIES

In the brief introduction to probabilistic analysis and in Example 6-1, the concept of a probability distribution was presented. Part a of Fig. 6-1 shows the smoothing of a sample distribution and Part b shows the integral of the free-form smoothing function. These two forms in Fig. 6-2 must be distinguished. The integral in Part b is usually called the cumulative probability function, while the distribution in Part a is usually called the probability density function and is obviously the differential of the function in Part b.

Probability density functions are often shown to illustrate the specific forms of distributions. The overlay plot, as in Part a, also tends to confirm or reject that a particular distribution is the correct one to use with the sample under study. The probability distribution in Part b, however, is the more usable form in decision making. Statements about the risks of certain magnitudes being exceeded a specific number of times can be made directly from this function. Also, it was shown in Example 6-1 that this is the function used in simulation to answer questions of risk and uncertainty.

We now introduce the concept that the probability distribution function can be obtained directly by smoothing. It is not necessary to first smooth the sample histogram and then integrate. The only requirement is that we first accumulate the sample histogram, constructing the subtotals of the classes. The division of these class subtotals by the total number of items in the sample yields a sample probability function with values from zero to 100 percent probability of exceedence and intermediate probability value for each class.

The calculation of sample probabilitics is illustrated in Table 6-4 using the data from Example 6-1 in Table 6-2. The class limits and class frequencies are shown at the left.

The number in the class is next accumulated, running from three to the sample total of 30. Each class subtotal is divided by 30 and multiplied by 100 to produce the sample probability values. We see that these numbers give the percent chance that any value of x selected at random will equal or fall below the upper limit of the class. For example, ten percent of the x's will equal or fall below a value of one. Also, 56.7 percent of the x's will be equal to or fall below a value of three. All values, 100 percent, will be equal to or fall below the upper limit of the last class. The sample probabilities, plotted against the upper limit of the class, form a graphical representation of the sample probabilities. Such graphs of the sample can be smoothed using sliding polynomials to yield directly the estimate of the population probability function.

6.3.1 Example 6.2: Calculation of a Free-form Probability Function

A sample of 45 items is shown in Table 6-5. These data were organized into classes and the sample probability function was calculated as shown in Table 6-6. This sample was plotted in Part a of Fig. 6-3. The next step is to smooth this sample probability function to obtain an estimate of the population probability function.

Two-arcs of the four-point sliding polynomials were used in smoothing. Data-boundary nodes were placed at the class limits of x = 1.2 and x = 3.0. A node midway between falls at x = 2.1, and outlier nodes fall at x = 0.3 and x = 3.9 to maintain uniform spacing. The sliding polynomial coefficients are shown for this full set of nodes at the top of Table 6-7. Boundary conditions were next imposed to enhance the smoothing operation. The node at x = 1.2 was set to zero probability, thus marking this as the left-hand boundary of all possible x-values. The node at x = 3.0 was set to 100 percent probability, making this the right-hand boundary of all possible x-values.

The reduced set of nodal coefficients is shown in the center of Table 6-7. The column of coefficients for the node at x = 1.2 is simply dropped because this node was set to zero. The column of coefficients for the node at x = 3.0 is multiplied by 100, the set value of the node, and moved over the implied equal sign of the observation equations to modify the sample probability values. The least-squares solution for the reduced set of nodes, plus imposed boundary nodes, is shown at the bottom of Table 6-7.

Values were interpolated for the two arcs of the four-point sliding polynomials running through the solution nodes. These values are plotted as the smoothing curve through the sample probabilities in Part a of Figure 6.3. Note that this smoothing curve closely represents the sample, and particularly, that it runs between the two values of 97.8 percent probability caused by the empty class of $2.6 \leq x \leq 2.8$ in Table 6-6.

This sliding polynomial probability function can be numerically differentiated using the first-derivative form of a six-point sliding polynomial. This differentiation will yield an estimate of the population probability density function. It is plotted in Part b of Fig. 6-3.

We conclude this example with an important remark about the imposed boundary values of zero percent probability and 100 percent probability. Note first that these boundaries were set subjectively, but obviously were conditioned strongly by the sample values. In practical probability analysis, we usually know something about the possible physical range of values of the variate under study. This knowledge should always be used to set the boundary conditions. If such boundaries cannot be set, we will see later how to set stochastic boundaries from properties of the smoothed sample.

6.4 TRANSFORMATION OF DATA

The sample probability and the smoothing curve in Part a of Figure 6-3 exhibit a shallow S-shape. This is a common shape of probability distribution functions, found to some degree in most sampling of stochastic processes. Most measurements from such a process will be centrally located near the mean of the measurements. Some few items will lie at greater distances from the mean. The familiar "bell-shape" of the normal probability density function is an example. When such samples are accumulated to form a cumulative probability distribution, the subtotals build slowly across the tails but build rapidly across the interior classes, which causes the S-shape.

When the S-shape of a sample becomes pronounced, it is difficult to smooth. Smoothing is always more efficient across shallow curvatures. It is customary to transform samples that show strong curvature before attempting a smoothing operation. Following smoothing the sample and the smoothing curve are "de-transformed" back to the original scale of the variate. Most samples of stochastic data have tails that gradually approach zero frequency of occurrences as the data values lie at increasing distances from the central values. This structure forms the basis for use of the descending exponential as the mathematical core of a data transform method. It forms the basis for the common practice of taking logarithms of data before fitting the classical distribution functions. We will develop a different form designed to pull in both the left and right tails of the samples.

The descending exponential may be written as:

$$v = K \exp(-g(m-x)) \qquad -\infty \le x \le m \tag{6-1}$$

In this equation, x is the measured variate, v is the transformed variate, m is a value near the mode of x, g is a shape parameter, and k is a scale parameter. Equation 6-1 forms one limb of the transform. A reversed image of this limb can be expressed as:

$$v = K(2 - \exp(g(m - x)) \qquad m \le x \le \infty \tag{6-2}$$

The definition of terms is the same as in Eq. 6-1.

Equations 6-1 and 6-2 are plotted as a continuous curve in Figure 6-4. For this plot, m was 70, g was 0.2, and k was 2.0. The transformed variate v has a total range from 0 to 4 as x goes from $-\infty$ to $+\infty$. The curve is symmetrical about the value m = 70. Equations 6-1 and 6-2 can be solved for x yielding the following:

$$x = \log_e(v/k)/g + m \qquad -\infty \le x \le m \tag{6-3}$$

$$x = m - (\log_e(2 - v/k))/g \qquad m \le x \le \infty \tag{6-4}$$

Now consider setting classes in the scale of the variate v into which to tally the sample. Figure 6-4 shows the range $0.2 \le v \le 3.8$ divided into nine equal classes. These class limits were then reflected into x-scale using the plotted transform curve. Alternatively, class limits in x-scale could be calculated from Eqs. 6-3 and 6-4. Note particularly the class widths in x-scale. The outer classes are wide, and the interior classes are narrow. The effect is to decrease the number of sample items that will tally into the narrow interior classes and increase the number of sample items that will tally into the wide outer classes. Since the tails have fewer items, the effect is compensating and tends to produce a more uniform, or more uniformly changing, number of items in each class. The accumulation by classes and the resulting sample probabilities are thus less curved in the v-scale than in the x-scale.

In Fig. 6-4, the value of 70 for m is centrally located. For skewed samples with one tail longer than the other, m is not central. If a value of m is maintained near the mode of the sample, the transformation curve will have one exponential limb longer than the other. This will serve to transform the long limb of the sample, the short limb being only moderately transformed.

6.4.1 Example 6-3: Smoothing Transformed Temperature Data with Boundary Values Fixed

Figure 6-5, Part a, shows a histogram of maximum daily temperature near Griffin, GA, for the years 1926 to 1981. This histogram was constructed using a uniform class width of two degrees Fahrenheit. Class limits are at half degrees since data were reported only to whole degrees (Kinard and Tippins, 1982).

The temperature data were transformed to the variate v using values of m = 70, g = 0.2, and k = 2.0 in Eqs. 6-1 and 6-2. The interval $0.15 \leq v \leq 3.75$ was then divided into 12 equal classes and the class limits in the v-scale were de-transformed to equivalent values of t (or x) using Eqs. 6-3 and 6-4. Corresponding class limits in v and t are shown in Table 6-8. The January maximum temperatures were then tallied into these classes, and a histogram was constructed. This histogram with its uniform class widths in the v-scale is shown in Part b of Figure 6-5.

Both histograms in Fig. 6-5 were accumulated and converted to sample probabilities. Both the t-scale and v-scale sample probabilities are plotted in Fig. 6-6. The S-shape of the t-scale probabilities is clearly evident in the data to the left in Fig. 6-6. The v-scale probabilities plot with much less curvature to the right in Fig. 6-6.

The sample probabilities on the right in Fig. 6-6 were smoothed using three spans of the four-point sliding polynomial. Nodes were located in v-values of 0.15, 1.35, 2.55, and 3.75. Outlier nodes are located at -0.15 and 4.05. The nodes at v = 0.15 and 3.75 were set to zero and 100 percent probability, respectively. Smoothing by least squares was then performed as illustrated in Table 6-7. The resulting smoothing curve is superimposed on the data to the right in Fig. 6-6. Correspondence between data and the curve is close. Note that smoothing filters the empty class for $0.75 \leq v \leq 1.05$, and also the relatively low occupancy class for $2.55 \leq v \leq 2.85$.

The smoothing, while derived in the v-scale, can also be displayed in the t-scale by calculating corresponding values of t for selected values of v where interpolation was performed. This de-transformed curve is shown superimposed on the t-scale probabilities to the left in Fig. 6-6. Again, the correspondence is good. However, we must remember that smoothing only took place over the slightly curved sample probability in the v-scale.

We should review the consequences of having set the node at $v = 0.15$ to zero probability of x and the node at $v = 3.85$ to 100 percent probability. These boundaries restrict the variate v to the range $0.15 \leq v \leq 3.85$ and variate t to the range $57.05°F \leq t \leq 80.40°F$ in simulation. Setting such boundaries can be a satisfactory procedure, if a particular problem is mainly concerned with mid-scale temperature variations only. However, any time that information is needed on variability of the extreme highs and lows of the data, such boundary values may not be set. In a broader context, boundaries should be set only when we have physically recognized limits. For example, on any day we may measure no rainfall, or we may measure some rainfall, but we cannot measure negative rainfall. Therefore, it is appropriate to set a zero boundary for a distribution of rainfall occurrences. In circumstances other than known physical boundaries, the highest and lowest values of a distribution are themselves stochastic variates and can only be sampled.

6.4.2 Example 6-4: Smoothing Transformed Temperature Data with Boundary Values Variable

The histograms of maximum daily temperature in July near Griffin, Ga., are shown in Fig. 6-7 (Kinard and Tippins, 1982). Class limits in the v- and the t-scale are shown in Table 6-9. Values of $m = 95.5$, $g = 0.18$, and $k = 2.0$ were used in Eqs. 6-1 and 6-2. Where class limits such as 93.06 and 93.86 occurred, making an exact tally impossible, the number of items was divided equally between the classes. The histogram in both the v- and t-scales were converted to sample probabilities and are plotted in Fig. 6-8. Again, we note the reduced curvature in the v-scale.

The sample probability in the v-scale was smoothed using four spans of the four-point sliding polynomial. All nodes were optimized by smoothing. Program B-6.3 was used. The results are shown in Table 6-10. Note that the node at v = 0.49 is not zero and the node at v = 3.69 is not 100 percent probability. The smoothing line crosses the zero percent axis to the right of v = 0.49 and crosses the 100 percent probability axis to the left of v = 3.69. All values of v must lie between these crossings, if the nodes are held constant. Now note that we have values of the standard deviations of all but the two exterior nodes. This means that we do not know the exact nodal values, and we regard the nodal values given in Table 6-10 as average values, not constant values.

The smoothing curve, which is passed through the average values of the nodes, is superimposed on the data in the v-scale in Fig. 6-8. As in Example 6-3, the correspondence between the data and the smoothing curve is good. The smoothing curve transformed to the t-scale is superimposed on the t-scale data to the left in Fig. 6-8. We see from this curve that the average lowest July maximum temperature is 88 degrees Fahrenheit. The average highest maximum temperature is 104.3 degrees Fahrenheit.

In Example 6-3, the boundaries for maximum January temperatures were set. The analysis, in effect, allowed no temperatures in the future to fall outside the set range. While this produced a simple analysis, it is not very realistic in that we do not know what the future temperature variations will be. We have only a sample of the temperature range, and we can never know the population from just one sample. Specifically, we can never know exactly what the highest and lowest maximum temperatures will be. We see now that our analysis of the July maximum temperatures is more realistic. We have not fixed a range, nor even fixed a distribution of the maximum temperature. We say only that the sliding polynomial smoothing curve is an average, but the standard deviation of the nodes allows us to vary them. In simulation, we calculate possible distributions around this average curve, not fixed on it. Two elements of variation in simulation now come into play. First, the nodes are varied with random normal numbers, and then these varied nodes are used to get a simulated value of v from a random probability of occurrence. Lastly, t is calculated from the simulated v.

The nodes are varied with random normal numbers because each node is a local average of the sample probabilities, and average values are known to vary approximately normally. To calculate a randomized nodal value, we simply equate the standardized variate of the distribution of the nodal values and the distribution of random normal numbers as in Eq. 6-5:

$$\frac{N - \bar{N}}{sd_N} = \frac{RNN - \overline{RNN}}{sd_{RNN}} \tag{6-5}$$

in which N is any value of a node, $\bar{N}$ is the mean value of the node found by least squares, sd_N is the standard deviation of the node, RNN is any random normal number, $\overline{RNN}$ is the mean of RNN, and sd_{RNN} is the standard deviation of RNN. Random normal numbers are usually generated with $\overline{RNN} = 0$ and $sd_{RNN} = 1$. With these values, we can solve Eq. 6-5 to yield N as:

$$N = \bar{N} + sd_N \cdot RNN \tag{6-6}$$

To use Eq. 6-6 in simulation, we generate RNN, multiply by sd_N, add to $\bar{N}$, and thereby get a randomized N.

Program B-6.2 in the Appendix was specifically prepared to generate synthetic samples of July maximum temperatures based on the analysis in this example. In order to show the effect of nodal variation, the program was used to generate 25 samples of 10 items each. Decisions as to the number of samples and number of items depend on the full scope of the problem under consideration. Synthetic samples of 10 items indicate we are interested in 10-year sequences of weather, perhaps to plan farming operations. We choose the number of samples, in this case 25, to establish the precision of our 10-year anticipations. For a highly precise analysis, many synthetic samples, say 500, are needed. A special feature is incorporated in Program B-6.2. After each synthetic sample is generated, it is arranged in order of magnitude of the individual items in the sample. We can then, secondarily, look at the distribution across the 25 samples of rank number one in ten, the distribution of the 25 values of rank number 10 in 10, or any other rank number.

In Fig. 6-9, the two total histograms of 250 items obtained by pooling the 25 synthetic samples are displayed. We note first that without nodal variation, the histogram is more compact. Adding nodal variation extends the tails and diminishes the peak classes. This is realistic since the "system variance" is added to the conventional simulation variance. We note further that the right tail is extended more than the left tail since the nodal standard deviations in Table 6-10 are much larger for higher temperatures.

The effect of nodal variation on the 25 individual samples of 10 items is shown using selected rank numbers in Fig. 6-10. In part a, the two histograms of the 25 values of rank number one in each sample of 10 items are displayed. The histogram for the synthetic samples with nodal variation lies to the right for that with fixed nodes. This is the same effect noted in Fig. 6-9. In Part b of Fig. 6-10, we see that rank number six in ten changes less than rank number one in ten. The nodal standard deviations are smaller, and typically, interior items of a sample are less variable than the tails. In Part c of Figure 6-10, the smallest values in each sample of 10 move slightly to the left, showing the same effect as shown by the total of all samples in Fig. 6-9.

Additional consequences of nodal variation in simulation are shown in Table 6-11. The highest and lowest values, the mean, and the standard deviation of the distributions of rank numbers 1-in-10, 6-in-10, and 10-in-10 are shown for samples with and without nodal variation. Note that nodal variation causes a consistent increase in the range between the highest and lowest values and in the standard deviation. The change in mean value is less pronounced except for rank number one.

As stated earlier, nodal variation produces more realistic simulations when the nodal standard deviations are significant. When the nodes are held constant, it appears that we have established our population values exactly, and we then underestimate variance of the simulations. When the nodes are varied, we simulated with greater accuracy the true variability to expect in the future. Sliding polynomials with nodal standard deviation are thus a simple and direct approach to the realistic simulation problem.

6.5 A GENERALIZED VARIATE TRANSFORM

The advantages of transforming sets of data for probabilistic analysis have been discussed earlier. Briefly, the process of smoothing the data is improved if the curvilinearity of the data is reduced, particularly in the tails of the distribution. Equations 6-2 through 6-4 developed a workable exponential form of data transform.

Thomas and Snyder (1986) published a more generalized form of transform that appears useful for many distributions. The method can be used for either symmetrical or skewed (positively or negatively) samples. The parent population for this transform must have an infinite boundary for the longer tail and a finite but indefinite boundary for the shorter tail. Most natural phenomena exhibit such characteristics, and many conventionally used distributions have such characteristics.

The core of the mathematical transform of the stochastic variate consist of two limbs of the common descending exponential, joined at a common point "C", as shown in Fig. 6-11. This compound curve is both shifted and shaped with the variation in skewness of the sample. The changes for positive skew are sketched in Fig. 6-11. Changes with negative skew are identical but reversed. The compound exponentials yield an intermediate variate, v, as a function of the historical variate, h. The major consequence of the transform is to cast small values of h with a finite lower bound, and larger values of h with an infinite bound, into the variate v with a finite range from 0 to 4. Since v is an internal variate, obviously any scale can be used. The range from 0 to 4 is a choice of convenience and has no effect on the analysis. The only requirement is consistency in the calibration of the following parameters.

The v_1 and v_2 limbs of the transform are given by

$$v_1 = VB - (VB - C)\exp(fh') \qquad v \geq C \tag{6-7a}$$

$$v_2 = C \exp(-d\,h') \qquad v < C \tag{6-7b}$$

where C is the common point, in v-scale, where the limbs join; VB is the asymptotic right boundary of the v_1 limb; f and d are the shape parameters of the limbs; and h' will be defined below.

C was made to shift linearly with the coefficient of skew, SK, of the sample being analyzed. The linear relationship was calibrated by setting C = 2 for zero skew and C = 3.5 for a coefficient of skew of 4. Symmetry requires that C be at the mid-range of v for zero skew. Also, it can only move from the mid-value of 2 to the boundary of 4 with increasing values of skew. While this calibration is judgmental, it is based on the geometrical restrictions of the v-scale in Fig. 6-11. This calibration yields the following for C:

$$C = 2 + 0.375\ SK \tag{6-8}$$

The right boundary of v, VB, shifts the same amount as C shifts, and starts from VB = 4.05 for zero skew. The value, 4.05, is an asymptotic boundary for the v_1 limb. It represents a rational value beyond the required finite value of 4.0 for the calibration point. The choice of 4.05 has exactly the same purpose as the accepted practice of adding a small value to each variate of a sample before taking logarithms in order to avoid the logarithm of zero. This calibration yields the following for VB:

$$VB = 4.05 + (C - 2.0) \tag{6-9}$$

The variate h' in Equation 6-7 is a re-scaled value of the variate h using the conventional rescaling:

$$h' = \frac{h - \bar{h}}{s_h} \tag{6-10}$$

where $\bar{h}$ is the mean of the sample under analysis, and s_h is its standard deviation.

Equation 6-10 was found to be insufficient re-scaling for the variate transform presented here. Equation 6-10 produces standardized deviates about the sample mean $\bar{h}$. For highly skewed samples the sample mean shifts significantly away from the median, or 0.5 probability value, in the direction of the skew. Too much of the sample then lies to one side of $\bar{h}$ where h' is zero, and where the two limbs of the compound curve join, as shown in Figure 6-11. The net result is insufficient reduction in curvilinearity of the sample when plotted against probability.

The solution to re-scaling skewed samples was to shift the h and h′ scales, Fig. 6-11 and Equation 6-1, relative to each other as the sample skew changes. The form of this shifting is:

$$h' = \frac{h - \bar{h} + a + b\ SK}{s_h} \tag{6-11}$$

Equation 6-11 retains Equation 6-10 as a component but adds a shift with skew. a and b are the calibrating parameters of the shift. They may be judgmentally evaluated by making the shift zero for zero skew, and one-half the standard deviation for SK = 4. This calibration is justified as follows. For a symmetrical sample with zero skew no shift of the h and h′ scales is desired. The tails of a symmetrical sample tend to be similar. For a skewed sample the tail for large sample items is longer than the tail for small items. The re-scaling of Eq. 6-11 has the effect of "pulling in" the long tail and creating a more symmetrical h′. The value of one-half of the standard deviation for SK = 4 was determined by trial analyses on various data sets. This calibration yields Eq. 6-12 for the operating equation to re-scale h to h′:

$$h' = \frac{h - \bar{h}}{s_h} + \frac{SK}{8} \tag{6-12}$$

The shape parameters of the two exponential limbs, f and d of Eq. 6-7, remain to be calibrated. This is done for each sample analyzed by specifying a value h_{min} as shown in Fig. 6-11. h_{min} is one coordinate of a calibrating point for the v_1 limb. The other coordinate of this point is v_1 = 4. Substituting these coordinate values in the v_1 limb of Eq. 6-7 yields:

$$f = \ln((4\text{-}VB/(C\text{-}VB))/h'_{min} \tag{6-13}$$

where h'_{min} is computed by setting h = h_{min} in Equation 6-12.

In addition to being common, the two limbs must be tangent at point C. Differentiating, and setting the derivatives equal at C, where h′ = 0, yields the following equation for d:

$$d = f(VB\text{-}C)/C \tag{6-14}$$

This completes the calibration of the compound curve for positively skewed samples. In summary, the shape of the curves is based on the mean, standard deviation, and coefficient of skew of the sample. Only the value of h_{min} needs to be an input to the calibration. The operating effect of h_{min} must be clearly visualized. It is used to set the boundary of the class containing the smallest events. It corresponds to setting the first class boundary when preparing sample histograms in simplistic data reduction. Any value can be used for h_{min} that is slightly less than the smallest item in the sample. The one exception to this selection is that h_{min} must be set to zero for variates such as precipitation that cannot have negative values. Specification of h_{min} does not establish a lower limit of the variate h.

As mentioned above, the transform for negatively skewed samples is identical but reversed. Small values of h cannot extend to negative infinity. The common point, C, moves to the left with increasing negative skew. The limiting boundary, VB, shifts from the upper right to the lower left in the lower portion of Fig. 6-11. The calibration point (v_1 = 4, h_{min}) changes to (v_2 = 0, h_{max}) and h_{max} is now an input to the analysis. The above considerations lead to Equation 6-15 for the two exponential limbs for negative skew.

$$v_1 = 4-(4-C)\ \exp(fh'), \qquad v \geq C$$
$$v_2 = (C+VB)\ \exp(-dh') - VB, \qquad v < C \tag{6-15}$$

The equation for the common point, Equation 6-8, is not changed but SK is now negative. Equation 6-9 changes to Equation 6-16 for VB.

$$VB = 0.05 + (2.0-C) \tag{6-16}$$

Equation 6-12 for re-scaling h to h' is unchanged, but SK values are now negative. Calibration for the parameter d in the v_2 limb of Equation 6-15 is given by

$$d = -1n(VB/(C+VB))/h_{max} \tag{6-17}$$

where h'_{max} is calculated from the input h_{max} using Equation 6-12. Mathematical continuity at the point C yields the following equation for parameter f:

$$f = d(C+VB)/(4-C) \tag{6-18}$$

The mathematics in the previous section allows transformation of any values of the variate, h, to values of the intermediate variate, v. Conversely, values of v can be converted to h. It is now possible to tally a sample of items, h, into classes that are linear in v-scale. The sample space in v, $0 < v \leq 4$, is divided into 40 equal classes, the class boundaries are reflected to h-scale, and the sample h is tallied into these 40 classes. The class subtotals are accumulated from largest to smallest h. The class accumulations are divided by the sample size, producing accumulated class totals as ratios running from zero to one.

The use of 40 classes represents both choice of convenience and the author's experience. No set rules can be established anymore than set rules can be established as to how many classes to use in preparing a simple histogram of a sample.

The accumulated class ratios are sample probabilities and can be plotted against the linearized intermediate variate v as indicated in the top portion of Fig. 6-11. Since accumulation is from large h to small h, the class subtotals of probability must be plotted at the right-hand class boundaries in v-scale. Not all of the classes of the 40 specified will necessarily be occupied, particularly for small samples. This only means that "plateaus" occur in the sample accumulated probabilities. The plotted sample, or its computerized counterpart, must now be smoothed, and this smoothing will eliminate the plateaus.

The sample-size requirements of the method are not stringent. Just as in the evaluation of distribution functions, larger samples produce better population estimates. Judgment as to sample sufficiency should be based on the information content of the sample, not solely on methodology. Thomas et al. (1977) have demonstrated in detail the smoothing capability of sliding polynomials with limited data.

Program B-6.3 in the Appendix will perform smoothing of samples using this generalized transform. The use of this program is illustrated by the result in Figures

6 – 12 and 6-13. In Figure 6-12 the results of probabilistic analyses of a 99-year record of November rainfall at Athens, GA, are shown. In Figure 6-13, the analysis of the November minimum temperature for 47 years at Watkinsville, GA, is shown. Brief characteristics of the samples and the parameters of the data transformation are given in Table 6-12. Note that the rainfall has a positive coefficient of skew and that the temperature has a negative coefficient. For the positively skewed November rainfall, h_{min} was set to zero since this is the lowest physically possible value, and h_{max} does not apply. For the negatively skewed November minimum temperature, h_{max} was set to 36 degrees Fahrenheit, a little higher than any value in the record. The parameter h_{min} does not apply.

The sample data and the smoothing curves are plotted in v-scale in Figures 6-12 and 6-13. With the parameters given in Table 6-12 and with Equations 6-7 through 6-18 the relationship between the historical variate, h, and the transformed variate, v, can be calculated for both positive and negative skew. This relationship is superimposed on the smoothed data in Figures 6-12 and 6-13.

Use of the dual plots is illustrated with the determination of the median value in both figures. The median values is the 50-percent probability value. One-half of the values in the parent populations are expected to fall above this value and one-half are expected to fall below. Entering the probability scale at 50 percent one moves horizontally to the smoothing curve. This establishes the transformed variate, v. From this intersection move vertically to an intersection with the v-h transform curve. This establishes a value of h, either rainfall or temperature. For comparison the arithmetic means of the samples are also shown.

Any value of probability can be transformed to a value of h. For example, if one obtained values of h for 25 and 75 percent, these with the medians would give the quartiles of the distribution. It should be noted that a value of probability can also be obtained for any value of h. The evaluation in Figures 6-12 and 6-13 is simply reversed.

The presentations in Figures 6-12 and 6-13 are convenient for establishing levels of risk, or conversely, establishing values of h for chosen values of risk, or the exceedance probability. The value of data transformation to enhance smoothing and the estimation of the population is illustrated. However, these graphs are not very useful in calculating the uncertainty of these risk values. When uncertainty is required, simulation with nodal variation should be used as illustrated in Example 6-4.

Note particularly that, when data transformation is used, the values of the sample probability are never transformed. The general principle of avoiding transformation of the dependent values, whether a property or a probability, is well known and has recently been discussed again (Koch and Smillie, 1986).

Program B-6.3 specifies that the boundary conditions that the probability distribution function derived by smoothing are zero and also have zero slope at v = 0. While the zero value of probability at v = 0 (h = + ∞) is required by the problem, a slope of zero in v-scale is not a requirement and could be modified.

6.6 SEASONALLY CONTINUOUS PROBABILITY

Our study of free-form probability analysis to this point has centered on the distributions of single variates. Analyses with more than one variate are also possible with sliding polynomials. Two general situations will be discussed. In this section we will present an analysis of a form of a bivariate distribution with one variate fixed and the other stochastic. In the next section, we will discuss a true bivariate distribution of paired stochastic variates.

Many historical data sets are constructed from observations or measurements at fixed time intervals. In order for such data to be useful, we must reduce them to workable averages or average relationships. We have looked previously at the expected values of the maximum or minimum temperature for a particular month. But for good planning over a longer period, we may need to know expected values for several months, perhaps for each month of the year. Now the climate changes in a regular and continuous manner from month to month. Therefore, good practice dictates that we pool all of our data before determining probabilities of expected events, but at the same time we must allow the probabilities to vary by months or by some other time period. And lastly, the transition of probabilities from time period to time period must be cyclically continuous so that the transition from December to January is just as smooth as, say, from January to February.

Snyder and Thomas (1986) developed a seasonally continuous probability distribution based on two-dimensional sliding polynomials. In this method the sample of data for each month is first transformed according to Equations 6-7 to 6-14. This brings the data sets for each month into close alignment of v-values even though the natural data may vary greatly by months. Smoothing for estimation of the population is then efficiently performed over a rectilinear space.

Using Equations 6-7 through 6-14, each monthly sample is tallied into one of 40 uniform classes in the transformed scale, v. Sample class probabilities are accumulated and 12 graphs of probability versus the abstract variate v can be constructed.

In order to produce continuous probability functions that follow the natural continuity of seasonal transitions, it is necessary to smooth simultaneously the 12 monthly sample probabilities and the month-to-month transition. Therefore, a surface of smoothing is required. The smoothing space is defined by the magnitude of events in v-scale versus the time by months. The height of the surface across this space is probability. The surface is constructed by using two-dimensional sliding polynomials as smoothing operators. Smoothing follows essentially the same methods as developed for the two-dimensional smoothing of properties in Chapter 5. Boundaries of the space in the v-scale follow the construction in the previous section. Boundaries of the space in the time scale have a particular requirement. Seasonal continuity requires that the smoothing surface pass through December at the end of the year and join with January at the beginning of the year with mathematical continuity. For this study a cyclically continuous form of sliding polynomials was constructed as described in the following paragraph.

The smoothing surface initially requires a 7 x 7 grid of nodes, as shown in Figure 6-14, in order to specify the two-dimensional sliding polynomial coefficient for any value of the transformed variate, v, and for any calendar month. Rows of nodes are located at January, April, July, October, and again in January. Rows of outlier nodes are located at an advanced October and at a following April. Columns of nodes are located at integral values of v from -1 to 5. This definitional set of nodes is reduced as follows. In order to produce a seasonal continuity of probability, the surface must "wrap around" to produce a cylinder function. The second January nodes must wrap around to overlap the initial January values. The rows of outlier nodes will at the same time wrap around to

October and April. Such cyclic boundary control was presented in more detail in Figure 5-11 and Table 5-11. The wrapping around is accomplished very simply by adding the coefficients for the same calendar months. For example, the advanced October coefficients are added to the October coefficients in the reduced set of nodes. In the v-scale the column of boundary nodes at v=0 is set to zero, and the column of nodes at v=-1 is algebraically eliminated by setting it equal to the column of nodes at v=1. The residual 4x5 grid of nodes, shown as solid dots in Figure 6-14, is then evaluated by least squares.

The method of deriving seasonally continuous probabilities is illustrated with three data sets. The first set is a 99-year record of monthly total rainfall at Athens, Georgia, for the years 1885 to 1983. The second example is a 44-year record of maximum daily streamflow for the Oconee River near Athens; the third is the record of the minimum daily flow at the same location. The record extended from 1938 to 1981. The sample characteristics for each month for the three data sets are given in Table 6-13. Note that rainfall is highly variable with a standard deviation several times the mean value. Skewness is slight to moderate, with the highest values being in summer and autumn. The maximum streamflow is less variable relative to the mean values. The skewness is moderate to high for most of the year, with peak values in June and November. The minimum streamflow is moderately variable and lightly skewed except for April. Program 6-4 in the Appendix was used to analyze these data sets.

The parameters of the transforming Equations 6-7 to 6-14 are given in Table 6-14. These parameters reflect the characteristics in Table 6-13. For example, the common point is near a v-value of 2.0 for the lightly skewed minimum flows, a little larger for the monthly rainfall, and exceeds a value of 3.0 for some months for the highly skewed maximum streamflows. The right boundary VB follows the same pattern of variation. The higher variability of the maximum flows causes a lower class limit of zero for h_{min}, even thought the less variable minimum flows have a lower class limit above zero.

The values of the 4 x 5 grid of nodes found by least squares are given in Table 6-15. These values are probabilities and give the height of the continuous probability surface at that point. The standard deviations of the nodes are also shown. These nodal standard deviations are measures of the statistical instability of the surface, or uncertainty, and can be taken as one measure of the goodness of fit. They are an alternative to the cumbersome errors of prediction from conventional regression analysis. Most of the standard deviations in Table 6-15 are less than one percent probability, with only two values being two percent or higher. We may judge, therefore, that the nodal values of probability are well defined, and consequently, that the surface through these nodes is also well defined.

The continuous seasonal probability contours for the three data sets are displayed in Figures 6-15, 6-16, and 6-17. Interpolation to construct these contours takes place in the uniform spacing of the v-scale, but the v-values are de-transformed back to the scales of the natural variates, h, prior to plotting. The de-transformed contours show the marked seasonal variations, but the seasonal transitions are smooth. In Figure 6-15 the high probability of high rainfall totals is sharply revealed. For example, the one percent probability contour shows that there is about 45 cm. of rain in August. In October, on the other hand, the one-in-one hundred risk level is less than 20 cm. March is probabilistically the second wettest month, and February, May, and September are relatively dry months.

The maximum daily streamflow, displayed in Figure 6-16, shows a seasonal pattern of three parts: a period of probability of high flows from January to July, low flows from June to October, and a recharge period from October to January. In April, the probability of getting a flow of 200 m^3/sec or higher is about five percent. In October, the probability of getting a maximum daily flow of 50 m^3/sec or higher is five percent.

The minimum daily streamflows are displayed in Figure 6-17. The probability of high minimum flows in April, at the end of the cool wet season, is strongly revealed. The 75 percent probability contour shows that in April the probability is 25 percent of getting flows of 9 m^3/sec or less. In September the probability is 25 percent for getting flows of only 2.5 m^3/sec or less.

6.7 BIVARIATE DISTRIBUTIONS

The surfaces for seasonal probabilities in the previous section were based on one stochastic variate, the magnitude of an event, and on one fixed variate, the time of the event. In this section we will discuss probability surfaces constructed on two stochastic variates. Such surfaces are conventionally called bivariate distributions. The particular topic here is the utilization of two-dimensional sliding polynomials to derive free-form bivariate surfaces from stochastic pairs of data points.

The development of a bivariate distribution is a simple extension of the single variate distributions in Figures 6-12 and 6-13. The same generalized variate transform is used, and each of the two variates is transformed according to its sample characteristics. h_{1i} and h_{2i}, $i=1$ to n, are defined as data pairs forming a sample of size n of a bivariate distribution. For each occurrence of the variate h_1 there is a corresponding or simultaneous occurrence of the variate h_2. In analysis, both variates are transformed using Equations 6-7 through 6-14. Assume that h_1 is transformed to v, as in Fig. 6-11, using the characteristics of its sample distribution. Assume also that h_2 is similarly transformed to z using the characteristics of its sample. We can then set uniform classes in the v- and z-scales and tally the sample into a two-dimensional grid of classes. The process is illustrated in Figure 6-18.

Part a of Figure 6-18 shows a completed tally of a bivariate sample into a 10x10 grid of classes. This is simply a two-dimensional version of a histogram, with the value shown representing the number of items of the sample that falls into each class. For example, there are four h_1-h_2 pairs with values of h_1 placing them in the largest v-class and at the same time with values of h_2 placing them in the largest z-class. Part b of Figure 6-18 shows the partial construction of the accumulated sample probabilities. The values in Part a have been summed horizontally producing a series of sample accumulations as in Part a, Figure 6-3.

Part c of Figure 6-18 shows the completion of the bivariate sample accumulation. The horizontal summations in Part b have now been summed vertically. We see that the sample size is 279 pairs of v-z. The accumulations start from zero at the lowest class boundaries along the bottom and left edges. The two-way accumulations are placed at the grid intersects since this corresponds to placing the single accumulations at the right-class limits as in Example 6-2. The column of accumulated values along the right edge can be called a marginal sample of z. This column includes all values of z, regardless of the value of v. The row of accumulated values across the top can be called a marginal sample of v. It includes all values of v regardless of the value of z.

Sample probabilities are calculated next by dividing each gridded accumulation in Part c by the sample size, 279. These sample probabilities are plotted in Part d and are shown as percentages. The right-hand column is a marginal sample probability of z, and the top row is a marginal sample probability of v. The grid of numbers in Part c can now be smoothed by a one-dimensional sliding polynomial analysis. Boundary conditions can be imposed to assure that the smoothing function is zero along both the left side and the bottom of the grid. Also, for infinite-tailed variates, the smoothing function must have zero slope along both the left side and the bottom of the grid.

The probabilities in Part d, Figure 6-18, must be read as dual statements of risk. The value of 10 percent at ($v=2$, $z=2$) means that only 10 percent of the data pairs had both values of $v=2$ or less and values of $z=2$ or less. In other words, only 10 percent of the pairs fall in the lower left quadrant of the grid. Also, 34 percent of the pairs fall in the left half of the grid. Therefore, (34-10), or 24 percent of the pairs fall in the top left quadrant. Such statements of probability, based on more than one variate, are called statements of joint probability.

We may also construct conditional probability distributions. Suppose we assume that some z falls in the sixth class from the bottom, which has lower and upper class limits of $z=2$ and $z=2.4$. Two ways to construct the probability of occurrences of v given this value of z are shown in Table 6-16. In the top part of the table, the probability values for the class boundaries of 2.4 and 2 are differenced and then re-scaled to read from zero to 100 percent. In the lower part of the table the class frequencies from Part a of Figure 6-18 for the z-class from 2.0 to 2.4 are given. These are accumulated and converted to probability values as for the distribution of a single variate. We see that within round-off error the two probability distributions of v are the same. Thus a conditional distribution is a distribution of either variate of a bivariate pair conditioned by a particular value of the other variate of the pair.

For simplicity the relevant definitions of a bivariate distribution were explained using sample values. In practice, we do not use sample values. Just as in single variate analysis, we use the smoothed surface as an estimate of the bivariate population after smoothing the sample. Marginal, conditional, and joint probabilities are evaluated from the smoothed surface. Smoothing a bivariate sample to estimate the population is not particularly complex, but it is a tedious and difficult calculation to perform manually. It is best performed by computer. Snyder and Thomas (1987) reported the development and use of such a program. They tested the program on eight data sets of varying degrees of correlation between the bivariate pairs. Two of these analyses are given below.

The first data were 480 pairs of monthly total rainfall and monthly total EI at Watkinsville, Georgia. EI is a composite rainfall amount-rainfall intensity index of the energy of a rainstorm and is used in the prediction of erosion from agricultural lands. The second data set was made up of 141 pairs of values of monthly total rainfall and monthly average temperature at Watkinsville. The square of the correlation coefficient between the paired variates was 0.43 for the first set and 0.11 for the second.

Following smoothing, contours of joint probability were drawn. These are shown in Figures 6-19 and 6-20. Note that in these figures the variates have been detransformed from the standardized v-z form in Figure 6-18 back to the scale of the original variates. These contours of joint probability are used as follows. In Fig. 6-19 note that the probability of the monthly total rainfall being 30 cm or greater and EI being 400 units or greater is about 3.3 percent. Also, the joint probability of rainfall being greater than 30 cm and EI being greater than 300 units is about 6.5 percent. Therefore, given the condition that rainfall is 30 cm or more, the probability that EI for the month will lie between 300 and 400 units is 6.5 minus 3.3 or 3.2 percent.

The patterns of contouring in Figures 6-19 and 6-20 are different, which indicates differing bivariate relationships. The temperature versus rainfall contours are more relaxed for middle and high probabilities, but more concentrated in the low probabilities, or for extreme values. This is the same as saying that the rainfall-temperature data indicate higher probabilities of extreme joint occurrences than is indicated by the rainfall–EI data. Figure 6-21, Parts a and b, shows sections through the form-free bivariate surface for each 10 classes. Sections in Part a are parallel to the rainfall axis. Sections in Part b are parallel to the temperature axis. The sections at class 40 in Part a is the marginal distribution of the monthly total rain, i.e., the distribution of rain without regard to temperature. The section at class 40 in Part b is the marginal distribution of temperature.

The sections through the surface show that the free-form sliding polynomials represent the data with little error, and also show that the surface conforms to the varying sectional shapes.

The greatest use of bivarate distributions presently is for generating synthetic pairs of the data. Such simulation is needed when both variates of the pair occur in some model for prediction. Separate simulation of the two variates is not possible because this would ignore the fact of the paired relationship. Bivariate simulation does take account of the relationship. The simulated variates must maintain the sample characteristics of each of the variates but must also maintain the relationship between the variates as exhibited by the original sample. Snyder and Thomas (1987) showed that the free-form bivariate surface did preserve the necessary characteristics when used in simulation.

6.8 SIMULATED STOCHASTIC INTEGRALS

With this section we begin the presentation of several specialized procedures for testing for significant differences. Many tests for difference exist in the statistical literature. They are part of a larger topic of decision theory. It is not our objective to review or comment on these tests. Rather, we wish to present the use of sliding polynomials with localized uncertainty to reduce non-conforming data sets to a form where the existing tests can be applied.

Snyder and Harper (1986) used the simulation of stochastic integrals to test for differences between the total flux of toxaphene to the air and to the soil after the chemical was applied to a cotton canopy. They also tested for the difference between toxaphene and DDT volatilization rates from the cotton canopy. They pointed out the difficulty in attempting to perform the test using conventional regression analysis.

For the first specialized situation we consider a problem where experimental data provide information about a rate of change, or a gradient, of some property, but a test must be performed on the total mass of property over some space or time. To illustrate this problem, suppose we have a set of measurements of the rate of transport of some chemical in water moving past an observation point along the stream channel. Such measurements can be made by taking samples of the water at selected times, finding the mass concentration of the chemical, and then multiplying the mass concentration by the mass flow rate of the stream. Since we must expect errors and lack of precision in representing a property of a stream by drawing a small sample from it, we immediately must face several questions. What are the probable ranges of the total mass of chemical transported? Because of the errors in sampling, can we show differences in sampling at two different points? The following examples will demonstrate an approach to answer these questions.

6.8.1 Example 6-5: Calculation of a Probable Amount of Mass Transport

The synthetic data set shown in Table 6-17 represents measurements of stream transport of a chemical past a measuring point A. We are interested in getting an estimate of the total amount carried past point A. The data were first smoothed with three spans of the four-point sliding polynomials using Program 5-3. The resulting values of the nodes and their standard deviations are shown in Table 6-18.

In Example 6-4 we used Equation 6-6 to randomize values of the nodes prior to simulation. In Chapter 4 we saw that the integral of the center span of a unit set of nodes of the four-point sliding polynomial is given by:

$$\int_{x=x_1}^{x=x_2} y\,dx = b \int_{z=-0.5}^{z=0.5} y\,dz \tag{6-19}$$

$$= b \left[-0.041667 \left(Y_i + Y_{i+3} \right) + 0.514667 \left(Y_{i+1} + Y_{i+2} \right) \right]$$

In this equation, b is the interval between nodes in x-scale (actually t-scale for this example), -0.041667 and 0.541667 are the sliding polynomial integral coefficients, and the Y_i are the unit set of nodes. If we use Equation 6-18 to integrate the rate of transport for each unit set of nodes, or each span, and then add the values for all spans, we obviously will get the total amount of chemical transported during the 30-hour period.

Our procedure is as follows. Each node in Table 6-18 is randomized with Equation 6-6. Then the randomized set is integrated by applying Equation 6-19 to the three spans and adding the span integrals. Equation 5-24 and 5-25 must be used to calculate the two outlier nodes for the first and third spans. This procedure was repeated 60 times, thus generating a 60-item sample of possible values of the amount of the chemical transported.

The histogram of simulated values is shown in Figure 6-23. All 60 values are possible amounts of transport, given the variability of the data in Figure 6-22. However, some values are much more likely than others. The mean of this simulated set is 626.16 kgm. The standard deviation is 20.80 kgm. Most values fall close to the mean.

Suppose we ask what range would encompass 90 percent of the simulated values. We can estimate this by dropping five percent of the lowest and five percent of the highest values, or three from each tail. This yields a range of about 589 to 660 kgm as the 90 percent likely range of the amount transported. If we assume that the sample in Figure 6-23 can be represented by the normal distribution, we can refine our expected range. Values of the standardize variate z of the normal distribution can be found in any standard statistical text. Values of the five percent point on the tails are 1.65. Then 626.16 ± 1.65*20.80 gives a range of 591.8 to 660.5 kgm as the 90 percent likely range.

6.8.2 Example 6-6: Testing for Significant Difference of Transported Amounts

Suppose we are interested in whether or not the amount of a chemical is changing as the mass is carried downstream. To this end, we have taken measurements at Point B downstream from the measurement Point A of Example 6.5. This downstream set of measurements is given in Table 6-19. The measurements have assumed that six hours is the travel time of the stream from Point A to Point B.

We proceed to analyze the Point B data as we analyzed the Point A data. Data are plotted and smoothed in Figure 6-24. Nodal values and standard deviations are given in Table 6-20. The histogram of simulated amounts transported past Point B is given in Part a of Figure 6-25. These values were actually simulated at the same time the values for Point A were calculated. For each paired simulation we can subtract the Point B simulated value from the Point A simulated value, thus getting a simulated difference of transport past the two measuring points. The histogram of 60 differences is shown as Part b of Figure 6-25.

The mean of the simulated distribution at Point B is 541.21 kgm and the standard deviation is 12.92 kgm, which is considerably smaller than at Point A. The mean of the simulated differences is 84.95 kgm and the standard deviation of the differences is

23.72 kgm. We are interested in the lowest likely value of the mean of the differences. Equation 6-20 can be used for such a calculation;

$$\mu = \bar{x} - t_{\alpha} \left(\frac{s}{\sqrt{n}} \right) \tag{6-20}$$

in which $\bar{x}$ is the mean of the sample, s is the standard deviation, n is the sample size, and t_{α} is the value of the "t-distribution" for a probability α and n-1 degrees of freedom. μ is that value of the mean so that α percent of values are larger or (100-α) percent are smaller. The value of t for a single-tailed test for 99.75 and n-1=59 is 2.92. This value, when used with the other elements in Equation 6-20, gives a value of μ of 76.0 kgm. There is almost no risk of a mean difference less than this value. We know then that the difference in chemical transport between the two sections is not zero. Therefore, some of the chemical is being lost between Points A and B. The material might be oxidizing, volatilizing, or settling.

6.9 TESTING FOR LOCAL DIFFERENCES

The analysis of experimental data sometimes requires determination of the significance of differences at selected points along two functions or of differences between a selected point on a function and the mean of some complementary data set at that point. Such situations arise, for example, in measuring concentrations of chemicals in the environment and in searching for significant pathways by which chemicals may move within the environment. Consider the situation where two methods of measurement are possible. One of these methods is tedious and expensive. Can acceptable results be obtained with the simple and more economical method of measurement?

Harper and Snyder (1987) studied such situations to document the uptake of nutrients during the stages of plant growth. In the following example, we will again use synthetic data in simpler situations to illustrate the procedure for localized testing.

6.9.1 Example 6-7: Testing for Difference Between a Function and an Auxillary Data Set

Suspended sediment carried by a stream was sampled at regular intervals for a 16-hour period, as shown in Table 6-21. This time span covered the rise, crest, and recession of the stream during a storm event. The sediment concentration is reported as parts-per-million mass ratio of sediment to water.

The timed samples are measured as averages along a vertical from the surface to the bottom at a fixed point in the cross section of the stream. At four hours and at twelve hours, seven extra samples were taken, each across additional verticals spaced across the stream. The 4-hour auxillary measurements are during the rising period of streamflow, when the stream tends to be more turbulent. The 12-hour auxillary measurements were during the recession when turbulence tends to diminish. Are there significant differences between the timed and the auxillary measurements?

The timed samples are plotted and smoothed in Figure 6-26. The auxillary samples at four hours and twelve hours, given at the bottom of Table 6-21, are also plotted. The values of the nodes at four hours and twelve hours, obtained by smoothing, are localized means of the timed data. The residual error variance, $E(e)^2$ in Equation 5-29, is known from the calculation of the standard deviation of the nodes. The sum of the coefficients weighting the errors to the nodes Σc in Equation 5-30, is the localized sample size. The mean, variance, and sample size of the auxillary data set are known or are readily calculated.

When the means, variances, and sample sizes are known, the significance of the difference between the means is easily tested. Natrella (1963, p 3-34) gives the following equations;

$$S_p = \sqrt{\frac{(n_A-1)\,s_A^2 + (n_B-1)\,s_B^2}{n_A + n_B - 2}} \tag{6-21}$$

and

$$u = t_{1-\alpha}\, S_p \sqrt{\frac{n_A + n_B}{n_A\, n_B}} \tag{6-22}$$

in which s_A^2 and s_B^2 are the error variances of the two sets, n_A and n_B are the sample sizes, s_p is the weighted standard deviation of the combined set, t is the t-statistic for significance level α with $(n_A + n_B - 2)$ degrees of freedom, and u is the largest mean difference at probability α. If the magnitude of the actual difference of means is greater than u, then the means of the two sets are significantly different.

The calculation for testing the difference between the timed data and auxillary data at four and at twelve hours is shown in Table 6-22. A significance level of 5 percent was used. At four hours the difference of 2.67 ppm is less than the statistic u of 23.52, so the measurements are not significantly different. At twelve hours the difference of 14.75 ppm is greater than the statistic u of 9.99, so the difference is significant. The auxillary data at four hours do not indicate a necessity for changing the position of the vertical where timed samples are taken. The twelve-hour data, however, indicate that the sampling vertical should be moved to one of the auxillary verticals where values are closer to the auxillary mean.

6.10 SIGNIFICANT DIFFERENCE OF SLOPES

In the previous sections, sliding polynomials with localized variances and localized degrees of freedom were used to test for differences between integrals of functions and between functions and auxillary data sets. They can also be used for testing differences between the slopes of two functions. Such problems may arise in water quality analyses. For example, we could measure the mass of some chemical in water at selected times as the mass changes due to oxidation. Suppose we had two series of such measurements. Can we decide if the rate of change, the rate of oxidation, is different in the two series?

6.10.1 Example 6-8: Testing for a Difference in BOD Changes

Biological oxygen demand (BOD) is a measure of the concentration of organic matter in water. The decay of material uses oxygen dissolved in the water. The greater the organic load, the greater the demand for dissolved oxygen in the decay process. Oxygen may be depleted to the point where fish and other organisms cannot survive. Reaeration through mechanical mixing can supply oxygen to the water and restore an oxygen balance.

Table 6-23 shows synthetic data for BOD over time under two different mixing rates. Is the decline of BOD significantly different under the two mixing rates? The data are plotted and smoothed with four-point sliding polynomials, as shown in Figure 6-27. The nodes and the error analysis of the smoothing, which are the elements of Equations 5-29 and 5-30, are shown in Table 6-24. The smoothing curves show that the levels of BOD are the same, 23 ppm, at 20 hours from the initial time of measurement with mixing rate #1, and six hours from the initial time for mixing rate #2. Are the slopes different for the two mixing rates when the BOD is 23 ppm in both runs?

In Chapter 3, we saw that the slope of the sliding polynomial is given by the calculation expressed by:

$$\text{slope} = dy/dt = dy/10dz = \frac{1}{10} \sum_{i=1}^{4} Y_i \; dc_i(z)/dz \qquad (6\text{-}23)$$

in which the Y_i are the nodes of any unit four-node set, the $c_i(z)$ are the sliding polynomial coefficients, and z ($-0.5 \le z \le 0.5$) is the unit variate. For a BOD of 23 ppm, z = -0.5 on the nodal set Y_{10}, Y_{20}, Y_{30}, Y_{40} for mixing rate #1 and z = 0.1 on the nodal set Y_{-10}, Y_0, Y_{10}, Y_{20}, for mixing rate #2. Y_{-10} is the outlier node algebraically eliminated by parabolic extension as in Equation 5-25. The values of $dc_i(z)/dz$ (i=1 to 4) are 0.5, 0.0, 0.5, 0.0 for z = -0.5 and 0.16, -1.38, 1.28, -0.06 for z = 0.1. The slope at z = -0.5 for mixing rate #1 is calculated by:

$$\frac{dy}{dt} = \frac{1}{10}(-0.5*32.104 + 0*23.123 + 0.5*20.406 + 0*17.647) = -0.5849 \text{ ppm/hr} \qquad (6\text{-}24)$$

The slope at z=0.1 for mixing rate #2 is calculated by:

$$\begin{aligned} dy/dt &= \frac{1}{10}\left(0.16 * Y_{-10} - 1.38 * Y_0 + 1.28\, Y_{10} - 0.06\, Y_{20}\right) \\ &= \frac{1}{10}\left\{\left(3 * 0.16 - 1.38\right) Y_0 + \left(-3 * 0.16 + 1.28\right) Y_{10} + \left(0.16 - 0.06\right) Y_{20}\right\} \\ &= \frac{1}{10}\left(-0.9 * 33.416 + 0.8 * 17.575 + 0.1 * 10.943\right) = -1.492 \text{ ppm/hr} \end{aligned} \qquad (6\text{-}25)$$

The slope variances are calculated using the conventional variance of a linear form:

$$S_1^2 = \frac{1}{100}\left[\left(-0.5 * 0.6738\right)^2 + \left(0.5 * 0.3719\right)^2\right] = 0.001481 \qquad (6\text{-}26a)$$

$$S_2^2 = \frac{1}{100}\left[\left(-0.9 * 1.6337\right)^2 + \left(0.8 * 0.8146\right)^2\right] = 0.025917 \qquad (6\text{-}26b)$$

The standard deviations are the square roots of these variances, 0.03848 and 0.12099 ppm/hr., for mixing rates #1 and #2, respectively.

The sample size can be estimated as a weighted average of the localized sample sizes:

$$n_1 = \frac{(-0.5)^2\,(9.3125) + (0.5)^2(9.3125)}{0.5^2 + 0.5^2} = 9.3125 \qquad (6\text{-}27a)$$

$$n_2 = \frac{(-0.9)^2\,(3.5156) + (0.8)^2\,(9.3125) + (0.1)^2\,(7.3438)}{0.9^2 + 0.8^2 + 0.1^2} = 6.0829 \qquad (6.27b)$$

Now given the standard deviations and sample sizes for the two slopes, we can use Equations 6-21 and 6-22 to calculate the probable largest slope difference for a chosen level of significance α. The standard deviation of the distribution of differences is

$$S_p = \left(\frac{(9.3125\text{-}1) * 0.001481 + (6.0829\text{-}1) * 0.025917}{9.3125 + 6.0829 - 2}\right)^{1/2} = 0.1037$$

For α of 1 percent $t_{1-\alpha}$ for 13 degrees of freedom is 2.650. Therefore, the probable largest difference of slopes is

$$u = 2.650 * 0.1037 * \left(\frac{9.3125 + 6.0829}{9.3125 * 6.0829}\right)^{1/2} = 0.1432$$

The magnitude of the sample difference of slopes is

$$\left| \; -1.492 + 0.585 \; \right| = 0.907$$

This sample slope difference far exceeds the probable slope difference and we conclude that the difference in mixing rates does have a significant effect on the rate that BOD diminishes.

6.10.2 Example 6-9: Confirmation of Slope Test

A critical feature of Example 6-8 is the use of Equation 6-26 to calculate the variance of the slope of a function. It is an interesting application of computer intensive statistics (Cover and Unny, 1986) to verify this method of slope variance calculation.

Assume the unit nodal set in Table 6-25. At z = 0 the values of $dc_i(z)/dz$ are 0.125, -1.375, 1.375, and -0.125. Thus the slope at z = 0 is

$$0.125 * 1.0 - 1.375 * 2.0 + 1.375 * 2.0 - 0.125 * 1.0 = 0$$

The nodal variance at z=0 is

$$(0.125 * 0.1)^2 + (-1.375 * 0.2)^2 + (1.375 * 0.2)^2 + (-0.125 * 0.1)^2 = 0.1516$$

The standard deviation is $\sqrt{0.15156} = 0.389$.

In order to confirm this value by simulation, we can randomize the four nodes using Equation 6-6, and for each randomized set of four nodes, we can calculate the slope at z=0. Program B-6.7 was prepared to simulate the slope 100 times. This program was run twice. The histograms of the 100 random slope values are displayed in Figure 6-28. These histograms show the wide variability of slopes due to the variability of the nodes, which in turn is due to the errors and uncertainties in the data. However, the mean slopes were -0.054 and 0.029 for the two runs, both of which are close to the known mean slope of 0.0. Also, the standard deviations of the simulated slopes in the two runs were 0.407 and 0.386, thus verifying the value of 0.389 obtained by direct calculation.

Programs such as B-6.6 and B-6.7 were constructed for specific problems. However, they should also serve as examples of approach for any problem where direct calculation of the necessary statistics is difficult.

6.11 A GENERALIZED PROCEDURE FOR SIGNIFICANCE TESTING

Examples 6-7, 6-8, and 6-9 illustrated significance testing based on sliding polynomials. The testing procedure was confirmed by computer simulations. We will now reduce significance testing to a more generalized procedure that can be used for functions, that can be used at the nodes or at interpolation points between nodes (Snyder and Harper, 1989).

Three specific items are needed for significance testing: (1) a value of some property, (2) the variance of that property, and (3) the degrees of freedom associated with the property. For our purposes we defined a property as: (1) the value of a function at a specific point, (2) the slope of the function at a point, or (3) the integral of a function between two points.

The general approximating equation for the property is given by:

$$P = \sum_{i=1}^{d} w_i \; Y_i \tag{6-28}$$

The property P may be the interpolated value of a function, its slope, or its integral as defined by the appropriate approximating equations in Chapters 1 through 4. The weights, w_i, are functions of the sliding polynomial coefficients $c_i(z)$ and the Y_i are nodal values. The index d is usually 2, 4, or 6, depending on the sliding polynomial system, but may be larger if P is an integral that covers more than one span.

The second item needed is the variance of the property. Note in Equation 6-28 that the weights, w_i, have no variance. They are single valued and calculated only on the specified values of the sliding polynomial coefficients, which in turn are dependent only on the values of the local variate z. Therefore, any variance in P must come only from the variances of the Y_i. Equation 6-28 is a linear form, and an advantage of the sliding polynomials is that they remain linear in integration and differentiation. This enables the variances of all properties P to be calculated using the standard calculation of variance of a linear form. Using the variance of a linear form the variance of P, S_P^2, can be written as

$$S_p^2 = \sum_{i=1}^{d} w_i^2 \; S_{Y_i}^2 \tag{6-29}$$

In Equation 6-29 the $S_{Y_i}^2$ are the variances of the individual nodes. The calculation of these nodal variances was given by Equations 5-29 and 5-30. We now introduce a slight modification in the calculation of the nodal variance. Equation 5-29 remains unchanged as the calculation of the sample variance of data at a node. The estimate of the population variance of data at the node is given by

$$\sigma_K^2 = E(e_i)^2 \sum_{m=1}^{n} c_K / \left[\sum_{m=1}^{n} c_K - 1 \right] \tag{6-30}$$

The variance of a node is then calculated as the variance of a mean value:

$$S_{Y_K}^2 = \sigma_K^2 / \sum_{m=1}^{n} c_K = \sum_{m=1}^{n} c_K^2 e_m^2 / \sum_{m=1}^{n} c_K{}^2 / \left[\sum_{m=1}^{n} c_K - 1 \right] \tag{6-31}$$

Equation 5-30 is the computation incorporated into the computer programs in the Appendix. It is an operationally "safe" approximation to use for small data sets where Σc_k -1 may approach zero; however, Equation 6-31 with its increase in precision, should be used for significance testing.

A comparison of Equations 5-30 and 6-31 shows that the ratio of variances is

$$\frac{S_K^2\ (6\text{-}30)}{S_K^2\ (5\text{-}30)} = \frac{(ND - NN)}{ND} \quad \frac{\sum c_K}{\left[\sum c_K - 1 \right]} \tag{6-32}$$

If one assumes he or she has a sample of 30 points smoothed using 5 nodes, and the data are uniformly distributed to the nodes, the ratio is

$$\frac{30 - 5}{30} \cdot \frac{6}{5} = 1$$

and the equations yield the same variance. A sample of 20 points smoothed using 5 nodes with uniform data distribution also gives

$$\frac{20 - 5}{20} \cdot \frac{4}{3} = 1$$

However, if the data are not uniformly distributed to the nodes, the correction is not unity. For example, for the 20 point sample, if Σc_k is 3, then the correction is

$$\frac{20 - 5}{20} \cdot \frac{3}{2} = 0.888$$

The third and last item needed for significance testing is an estimate of the degrees of freedom. As explained in Chapter 5 the term Σc_k is the "portion of the sample" weighted to node K. This is because the sliding polynomial system is unbiased. The sum of the Σc_k across all nodes is equal to the sample size. Then the degrees of freedom of a node is simply $\Sigma c_k - 1$ since one degree of freedom is used in computing the nodal value as a local mean value. It follows that the degrees of freedom of a predicted value is a weighted mean of the degrees of freedom of the nodes used in the prediction. The w_i of Equation 6-28 and 6-29 can be used as weights:

$$\text{d.f.(P)} = \sum_{i=1}^{d} w_i^2 \left[\sum_{m=1}^{n} c_i \ -1 \right] / \sum_{i=1}^{d} w_i^2 \qquad (6\text{-}33)$$

In Equation 6-28 through 6-32 "i" has been used as an index to refer to nodes in a sliding polynomial unit set of nodes; "K" has been used to refer to nodes in the entire problem space.

All elements are now available for use in Equations 6-21 and 6-22 for significance testing. Note that n_A and Σc_k are sample sizes while $(n_A - 1)$ and $(\Sigma c_k - 1)$ are degrees of freedom.

6.11.1 Example 6-10: Test for a Significant Difference of Slopes

Two data sets of 16 points each are plotted in Figure 6-29. Both sets have been smoothed using four-point sliding polynomials with a parabolic extension at the boundaries. The results of smoothing are given in Table 6-26. Do the two curves have different slopes at x=1.35?

For this problem, P, which is the property in Equation 6-28, is $dy/dx|_{x=1.35}$. The weights, w_i, are the differentials of the coefficients, as given in Equations 3-9. These coefficients are divided by 0.25 for the scale change from global variate x to local variate z. The slopes are:

$$P_1 = \frac{0.06}{0.25} * 1.476 - \frac{1.28}{0.25} * 2.594 + \frac{1.38}{0.25} * 3.081 - \frac{0.06}{0.25} * 3.147 = 3.325$$

$$P_2 = 0.24 * 1.391 - 5.12 * 2.562 + 5.52 * 3.185 - 0.24 * 3.499 = 3.958$$

The variances are calculated from Equation 6-29 as:

$$S^2_{P_1} = 0.24^2 * 0.0602 + 5.12^2 * 0.0187 + 5.52^2 * 0.0297 + 0.24^2 * 0.0963 = 1.404$$

$$S^2_{P_2} = 0.24^2 * 0.0181 + 5.12^2 * 0.0065 + 5.52^2 * 0.0409 + 0.24^2 * 0.2346 = 1.431$$

The degrees of freedom are calculated from Equation 6-33 as:

$$df_1 = \left[\frac{0.24^2 * 0.3962 + 5.12^2 * 5.5683 + 5.52^2 * 4.8349 + 0.24^2 * 1.2006}{0.24^2 + 5.12^2 + 5.52^2 + 0.24^2} \right] = 5.165$$

$$df_2 = \left[\frac{0.24^2 * 0.7391 + 5.12^2 * 6.5476 + 5.52^2 * 4.3676 + 0.24^2 * 0.3457}{56.8} \right] = 5.366$$

The standard deviation of the difference, given by Equation 6-21, is:

$$S_P = \left[\frac{5.165 * 3.325 + 5.366 * 3.958}{5.165 + 5.366} \right]^{1/2} = 1.910$$

The least significant difference, given by Equation 6-22, is:

$$u = 1.812 * 1.910 * \left[\frac{6.165 + 6.366}{6.165 * 6.366} \right]^{1/2} = 1.955$$

The actual difference of slopes is 3.958 - 3.325 = 0.633. Therefore, the slopes of the two sliding polynomials at $x=1.35$ are not significantly different.

6.11.2 Example 6-11: Test for Significant Difference of Integrals

With reference to the data in Figure 6-29, we can ask: Are the integrals of the two curves in the interval $1.25 \le x \le 1.70$ significantly different?

For this example the integrals lie in two spans; therefore, the P of Equation 6-28 is

$$\int ydx = 0.25 \left[\int ydz|_{1.25 \le x \le 1.50} + \int ydz|_{1.50 \le x \le 1.70} \right]$$

The w_i of Equation 6-28 are computed from the integral coefficients given in Equation 4-5. For the full span, $1.25 \le x \le 1.50$, the integral is:

$$\int y dx = 0.25 \left[-0.04167\ Y_{1.0} + 0.54167\ Y_{1.25} + 0.54167\ Y_{1.5} - 0.04167\ Y_{1.75} \right]$$

For the partial span, $1.50 \le x \le 1.70$, the integral is:

$$\int y dx = 0.25 \left[-0.04053\ Y_{1.25} + 0.52693\ Y_{1.5} + 0.34773\ Y_{1.75} - 0.03413\ Y_{2.0} \right]$$

The outlier node $Y_{2.0}$ was eliminated during smoothing by a parabolic extension $Y_{2.0} = 3Y_{1.75} - 3Y_{1.5} + Y_{1.25}$. Substituting this expression for $Y_{2.0}$ yields, for the partial span:

$$\int y dx = 0.25 \left[-0.07467\ Y_{1.25} + 0.62933\ Y_{1.5} + 0.24533 Y_{1.75} \right]$$

The weights w_i of the total integral of the full and partial spans are

$$\begin{aligned} w_1 &= 0.25\ (-0.04167) &&= -0.01042 \\ w_2 &= 0.25\ (0.54167 - 0.07467) &&= 0.11675 \\ w_3 &= 0.25\ (0.54167 + 0.62933) &&= 0.29275 \\ w_4 &= 0.25\ (-0.04167 + 0.24533) &&= 0.05092 \end{aligned}$$

Then the integrals are:

$$P_1 = -0.01042*1.467 + 0.11675*2.594 + 0.29275*3.081 + 0.05092*3.147 = 1.3497$$

$$P_2 = -0.01042*1.391 + 0.11675*2.562 + 0.29275*3.185 + 0.05092*3.499 = 1.3952$$

The variances of the integrals are calculated as:

$$S^2_{P_1} = 0.01042^2*0.0602 + 0.11675^2*0.0187 + 0.29275^2*0.0297 + 0.05092^2*0.0963 = 0.003056$$

$$S^2_{P_2} = 0.01042^2*0.0181 + 0.11675^2*0.0065 + 0.29275^2*0.0409 + 0.05092^2*0.2346 = 0.004204$$

The degrees of freedom are the weighted values:

$$df_1 = (0.01042^2*0.3962 + 0.11675^2*5.5683 + 0.29275^2*4.8349 + 0.05092^2*1.2006)/0.10203 = 4.8356$$

$$df_2 = (0.01042^2*0.7391 + 0.11675^2*6.5476 + 0.29275^2*4.3676 + 0.05092^2*0.3457)/\ 0.10203 = 4.5530$$

The standard deviation of the difference is

$$S_P = \left[\frac{4.8358*0.003056 + 4.5530*0.004204}{4.8358 + 4.5530}\right]^{1/2} = 0.06010$$

and the least significant difference is

$$u = 1.833 * 0.06010 * \left[\frac{5.8358 + 5.5530}{5.8358 * 5.5530}\right]^{1/2} = 0.0653$$

Since the real difference, 1.3952 - 1.3497 = 0.0455, is smaller than the least significant difference of 0.0653, the integrals are not significantly different.

EXERCISES

6-1. Six stages of data processing are discussed in Section 6.2. Discuss these stages with respect to the problem of earthquake prediction.

6-2. The following data are based on measurements of the sediment yield (acre-feet per square mile per year) for 37 watersheds:

0.14	0.61	1.65	0.04	0.16	0.17
0.12	0.69	0.14	0.99	0.37	0.17
0.21	2.31	0.07	0.25	0.21	
0.15	2.65	0.22	2.20	0.35	
0.076	2.37	0.20	0.02	0.64	
0.66	0.51	0.18	0.03	1.55	
2.67	1.42	0.02	0.036	0.09	

Derive a free-form cumulative probability distribution for a histogram based on the above data. Smooth the cumulative histogram using a two-point sliding polynomial. Compute the probability density function by numerical differentiation. What magnitude would be greater than 95 percent of the expected population? What exceedence probability would be associated with a magnitude of 2.0?

6-3. The following data are measurements of carbon monoxide (CO) in grams/mile for vehicles chosen at random:

7.2	11.6	22.3	15.1	10.3	14.4	21.7	8.1
8.5	17.9	12.6	8.3	12.8	9.1	25.0	13.3
17.4	5.8	6.7	2.6	5.3	18.9	3.9	5.5
19.6	8.0	16.3	24.5	20.7	7.5	23.5	17.8
4.7	13.4	9.8	11.4	13.0	20.6	14.6	23.1
3.4	21.1	6.2	5.3	19.1	12.2	6.2	7.9
24.6	9.6	25.4	17.9	7.7	23.8	15.6	16.2

Derive a free-form cumulative probability distribution. Smooth the cumulative distribution using a four-point sliding polynomial and compute the probability density function by numerical differentiation. What magnitude can we expect 75 percent of all automobiles to exceed? What exceedence probability is associated with a CO level of 20 grams/mile?

6-4. Use Program B-6.3 to determine a free-form probability distribution of the monthly rainfall data of Problem 5-10. What are the 1-in-100 and 1-in-50 risk levels of monthly rainfall?

6-5. Use Program B-6.3 to determine a free-form probability distribution of the monthly runoff data of Problem 5-10. What are the 1-in-100 and 1-in-50 risk levels of monthly runoff?

6-6. Use Program B-6.2 to generate 10 samples of 25 items each using the free-form distribution derived in Problem 6-4. Find the second highest simulated monthly rainfall in each of the 10 samples. Find the highest of these ten values. What is its meaning?

6-7. Repeat Problem 6-7 using the free-form monthly runoff distribution derived in Problem 6-5.

6-8. What are the 95 percent confidence limits off the nodal values derived in Problem 5-9? Use tables of the normal curve.

6-9. Find the values of the four nodes at x,w values of (0,1), (0.1,1), (0,2), and (0.1,2). Compute these random values using Eq. 6-6. Using the randomized nodes predict (interpolate) a value of y at (0.05, 1.5). What is the meaning of this predicted value as compared to a value of y at the same point using the original nodes?

6-10. How does the procedure of Problem 6-9 compare with "error of prediction" in conventional regression analysis?

Table 6-1

Data For Example 6-1

Event Number	Magnitude	Ordered Sequence
1	4.506	5.974
2	4.390	5.455
3	5.974	4.953
4	0.796	4.506
5	1.921	4.390
6	2.426	4.145
7	1.295	4.003
8	2.126	3.978
9	1.852	3.737
10	4.145	3.516
11	0.618	3.287
12	5.455	3.182
13	0.562	3.091
14	4.003	2.989
15	4.953	2.717
16	2.989	2.692
17	3.978	2.541
18	1.971	2.492
19	2.492	2.426
20	2.272	2.404
21	2.541	2.272
22	3.737	2.126
23	2.692	1.971
24	3.287	1.921
25	3.516	1.852
26	1.571	1.571
27	2.717	1.295
28	3.091	0.796
29	3.182	0.618
30	2.404	0.562

Table 6-2

Smoothing The Histogram For Example 6-1

Class Limits	Number of Events in Class	Midpoint of Class	z	Coefficients for Nodes at x: 0	1.5	3.0	4.5	6.0
0-1	3	0.5	-0.167	0.667	0.333			
1-2	5	1.5	0		1.0			
2-3	9	2.5	+0.167		0.333	0.667		
3-4	6	3.5	-0.167			0.667	0.333	
4-5	5	4.5	0				1.000	
5-6	2	5.5	0.167				0.333	0.667

Reduced Matrix for Least Square Solution

Midpoint of Class	Coefficients for Nodes at x: 1.5 & 4.5	3	Number in Class
0.5	0.333	0	3
1.5	1.000	0	5
2.5	0.333	0.667	9
3.5	0.333	0.667	6
4.5	1.000	0	5
5.5	0.333	0	2

Solution Node

x = 1.5	x = 3	x = 4.5
5.250	8.623	5.250

Table 6-3

Construction of the Probability Function of Example 6-1

Equations of the Segments

Segment	Segment Function	Segment Integral
0.0-1.5	$f = 3.500x$	$\int f dx = 3.500x^2/2$
1.5-3.0	$f = 1.877 + 2.249x$	$\int f dx = -5.346 + 1.877x + 2.249x^2/2$
3.0-4.5	$f = 15.370 - 2.249x$	$\int f dx = -35.990 + 15.370x - 2.249x^2/2$
4.5-6.0	$f = 21.000 - 3.500x$	$\int f dx = -59.062 + 21.000x - 3.500x^2/2$

Integrals and Probability

x	Segment Integrals	Total Integral	Probability =Total/28.686
0	0	0	0
0.5	0.438	0.438	1.5
1.0	1.750	1.750	6.1
1.5	3.938	3.938	13.7
2.0	2.906 + 3.938 =	6.844	23.9
2.5	6.375 + 3.938 =	10.373	36.2
3.0	10.405 + 3.938 =	14.344	50.0
3.5	4.030 + 14.344 =	18.374	64.1
4.0	7.498 + 14.344 =	21.842	76.1
4.5	10.405 + 14.344 =	24.748	86.3
5.0	2.188 + 24.748 =	26.936	93.9
5.5	3.500 + 24.748 =	28.248	98.5
6.0	3.938 + 24.748 =	28.686	100.0

Table 6-4

Calculation of Sample Probability

Class Limits	Class Frequency	Accumulated Number	Sample Probability
0-1	3	3	10.0
1-2	5	8	26.7
2-3	9	17	56.7
3-4	6	23	76.7
4-5	5	28	93.3
5-6	2	30	100.0

Table 6-5

Data For Example 6-2

Number	x	Number	x
1	2.05	24	2.01
2	1.28	25	1.49
3	1.61	26	1.89
4	1.91	27	1.41
5	1.54	28	1.45
6	2.55	29	1.61
7	2.24	30	1.89
8	1.80	31	1.80
9	1.81	32	1.83
10	1.78	33	1.59
11	2.29	34	1.95
12	1.38	35	1.94
13	2.13	36	2.48
14	2.14	37	2.13
15	2.16	38	1.79
16	1.39	39	1.70
17	1.47	40	1.79
18	1.87	41	1.61
19	2.41	42	1.47
20	1.93	43	1.60
21	1.69	44	1.82
22	2.46	45	1.50
23	2.93		

Table 6-6

Sample Probabilities For Example 6-2

Class Limits	Frequency	Accumulated	Sample Probability
1.2 - 1.4	3	3	6.7
1.4 - 1.6	9	12	26.7
1.6 - 1.8	10	22	48.9
1.8 - 2.0	10	32	71.1
2.0 - 2.2	6	38	84.4
2.2 - 2.4	2	40	88.9
2.4 - 2.6	4	44	97.8
2.6 - 2.8	0	44	97.8
2.8 - 3.0	1	45	100.0

Table 6-7

Sliding Polynomial Coefficients For Example 6-2

Full Set of Nodes

Class Number	Coefficients for Node at x = 0.3	1.2	2.1	3.0	3.9	P
1	-.0672	.8930	.1934	-.0192		6.7
2	-.0686	.6379	.4856	-.0549		26.7
3	-.0370	.3333	.7778	-.0741		48.9
4	-.0055	.0782	.9712	-.0439		71.1
5		-.0439	.9712	.0782	-.0055	84.4
6		-.0741	.7778	.3333	-.0370	88.9
7		-.0549	.4856	.6379	-.0686	97.8
8		-.0192	.1934	.8930	-.0672	97.8
9		0	0	1.0000	0	100.0

Reduced Set of Nodes

Class Numbers	Coefficients for Node at x = 0.3	2.1	3.9	Reduced P
1	-.0672	.1934		6.7 + 1.92 = 8.62
2	-.0686	.4856		26.7 + 5.49 = 32.19
3	-.0370	.7778		48.9 + 7.41 = 56.31
4	-.0055	.9712		71.1 + 4.39 = 75.49
5		.9712	-.0055	84.4 - 7.82 = 76.58
6		.7778	-.0370	88.9 - 33.33 = 55.57
7		.4856	-.0686	97.8 - 63.79 = 34.01
8		.1934	-.0672	97.8 - 89.30 = 8.50
9		0	0	100.0 - 100.00 = 0.00

Solution

Node Location	0.3	1.2	2.1	3.0	3.9
Node	93.99	0	78.13	100.0	84.87

Table 6-8
Class-Limit Correspondence For Example 6-3

Limits in v-scale	Limits in Temperature (t)-scale Degrees F
0.15	57.05
0.45	62.54
0.75	65.10
1.05	66.78
1.35	68.03
1.65	69.04
1.95	69.87
2.25	70.67
2.55	71.61
2.85	72.77
3.15	74.28
3.45	76.45
3.75	80.40

Table 6-9
Class-Limit Correspondence For Example 6-4

Limits in v-scale	Limits in Temperature (t)-Scale Degrees F
0.49	87.69
0.69	89.59
0.89	91.00
1.09	92.13
1.29	93.06
1.49	93.86
1.69	94.56
1.89	95.19
2.09	95.76
2.29	96.37
2.49	97.06
2.69	97.85
2.89	98.77
3.09	99.87
3.29	101.25
3.49	103.09
3.69	105.86

Table 6-10

Results of Smoothing Sample Probabilities For Example 6-4

Node Location in v-scale	Nodal Value (%)	Standard Deviation of Node (%)
-0.31	3.8803	----
0.49	-0.2150	0.5486
1.29	16.1633	0.9789
2.09	53.0150	1.4251
2.89	77.8876	1.6728
3.69	103.0097	2.3483
4.49	128.3813	----

Table 6-11

Comparison Of Simulations With And Without Variation of Nodes (Values in Degrees Fahrenheit)

Rank Number in Ten

1				6				10			
High	Low	Mean	Stnd. Dev.	High	Low	Mean	Stnd. Dev.	High	Low	Mean	Stnd. Dev.
Without											
103.9	96.5	100.9	2.04	99.3	93.7	96.0	1.42	93.7	88.4	91.0	1.47
With											
105.7	97.1	102.4	2.18	101.5	94.4	96.5	1.86	94.4	88.2	90.7	1.67

Table 6-12
Characteristics of Test Data

	November Total Monthly Rainfall	November Minimum Temperature
Mean*	3.95 (7.77)	23.87 (4.52)
Std. Dev.*	2.33 (5.92)	5.08 (2.82)
Coef. of Skew	2.06	-0.99
Range*	15.78- (40.08- 0.33 0.84)	35.- (1.67- 5. -15.00)
C**	2.77	1.63
VB**	4.82	0.42
f**	0.86	0.61
d**	0.64	0.70
h_{min}*	0 (0)	---
h_{max}*	---	36.00 (2.22)

*Units: rainfall, inch (cm.); temperature, ^{o}F (^{o}C).

**Definitions: C - common point, VB - asymptotic boundary, f and d - shape parameters

Table 6-13
Sample Characteristics of the Test Data For Seasonally Continuous Probabilities

	Jan	Feb	Mar	Apr	May	Jun	Jul	Aug	Sep	Oct	Nov	Dec
Athens Monthly Total Rainfall												
M, cm x 10^{-1}	1.19	1.22	1.37	1.00	0.98	1.00	1.26	1.08	0.86	0.76	0.78	1.14
s, cm	5.94	5.85	7.04	5.74	5.55	5.75	6.85	7.67	6.00	5.62	5.91	5.92
SK	0.84	0.30	0.83	0.73	0.74	1.39	1.18	2.15	0.96	0.80	2.06	0.86
R, cm x 10^{-1}	3.26	2.86	3.59	2.95	2.88	3.38	3.90	4.68	2.69	2.44	4.01	2.87
	0.19	0.16	0.25	0.02	0.03	0.18	0.10	0.05	0.08	0.00	0.06	0.16
Oconee River Daily Maximum Streamflow												
M, m^3 x 10^{-1}	8.63	7.12	9.36	7.67	5.78	4.39	2.91	3.39	2.15	2.17	3.96	4.71
s, m^3/s x 10^{-1}	7.50	5.65	6.83	6.59	6.84	5.32	2.68	5.10	1.85	2.04	6.62	4.69
SK	1.21	1.64	0.90	1.76	2.42	3.26	2.10	2.96	2.26	1.85	3.66	2.30
R, m^3/s x 10^{-2}	3.03	2.86	2.72	3.00	3.29	2.97	1.35	2.40	1.04	0.92	3.46	2.23
	0.08	0.14	0.18	0.13	0.09	0.06	0.05	0.03	0.02	0.02	0.05	0.05
Oconee River Daily Minimum Streamflow												
M, m^3/s x 10^{-1}	0.98	1.13	1.23	1.19	0.87	0.63	0.53	0.41	0.38	0.45	0.59	0.73
s, m^3/s	3.45	3.57	3.04	4.14	2.73	2.44	2.33	1.75	1.83	2.08	2.17	2.75
SK	0.68	0.40	0.19	1.77	0.32	0.33	0.74	0.44	0.62	0.60	0.49	1.25
R, m^3/s x 10^{-1}	1.83	1.98	2.12	2.95	1.63	1.16	1.16	0.86	0.96	1.08	1.12	1.71
	0.29	0.50	0.71	0.62	0.31	0.22	0.17	0.10	0.07	0.08	0.16	0.30

M - mean, s - standard deviation, SK - coefficient of skew, R - range.

Table 6-14
Transform Parameters of The Test Data For Seasonally Continuous Probabilities

	Jan	Feb	Mar	Apr	May	Jun	Jul	Aug	Sep	Oct	Nov	Dec
	Athens Monthly Total Rainfall											
C	2.32	2.11	2.31	2.27	2.28	2.52	2.44	2.81	2.36	2.30	2.77	2.32
VB	4.37	4.16	4.36	4.32	4.33	4.57	4.49	4.86	4.41	4.35	4.82	4.37
f	0.97	1.29	1.04	1.12	1.09	0.88	0.84	0.77	1.22	1.42	0.86	0.97
d	0.86	1.26	0.92	1.01	0.98	0.72	0.71	0.56	1.06	1.27	0.64	0.86
h'_{min}, cm	0.74	0.46	1.12	0.00	0.00	0.61	0.00	0.00	0.00	0.00	0.00	0.41
	Oconee River Daily Maximum Streamflow											
C	2.45	2.61	2.34	2.66	2.91	3.22	2.79	3.11	2.85	2.70	3.37	2.86
VB	4.50	4.66	4.39	4.71	4.96	5.27	4.84	5.16	4.90	4.75	5.42	4.91
f	1.41	1.12	1.40	1.12	1.40	1.14	1.09	1.94	0.94	1.21	2.60	1.13
d	1.18	0.88	1.23	0.87	0.99	0.73	0.80	1.28	0.68	0.92	1.58	0.81
h'_{min}, m^3/s	0.00	2.78	4.60	0.00	0.00	0.00	0.00	0.00	0.00	0.00	0.00	0.00
	Oconee River Daily Minimum Streamflow											
C	2.26	2.15	2.07	2.66	2.12	2.12	2.28	2.17	2.23	2.23	2.18	2.47
VB	4.31	4.20	4.12	4.71	4.17	4.17	4.32	4.22	4.28	4.28	4.23	4.52
f	0.90	1.22	1.52	0.78	1.12	1.33	1.11	1.17	1.12	1.07	1.04	0.86
d	0.81	1.16	1.50	0.60	1.08	1.29	1.00	1.11	1.03	0.98	0.98	0.71
h'_{min}, m^3/s	2.20	4.27	6.52	5.37	2.57	1.69	1.26	0.61	0.37	0.38	1.18	2.48

C - common point, VB - asymptotic boundary, f and d - shape parameters.

h'_{min} - class boundary.

Table 6-15
Derived Nodes and Their Standard Deviations in Probability Scale

	Node Number	Data Set: All values are percent probability					
		I		II		III	
		Node	S.D.	Node	S.D.	Node	S.D.
January	1	18.2	0.7	20.6	1.0	23.0	1.2
	2	40.7	0.9	32.4	0.8	40.5	0.9
	3	67.6	0.6	50.4	0.8	63.3	0.7
	4	100.3	0.7	102.1	1.2	98.3	0.9
	5	94.8	—	164.2	—	32.8	—
April	1	24.2	0.6	18.1	0.9	21.9	2.1
	2	46.8	0.6	30.6	0.9	46.2	2.0
	3	60.3	0.5	50.8	1.0	67.0	0.9
	4	100.1	0.6	103.3	1.0	100.2	1.2
	5	112.6	—	170.8	—	173.0	—
July	1	10.1	0.8	11.4	0.9	26.3	0.6
	2	33.6	0.8	23.5	0.9	41.6	0.6
	3	63.7	0.6	44.9	0.9	64.6	0.7
	4	100.9	0.3	101.8	0.7	99.0	0.6
	5	73.3	—	79.9	—	101.7	—
October	1	21.5	1.4	12.1	0.5	18.3	0.6
	2	36.7	1.0	30.1	0.6	41.0	0.5
	3	55.4	0.5	55.8	1.1	75.4	0.6
	4	99.3	0.8	101.8	1.5	100.5	0.5
	5	162.7	—	121.0	—	178.8	—

I - Athens monthly total rainfall.

II - Oconee River daily maximum streamflow.

III - Oconee River daily minimum streamflow.

Table 6-16

Conditional Probabilities of v

Calculation from Part d

Value of z	v-Probabilities										
2.4	0	1.8	3.6	6.5	9.4	14	20	27	35	43	49
2.0	0	1.1	2.9	5.0	6.8	10	15	21	28	34	39
Difference	0	0.7	0.7	1.5	2.6	4	5	6	7	9	10
Rescaled to 100 %	0	7	7	15	26	40	50	60	70	90	100

Calculation from Part a

Class Freq.		2	0	2	3	4	2	3	4	3	3
Accumulation	0	2	2	4	7	11	13	16	20	23	26
Probability	0	8	8	15	27	42	50	61	77	88	100

Table 6-17

Data For Example 6-5

Data Point	Time Hrs.	Transport Rate Kgm/hr.	Data Point	Time Hrs.	Transport Rate Kgm/hr.
1	0	28	7	14.0	24
2	1.5	33	8	16.0	17
3	4.0	31	9	19.0	17
4	5.5	34	10	23.0	13
5	7.5	34	11	26.5	6
6	9.5	29	12	30.0	5

Table 6-18

Nodes and Standard Deviations From Smoothing in Example 6-5

Node	Nodal Value Kgm/hr.	Nodal Standard Deviation Kgm/hr.
1	30.0770	1.4379
2	30.1769	1.0575
3	13.9263	1.6005
4	5.0156	0.7576

Table 6-19

Data Points for Example 6-6

Data Point	Time Hrs.	Transport Rate Kgm/hr.	Data Point	Time Hrs.	Transport Rate Kgm/hr.
1	6.0	19	8	17.0	24
2	7.0	22	9	20.0	20
3	8.5	20	10	23.0	16
4	9.5	24	11	26.5	16
5	11.0	24	12	30.0	12
6	12.0	21	13	33.0	14
7	14.5	22	14	36.0	11

Table 6-20
Nodes and Standard Deviation From Smoothing Point B Data in Example 6-6

Node	Nodal Value Kgm/hr.	Nodal Standard Deviation Kgm/hr.
1	19.9323	0.8311
2	22.7617	0.7257
3	15.0663	0.8595
4	11.8070	0.9911

Table 6-21
Data For Example 6-7, Timed Samples

Point	Time (Hrs)	Concentration (ppm)	Point	Time (Hrs)	Concentration (ppm)
1	0	25	12	8.8	125
2	0.8	60	13	9.6	105
3	1.6	75	14	10.4	85
4	2.4	80	15	11.2	80
5	3.2	110	16	12.0	60
6	4.0	105	17	12.8	55
7	4.8	130	18	13.6	45
8	5.6	165	19	14.4	35
9	6.4	185	20	15.2	35
10	7.2	165	21	16.0	30
11	8.0	155			

Auxiliary Samples

Point	4-Hours Concentration (ppm)	12-Hours Concentration (ppm)
1	155	85
2	145	80
3	140	75
4	125	75
5	110	73
6	95	65
7	75	50

Table 6-22

Significance Tests For Example 6-7

Test at Four Hours

Timed Data			Auxiliary Data		
Node at 4 hrs.	Error Variance at Node	Sample Portion	Mean	Variance	Sample Size
123.387	172.281	5.8	120.714	28.929	7

$$S_p = \sqrt{\frac{4.8 * 172.281 + 6 * 836.887}{4.8 + 6}} = 23.27 \qquad t_{.05} \approx 1.8$$

$$u = 1.8 * 23.27 * \sqrt{\frac{5.8 + 7}{5.8 * 7}} = 23.52 \qquad |123.39 - 120.71| = 2.67$$

Test at Twelve Hours

Timed Data			Auxillary Data		
Node at 12 hrs.	Error Variance at Node	Sample Portion	Mean	Variance	Sample Size
57.108	56.145	5.8	71.857	130.805	7

$$S_p = \sqrt{\frac{4.8 * 56.145 + 6 * 130.805}{4.8 + 6}} = 9.88 \qquad t_{.05} \approx 1.8$$

$$u = 1.8 * 9.88 * \sqrt{\frac{5.8 + 7}{5.8 * 7}} = 9.99 \qquad |57.11 - 71.86| = 14.75$$

Table 6-23

Data For Example 6-8

Sample	Sample Time Hrs.	*Mixing* Rate 1 BOD ppm	*Mixing* Rate 2 BOD ppm
1	0	49	36
2	1.25	51	34
3	2.5	48	25
4	3.75	40	26
5	5.0	39	22
6	6.25	35	18
7	7.5	34	22
8	8.75	36	18
9	10.0	32	16
10	11.25	33	19
11	12.5	29	16
12	13.75	29	16
13	15.0	27	17
14	16.25	23	15
15	17.5	26	13
16	18.75	24	12
17	20.0	23	9
18	21.25	21	10
19	22.50	23	7
20	23.75	22	7
21	25.0	21	9
22	26.25	22	8
23	27.5	21	8
24	28.75	19	8
25	30.0	19	7
26	31.25	21	7
27	32.5	20	9
28	33.75	21	8
29	35.0	19	6
30	36.25	19	8
31	37.5	18	8
32	38.75	17	6
33	40.0	18	6

Table 6-24

Results of Smoothing For Example 6-8

Time	Node (ppm)	$\Sigma c^2 e^2$	Σc^2	Residual Variance $E(e^2)$ (ppm^2)	Portion of N Σc	Nodal Std. Dev. S_K (ppm)
			Mixing Rate 1			
0	50.674	15.755	2.6169	6.020	3.5156	1.4207
10	32.104	26.987	7.5238	3.587	9.3125	0.6738
20	23.123	8.691	6.6097	1.315	7.3438	0.4594
30	20.406	8.220	7.5238	1.093	9.3125	0.3719
40	17.647	1.059	2.6169	0.405	3.5156	0.3682
					33.0000	
			Mixing Rate 2			
0	33.416	20.834	2.6169	7.961	3.5156	1.6337
10	17.575	39.453	7.5238	5.244	9.3125	0.8146
20	10.943	20.915	6.6097	3.164	7.3438	0.7126
30	7.505	6.955	7.5238	0.924	9.3125	0.3420
40	6.998	2.741	2.6169	1.047	3.5156	0.5925
					33.0000	

Table 6-25

Data For Example 6-9

Unit Variate z	Value of Node	Nodal Standard Deviation
-3/2	-	-
1.0	1.0	0.1
-1/2	2.0	0.2
1/2	2.0	0.2
3/2	1.0	0.1

Table 6-26

Results of Smoothing Data in Example 6-10

	Set 1			Set 2		
Node	Nodal Value	Nodal Variance	Nodal Sample	Nodal Value	Nodal Variance	Nodal Sample
1	1.476	0.0602	1.3962	1.391	0.0181	1.7391
2	2.594	0.0187	6.5683	2.562	0.0065	7.5476
3	3.081	0.0297	5.8349	3.185	0.0409	5.3676
4	3.147	0.0963	2.2006	3.499	0.2346	1.3457
			16.0000			16.0000

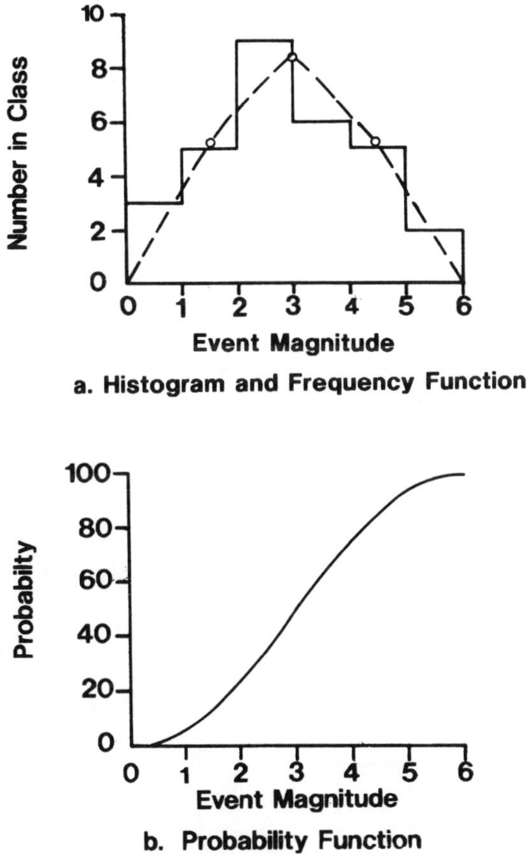

Fig. 6-1. Probability Analysis for Example 6-1.

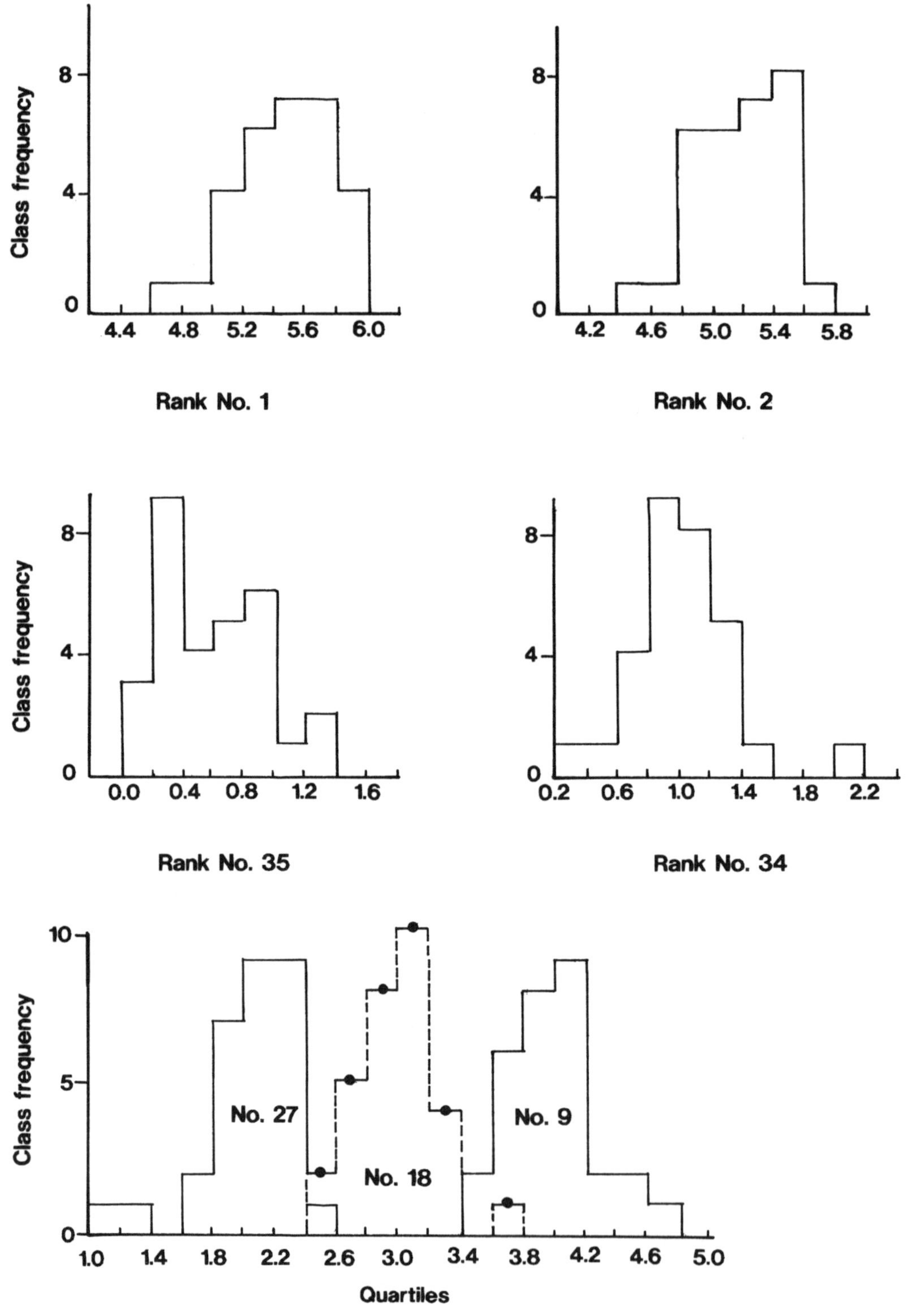

Fig. 6-2. **Sampling Distributions of Ranked Items**

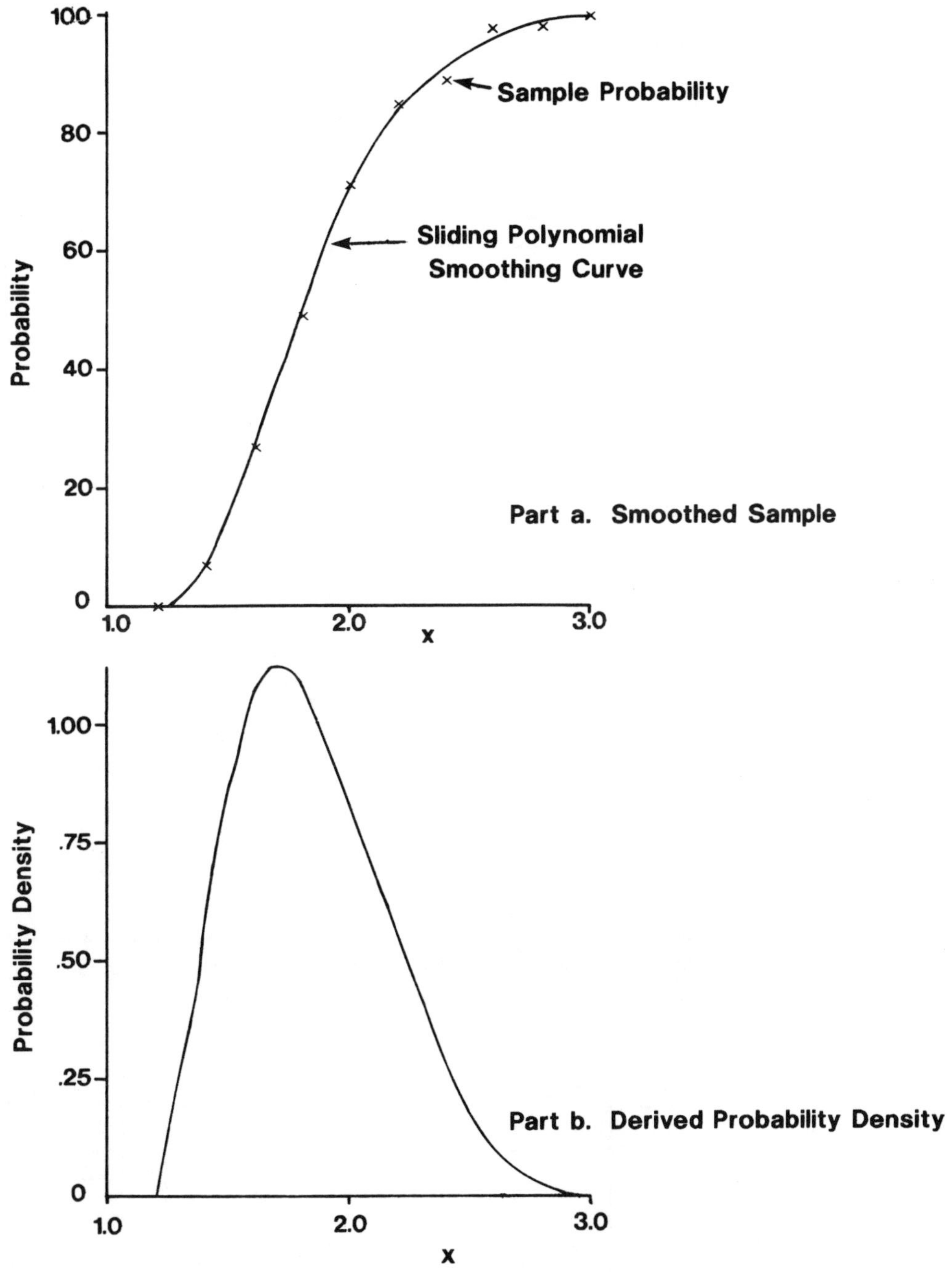

Fig. 6-3. Smooth Sample Probability, Example 6-2

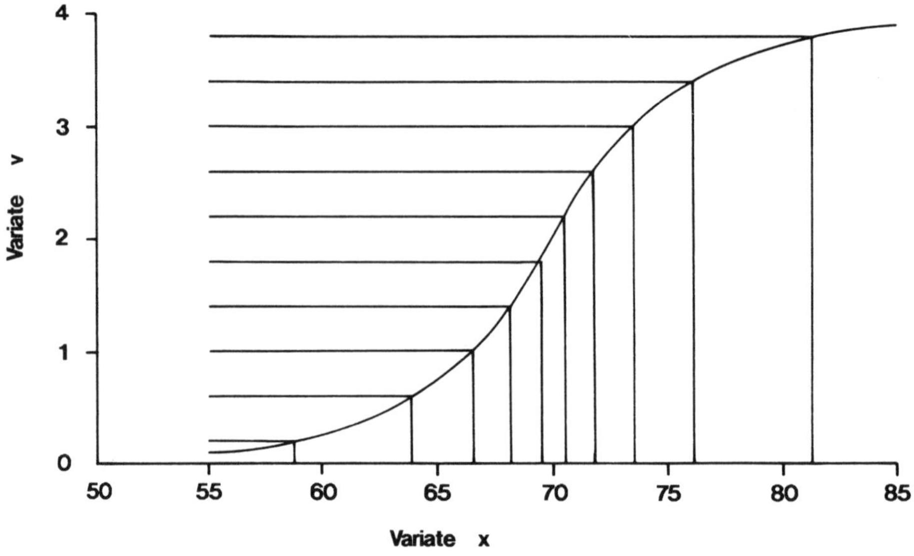

Fig. 6-4. Variate Transform

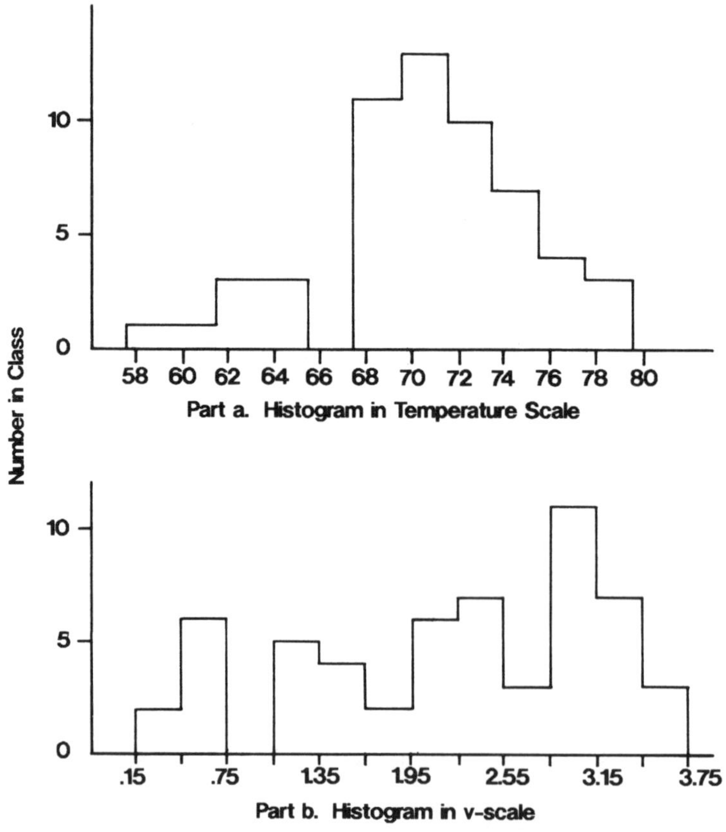

Fig. 6-5. Maximum Daily Temperature for January Near Griffin, GA. 1926-1981. Example 6-3.

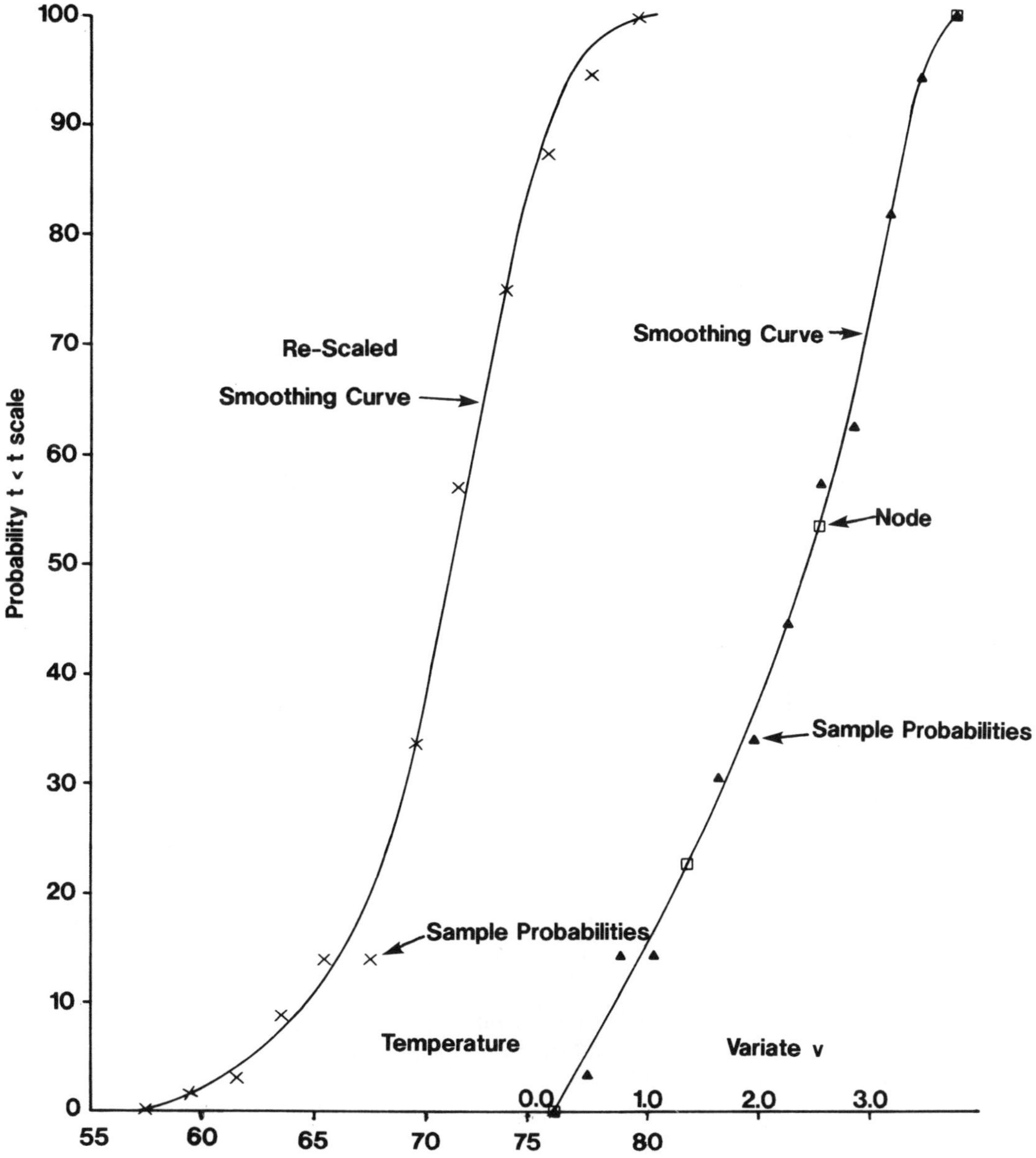

Fig. 6-6. Sample Probabilities, January Maximum Temperatures, Griffin, GA 1926-1981

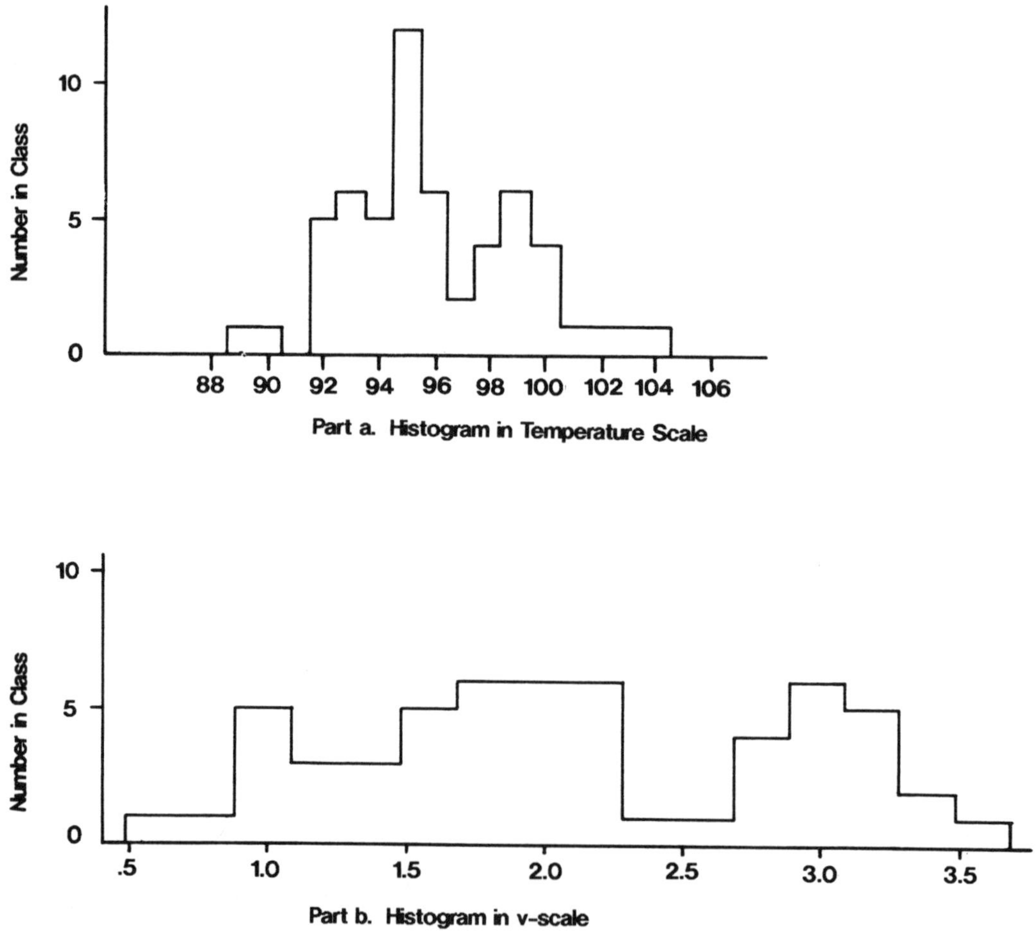

Fig. 6-7. Maximum Daily Temperature for July Near Griffin, GA 1926-1981, Example 6-4

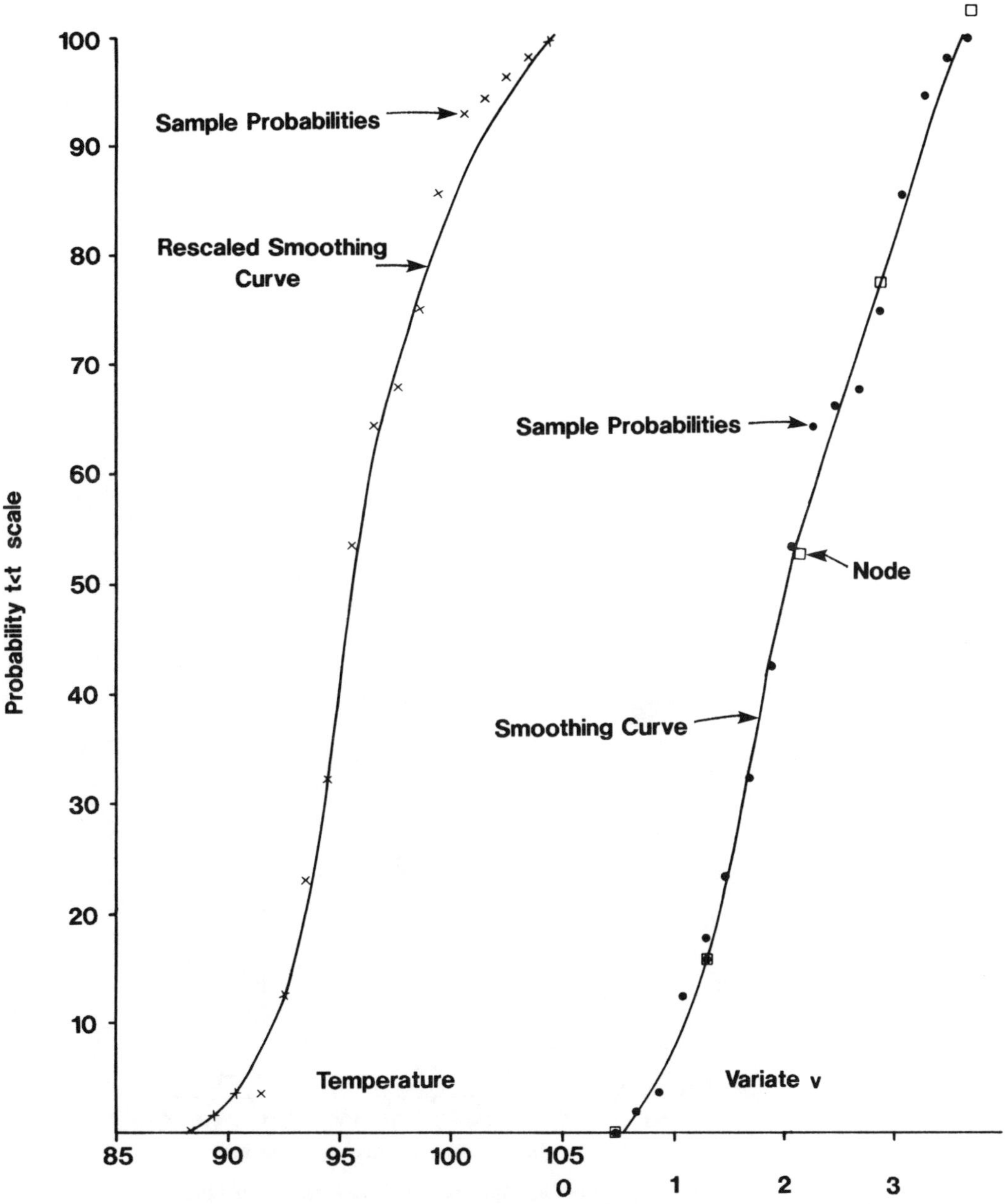

Fig. 6-8. Sample Probabilities, July Maximum Temperatures, Griffin, GA 1926-1981

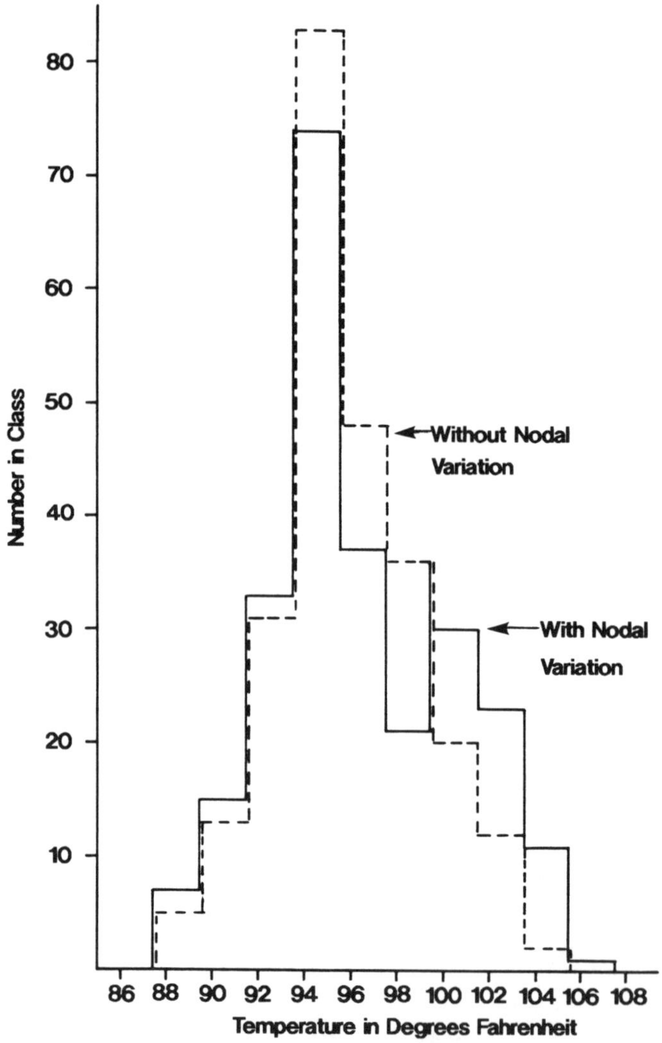

Fig. 6-9. Total Histograms of Simulated July Maximum Daily Temperature, Griffin, GA

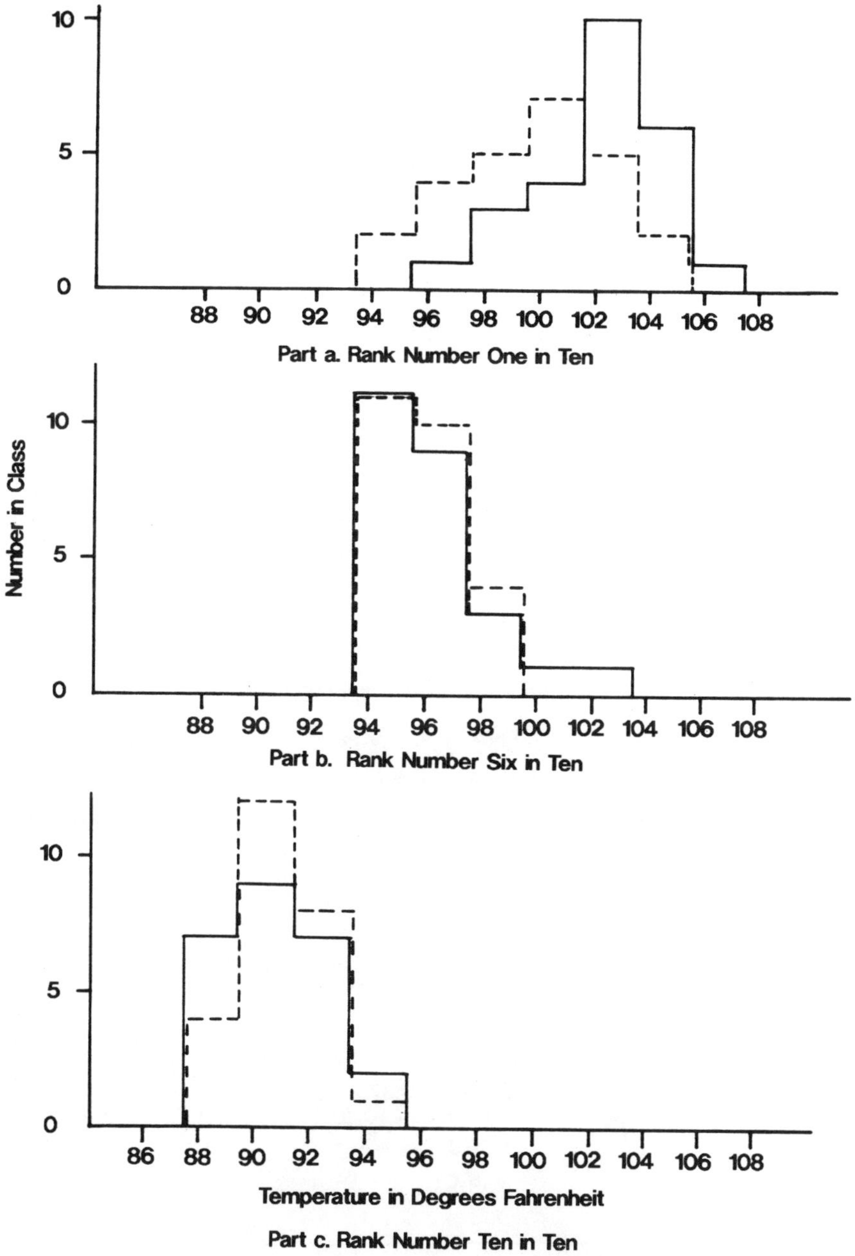

Fig. 6-10. Histograms of Selcted Rank Numbers of Simulated July Maximum Temperatures, Griffin, GA

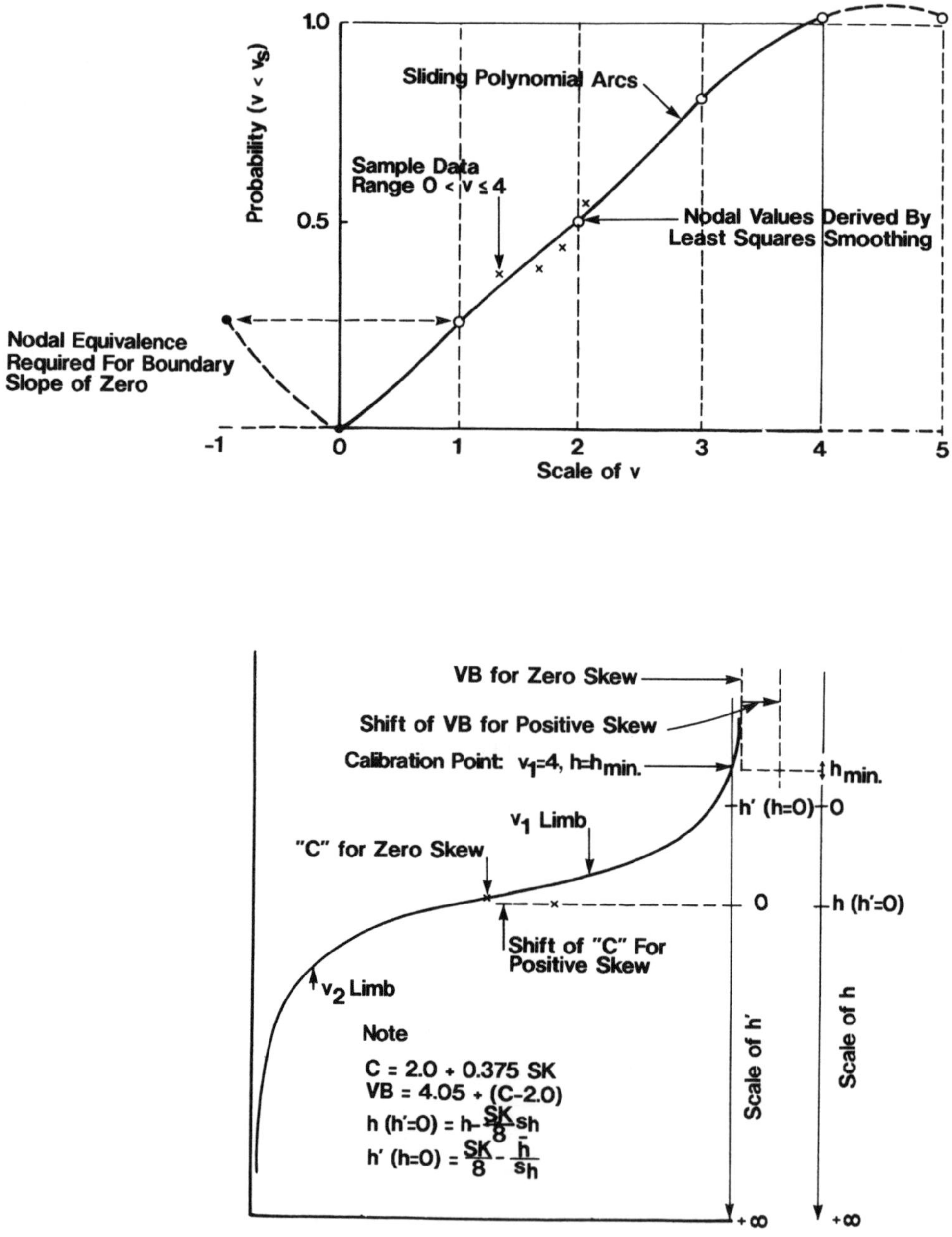

Fig. 6-11. Schematic for Variate Transformation and Smoothing

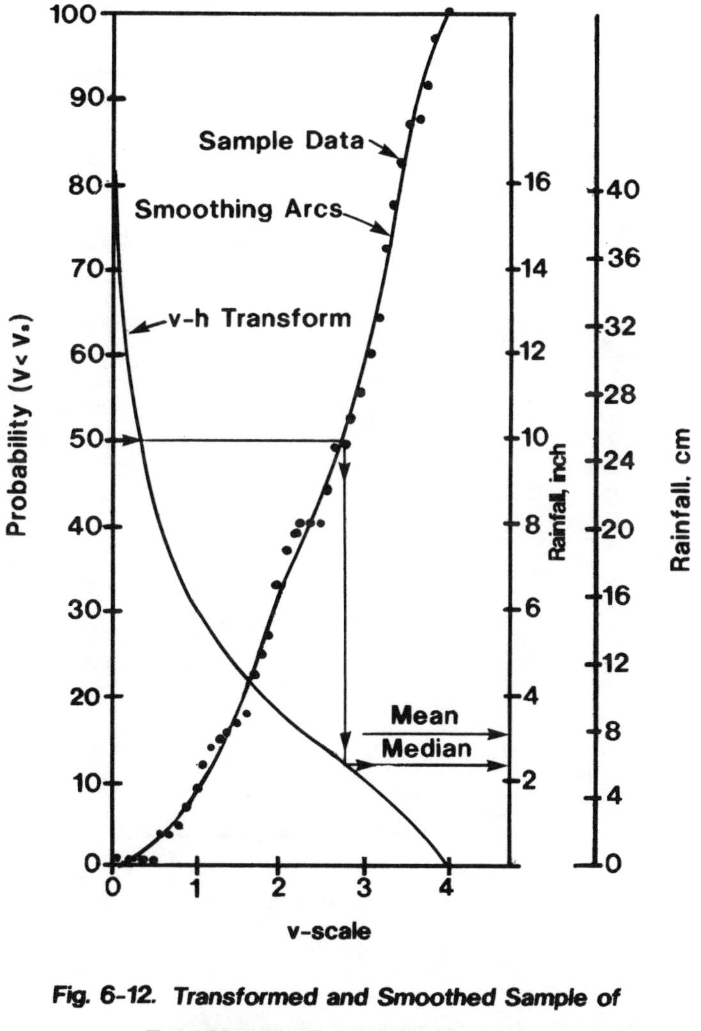

Fig. 6-12. Transformed and Smoothed Sample of Total Monthly Rainfall for November at Athens, GA.

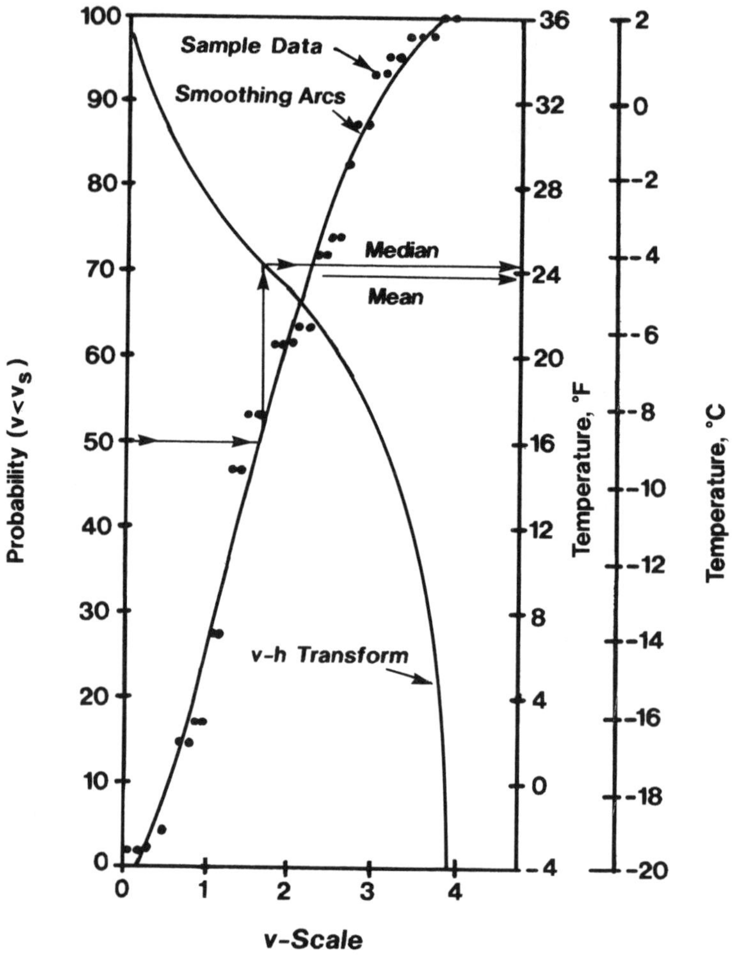

Fig. 6-13. Transformed and Smoothed Sample of Minimum Monthly Temperature for November at Watkinsville, GA.

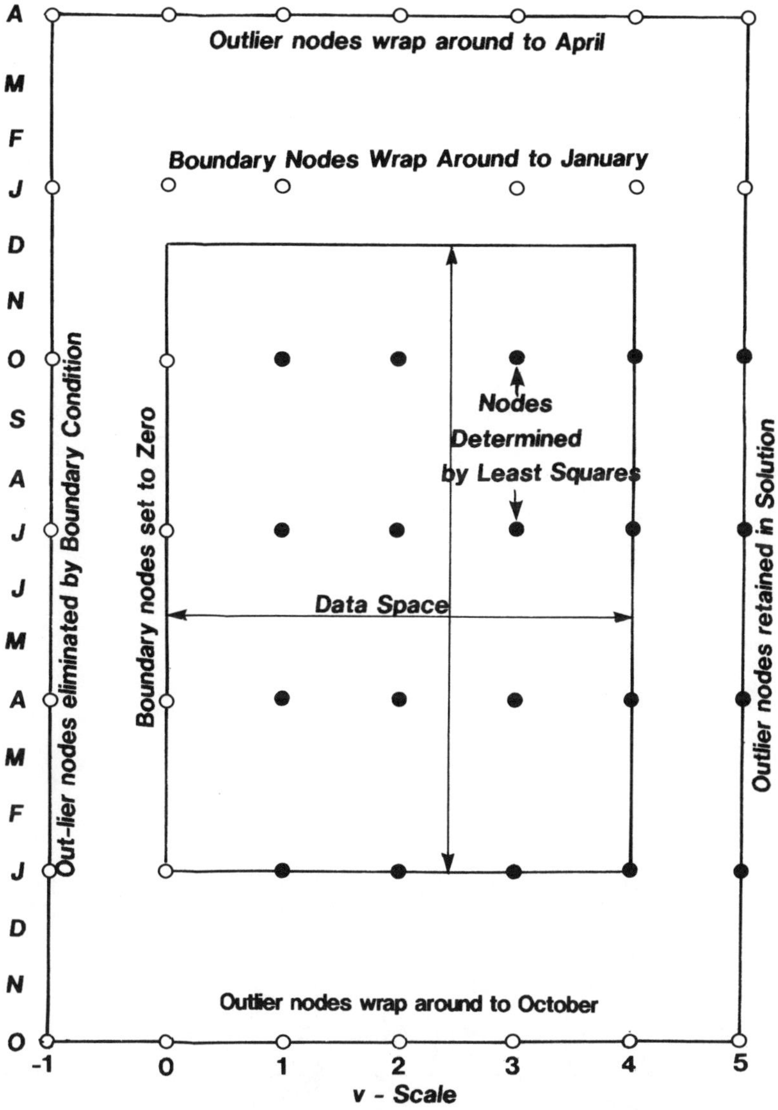

Fig. 6-14. Schematic Arrangement of Nodes for Seasonally Continuous Probability Analysis

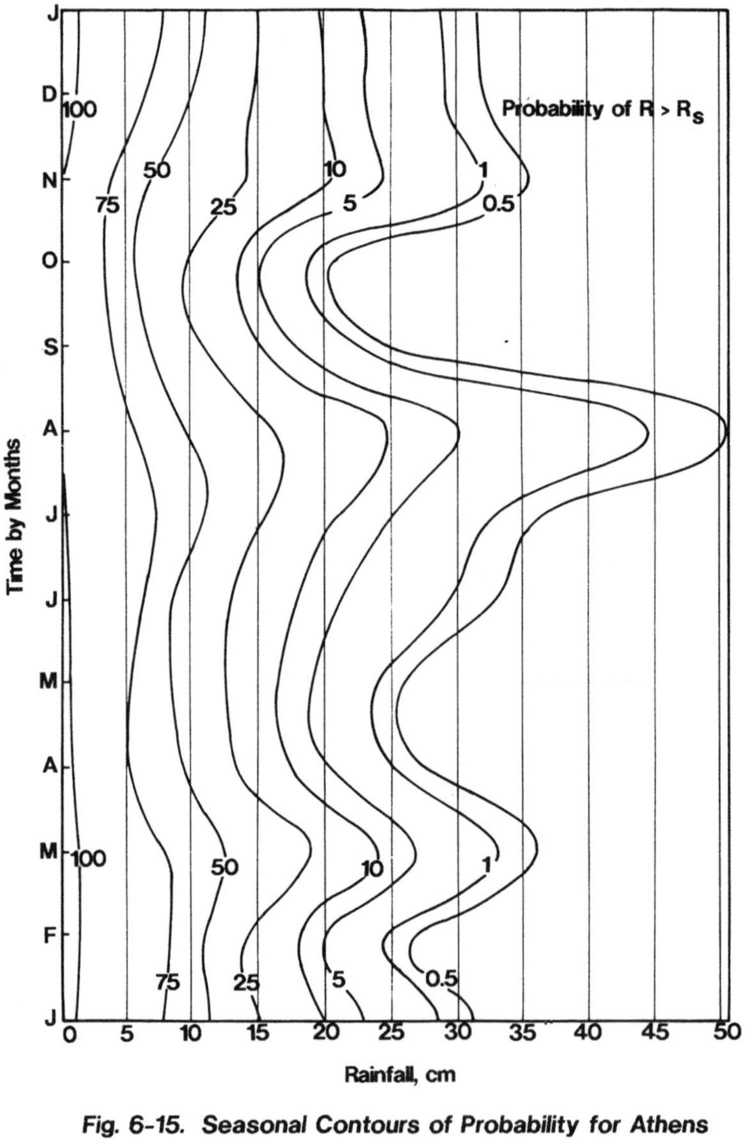

Fig. 6-15. Seasonal Contours of Probability for Athens Monthly Total Rainfall

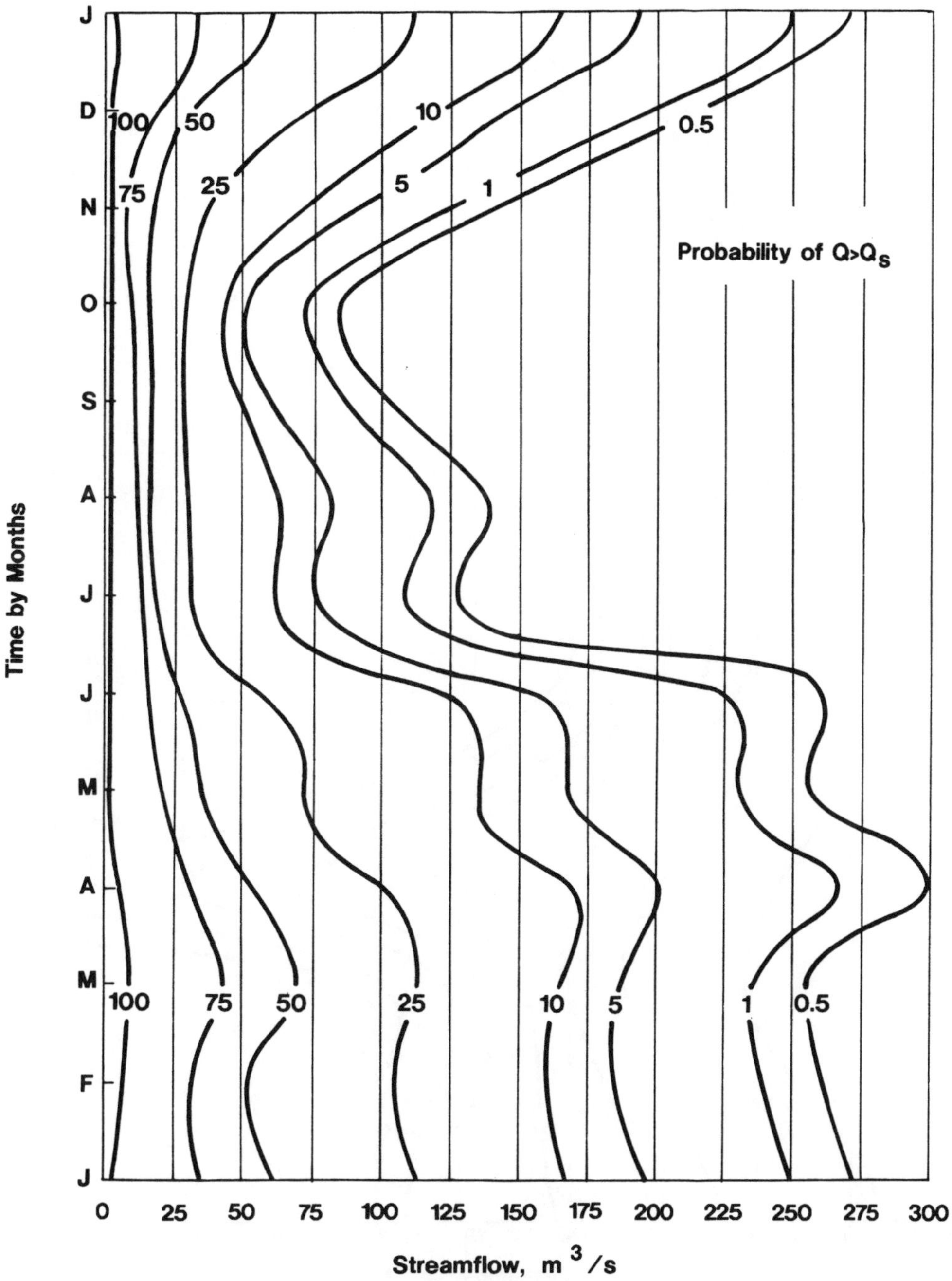

Fig. 6-16. Seasonal Contours of Probability for Oconee River Daily Maximum Streamflow

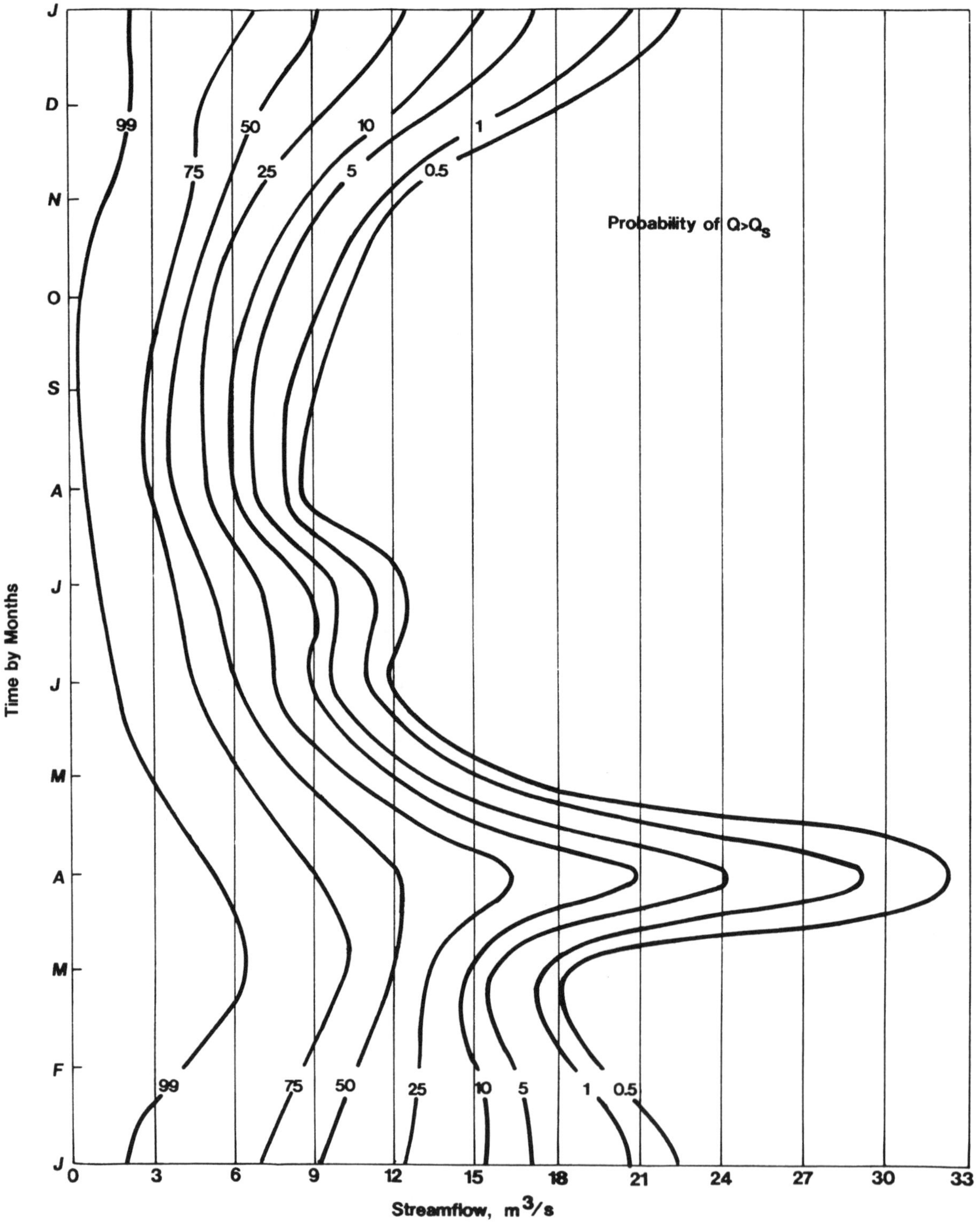

Fig. 6-17. Seasonal contours of probability for Oconee River Daily Minimum Streamflow

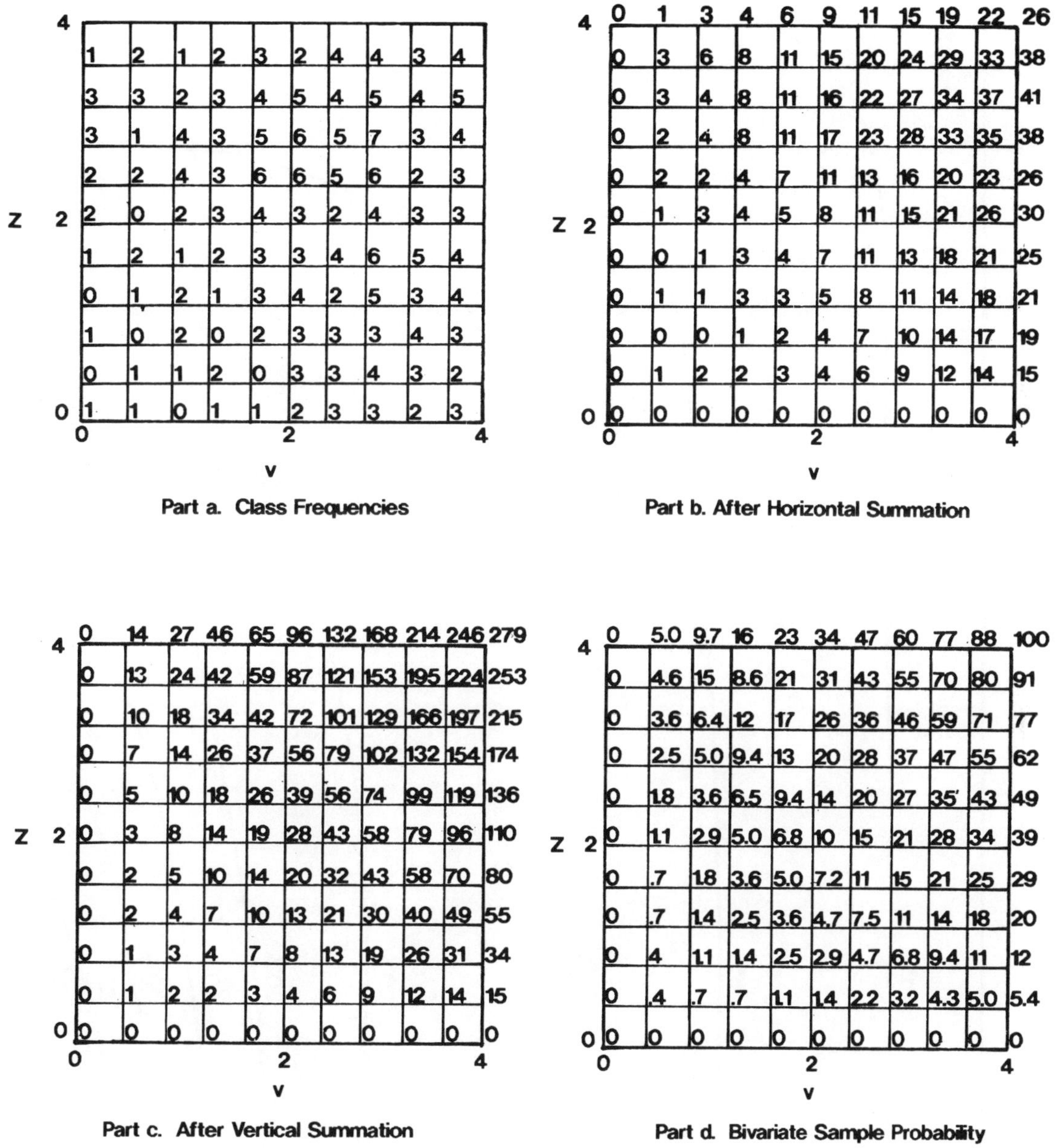

Fig. 6-18. Construction of a Bivariate Probability

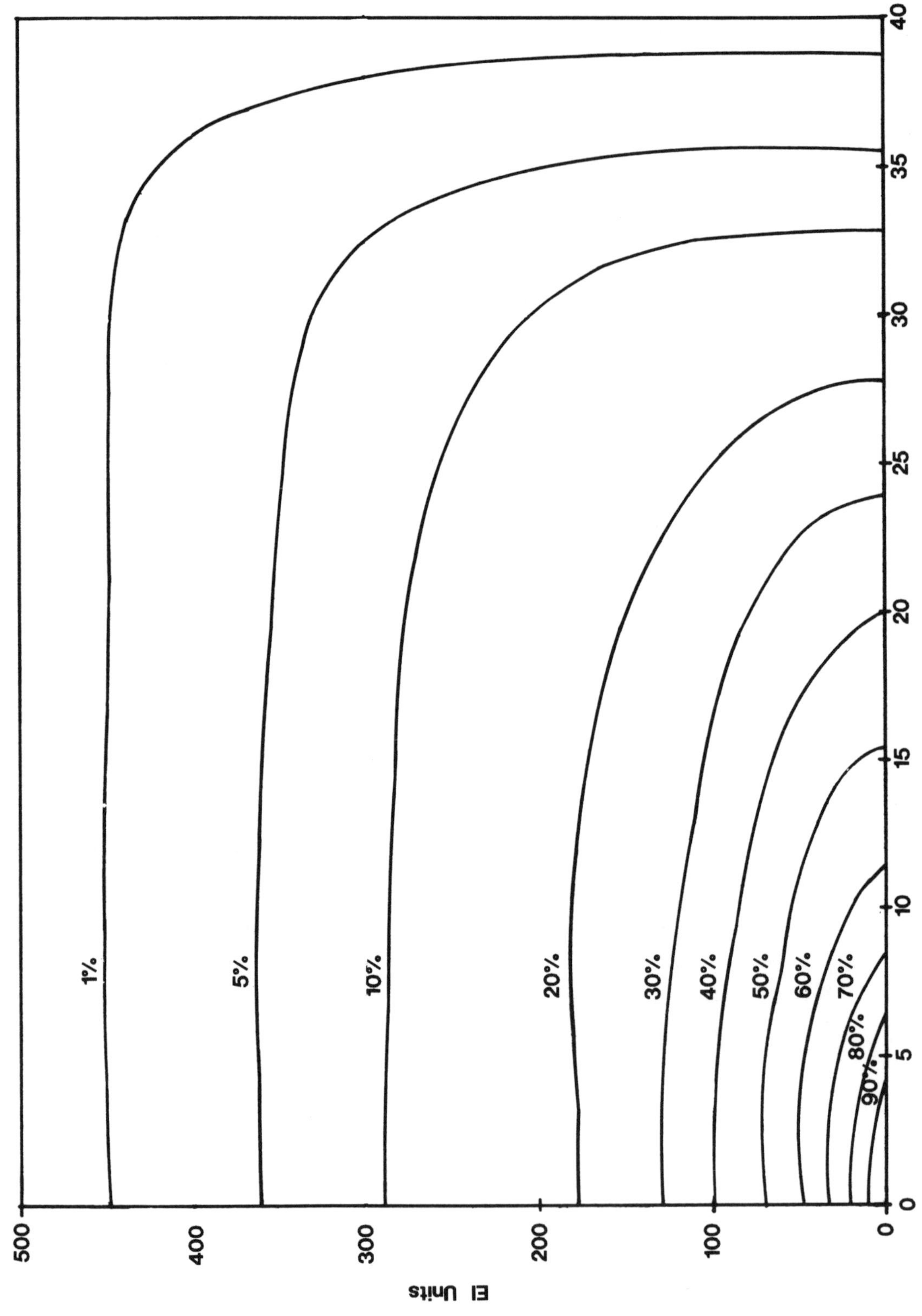

Fig. 6-19. Contours of Joint Probability of Monthly Total Rainfall and Monthly Total *EI*

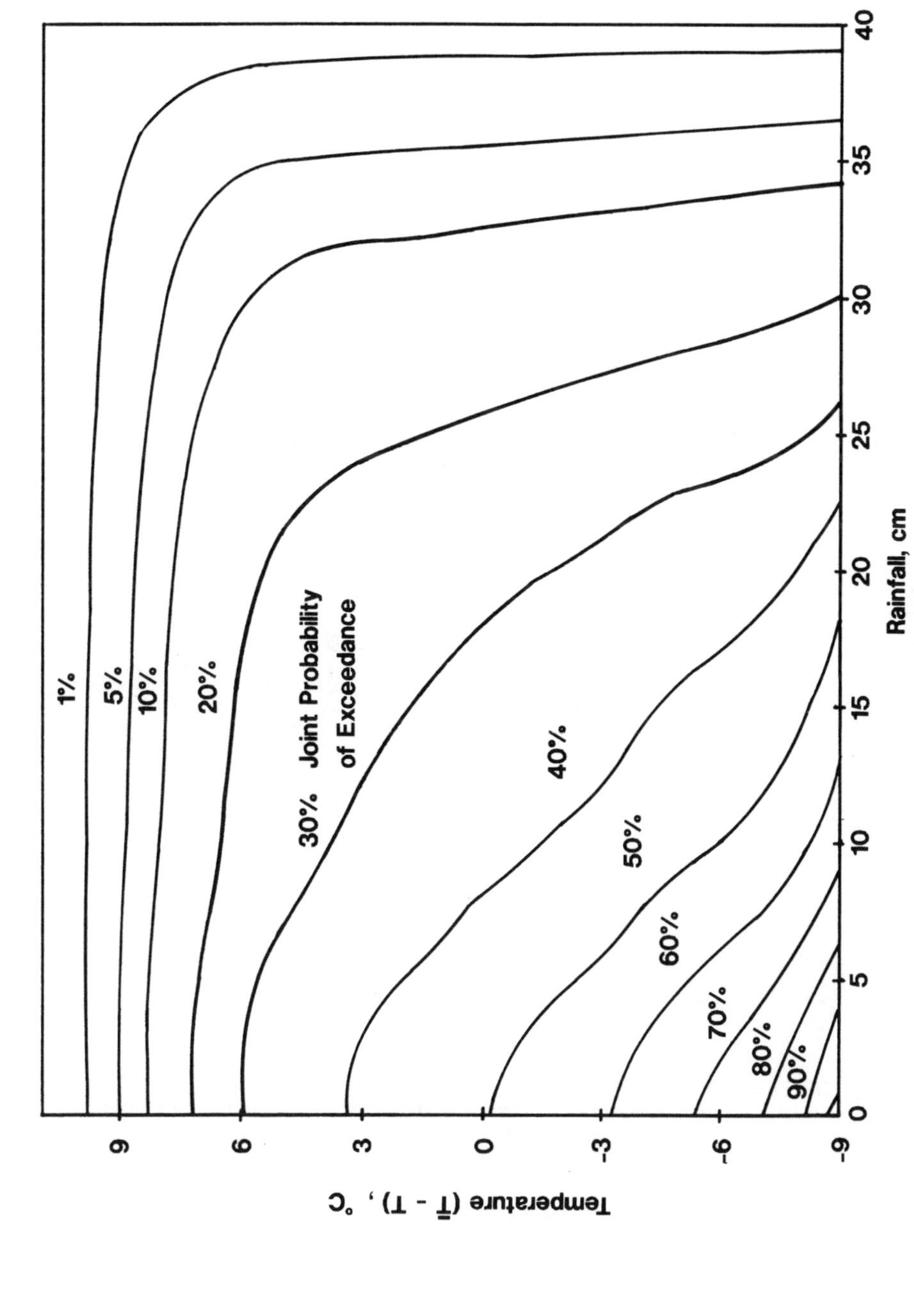

Fig. 6-20. Contours of Joint Probability of Monthly Total Rainfall and Monthly Average Temperature.

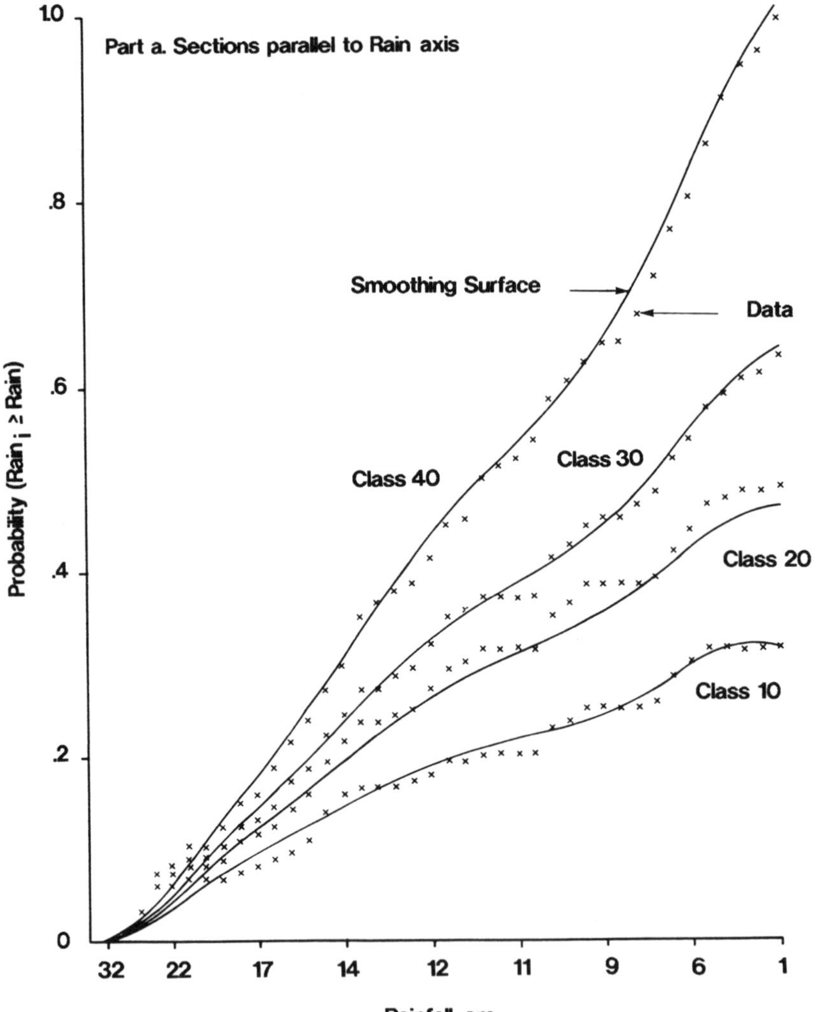

Fig. 6-21a. Cross Sections Through The Joint Probability Surface of Monthly Total Rainfall and Monthly Average Temperature

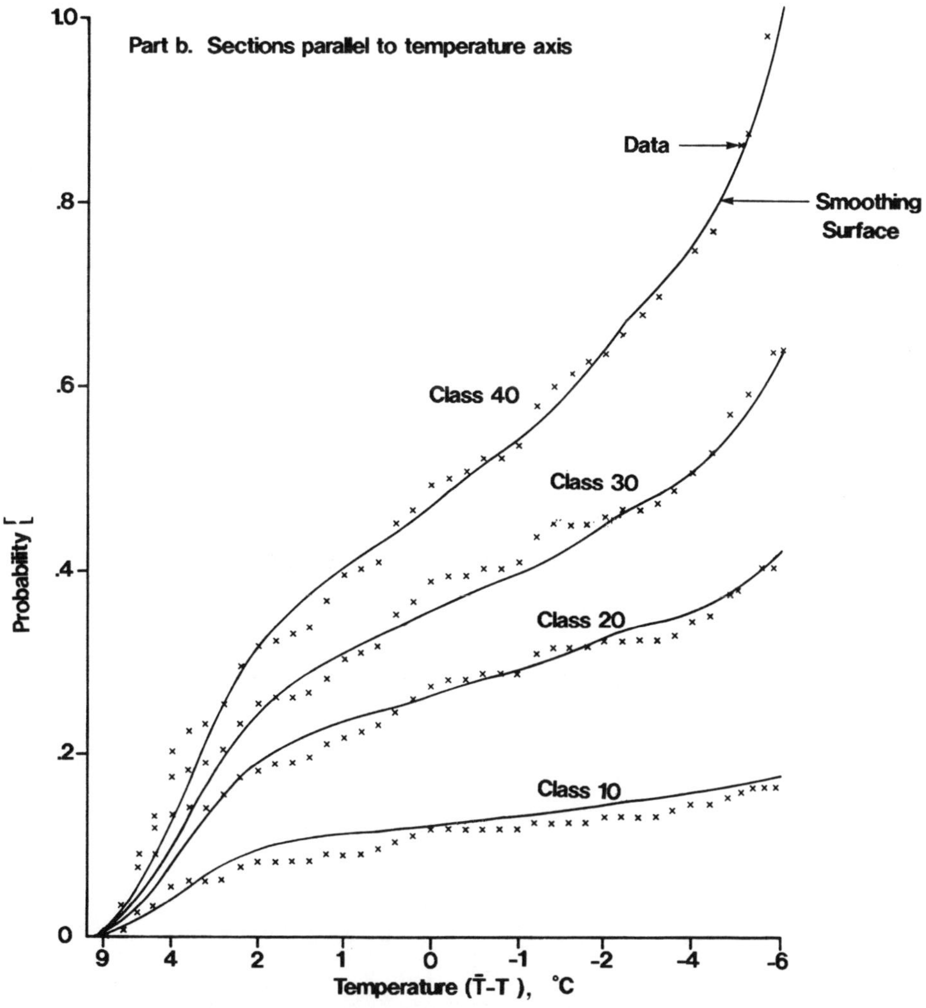

Fig. 6-21b. Cross Sections Through The Joint Probability Surface of Monthly Total Rainfall and Monthly Average Temperature

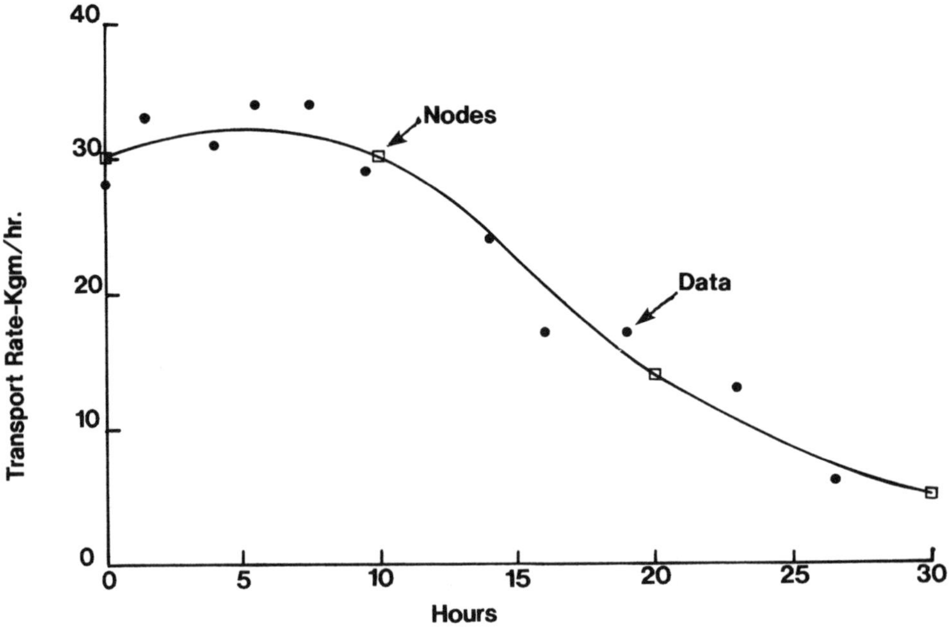

Fig. 6-22. Smoothed Data for Example 6-5

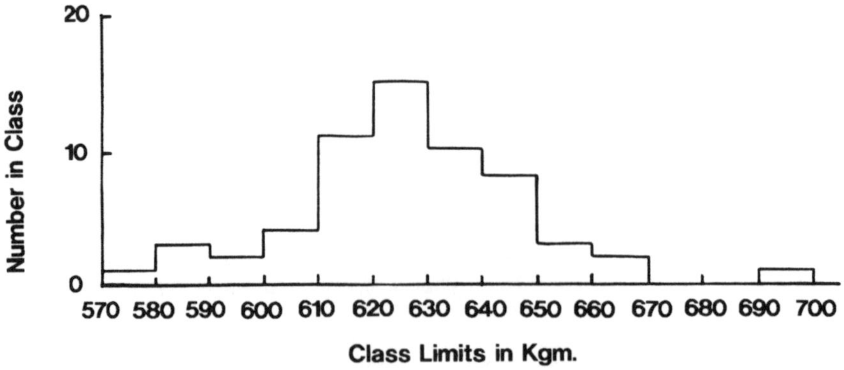

Fig. 6-23. Distribution of Transported Amounts of Point A.

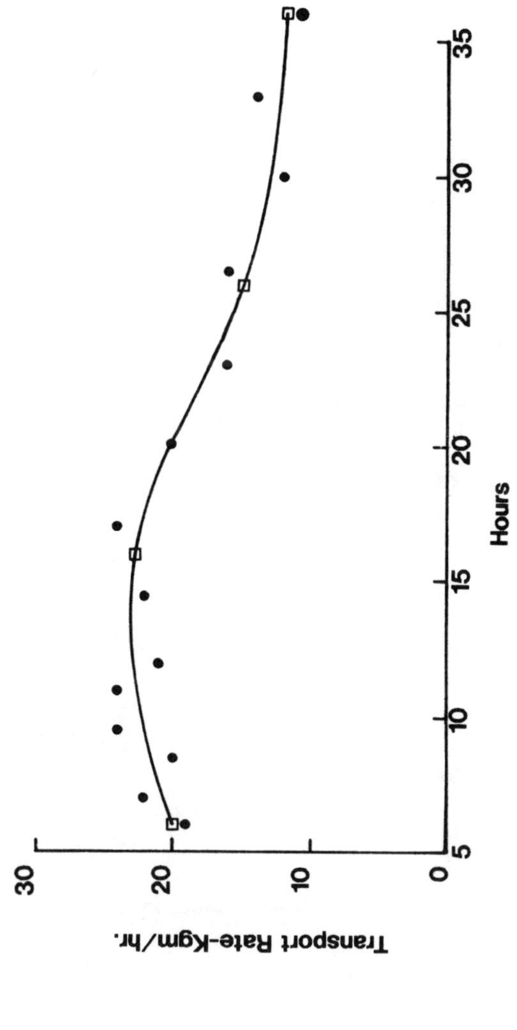

Fig. 6-24. Smoothed Data for Example 6-6

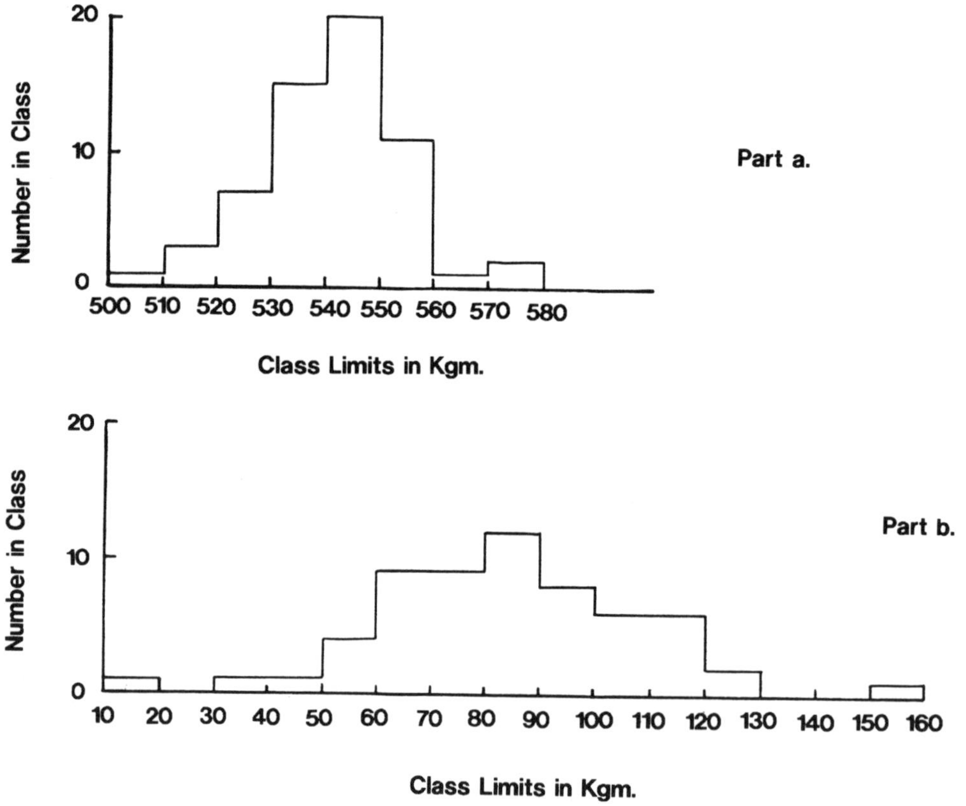

Fig. 6-25. **Distribution of Transported Amounts of Point B and Difference from Point A.**

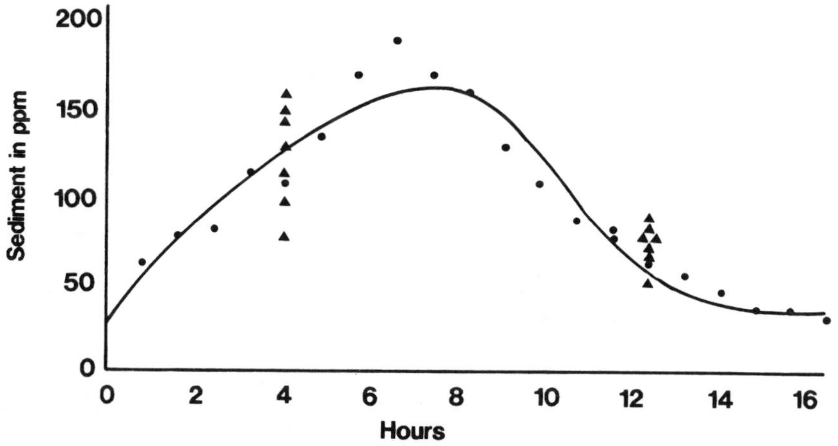

Fig. 6-26. Smoothed Data for Example 6-7

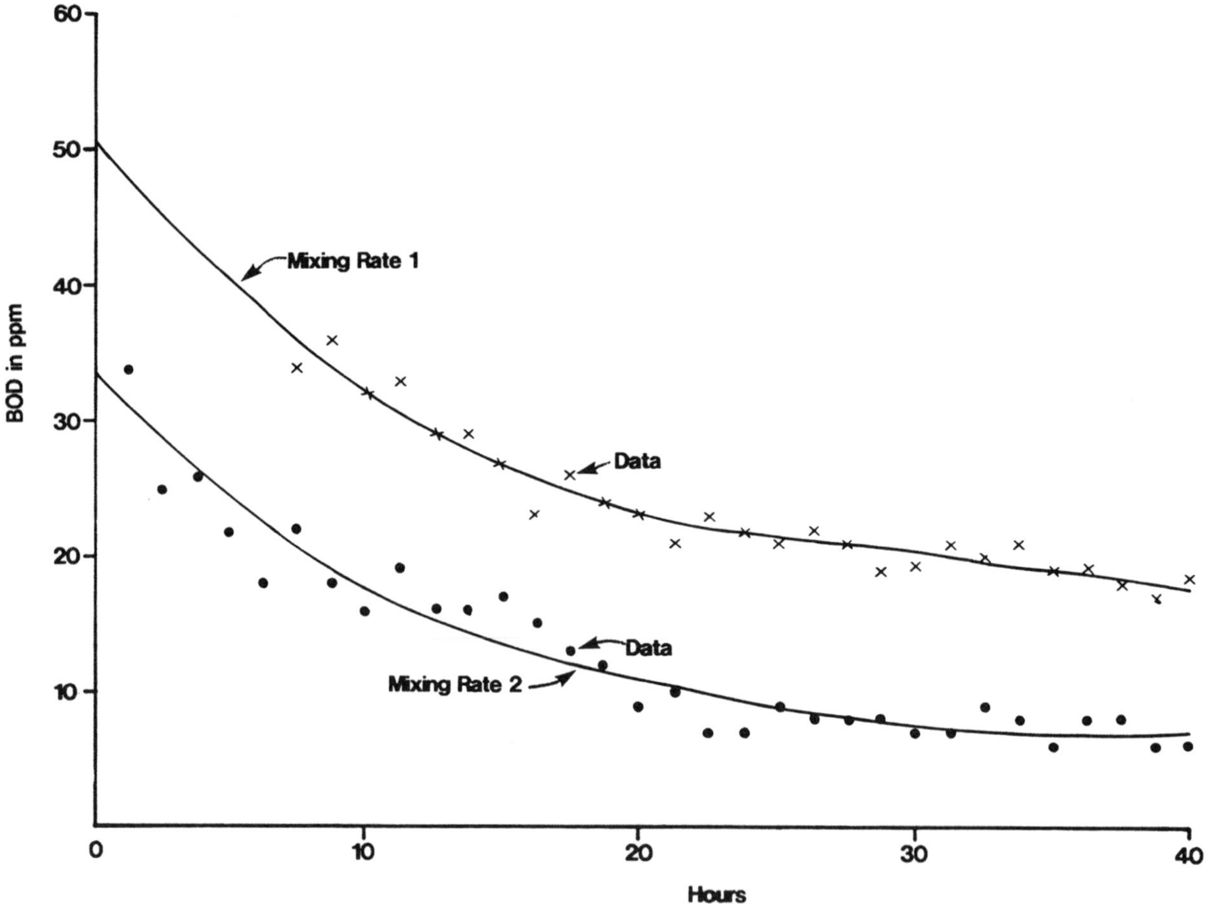

Fig. 6-27. Smoothed Data for Example 6-8

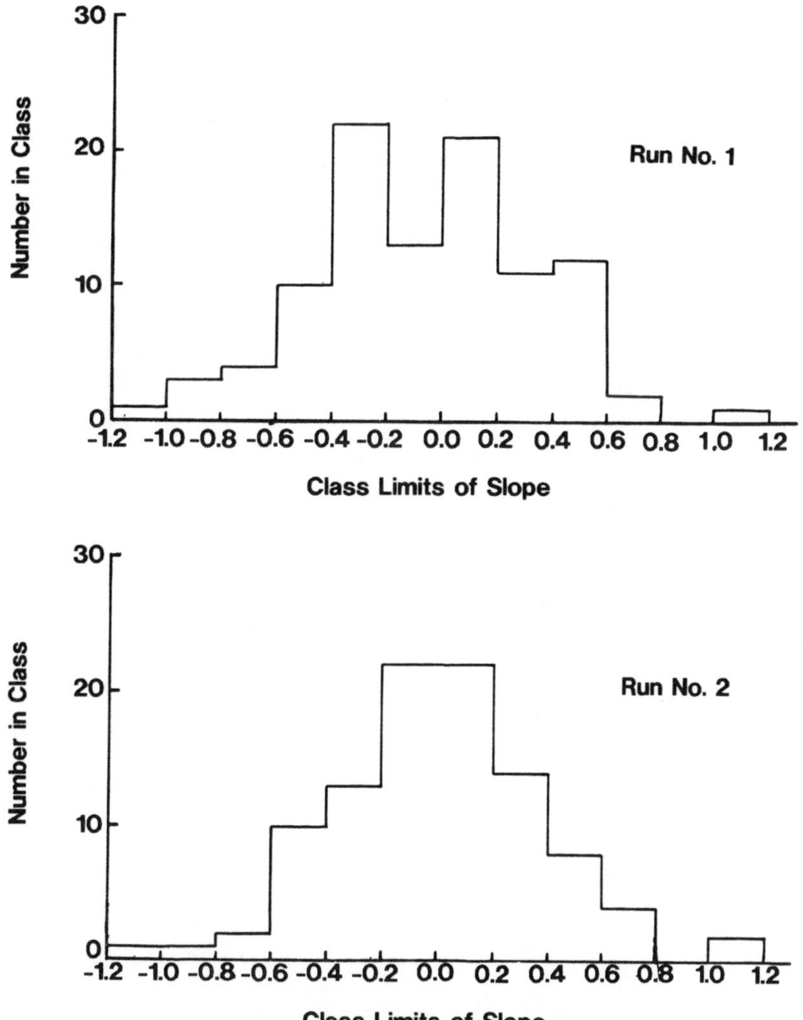

Fig. 6-28. Histograms of Simulated Slope

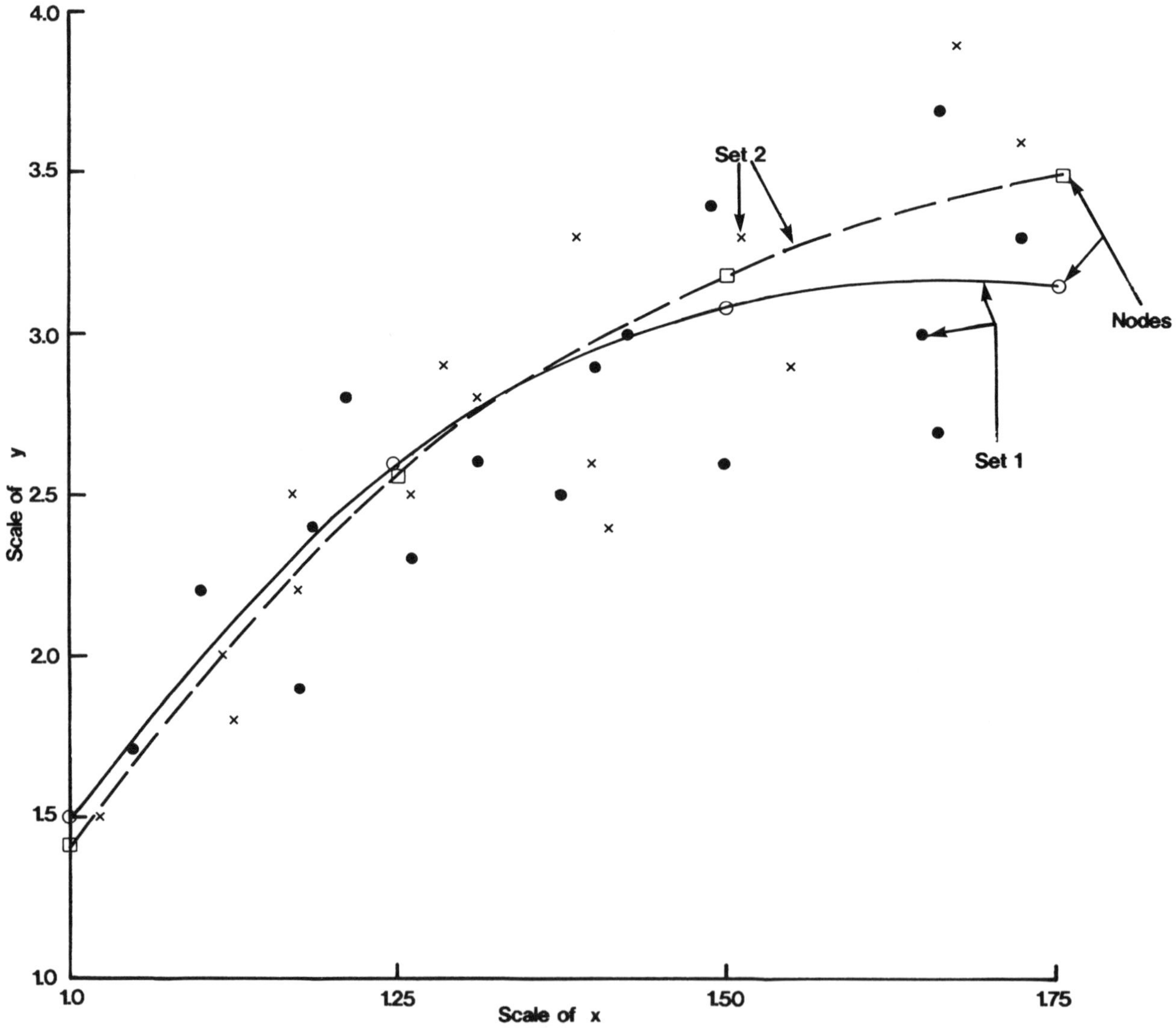

Fig. 6-29. Smoothed Data for Example 6-10

CHAPTER 7

Solution of Differential And Integral Equations By Numerical Approximation With Least Squares

7.1 INTRODUCTION

In Chapters 5 and 6 we explored ways to approximate functions from a series of observations or measurements. Particularly in Chapter 5 we used the piecewise sliding polynomials to smooth experimental data. Such smoothing operators both represent the data and become substitutes for the data. In addition, they are customarily considered to be approximations for the physical-chemical-biological processes that produced the experimental data. In Examples 5-5 and 5-6 we combined smoothing with a change of functional hierarchy, in one case smoothing and differentiating and in the other case smoothing and integrating simultaneously.

In this chapter and the next we will also use numerical information, but this information will usually come from algebraic statements about a known process instead of from observation. Operationally, functional approximation here reduces to the numerical solution of integral and differential equations. Such numerical techniques must be used when we have complex equations that cannot be readily treated through classical calculus. In this chapter we will use least squares to achieve optimal correspondence between the function and its approximation. In Chapter 8 we will use finite elements based on sliding polynomials. In both chapters the analytical unit in the approximation will be the sliding polynomials.

7.2 ONE-DIMENSIONAL ANALYSIS

By one-dimensional analysis we mean a dependent variable functionally dependent on one independent variable. We need to review briefly the functional concepts underlying Eqs. 5-7 through 5-12. The hierarchy of the following equations will form the beginning point in functional approximation:

$$\sum_{i=1}^{n} \int_{-1/2}^{1/2} C_i(z)\, dz\, \frac{dx}{dz}\, Y_i = \int_{x_1}^{x_2} \hat{y}dx \approx \int_{x_1}^{x_2} ydx = \int_{x_1}^{x_2} f_1(x)\, dx \tag{7-1}$$

$$\sum_{i=1}^{n} C_i(z)\, Y_i = \hat{y} \approx y = f_1(x) \tag{7-2}$$

$$\sum_{i=1}^{n} C_i'(z)\, \frac{dz}{dx}\, Y_i = \frac{d\hat{y}}{dx} \approx \frac{dy}{dx} = f_1'(x) \tag{7-3}$$

$$\sum_{i=1}^{n} C''(z)\, \frac{d^2z}{dx^2}\, Y_i = \frac{d^2\hat{y}}{dx^2} \approx \frac{d^2y}{dx^2} = f_1''(x) \tag{7-4}$$

In this set of equations, Eq. 7-2 is usually considered the fundamental functional expression of interest. We can move from Eq. 7-1 to Eq. 7-2 by a process of differentiation. We will call this the solution of an integral equation. We can move from either Eq. 7-3 or 7-4 to Eq. 7-2 by a process of integration. We will call this the solution of differential equations.

Equations 7-1 to 7-4 are usually simplified for practical work. For example, the full expression of Eq. 7-2 is as follows. y is a function of x. It is approximated by $\hat{y}$. This approximation is the sliding polynomial form of a product-summation of coefficients and nodes. The simplification is to set the approximating form equal to the function of x as:

$$\sum_{i=1}^{n} C_i(z)\, Y_i \approx f_1(x) \tag{7-5}$$

in which n is 2, 4, or 6 depending upon which system of sliding polynomials is used. In all of the equations, we also imply a second function, $z = f_2(x)$. With this scale-change function, the equations can be cleared to a common and standardized independent variable, z, so that the sliding polynomial coefficients that are expressed as a function of z can be used.

The right-hand sides of Eqs. 7-1 to 7-4 must provide a vector of numerical values. Such numerical values result either from stated boundary conditions, or initial values, or from known values of the process expressed by the function of x. The vector of numeric values provides the basis for the solution for a set of nodes Y_i. This vector of calculated values simply replaces the vector of measured values we used in Chapter 5.

In Eqs. 7-1 to 7-4 it must be clearly understood that the nodes Y_i in all four hierarchial levels are the same. They express an approximation of a function y at the level of Eq. 7-2. Given a numeric right-hand side and the coefficients, C_i, we can solve for the nodal set Y_i to approximate y, and this $\hat{y}$ is always at the level of Equation 7-2. The method is identical to the method of smoothing as discussed in Chapter 5. The only difference is the right-hand numerical vectors. While in Chapter 5 we used observations, in this chapter we will use calculated values.

The use of least squares for the solution of differential and integral equations will be further explained using a series of examples. Since an equation may have a form such as

$$\frac{dy}{dx} + y = f(x)$$

We will illustrate the approximation of y, as well as forms such as dy/dx. In order to illustrate, as well as test the least squares methodology, we will operate on a known function. The four levels of the chosen function are given by:

$$\int_0^x y dx = x \exp(-0.3x) \tag{7-6}$$

$$y = (1 - 0.3x) \exp(-0.3x) \tag{7-7}$$

$$\frac{dy}{dx} = (-0.6 + 0.09x) \exp(-0.3x) \tag{7-8}$$

$$\frac{d^2y}{dx^2} = (0.27 - 0.027x) \exp(-0.3x) \tag{7-9}$$

7.2.1 Example 7-1: Two-Point Functional Approximation

The two-point sliding polynomial will be used to approximate Eq. 7-7 for the range $0 \leq x \leq 6$. Nodes will be placed at x = 0, 2, 4, and 6 to break this range into three spans.

In order to use the method of least squares, many points in the range $0 \leq x \leq 6$ must be specified where the difference between the function and the sliding polynomial approximation is minimized. We will call these points match points, and specify that a match point be located at the end and quarter points of each of three spans. At each of these points, the sliding polynomial coefficients and the value of the function must be calculated. In brief summary, the function y(x) will be approximated by three linear segments. The approximation will be a "best fit" between the $\hat{y}$ of the two-point sliding polynomial and the y(x) of Eq. 7-7 at the generated match points.

The data are organized for a least-squares solution in Table 7-1. The match points are numbered sequentially from 1 to 13 at the left. Values of x for each of these match points are shown next, then the values of the local unit variable z are shown. With these values of z, the coefficient equations (Eqs. 1-3) can be evaluated for each point, and the coefficients assigned to the appropriate nodes for the span. These coefficients are shown under the appropriate nodes in the center of the table. These four columns of coefficients are the numerical expression of the left-hand side of Eq. 7-5. The right-hand side is expressed numerically by evaluating Eq. 7-7 at each of the match points, as shown in the right-hand column of Table 7-1.

The five columns on the right of Table 7-1 are the basis for a straight-forward least squares solution, following exactly the methods used in smoothing in Chapter 5. The only difference is that in smoothing we computed sliding polynomial coefficients for data points, which were observed values. In functional analysis we compute the coefficients for a series of match points. The right-hand vectors in smoothing are data points, while in functional analysis they are calculated match-point values. The solution for the four nodal values is shown at the bottom of Table 7-1.

The function and its approximation are shown in Figure 7-1. Overall, the approximation is good. It overlays the function with a balancing of plus and minus departures at the match points across the three spans. However, it suffers from the limitations imposed by the linear segmented form.

In summary, we have presented a simple approach to approximation of a known function. This example has little intrinsic value but it is important in that it illustrates the basic methodology to be used in more complex problems. Specifically, it demonstrates how the y in a differential equation such as

$$\frac{dy}{dx} + y = f(x)$$

is approximated. Further examples will show how to approximate other terms and combine them for solution of a complete differential or integral equation.

7.2.2 Example 7-2: Two-Point Integral Approximation

In this example we will derive an approximation of the function in Eq. 7-7 from the integral in Eq. 7-6 using the two-point sliding polynomial. This is the solution of an integral equation. For this solution we must use the integrals of the sliding polynomial coefficients as indicated in Eq. 7-1. Additionally, we must take account of the scale change from x to the local unitized variable z. With nodes at 0, 2, 4, and 6, dx/dz is 2.

The process of integration expressed by Eq. 7-1 is within one span. Therefore, for this example, Eq. 7-1 covers directly the integral for the first span, where $0 \leq x \leq 2$. For the second span the process of integration must be the summation of the integral for the full span $0 \leq x \leq 2$ plus the within-span integral for the span $2 \leq x \leq 4$. For the third span the full integral from $x = 0$ is the summation of the integral for the two full spans $0 \leq x \leq 4$ plus the within-span integral for the span $4 \leq x \leq 6$.

The organization of the data for least squares is shown in Table 7-2. Match points 1 through 5 are within the first span; therefore, Eqs. 4-4 can be used directly to calculate the integral coefficients for nodes Y_0 and Y_2 in this span. Match points 6, 7, 8, and 9 are within the second span. Therefore, the coefficients for these within-span integrals are added to the coefficients for the full span $0 \leq x \leq 2$ to the left in order to express the full integral from $x = 0$ to the point within the second span. Match points 10, 11, 12, and 13 are within the third span. Therefore, the integral coefficients for two full spans to the left are added to the within-span coefficients. Values of the integral of the function, Eq. 7-6, are shown in the column at the right in Table 7-2. The process of least squares uses those rows of Table 7-2 that have match point numbers and values of x exp(-0.3x).

The solution by least squares is shown at the bottom of Table 7-2. It should be noted that the area under the lines connecting the nodes is given by the following simple geometric equation:

$$\left(\frac{Y_0 + Y_2}{2}\right) 2 + \left(\frac{Y_2 + Y_4}{2}\right) 2 + \left(\frac{Y_4 + Y_6}{2}\right) 2 = Y_0 + 2Y_2 + 2Y_4 + Y_6 \qquad (7\text{-}10)$$

This agrees with the values of the coefficient integrals for point 13 in Table 7-2.

The integral equation, Eq. 7-6, and its two-point sliding polynomial approximate solution are both shown in Fig. 7-2. Note that we got the same set of nodes by least squares as we did in the direct function approximation in Example 7-1. Also note that this set of nodes is a good approximation to (1-0.3x) exp(-0.3x) which is the first derivative of x exp(-0.3x). In summary, we have demonstrated the solution of the integral Eq. 7-6 to obtain the function Eq. 7-7.

7.2.3 Example 7-3: Four-Point Integral Approximation

The solution of Example 7-2 can also be derived using four-point sliding polynomials. Table 7-3 shows the four-point coefficients and the right-hand side values for least squares. The structure is similar to that used in Table 7-2, except that the four-point system adds the outlier nodes at x = -2 and at x = 8. Four coefficients cover each match point compared to the two coefficients in Table 7-2. The function is again approximated using calculated values at the quarter-points of each span.

The solution is shown at the bottom of Table 7-3, with the solution plotted in Fig. 7-3, along with the integral function. Note the improvement in the approximation in Fig. 7-3 over that in Fig. 7-2. The four-point sliding polynomial approximation closely overlays the test function of Eq. 7-7.

7.3 SOLUTION OF DIFFERENTIAL EQUATIONS

The solution of differential equations requires consideration of two elements that are not necessary either in the solution of integral equations or in direct function approximation. First, boundary values must be added to the solution. Boundary values serve the same purpose as constants of integration in simple classical calculus. Stated very simply, the solution of a differential equation is the process of integration or summation, with the summation starting from a known point or points. Boundary conditions in smoothing were introduced in Chapter 5.

The second element is a requirement for mathematical continuity of the approximating function. We saw in Chapter 1 that the two-point sliding polynomial is continuous at the hierarchial level of Eqs. 7-1 and 7-2. The four-point sliding polynomial is continuous down to the level of Equation 7-3. The six-point sliding polynomial adds continuity at the level of Equation 7-4. It was for this reason that we did not attempt to solve Eq. 7-8 to approximate Eq. 7-7 with the two-point system. However, we can use the four-point system for this approximation. We cannot use the four-point system to solve Eq. 7-9, but we can use the six-point sliding polynomials.

7.3.1 Example 7-4: Four-Point Differential Approximation

The four-point system of sliding polynomials can be used to solve Eq. 7-8, with the boundary condition that $Y_0 = 1$. The arrangement of the data for solution is shown in Table 7-4. Note the following change of scale in the coefficients:

$$\frac{d\hat{y}}{dx} = \sum_{i=1}^{4} \frac{dC_i(z)}{dx} Y_i = \sum_{i=1}^{4} \frac{dC_i(z)}{dz} \frac{dz}{dx} Y_i = \frac{1}{2} \sum_{i=1}^{4} \frac{dC_i(z)}{dz} Y_i \qquad (7\text{-}11)$$

The four-point differential coefficients were calculated using Eqs. 3-9. Since we use a boundary value of $Y_0 = 1$, the coefficients for Y_0 must be moved across the equal sign of the implied equation and must be combined with the values of dy/dx. The Y_0 column is dropped from the matrix of coefficients and only Y_{-2}, Y_2, Y_4, Y_6, and Y_8 are evaluated by least squares.

Values of the nodes derived by least squares are shown at the bottom of Table 7-4. Since Y_0 is set to 1, the values differ slightly from those in Table 7-3, but still form a good approximation to Eq. 7-7, which is the test function. Equation 7-8 and $\hat{y}$, which is the approximation to Eq. 7-7, are both shown in Fig. 7-4. In summary, we solved the differential Eq. 7-8 to get the function of Eq. 7-7.

7.3.2 Example 7-5: Six-Point Differential Approximation

The six-point sliding polynomial can be used to solve Eq. 7-9, with the boundary conditions $Y_0 = 1$ and $Y_6 = -0.1322$. Two boundary conditions are necessary since this is a second-order differential equation.

Data for the least squares solution are shown in Table 7-5. The following change of the scale is made in the coefficients:

$$\frac{d^2y}{dx^2} = \sum_{i=1}^{6} \frac{d^2C_i(z)}{dx^2} Y_i = \sum_{i=1}^{6} \frac{d^2C_i(z)}{dz^2} \frac{d^2z}{dx^2} Y_i = \frac{1}{4} \sum_{i=1}^{6} \frac{d^2C_i(z)}{dz^2} Y_i \tag{7-12}$$

Eqs. 3-12 were used to evaluate the coefficients. Y_0 and Y_6 are algebraically eliminated by moving the product with their coefficients across the implied equal sign in the table and combining this product with the derivative d^2y/dx^2 to form the right-hand vector. The Y_0 and Y_6 columns are then dropped from the coefficient matrix. The least squares solution is shown at the bottom of the table.

The second derivative of the test function, Eq. 7-8, and the approximation of Eq. 7-7 derived from it, are both shown in Fig. 7-5. Thus, we have demonstrated that all changes in the hierarchy expressed by Eqs. 7-6 through 7-9 can be accomplished by use of sliding polynomials and least squares. The approximations are good, as shown by the various sets of nodal values derived by the least squares.

We should note in the conclusion of this series of examples that six-point sliding polynomials can be used for the entire range of Eq. 7-6 through Eq. 7-9. The four-point system can be used for the range from Eq. 7-6 through Eq. 7-8. The two-point system can only be used with Eqs. 7-6 and 7-7 since this system has no mathematical continuity in its derivatives.

7.3.3 Example 7-6: Formulation and Solution of an Elementary Differential Equation

Examples 7-1 through 7-5 served mainly to demonstrate the method of solution of differential and integral equations by least squares using sliding polynomials as the core analytical device. All of these examples tested the method against a known function. In this example the function is unknown. This is the usual situation in practical problem solving.

Figure 7-6 shows a two-compartment tank, with a length of 12 meters and a width of w meters. The top compartment contains some mass of chemical, M, diffused through a layer of sand. The bottom compartment shows a concentrated mass of the chemical. The task is to derive a curve that approximates the total amount of chemical measured as some sensing equipment moves from left to right over the full length of the tank. The accumulation process is started at zero at x = 0.

In the diffused chemical in the top of the tank, a differential volume of h w dx meters cubed would hold a differential mass of chemical dM_d = c h w dx grams, where c is the concentration in gm/m^3. Suppose h, the depth of the sand, varies as h = f exp(gx), then dM_d/dx = cf exp(gx) dx. The concentrated material in the bottom of the tank is assumed to have a parabolic distribution of mass of chemical $M_c = (a_1 + a_2x + a_3x^2)\, w$, or a rate of increase per unit length of $dM_c = (a_2 + 2a_3x)\, wdx$. The total change in mass per unit length of tank is given by:

$$\frac{dM_T}{dx} = \frac{dM_d}{dx} + \frac{dM_c}{dx} = (cf\ \exp(gx) + a_2 + 2a_3\ x)\ w \tag{7-13}$$

For consistency of units c is gm/m^3, f is m, g is 1/m, a_2 is gm/m^2, a_3 is gm/m^3, and w is m.

Table 7-6 shows the arrangement of data for the solution with nodes at x equal to 0, 4, 8, and 12 meters, and outlier nodes for the four-point sliding polynomial at x = -4 and x = 16. Note that dx = 4 dz. The scale-change parameter 4 is used with the right-hand vector in Table 7-6. Numerical values of $c = 0.2$, $f = 2$, $g = 0.1$, $a_2 = 2.0$, $a_3 = 0.3$, and $w = 1$ were used to compute the right-hand vector of Eq. 7-13. The solution is also plotted in Fig. 7-7. This curve represents the integral of an exponential and a parabola. We have no direct algebraic expression of this curve.

7.3.4 Example 7-7: Solution of a Differential Equation with Discrete Variation

For this example we will consider a slight modification of Example 7-6. Imagine that some of the concentrated chemical has been removed from the lower compartment of the tank. The location and pattern of this removed material, ΔM, is shown to the right of Table 7-6. The amount removed is shown multiplied by 4 to make it consistent with the right-hand vector of Example 7-6. The removal has caused a peak reduction of 2 gm/m from the differential of mass in Example 7-6. There has been no removal between zero and four meters and none at 12 meters. We want to find the curve that now approximates the accumulating amount of material as we measure from left to right along the tank.

If we use the same criteria as for Example 7-6, we can simply reduce the right-hand vector by the amount removed, as shown in Table 7-6. The least squares solution uses the same coefficient matrix as Example 7-6.

The solution is shown at the bottom of Table 7-6 and is also plotted in Fig. 7-7. The comparison with the curve for Example 7-6 is interesting. The two curves are identical from $x = 0$ to $x = 4$, the beginning of the depleted zone. The curve for Example 7-7 draws away from that for Example 7-6 through the depleted zone, but then tends to become parallel again as the depleted zone is exited.

There is no simple way to calculate the effect of the removed material on the integral using basic calculus. Note that what we have done is study the effect of discrete variation of a_2 in Eq. 7-13.

7.3.5 Example 7-8: Solution of a Differential Equation of Two Dependent Terms

The equations solved in examples to this point have been simple. For example, in Eq. 7-13 there is only one term involving the dependent variable. Most differential and integral equations used in practical problem solving will have more than one term involving the dependent variable. Example 7-6 will again be modified to illustrate the methodology required to solve such equations.

In Example 7-6 we proposed the problem of approximating the curve giving the mass of a chemical as sensing equipment scanned the concentration from left to right. We now propose a more complex problem. While the sensing equipment scans from left to right, we will continuously remove some of the chemical. The rate of removal will depend on the mass of chemical accumulated to that point. The total change in mass per unit length of the tank is then given by:

$$\frac{dM_T}{dx} = \frac{dM_0}{dx} + \frac{dM_c}{dx} - kM_T \qquad (7\text{-}14)$$

Rearranging and adding Eq. 7-13:

$$\frac{dM_T}{dx} + kM_T = cf\ \exp(gx) + a_2 + 2a_3x \qquad (7\text{-}15)$$

In these equations k is a proportionality factor. There are now two terms on the left-hand side involving the dependent variable M_t. For computation Eq. 7-15 is rescaled to the following form:

$$\frac{dM_T}{dz} + 4kM_T = 4\ (cf\ \exp(gx) + a_2 + 2a_3\ x) \qquad (7\text{-}16)$$

Table 7-7 shows the calculation of the composite coefficient matrix indicated by the left-hand side of Eq. 7-16. Only one span is shown; additional spans must be added, as needed, as in the earlier examples. A value of 0.05 was used for k. Table 7-7 shows that the values of the different coefficients, function or derivative, are simply added, since the terms are added in Eq. 7-16. The summation is performed at each quarter-point of the span, this being the point density used to define the left-hand vectors.

The composite coefficients from Table 7-7 are used to construct a three-span approximation in Table 7-8. The values are arranged simply in a band matrix previously illustrated in tables such as Tables 7-5 and 7-6. The least squares solution is shown at the bottom of Table 7-7, and the solution is plotted in Fig. 7-7. This curve differs from that in Example 7-7 in that it shows a steady deviation from the curve of Example 7-6. Example 7-7 showed the removal of a discrete limited amount of chemical. Example 7-8 shows a steady removal at a rate that is proportional to the amount present.

7.4 TWO-DIMENSIONAL ANALYSIS

In one-dimensional analysis the problems consisted of one independent variable related to a dependent variable. In two-dimensional analysis the problems will consist of two independent variables related to a dependent variable. Solutions in one-dimensional analysis are lines or curves. Solutions in two-dimensional analysis are surfaces. In one-dimensional analysis we could use the two-point, four-point, or six-point system of sliding polynomials. We have these same options in two-dimensional analysis.

One-dimensional functional analysis using least squares was numerically the same as one-dimensional smoothing in Chapter 5. The only difference was that we worked with values calculated from algebraic expressions, rather than from measurements or observations. Two-dimensional analysis using least squares is the same numerically as two-dimensional smoothing in Chapter 5, but again we work with calculated values rather than observations.

7.4.1 Example 7-9: Approximate a Simple Polynomial Surface

The following two-point, two-dimensional sliding polynomial surface will be used to illustrate the general concept:

$$y = a + bwx + cw^2x \tag{7-17}$$

Values of the coefficients are $a = 2$, $b = 3$, and $c = 4$. One span in the w-dimension with $1 \leq w \leq 2$ will be used. Two spans in the x-dimension with $0 \leq x \leq 2$ will be used. The values of the function, Eq. 7-17, are shown in Fig. 7-8. The equation was evaluated at all half-points of the two elements.

Any one element has nine match points based on the half-points of the element. Other spacings for evaluation could by used, e.g., the third-points or quarter-points. The half-points appear to be sufficient for many problems. Each of the nine match points in a unit surface has coefficients weighting it to the four nodes at the corners of that surface. The two-point sliding polynomial coefficients for the nine positions are shown in Fig.7-9. Note that these are products of the one-dimension coefficients in local unit variates z and u. Such products are based on the "function-of-functions" structure described in Chapter 1. The coefficients in Fig. 7-9 were arranged with the function values of Fig. 7-8 in Table 7-9 for the least squares solution. The optimal values of the six nodes are shown at the bottom of the table.

The surface defined by the six optimized nodes is an approximation of the surface of Eq. 7-17. This approximating surface is shown as a set of contour lines in Fig. 7-10. Note that the contours kink as they cross the boundary between the two elements. Such kinking is caused by the discontinuity of the first and higher derivatives in the two-point system. The nature of the break in the contours and the correspondence between the true surface and the approximating surface is more fully revealed in Fig. 7-11. Three cross-sections are shown parallel to the x-dimension for values of w of 1.0, 1.5, and 2, which corresponds to the alignment of points in Fig. 7-8. The approximating linear segments are shown, along with the 15 calculated values on which the approximation is based.

We can see that the two-point system forms a useful approximation when high precision is not required. Its advantage lies in its simplicity, and this simplicity was used here for the introduction to two-dimensional analysis. We should reemphasize that in most practical problems the true function is not known. We assumed Eq. 7-17 only to provide numerics to illustrate and test the method. Also, most problems will require a change in the mathematical hierarchy in Eqs. 7-1 through 7-4 to be performed simultaneously with the approximating solution. However, direct functional approximation is frequently a part of the required solution, just as in the illustration with the two dependent terms in Example 7-8.

7.4.2 Example 7-10: Solution of a Two-Dimensional Integral Equation

The two-point system can be used to solve the following integral:

$$y = w^{0.3}x \exp(-0.3w^{0.3}x) \tag{7-18}$$

This equation is to be considered an integral only in the x-dimension and is to be considered of proper order in the w-dimension. It is nearly the same as Eq. 7-6 except that the second independent variable has been added.

The solution will cover the range $1 \leq w \leq 3$, $0 \leq x \leq 6$, with two spans in w and three spans in x. The arrangement of the grid elements and the numbering convention for the 12 nodes is shown in Fig. 7-12.

As in Example 7-9 we will use calculated values at the half-points of the elements. Figure 7-13 shows the coefficients for the four nodes of an element for the nine evaluation positions. These coefficients are products of the one-dimensional coefficients in the w-dimension and of the one-dimensional integral coefficients in the x-dimension. The assignment of coefficient-values to the 12 nodes is shown in Table 7-10. Remember that in elements two and three, counting from the left, the coefficients must be added to the coefficients of the full elements to the left in order to express the total integral from $x = 0$. To the right in Table 7-10 are the calculated values of Equation 7-18 scaled by dx/dz. Note that all calculated values for points along the $x = 0$ vertical are omitted since the surface integral is zero along this boundary.

The solution for the 12 nodes is shown at the bottom of Table 7-10. Contours across the six elements are plotted in Figure 7-14. A section of this surface along the $w = 1$ boundary is the same as the solution of Example 7-2 shown in Fig. 7-2. The surface in Fig. 7-14 is in reality the partial derivative, $\partial y/\partial x$, of Eq. 7-18. The zero contour in Fig. 7-14 shows how the peak of the function, Eq. 7-18, moves to the left with increasing values of w.

7.5. SOLUTION OF NONLINEAR DIFFERENTIAL EQUATIONS BY NON-LINEAR LEAST SQUARES

All of the differential equations that have been discussed in this chapter have been linear in the structure of their nodes for the approximating sliding polynomials. Even the two-dimensional solutions have been linear in their nodal structure. Because of this linearity, ordinary least squares could be used to force maximum correspondence between the approximating function and the match points.

Some differential equations may not be linear in the structure of the nodes. For example, the following is not linear:

$$y \frac{dy}{dx} = f(x, a_1, a_2 \ldots) = R \tag{7-19}$$

Consider one span of a possible sliding polynomial approximate solution. In this span

$$y \approx \hat{y} = C_1Y_1 + C_2Y_2 + C_3Y_3 + C_4Y_4$$

for a solution based on the four-point system. Also,

$$\frac{dy}{dx} \approx b \frac{dy}{dz} = b\ (dC_1Y_1 + dC_2Y_2 + dC_3Y_3 + dC_4Y_4)$$

and the product y dy/dx will produce forms such as $C_1\ dC_2\ Y_1\ Y_2$. These products Y_iY_j are obviously not linear. In Eq. 7-19, R is introduced to be the symbolic representation of the right-hand side of the equation.

A technique known as non-linear least squares can be used to derive values of the nodes of the sliding polynomial approximations when the differential (or integral) equations are non-linear. This technique is also known as the method of differential correction (Nielson, 1956, p. 309). The method is an iterative one. We start with an estimated set of nodes, and then we use least squares to get corrections to this set that will reduce the error between the stated right-hand side of the function and our prediction of it. We make the correction on the nodes, and then perform the operation again, this time using the updated nodes. This process continues until the corrections approach zero.

In further explanation of the method consider the following:

$$\frac{\hat{R}_1-\hat{R}}{\Delta Y_1}\,dY_1 + \frac{\hat{R}_2-\hat{R}}{\Delta Y_2}\,dY_2 + \frac{\hat{R}_3-\hat{R}}{\Delta Y_3}\,dY_3 + \text{----} = R-\hat{R} = E \tag{7-20}$$

In the equation, R is the calculated or stated right-side value at a match point. It may be zero or some other constant value. $\hat{R}$ is the estimate of R using sliding polynomial operations on the initial estimated set of nodes. Call these initial nodes Y_{0-1}, Y_{0-2}, Y_{0-3}, and so on, with the zero subscript indicating the initial value and the second integer being the node number. $\hat{R}_1$ is the value of $\hat{R}$ obtained by incrementing node Y_{0-1} by the small quantity ΔY_1. Then $(\hat{R}_1-\hat{R})/\Delta Y_1$ is approximately the partial derivative, the slope of $\hat{R}$ with respect to Y_{0-1}. $(\hat{R}_2-\hat{R})/\Delta Y_2$ is the slope of the right-hand side $\hat{R}$ with respect to node Y_{0-2}. We calculate these slopes for every match point for every node that we wish to find. The only unknowns in Eq. 7-20 are the nodal corrections dY_1, dY_2, dY_3, etc. We use least squares to solve for these corrections by treating them as the regression coefficients of any least-squares solution.

For round two of the iterative scheme, we update the nodes as follows: $Y_{1-1} = Y_{0-1} + dY_1$, $Y_{1-2} = Y_{0-2} + dY_2$, $Y_{1-3} = Y_{0-3} + dY_3$, and so on. These nodes are used to compute new values of $\hat{R}$ and new values of the slopes $(\hat{R}_i-\hat{R})/\Delta Y_i$. Then a second application of least squares gives a new set of corrections dY_i. We continue until all of the dY_i become very small.

The computational process for non-linear least squares will be presented in detail using an exceedingly simple problem. Consider the following simple non-linear differential equation:

$$y \cdot \frac{dy}{dx} = \frac{x}{2} \tag{7-21}$$

where

$$y \cdot \frac{dy}{dx} \approx y \cdot \frac{dy}{dz}\frac{dz}{dx} = (C_1Y_1+C_2Y_2+C_3Y_3+C_4Y_4)\;(dC_1Y_1+dC_2Y_2+dC_3Y_3+dC_4Y_4)\;\frac{dz}{dx}$$

We will calculate an approximate solution for this equation for the range $0 \leq x \leq 4$. This range will be covered using one span of the four-point sliding polynomials. The interior nodes will be at $x = 0$ and $x = 4$. The outlier nodes will be at $x = -4$ and x = 8. The node at x = 0 will be set to unity as a boundary condition. Five match points, which are located at the edge and the quarter points of the span, will be used.

Details of the first round of the computation are shown in the top portion of Table 7-11. The R of Eq. 7-20 is computed first. R is x/2 for our simple illustration. An initial set of nodes is estimated as 0, 1, 2, and 3. The values of y and dy/dx are calculated on these nodes for each match point, using the four-point sliding polynomial. The $\hat{R}$ of Eq. 7-20 is computed as y • dy/dx using the form of our simple Eq. 7-21. As the next step, the outlier node at x = -4 is incremented from 0 to 0.05. This means that 0.05 was used for ΔY_1, as in Eq. 7-20. Now y, dy/dx, and y • dy/dx = $\hat{R}$ are computed for the nodal values of 0.05, 1, 2, and 3. These calculations are again performed for each match point. As the next step in these initial calculations, $\hat{R}_3$ is calculated using nodal values of 0, 1, 2.05, and 3; then $\hat{R}_4$ is calculated using nodal values of 0, 1, 2, and 3.05. Note that the second node, with a value of 1, is not incremented. This node is to be maintained at a value of 1 as the boundary condition. Only the first, third, and fourth nodes can change during iteration.

Following the computation of $\hat{R}$, $\hat{R}_1$, $\hat{R}_3$, and $\hat{R}_4$, the slopes, or partial derivatives, can be computed. These values are shown in the center portion of Table 7-11. For example, for match point number 1, $(\hat{R}_1-\hat{R})/0.05$ =(0.24375-0.25)/0.05 = -0.125. Also $(\hat{R}_4-\hat{R})/0.05$ = (0.25-0.25)/0.05 = 0. The error terms, R - $\hat{R}$, are also computed for each match point. All terms in Eq. 7-20 have now been calculated except dY_1, dY_3, and dY_4. These are now computed by least-squares operation on the data for the five match points in the center portion of Table 7-11. The values given by the least-squares solution are shown in the lower part of Table 7-11. When these corrections are added to the initial estimates of the nodal values, an updated set of nodal values results.

Another round of iteration can now be performed by simply using the updated nodes in place of the initial estimated nodes. Aside from this change of nodal values, the calculations are identical to the first-round calculations. A second round of calculations was performed as shown by the abbreviated Table 7-12. For this round the nodes were incremented by 0.025, instead of 0.05 as in the first round. The new partial derivatives and errors are shown in the center portion of Table 7-12. Note particularly the significant reduction in errors caused by the first-round corrections. The least-squares solution and the updated nodes after the second round corrections are shown in the lower portion of Table 7-12.

A third round of iteration was performed. Details of these calculations are not shown. However, the results of the least-squares solution and the updated nodes after the third round are included in the bottom portion of Table 7-12. The corrections were very small after three rounds and this simple illustrative problem was halted here. The approximating solution of Eq. 7-21 is shown in Fig. 7-15. The values of R and its prediction by the sliding polynomial, $\hat{R}$, are plotted in the top part of the figure. Correspondence between R and $\hat{R}$ is excellent for the five match points. The approximating four-point sliding polynomial is shown in the bottom portion of the figure. This polynomial times its first derivative predicts the $\hat{R}$ values in the top portion. Therefore, it is an approximate solution of Eq. 7-21.

We may not assume that a solution of a non-linear equation is a unique and only solution. We should remember the elementary example based on simple square roots. Suppose we say $x^2 = 4$. Engineers and scientists usually assume $x = 2$ because we need "real world" solutions. However, $x = -2$ is also a solution. We must always examine our solution for physical rationality and plausibility. In new work, or in questionable situations, it is advisable to re-solve the differential equation using different initial estimates of the nodes.

The approximate solution of Eq. 7-21 carried out in detail in Tables 7-11 and 7-12 is both laborious and error-prone when performed manually. However, because of the repetitive nature of the calculations the non-linear least-squares technique produces highly efficient computer programs. Several examples follow that serve to illustrate and develop further the utility of the method. All of these examples use the six-point sliding polynomial so that differential equations of the second order can be approximated by a system with a continuous second derivative. The examples are based on the schematic definition of problem geometry as shown in Fig. 7-16. The ranges of solution of the equations are set by variables xL and xR and the range of solution is $xL \leq x \leq xR$. This chosen range in x is divided into 12 spans of the six-point sliding polynomial. A total of 17 nodes is specified. Node number 3 is fixed as a boundary condition for a first-order differential equation. Node number 15 is also fixed as a right boundary condition for second-order differential equations. Nodes numbered 1, 2, 16, and 17 are outlier nodes required by the six-point sliding polynomial. Therefore, 16 nodes are iteratively evaluated for first-order equations, while 15 nodes are evaluated for second-order equations. A total of 49 match points are located at the edges and quarter points of each of the 12 spans.

7.5.1 Example 7-11: Demonstration with a Known Function

In this example, we will assume a simple function, construct a first-order non-linear differential operation on this function, and demonstrate that the non-linear least-squares solution does in fact approximate the simple assumed function.

First, assume the simple function

$$y = axe^{-bx} \tag{7-22}$$

The first derivative of this function is

$$\frac{dy}{dx} = ae^{-bx}\ (1 - bx) \tag{7-23}$$

Now construct the differential operation

$$y \bullet \frac{dy}{dx} = a^2xe^{-2bx}\ (1 - bx) \tag{7-24}$$

Our problem is to use non-linear least squares to get a solution to Eq. 7-24 and to check that this solution is an acceptable approximation to Eq. 7-22.

Referring to Eq. 7-20, the expression to the right of the equal sign in Eq. 7-24 is R. The left side, $y \cdot dy/dx$, with the terms approximated by the six-point sliding polynomial through any set of 17 nodes, is the predicted value, $\hat{R}$. Program B-7.1 was prepared to solve Eq. 7-24. The results of two runs of the program are shown in Fig. 7-17. Remember, these runs should yield approximate solutions to Eq. 7-22 for the given problem range and parameter values used. For the first run, a = 4.0, b = 0.02, and $0 \leq x \leq 6.0$. For the second run a = 4.0, b = 0.05, and $0 \leq x \leq 12.0$. The 13 interior nodes, Fig. 7-16, are shown for the two runs in Figure 7-17. A left bound of zero was used for both runs. R and $\hat{R}$ agreed to within three decimal points for the first run. For example, for match point number 49, at xR, R was 66.454 and the predicted $\hat{R}$ was also 66.454. The nodal values are nearly an exact expression of Equation 7-22. Again as an example, the value of y from Eq. 7-22 at x = 6.0 is 21.286. The value of node 15, at xR = 6.0 is also 21.286. Other points agree to the same precision. The second run, across twice the problem range, shows slightly less precision. The greatest error in match points is at number 22 where R is 36.647 and $\grave{R}$ is 36.633. However, at x = 12, y from Eq. 7-22 is 26.342. This is also the value of node number 15. Only eight rounds of iteration were necessary to make the corrections to the nodes, dY_i, essentially zero.

We have thus demonstrated that non-linear least squares did, in fact, yield a set of nodes that closely approximates a known solution to a non-linear differential equation. Caution is necessary, since we know that y · dy/dx = -y (-dy/dx). Therefore, at least one other solution is mathematically possible.

7.5.2 Example 7-12: A Non-linear Differential Equation with an Empirical Component

For this example we will operate on a variant of the common descending exponential:

$$y = ce^{-bt} \qquad (7\text{-}25)$$

The first derivative of this equation is given by:

$$\frac{dy}{dt} = -cbe^{-bt} = -by \qquad (7\text{-}26)$$

Equation 7-25 is often used to express decay of material over time. The rate of decay, dy/dt, in say, gms/hr is proportional to the amount of material present, y. The proportionality constant, or decay constant, has units of 1/hrs, and y is in gms. We can rearrange Eq. 7-26 to a form similar to Eq. 7-19 by writing

$$\frac{dy}{dt} + by = 0 \qquad (7\text{-}27)$$

Now suppose we have empirical information that b is not actually constant in a particular problem but varies with y. Suppose further we have made measurements to show how b varies with y and from this developed an empirical function b(y). Consider further that Eq. 7-27 allows only for decay. We can make the equation more versatile if we allow material to be added to the system while decay is taking place. This can be done by changing the zero value on the right-hand side of Eq. 7-27 to a(t). Now a(t) is a discrete function of t and can vary for each match point. In other words, we can add (or subtract) material to the system match point by match point. We then have as the differential equation

$$\frac{dy}{dt} + b(y)y = a(t) \tag{7-28}$$

The discrete values of a(t) must have units gms/hr.

Computer program B-7.2 was prepared to solve this equation in the geometry of Fig. 7-16. The equation was solved four times, varying b(y), a(t), and the left boundary value. Values of b(y) for the four runs are shown in Table 7-13. Run number 2 has a constant value of b(y) and, therefore, reduces to the simple solution, Eq. 7-25. Values of b(y) for Run numbers 1, 3, and 4 have low values of b(y) for low y. Therefore, the rate of decay will be more rapid for high y and will diminish rapidly with low y. These statements are relative to normal exponential decay. Runs number 3 and 4 will have more rapid decay than Run number 1.

Values of the system input function, a(t), are given in Table 7-14. The 49 match points span six hours of time. Therefore, any match point can be converted to hours by $(mp-1)/4$.

Runs number 1 and 2 of Eq. 7-28 are displayed in Fig. 7-18. Runs number 3 and 4 are displayed in Fig. 7-19. For Runs number 1 and 2, node number 3 was set to a value of 4.0 for the boundary condition. For Runs number 3 and 4, the value was set to 20.0. The values of R and $\hat{R}$ for the 49 match points for Run number 2 are shown at the top of Fig. 7-18. The close agreement between R and $\hat{R}$ is about the same as it was for Run number 1. At the bottom of Fig. 7-18 the solutions of Eq. 7-28 show a sluggish decay for Run number 1. This decay curve is interrupted by the response to the input values of a(t) for match points 15 through 25. The more rapid decay for Run number 2 exhibits only a slight response to the inputs at about 2.5 hours.

The agreement between R and $\hat{R}$ for the 49 match points for Run number 4 is shown at the top of Figure 7-19. Agreement is good when one realizes that the continuous values of $\hat{R}$ must necessarily "round the corners" of the discrete values of R. Inputs for Run number 3 are zero; therefore, R is zero for all match points and $\hat{R}$ agrees closely. At the bottom of Fig. 7-19, the two solutions are similar. Decay is initially rapid, but becomes very slow below a value of about 5.0 for y. Run number 4 responds only slightly to the input values a(t) for match point numbers 11 through 22.

7.6 SECOND ORDER NON-LINEAR DIFFERENTIAL EQUATIONS

Examples 7-11 and 7-12 illustrated the solution of first-order non-linear differential equations. Both of these examples were based on the geometry of Fig. 7-16. Node number 3 was fixed for the single boundary condition required for a first-order equation.

Second-order differential equations can also be solved by non-linear least squares. The geometry of Fig. 7-16 will be retained in the following two examples; however, node number 15 will now also be fixed, thus providing two boundary conditions for the second-order equations. The number of nodes to be found by iteration is decreased from 16 to 15.

It must be made clear that other geometries and other boundary conditions could be used. In practice, the boundary conditions will be determined by known characteristics of the problem. This is true for any method of solution. The number of nodes necessary to define the solution within the chosen problem space can also be changed from the 17 nodes of Fig. 7-16. This figure serves only to illustrate methodology.

7.6.1. Example 7-13: Demonstration of Solution of a Second-Order Equation

For this example we will again assume an equation and operate on it to produce a second-order differential equation. We will then solve the differential equation by non-linear least squares to demonstrate that we do, in fact, approximate the assumed function that was the starting point. The six-point sliding polynomial will be used, taking advantage of its continuous second derivative.

The assumed function is

$$y = a \sin(bx) \exp(-fx) \tag{7-29}$$

This is the equation of damped harmonic motion, a sine wave that gradually diminishes in amplitude. The equation with a selected set of parameters is illustrated as a connected set of points in Fig. 7-20. These points will be check points of the test. Parameter values used are noted in the figure.

We specify the differential equation as:

$$\frac{d^2y}{dx^2} + gy\,\frac{dy}{dx} = R \tag{7-30}$$

From Eq. 7-29 we have

$$\frac{dy}{dx} = a\ [-f\ \sin(bx) + b\ \cos(bx)]\ \exp(-fx) \tag{7-31}$$

and

$$\frac{d^2y}{dx^2} = a\ [(f^2-b^2)\ \sin(bx)\ -\ 2bf\ \cos(bx)]\ \exp(-fx) \tag{7-32}$$

From a set of parameters and Eqs. 7-29, 7-31, and 7-32, we can compute R in Eq. 7-30. A value of 0.2 is used for g. Also, given a set of 17 nodes, as in Fig. 7-16, we can approximate $d^2y/dx^2 + gy\ dy/dx$ using the six-point sliding polynomial. Both R and $\hat{R}$ must be evaluated for the 49 match points.

Program B-7.3 was prepared to solve this differential equation. The program was run three times as a test. The positioning of the runs for values of the right side R are shown in Fig. 7-20.

The confirmation of Run number 1 is shown in Fig. 7-21. The required left and right boundary values were taken as the true values from Eq. 7-29. The correspondence between R and $\hat{R}$ is so close for the 49 match points that they cannot be distinguished in the plot. Also the correspondence between the approximating sliding polynomial and the true function, Eq. 7-29, is so close as to make them indistinguishable. For example, the value of the node at x = 110 is 1.343. The true value of Eq. 7-29 at this point is 1.331.

The confirmation of Run number 2 is shown in Fig. 7-22. The left and right boundary values again were set with values of R and $\hat{R}$ and between true values and approximating values of the function is still very good relative to the scales involved, but is less precise than for Run number 1. For example, at x = 22, y from Eq. 7-29 is 1.000, while $\hat{y}$ from the approximating six-point sliding polynomial is 1.251. We should note that the main difference between Run number 1 and Run number 2 is the amount of detail we attempt to approximate. Run number 1 in Fig. 7-21 shows a configuration with little detail and a clear definition of features by the number of nodes used. Run number 2, on the other hand, shows more detail, particularly in the match points, and this degree of detail cannot be precisely approximated by the number of nodes for the specified range of x.

The match-point and function approximations for Run number 3 are shown in Fig. 7-23. The lack of close correspondence between R and $\hat{R}$ and between y and $\hat{y}$ shows that this is an unacceptable solution. The specified problem range of $120 \leq x \leq 192$ contains a great amount of detail. This detail cannot be approximated with sufficient precision with the number and spacing of nodes given by the geometry of Fig. 7-16.

In summary, Runs number 1 and 2 of this example show that non-linear least-squares optimization of the six-point sliding polynomial will provide a satisfactory solution of the second-order non-linear differential equation of Eq. 7-29 when nodal spacing is close enough to define the detail of the sine waves. Run number 3 shows that the method will not produce an acceptable solution when the nodes are not spaced closely enough to define the sine wave.

The observation in the preceding paragraph should be generalized. The spacing of nodes must always be close enough to define features under study, whether these are sine waves or other functional variations. If features are not defined with sufficient detail, one can either shorten the problem range in x and maintain a geometry such as in Fig. 7-16 or one can change the geometry, incorporating additional nodes.

7.6.2 Example 7-14: A Second-Order Differential Equation with Unknown Solution

In Example 7-13 we assumed a function to be a solution to an equation, and then applied non-linear least squares to test whether we would in fact, approximate this known solution. Within the limits governed by detail in definition of the functional features, the method tested satisfactory. In practice, of course, we do not known the solution. There would be no need to approximate it if we did. In this example we will not know an explicit solution. We will derive approximations to an unknown function.

For our differential operation we will keep the same form as in Example 7-13 but use a different right side:

$$\frac{d^2y}{dx^2} + gy\frac{dy}{dx} = a(x_m) + bx \tag{7-33}$$

in which $a(x_m)$ is a discrete function of x by match points as already introduced in Eq. 7-28 in Example 7-12. Following the problem geometry of Fig. 7-16, 49 match-point values of $a(x_m)$ are needed.

If one retains the geometry of Fig. 7-16, then only a slight change is necessary in the computer program from Example 7-13. Only the calculation of R, the equation right side, must be changed. In Example 7-13, R was calculated as the differential operation on Eq. 7-29. In this example R is the right side of Eq. 7-33. This program is included as Program B-7.4 in the Appendix.

Three runs of the computer program for Eq. 7-33 were performed. The equation parameters are given in Table 7-15. Values of $a(x_m)$ for Runs number 2 and 3 are given in Table 7-16. For Run number 1 $a(x_m)$ was held at a constant value of 0.03.

The approximate solution of Eq. 7-33 for Run number 1 is shown in Fig. 7-24. The 49 match points are shown at the top of the figure. The agreement between R and $\hat{R}$ is very close and only distinguishable at the right side. The functional approximation is shown at the bottom of the figure. Left and right boundary values of -1.0 and 2.8, respectively, were used. The solution is a smoothly curving set of values with a minimum near $x = 5$, a relatively simple feature. We now argue that the six-point sliding polynomial differential operation on this set of nodes produces the match-point agreement at the top. Therefore, it is a possible solution of Eq. 7-33.

Run number 2 is displayed in Fig. 7-25. The match points at the top show values of R with one major break in slope at match point number 20. This is a slightly more complex pattern than in Fig. 7-24. Any differences between R and $\hat{R}$ are too small to show on the graph. The approximate solution at the bottom of the figure falls from the left boundary value of zero to a minimum near $x = 3.0$, and then climbs to the right boundary value of 1.6. As in the previous example, this is a relatively simple feature. Again, we accept the functional approximation as a possible solution for Eq. 7-33 for the given boundary values and parameters.

The third run in solution of Eq. 7-33 is shown in Fig. 7-26. The match points at the top show a further increase in complexity of pattern. The agreement of R and $\hat{R}$ is generally good. A bounce in $\hat{R}$ occurs during the rapid rise and fall at match point numbers 24 and 33. The functional approximation at the bottom is a simple monotonically rising set of values. The nodal values rise fairly steeply from the left boundary value of 0 and then bend and pass through the right boundary value of 4.0. Again we judge that the approximation is a possible solution of Eq. 7-33 for the given boundaries and parameters.

7.7 SUMMARY OF THE NON-LINEAR LEAST-SQUARES METHOD

The discussion of the non-linear least-squares approach to a solution of non-linear differential equations is intended to instruct and demonstrate. The examples were developed to show varieties of possible forms and solutions rather than as solutions to particular problems.

Many assumptions were made for the development of demonstrations for this methodology. Perhaps most important is that the forms of the non-linear differential equations were assumed. It must be thoroughly understood that in solution of real problems the differential equation is the result of rigorous statements of the processes controlling the problem. Also we assumed equation parameters, we assumed problem ranges in x or t, and we assumed the boundary values. Again, in solution of real problems appropriate values come from the problem environment.

Lastly, the demonstrations were based on the problem geometry of Fig. 7-16. A fixed number of nodes of the six-point sliding polynomial were used. Boundaries were set by fixing nodal values. This procedure was used only for simplicity and ease in programming. Many other configurations are possible. The number of nodes can be changed, or the boundary conditions can be changed to reflect real problems. For example, boundary condition may be required to be fixed by the slope of the function rather than by its value. Lastly, the six-point sliding polynomial was used to take advantage of its continuous second derivative. For a first-order differential equation the four-point sliding polynomial could be used.

These demonstrations concluded with Example 7-14. In this example we did not know a specific solution to the problem. We only have the derived functional approximation. This is also the case with real problems. We do not know a specific solution. If we can get good agreement between R and $\hat{R}$ for the match points, then we have a mathematically possible solution. However, the scientist must judge, based on his or her knowledge of the processes and their environment, whether it is a rational solution. In critical work the effect of changing the initial estimate of the nodes should be explored. This is feasible because of the extremely rapid decay of the nodal corrections during successive rounds of the iterative solution. This rapid convergence of the corrections to 0 is illustrated in Table 7-17. Three selected nodes from Run Number 3 of Example 7-14 reached stable values in just three rounds.

EXERCISES

7-1. Steady flow occurs in a porous medium between two water bodies. Assuming a free surface without infiltration, the velocity of flow at a location where the depth of flow is h is given by:

$$V = \frac{K\left(h_0^2 - h_1^2\right)}{2Lh}$$

in which h_0 and h_1 are the water surface depths for the upper and lower water bodies, respectively; K is the hydraulic conductivity; and h is the horizontal distance between the two water bodies. Assume K = 6, L = 600, h_0 = 15, h_1 = 8. Using a two point approximation with four nodal points at 0, 200, 400, and 600, find the least squares approximation of the above function. Generate data points at the quarter-points of each of the three spans.

7-2. The elevation y of the bottom of the reservoir above some datum is defined by the function $y = 40e^{-x/150}$. Determine the volume of water per unit width as a function of x when the depth at x = 0 is 40 feet. Divide the length (x-scale) into three spans by the nodal points x = 0, 100, 200, and 300. Use a two-point approximation, with match points at the quarter-points of each span.

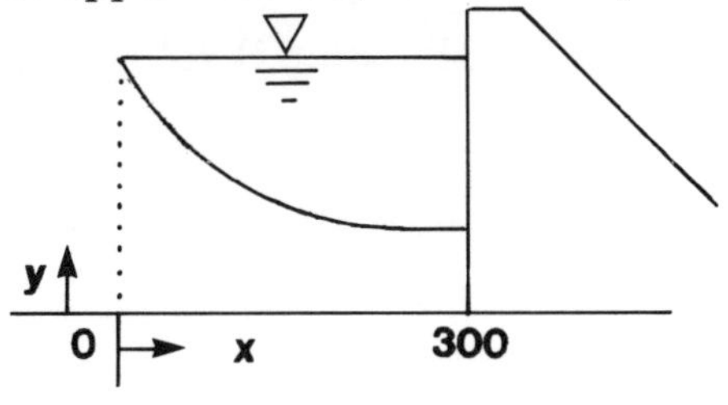

7-3. Estimate the volume for the case of problem 7-2 using a four-point approximation. As outlier nodes, use y = 60 at x = -100 and y = 0 at x = 400.

7-4. For steady unconfined radial flow toward a well, the slope of the water surface is defined by

$$\frac{dy}{dx} = \frac{Q}{2\pi xyK}$$

in which Q is the discharge rate, $2\pi xy$ is the area of a cylindrical surface with the well at its central point, x is the distance from the center point of the well, y is the distance above the impervious stratum, and K is the hydraulic conductivity. Using a four-point approximation compute the water table gradient for distances up to 300 feet from the well. Assume K = 1000 gpd/ft^2, Q = 500 gpm. The 12-inch well fully penetrates the unconfined aquifer of 80 feet depth. Observation wells at distances of 100, 200, and 300 feet from the discharging well have surfaces of 60, 63, and 65 feet, respectively, with the depth of 55 feet in the discharging well.

7-5. Following the procedures of Section 7.5 solve the non-linear differential equation

$$y \cdot \frac{dy}{dx} = 3x^2$$

Use the same specifications as for Eq. 7-21.

7-6. Expand the solution of Eq. 7-21 (Fig. 7-15) to two spans by adding a node at x = 8 and moving the outlier node to x = 12.

7-7. Program B-7.1 was used to solve iteratively Eq. 7-24. What changes would have to be made in this program to solve the equation

$$y \cdot \frac{dy}{dx} = a + bx + cx^2$$

7-8. Change the values for b(y) and a(t) in Tables 7-13 and 7-14 and run Program B-7.2. Compare with results of Example 7-12 in Figs. 7-18 and 7-19.

7-9. Example 7-9 demonstrated use of the two-point sliding polynomials to approximate Eq. 7-17. Re-calculate this example using the four-point sliding polynomials.

7-10. Left and right boundary values were used in solution of the second-order differential equations in Examples 7-13 and 7-14. Why are two boundary values used? Consider changing the two boundary conditions to one boundary value and a boundary slope at the left edge. Can this problem be solved? Would the solution at the right edge be sensitive to errors made in boundary and boundary slope at the left?

Table 7-1
Least-Squares Solution For Example 7-1

Match Point	x	z for span	c(z) Y_0	Y_2	Y_4	Y_6	y = (1-.3x)exp(-.3x)
1	0.0	-0.5	1.0	0	0	0	1.0
2	0.5	-0.25	0.75	0.25	0	0	0.7316
3	1.0	0.0	0.5	0.5	0	0	0.5186
4	1.5	0.25	0.25	0.75	0	0	0.3507
5	2.0	0.5	0	1.0	0	0	0.2195
6	2.5	-0.25	0	0.75	0.25	0	0.1181
7	3.0	0.0	0	0.5	0.5	0	0.0407
8	3.5	0.25	0	0.25	0.75	0	-0.0175
9	4.0	0.5	0	0	1.0	0	-0.0602
10	4.5	-0.25	0	0	0.75	0.25	-0.0907
11	5.0	0.0	0	0	0.5	0.5	-0.1116
12	5.5	0.25	0	0	0.25	0.75	-0.1248
13	6.0	0.5	0	0	0	1.0	-0.1322
Least-squares Solution:			0.9529	0.1743	-0.0732	-0.1396	

Table 7-2
Least-Squares Solution For Example 7-2

Data Point	x	z for Span	$2\int_{-0.5}^{z} C(z)dz$				$\int_{o}^{x} y\,dx =$ $x\exp(-.3x)$
			Y_0	Y_2	Y_4	Y_6	
1	0.0	-0.5	0	0			0
2	0.5	-0.25	0.4375	0.0625			0.4304
3	1.0	0.0	0.75	0.25			0.7408
4	1.5	0.25	0.9375	0.5625			0.9564
5	2.0	0.5	1.0	1.0			1.0976
			1.0	1.0			
		-0.25		0.4375	0.0625		
6	2.5		1.0	1.4375	0.0625		1.1809
			1.0	1.0			
		0.0		0.75	0.25		
7	3.0		1.0	1.75	0.25		1.2197
			1.0	1.0			
		0.25		0.9375	0.5625		
8	3.5		1.0	1.9375	0.5625		1.2248
			1.0	1.0			
		0.5		1.0			
9	4.0		1.0	2.0	1.0		1.2048
			1.0	2.0	1.0		
		-0.25			0.4375	0.0625	
10	4.5		1.0	2.0	1.4375	0.0625	1.1666
			1.0	2.0	1.0		
		0.0			0.75	0.25	
11	5.0		1.0	2.0	1.75	0.25	1.1157
			1.0	2.0	1.0		
		0.25			0.9375	0.5625	
12	5.5		1.0	2.0	1.9375	0.5625	1.0563
			1.0	2.0	1.0		
		0.5			1.0	1.0	
13	6.0		1.0	2.0	2.0	1.0	0.9918
Least-Squares Solution			0.9212	0.1800	-0.0749	-0.1400	

Table 7-3
Least-Squares Solution For Example 7-3

Data Point	x	Z for Span	$2\int_{-1/2}^{Z} C(Z)\,dZ$ Y_2	Y_o	Y_2	Y_4	Y_6	Y_8	$\int_0^x y\,dx =$ $x\,\exp(-.3x)$
1	0.0	-0.5	0	0	0	0			0
2	0.5	0.25	-.0219	.4768	.0492	-.0042			0.4304
3	1.0	0.0	-.0572	.8386	.2448	-.0260			0.7408
4	1.5	0.25	-.0792	.0342	.6064	-.0616			0.9564
5	2.0	0.5	-.0834	.0834	1.0834	-.0834			1.0976
			-.0834	.0834	1.0834	-.0834			
		-0.25		-.0219	.4768	.0492	-.0042		
6	2.5		-.0834	1.0615	1.5602	-.0342	-.0042		1.1809
			-.0834	1.0834	1.0834	-.0834			
		0.0		-.0572	.8386	.2448	-.0260		
7	3.0		-.0834	1.0262	1.9220	.1614	-.0260		1.2197
			-.0834	1.0834	1.0834	-.0834			
		0.25		-.0792	1.0342	.6064	-.0616		
8	3.5		-.0834	1.0042	2.1176	.5230	-.0616		1.2248
			-.0834	1.0834	1.0834	-.0834			
		0.5		-.0834	1.0834	1.0834	-.0834		
9	4.0		-.0834	1.0	2.1668	1.0	-.0834		1.2048
			-.0834	1.0	2.1668	1.0	-.0834		
		-.25			-.0219	.4768	.0492	-.0042	
10	4.5		-.0834	1.0	2.1449	1.4768	-.0342	-.0042	1.1666
			-.0834	1.0	2.1668	1.0	-.0834		
		0.0			.0572	.8386	.23448	-.0260	
11			-.0834	1.0	2.1668	1.0	-.0834	-.0260	1.1157
			-.0834	1.0	2.1	2.1668	1.0	-.0834	
		0.25	-	-	-.0792	1.0342	.6064	-.0616	
12	5.5		-.0834	1.0	2.0876	2.0342	.5230	-.0616	1.0563
			-.0834	1.0	2.1668	1.0	-.0834		
		0.5			1.0834	1.0834	1.0834	-.0834	
13	6.0		-.0834	1.0	2.0834	2.0834	1.0	-.0834	0.9918
Least-Squares Solution:			2.9258	1.0139	0.2226	-.0586	-0.1353	-0.1834	

Table 7-4 Least-Squares Solutions for Example 7-4

Data Point	x	Z for Span	1/2 d C(Z)/dZ Y_2	Y_o	Y_2	Y_4	Y_6	Y_8	dy/dx = (-.6+.09x)exp(.3x)	For Y_o = 1	Right hand Vector
1	0.0	-0.5	-.25	0	-.25	0			-.6	0	-.6
2	0.5	-0.25	-.0468	-.4844	.6094	-.0781			-.4777	.4844	.0067
3	1.0	0.0	.0625	-.6875	.6875	-.0625			-.3778	.6875	.3097
4	1.5	0.25	.0781	-.6094	.4844	.0468			-.2965	.6094	.3129
5	2.0	0.5	0	-.25	0	.25			-.2305	.25	.0197
6	2.5	-0.25		-.0468	.4844	.6084	-.0781		-.1771	.0468	-.1303
7	3.0	0.0		.0625	-.6875	.6875	-.0625		-.1342	-.0625	-.1967
8	3.5	0.25		.0781	-.6094	.4844	.0468	-.0997	-.0781	-.0781	-.1778
9	4.0	0.5		0	-.25	0	.25		-.0723	0	-.0723
10	4.5	-0.25			.0468	.4844	.6094	-.0781	-.0506	0	-.0506
11	5.0	0.0			.0625	-.6875	.6875	-.0625	-.0335	0	-.0335
12	5.5	10.25			.0781	-.6094	.4844	.0468	-.0202	0	-.0202
13	6.0	0.5			0	-.25	0	.25	-.0099	0	-.0099
Least-Squares Solution:			2.6229	1.0*	0.2219	-0.0549	-0.1261	-0.0972			

* Set boundary value

Table 7-5 Least-Squares Solution for Example 7-5

Data Point	x	Z for Span	1/4 $d^2C(Z)/dZ^2$ Y_4	Y_2	Y_o	Y_2	Y_4	Y_6	Y_8	Y_{10}	$d^2y/dx^2 =$ (27-.027x)exp(-.3x)	For Y_o = 1	For Y_6 = .1322	Right hand Vector
1	0.0	-0.5	-.0208	0.3333	-.6250	.3333	.0208	0			.27	.6250	0	.8950
2	0.5	-0.25	-.0762	.5241	-.8971	.5586	-.1413	.0379			.2208	.8971	.0050	1.1229
3	1.0	0.0	-.0260	.2031	-.1771	-.1771	.3031	-.0260			.1800	.1771	-.0034	.3537
4	1.5	0.25	.0379	-.1413	.5586	-.8971	.5241	-.0762			.1463	-.5586	-.0101	-.4224
5	2.0	0.5	0	-.0208	.3333	-.6250	.3333	-.0208	0		.1185	-.3333	-.0028	-.2176
6	2.5	-0.25		-.0762	.5241	-.8971	.5586	-.1413	.0379		.0957	-.5241	-.0187	-.4471
7	3.0	0.0		-.0260	.2031	-.1771	-.1771	.2031	-.0260		.0768	-.2031	.0268	.0995
8	3.5	0.25		.0379	-.1413	.5586	-.8971	.5241	-.0762		.0614	.1413	.0693	.2720
9	4.0	0.5		0	-.0208	.3333	-.6250	.3333	-.0208	0	.0488	.0208	.0441	.1137
10	4.5	-0.25			-.0762	.5241	-.8871	.5586	-.1413	.0379	.0385	.0762	.0738	.1885
11	5.0	0.0			-.0260	.2031	-.1771	-.1771	.2031	-.0260	.0301	.0260	-.0234	.0327
12	5.5	0.25			.0379	-.1413	.5586	-.8971	.5241	-.0762	.0233	-.0379	-.1186	-.1332
13	6.0	0.5			0	-.0208	.3333	-.6250	.3333	-.0208	.0179	0	-.0826	-.0674
Least-Squares Solution:			6.7064	2.8687	1.0*	0.2191	0.0597	-0.1322*	0.1052	0.1120				

*Set boundary value

Table 7-6
Least-Squares Solutions For Examples 7-6 and 7-7

x	dc(z)/dz						Example 7-6	Example 7-7	
	Y_{-4}	Y_o	Y_4	Y_8	Y_{12}	Y_{16}	4* dm/dx	4* M	4*dM/dx-4* M
m.							gm/m	gm/m	gm/m
0	-.5	0	.5	0			9.6		9.6
1	-.09375	-.96875	1.21875	-.15625			12.2		12.2
2	.125	-1.375	1.375	-.125			14.8		14.8
3	.15625	-1.21875	.96875	.09375			17.4		17.4
4	0	-.5	0	.5	0		20.0		20.0
5		-.09375	-.96875	1.21875	-.15625		22.6	1	21.6
6		.125	-1.375	1.375	-.125		25.3	2	23.3
7		.15625	-1.21875	.96875	.09375		28.0	4	24.0
8		0	-.5	0	.5	0	30.8	8	22.8
9			.09375	-.96875	1.21875	-.15625	33.5	4	29.5
10			.125	-1.375	1.375	-.125	36.3	2	34.3
11			.15625	-1.21875	.96875	.09375	39.2	1	38.2
12			0	- .5	0	.5	42.1		42.1
Example 7-6 Solution	-4.41	0.0*	14.80	40.11	76.47	124.34			
Example 7-7 Solution	-4.27	0.0*	14.84	37.38	70.89	121.51			

Table 7-7
Calculation Of Composite Matrix

Unit Variate z	Unit Set Node				
	Y_1	Y_2	Y_3	Y_4	
-.5	-.5	0	.5	0	dc(z)/dz
	0	1.0	0	0	c(z)
	-.5	0.2	.5	0	dc(z)/dz+0.2c(z)
-.25	-.09375	-.96875	1.21875	-.15625	dc(z)/dz
	-.07031	.86719	.22656	-.02344	c(z)
	-.10781	-.79531	1.25406	-.16094	dc(z)/dz+0.2c(z)
.0	.125	-1.375	1.375	-.125	dc(z)/dz
	-.0625	.5625	.5625	-.0625	c(z)
	.1125	-1.2625	1.4875	-.1375	dc(z)/dz+0.2c(z)
.25	.15625	-1.21875	.96875	.09375	dc(z)/dz
	-.02344	.22656	.86719	-.07031	c(z)
	.15156	-1.17344	1.14219	.07969	dc(z)/dz+0.2c(z)
.5	0	-.5	0	.5	dc(z)/dz
	0	0	1.0	0	c(z)
	0	-.5	.2	.5	dc(z)/dz+0.2c(z)

Table 7-8
Least-Squares Solution For Example 7-8

x	dc(z)/dz + .2c(z)						
x	Y_{-4}	Y_0	Y_4	Y_8	Y_{12}	Y_{16}	4*dm/dx
m.							
0	-.5	.2	.5	0			9.6
1	-.10781	-.79531	1.25406	-.16094			12.2
2	.1125	-1.2625	1.4875	-.1375			14.8
3	.15156	-1.17344	1.14219	.07969			17.4
4	0	- .5	.2	.5	0		20.0
5		- .10781	- .79031	1.25406	-.16094		22.6
6		.1125	-1.2625	1.4875	-.1375		25.3
7		.15156	-1.17344	1.14219	.07969		28.0
8		0	- .5	.2	.5	0	30.8
9			- .10781	-.79531	1.25406	-.16094	33.5
10			.1125	-1.2625	1.4875	-.1375	36.3
11			.15156	-1.17344	1.14219	.07969	39.2
12			0	- .5	.2	.5	42.1
Least-Squares Solution:	-5.66	|0.0|**	13.62	34.36	61.38	93.84	

** Set Boundary Condition

Table 7-9
Least-Squares Solution For Example 7-9

Data Point	Coefficients for Node Y_1	Y_2	Y_3	Y_4	Y_5	Y_6	$y = 2 + 3wx + 4w^2x^2$
1	1.0	0		0	0		2.0
2	0.5	0.5		0	0		4.5
3	0	1.0		0	0		9.0
4		0.5	0.5	0	0		15.5
5		0	1.0		0	0	24.0
6	0.5	0		0.5	0		2.0
7	0.25	0.25		0.25	0.25		6.5
8	0	0.5		0	0.5		15.5
9		0.25	0.25		0.25	0.25	29.0
10		0	0.5		0	0.5	47.0
11	0	0		1.0	0		2.0
12	0	0		0.5	0.5		9.0
13	0	0		0	1.0		24.0
14		0	0		0.5	0.5	47.0
15		0	0		0	1.0	78.0
Least-Squares Solution:	1.738	8.143	22.405	0.881	21.429	75.548	

Table 7-10 Least-Squares Solution For Example 7-10

TABLE 7-10. LEAST SQUARES SOLUTION FOR EXAMPLE 7-10

Verticle Span	Horizontal Span	Point No.	Coefficients for Node Y_1	Y_2	Y_3	Y_4	Y_5	Y_6	Y_7	Y_8	Y_9	Y_{10}	Y_{11}	Y_{12}	y/2*
1	1	1	0.375	0.125	0	0	0	0	0	0	0	0	0	0	0.370
1	1	2	0.5	0.5			0	0							.549
1	1	3	0.1875	0.0625			0.1875	0.0625							.402
1	1	4	0.25	0.25			0.25	0.25							.573
1	1	5	0	0			0.375	0.125							.425
1	1	6	0	0			0.5	0.5							.588
1	2	7	0.5	0.875	0.125		0	0	0						.610
1	2	8	0.5	1.0	0.5		0	0	0						.603
1	2	9	0.25	0.4375	0.0625		0.25	0.4375	0.0625						.613
1	2	10	0.25	0.5	0.25		0.25	0.5	0.25						.582
1	2	11	0	0	0		0.5	0.875	0.125						.610
1	2	12	0	0	0		0.5	1.0	0.5						.562
1	3	13	0.5	1.0	0.875	0.125	0	0	0	0					.558
1	3	14	0.5	1.0	1.0	0.5	0	0	0	0					.496
1	3	15	0.25	0.5	0.4375	0.0625	0.25	0.5	0.4375	0.0625					.519
1	3	16	0.25	0.5	0.5	0.25	0.25	0.5	0.5	0.25					.444
1	3	17	0	0	0	0	0.5	1.0	0.875	0.125					.486
1	3	18	0	0	0	0	0.5	1.0	1.0	0.5					.403
2	1	19					0.1875	0.0625	0	0	0.1875	0.0625			.443
2	1	20					0.25	0.25	0	0	0.25	0.25			.597
2	1	21					0	0	0	0	0.375	0.125			.458
2	1	22					0	0	0	0	0.5	0.5			.602
2	2	23					0.25	0.4375	0.0625	0	0.25	0.4375	0.0625		.604
2	2	24					0.25	0.5	0.25	0	0.25	0.5	0.25		.542
2	2	25					0	0	0	0	0.5	0.875	0.125		.597
2	2	26					0	0	0	0	0.5	1.0	0.5		.524
2	3	27					0.25	0.5	0.4375	0.0625	0.25	0.5	0.4375	0.0625	.457
2	3	28					0.25	0.5	0.5	0.25	0.25	0.5	0.5	0.25	.369
2	3	29					0	0	0	0	0.5	1.0	0.875	0.125	.432
2	3	30	0	0	0	0	0	0	0	0	0.5	1.0	1.0	0.5	.341
Least-Squares Solution:			0.952	0.147	-0.041	-0.185	1.093	0.097	-0.167	-0.154	1.203	0.016	-0.185	-0.183	

* $y/2 = w^{0.3}\ x \exp(-0.3\ w^{0.3}\ x)/2$

Table 7-11 Illustration of Non-linear Least Squares Details of Computation

			Initial Nodes: 0, 1, 2, 3			Nodes: 0.05, 1, 2, 3			Nodes: 0, 1, 2.05, 3			Nodes: 0, 1, 2, 3.05		
Match Point	x	R=x/2	x	dy/dx	$\hat{R} = y \cdot dy/dx$	y	dy/dx	$\hat{R}_1 = y \cdot dy/dx$	y	dy/dx	$\hat{R}_3 = y \cdot dy/dx$	y	dy/dx	$\hat{R}_4 = y \cdot dy/dx$
1	0	0	1.0	0.25	.25	1.0	.24375	.24375	1.0	.25625	.25625	1.0	.25	.25
2	1	0.5	1.25	0.25	.3125	1.24648	.24883	.31016	1.26133	.26523	.33455	1.24883	.24805	.30977
3	2	1.0	1.5	0.25	.375	1.49688	.25156	.37656	1.52812	.26719	.40830	1.49688	.24844	.37188
4	3	1.5	1.75	0.25	.4375	1.74883	.25195	.44062	1.79336	.26211	.47006	1.74648	.25117	.43867
5	4	2.0	2.0	0.25	.5	2.0	.25	.5	2.05	.25	.5125	2.0	.25625	.5125

Partial Derivatives and Error

Match Point	$\frac{\hat{R}_1 - \hat{R}}{.05}$	$\frac{\hat{R}_3 - \hat{R}}{.05}$	$\frac{\hat{R}_4 - \hat{R}}{.05}$	$\varepsilon = R - \hat{R}$
1	-.125	.125	0	-.125
2	-.0468	.441	-.0546	.1875
3	.0312	.666	-.0624	.625
4	.0629	.6512	.0234	1.0625
5	0	.25	.25	1.5

Least-Squares Solution:	$DY_1 = 2.9757$,	$DY_3 = 1.2294$,	$DY_4 = 4.7132$
Initial Nodes:	0	2.0	3.0
Updated Nodes:	2.9757	3.2294	7.7132

Table 7-12
Continuation Of Non-Linear Least Squares

Computed Values of R

Match Point	$\hat{R}_1$	$\hat{R}_3$	$\hat{R}_4$	$\hat{R}$
1	.02859	.03484	.03171	.03171
2	.44680	.45951	.44676	.44815
3	1.05829	1.08144	1.05561	1.05792
4	1.87218	1.90182	1.87033	1.87028
5	2.70995	2.73093	2.72004	2.70995

Partial Derivatives and Error

Match Point	$\frac{\hat{R}_1 - \hat{R}}{.025}$	$\frac{\hat{R}_3 - \hat{R}}{.025}$	$\frac{\hat{R}_4 - \hat{R}}{.025}$	$E = R - \hat{R}$
1	-.1248	.1252	0	-.03171
2	-.0540	.4544	-.0556	.05185
3	.0148	.9408	-.0924	-.05792
4	.0760	1.2676	.0020	-.37028
5	0	.8392	.4036	-.70995

Least-Squares Solution:	DY_1 = -.3975,	DY_3 = -.2243	DY_4 = -1.3334
First-Round Nodes:	2.9757	3.2294	7.7132
Updated Nodes:	2.5782	3.0051	6.3798

Round Three Solution	DY_1 = .0020,	DY_3 = -.0086,	DY_4 = -.0861
Updated Nodes	2.5802	2.9965	6.2937

Table 7-13
Empirical Functions For Example 7-12

y	Run #1	Run #2	Run #3 and #4
2	0.01	0.2	0.01
1	0.02	0.2	0.02
0	0.03	0.2	0.04
1	0.04	0.2	0.07
2	0.05	0.2	0.11
3	0.06	0.2	0.16
4	0.07	0.2	0.19
5	0.08	0.2	0.23
6	0.09	0.2	0.27
7	0.10	0.2	0.31
8	0.12	0.2	0.35
9	0.14	0.2	0.39
10	0.16	0.2	0.43
11	0.18	0.2	0.46
12	0.20	0.2	0.49
13	0.22	0.2	0.52
14	0.24	0.2	0.54
15	0.26	0.2	0.56
16	0.28	0.2	0.58
17	0.30	0.2	0.59
18	0.31	0.2	0.60
19	0.32	0.2	0.61
20	0.33	0.2	0.62
21	0.35	0.2	0.63
22	0.36	0.2	0.64

Table 7-14
Values Of a(t) - Example 7-12

Units are gms/hr

Match Points	Runs #1 and 2	Match Points	Run #4
0-14	0	0-10	0
15	.05	11	0.2
16	.10	12	0.6
17	.15	13	1.2
18	.20	14	1.8
19	.20	15	2.4
20	.20	16	3.0
21	.20	17	3.0
22	.15	18	2.4
23	.10	19	1.8
24	.05	20	1.2
25-49	0	21	0.6
		22	0.2
		23-49	0

Run # 3 - All Zero

Table 7-15
Parameters For Example 7-14

Parameter	Run 1	Run 2	Run 3
g	-0.25	0.30	0.40
b	0.20	0.10	0.0

Table 7-16

Values of Discrete Function $a(x_m)$ in Example 7-14

Run No. 2				Run No. 3			
x_m	$a(x_m)$	x_m	$a(x_m)$	x_m	$a(x_m)$	x_m	$a(x_m)$
1	.0	26	.26	1	.0	26	.32
2	.02	27	.24	2	.0	27	.40
3	.04	28	.22	3	.0	28	.44
4	.06	29	.20	4	.0	30	.46
5	.08	30	.18	5	.0	30	.46
6	.10	31	.16	6	.0	31	.44
7	.12	32	.14	7	.0	32	.40
8	.14	33	.12	8	.0	33	.32
9	.16	34	.10	9	.0	34	.16
10	.18	35	.08	10	.0	35	.08
11	.20	36	.06	11	.0	36	.04
12	.22	37	.04	12	.0	37	.02
13	.24	38	.02	13	.0	38	.01
14	.26	39	0	14	.0	39	0
15	.28	40	0	15	.0	40	0
16	.30	41	0	16	.0	41	0
17	.32	42	0	17	.0	42	0
18	.34	43	0	18	.0	43	0
19	.36	44	0	19	.0	44	0
20	.38	45	0	20	.0	45	0
21	.36	46	0	21	.01	46	0
22	.34	47	0	22	.02	47	0
23	.32	48	0	23	.04	48	0
24	.30	49	0	24	.08	49	0
25	.28			25	.16		

Table 7-17
Successive Values Of Selected Nodes
In Run #3 Of Example 7-14

Round	Node Number 6	9	12
initial	2.0	3.0	3.4
1	1.680143	2.897338	3.625725
2	1.671923	2.888074	3.621856
3	1.671923	2.888074	3.621856
4	1.671923	2.888074	3.621856

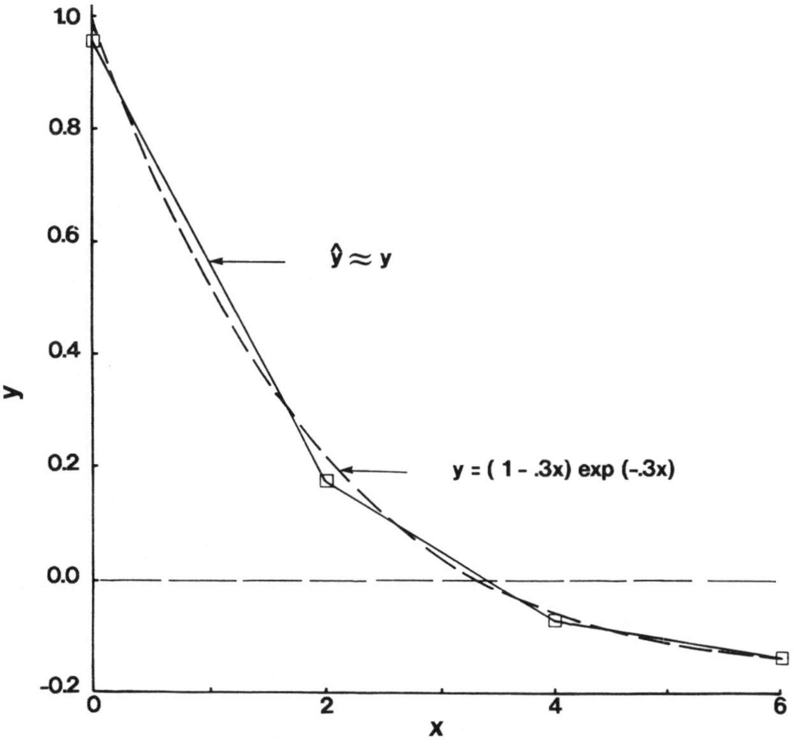

Fig. 7-1. The Functional Approximation of Example 7-1

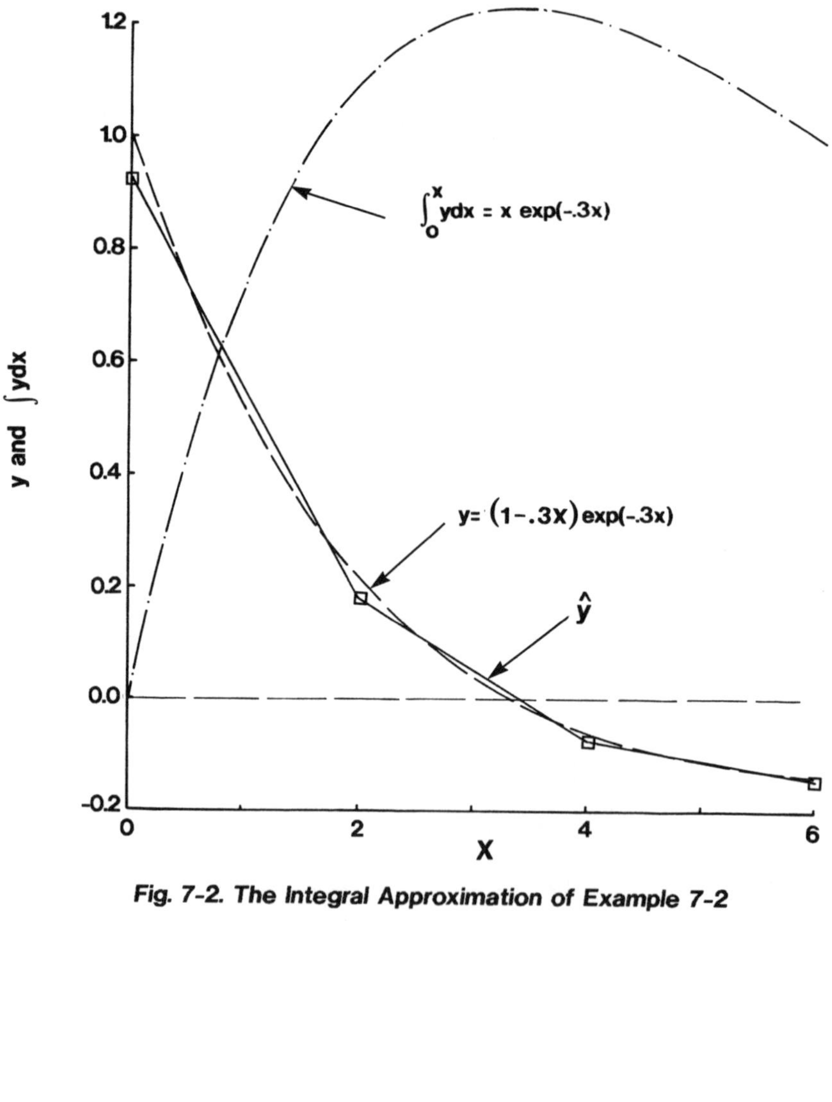

Fig. 7-2. The Integral Approximation of Example 7-2

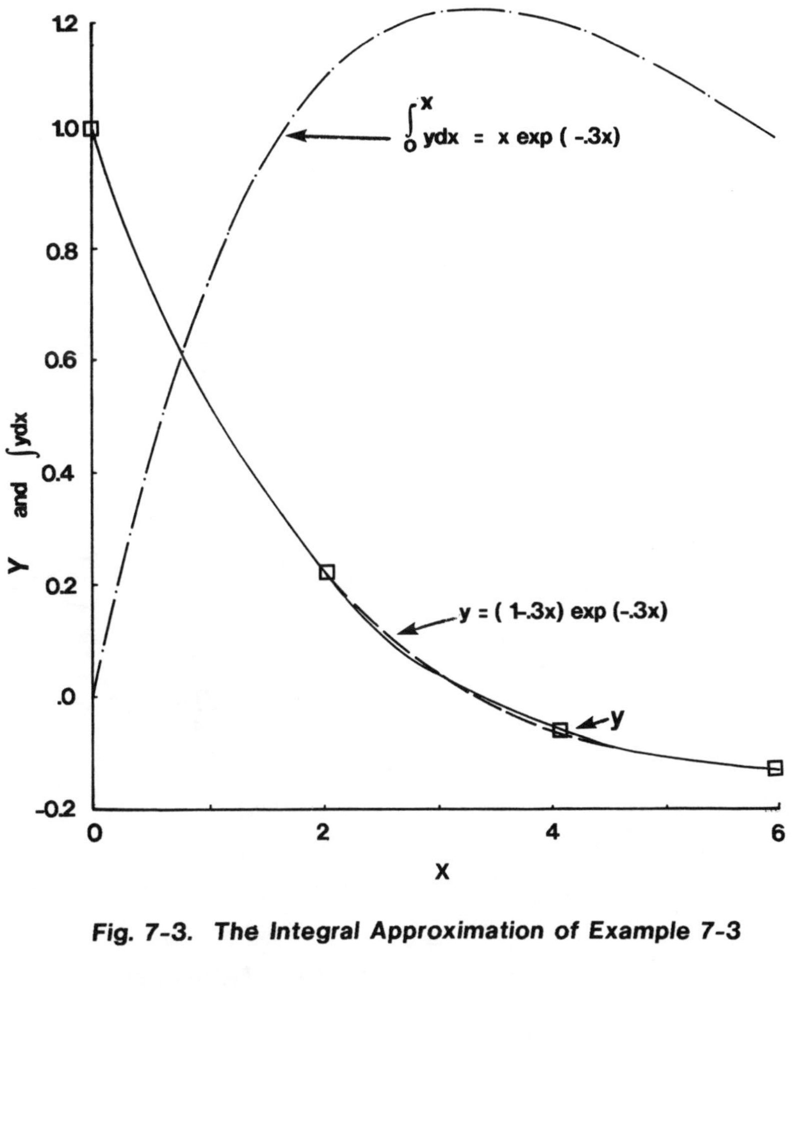

Fig. 7-3. The Integral Approximation of Example 7-3

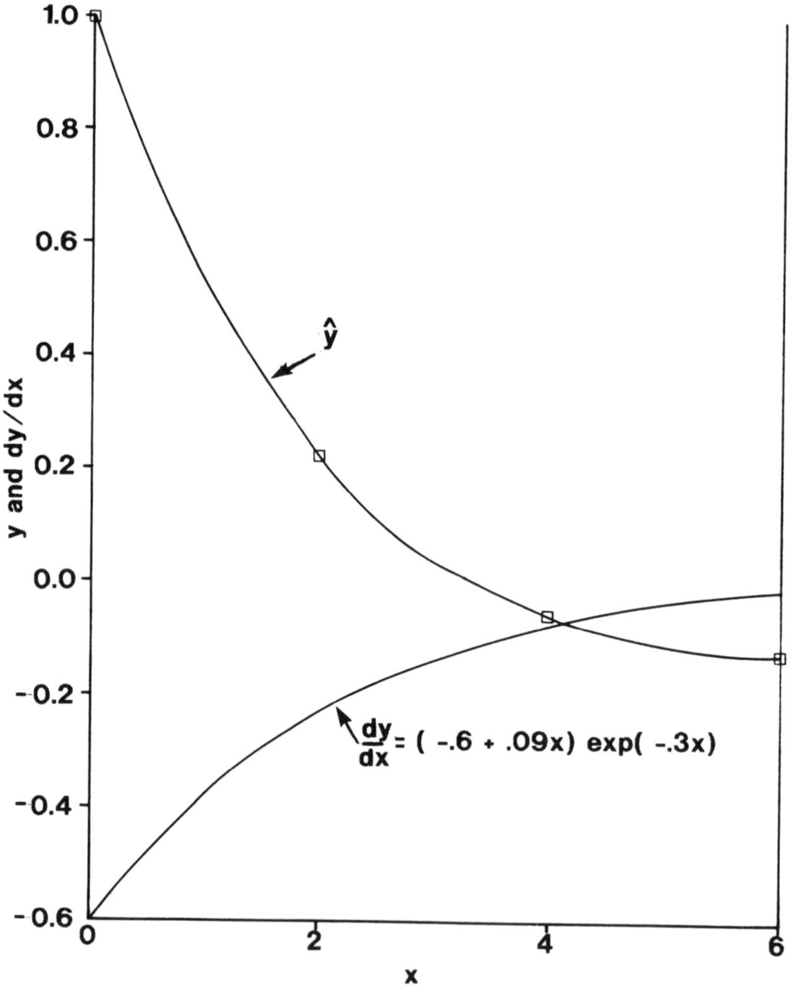

Fig. 7-4. The Differential Approximation of Example 7 - 4

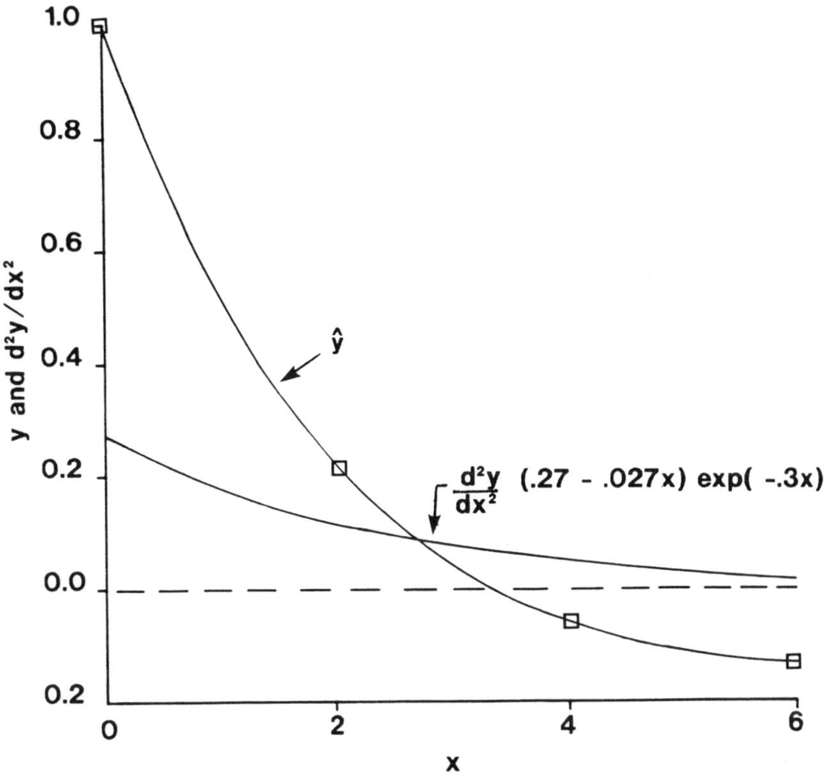

Fig. 7-5. The Differential Approximation of Example 7-5

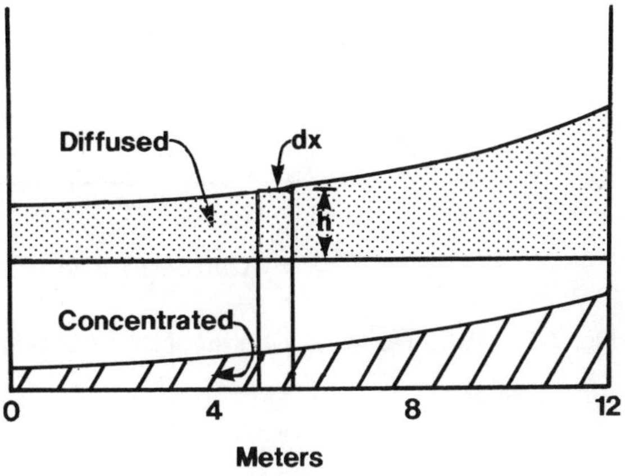

Fig. 7-6. Sketch for Example 7-6

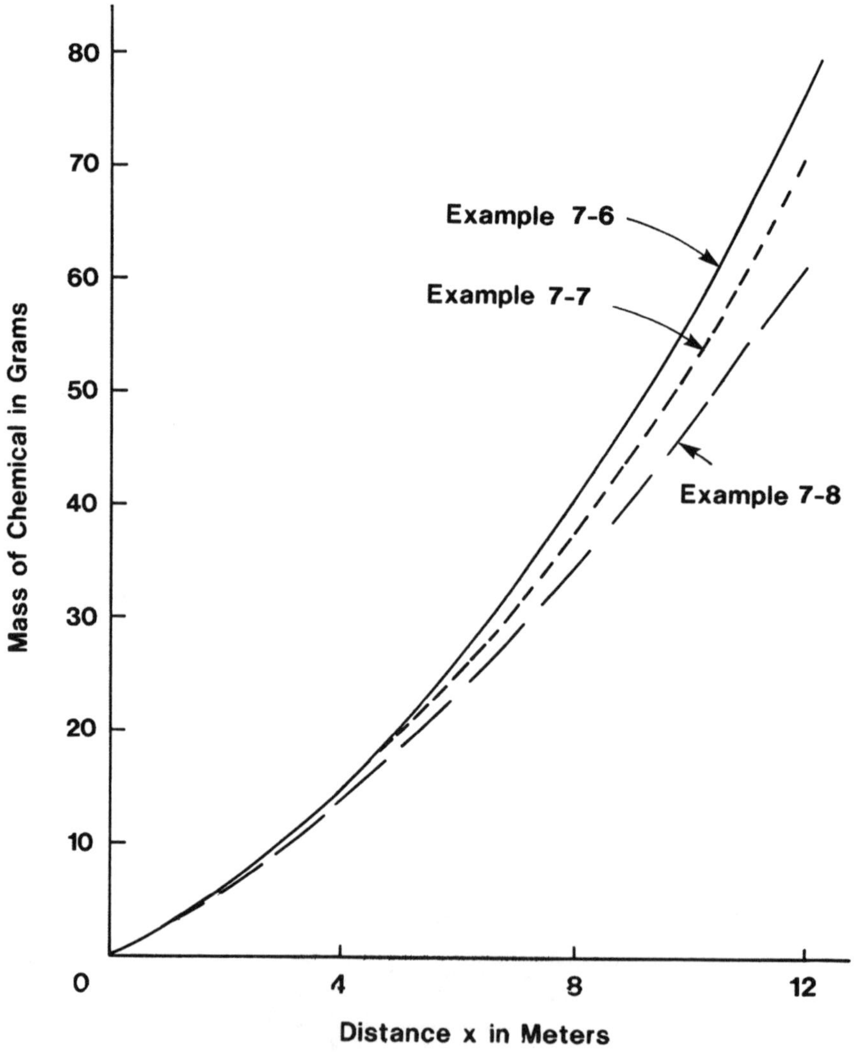

Fig. 7-7. Total Mass of Chemical as Found in Examples 7-6, 7-7, and 7-8

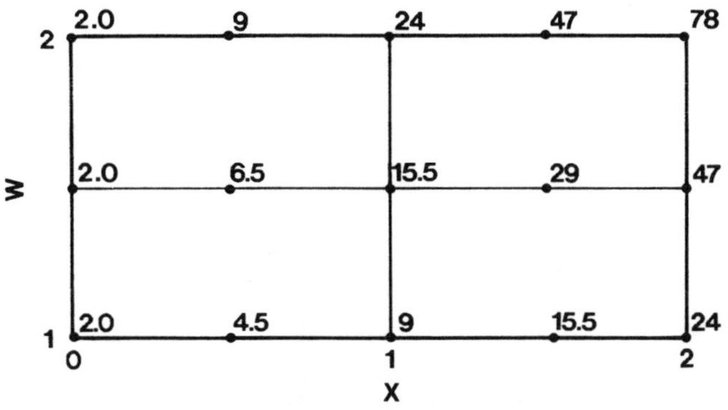

Fig. 7-8. **Values of Function to be Approximated in Example 7-9**

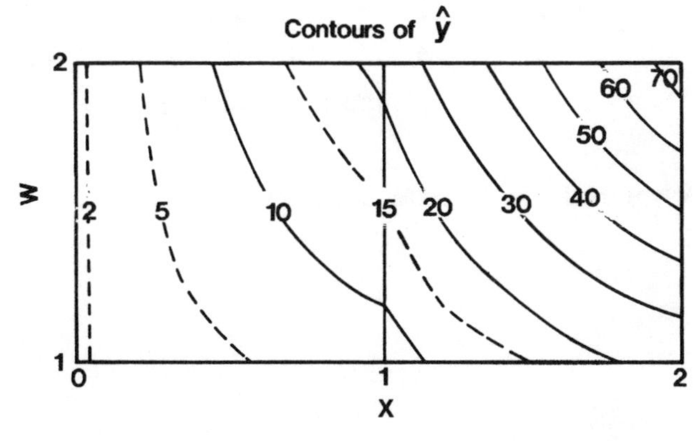

Fig. 7-10. **Contours of the Approximating Surface of Example 7-9**

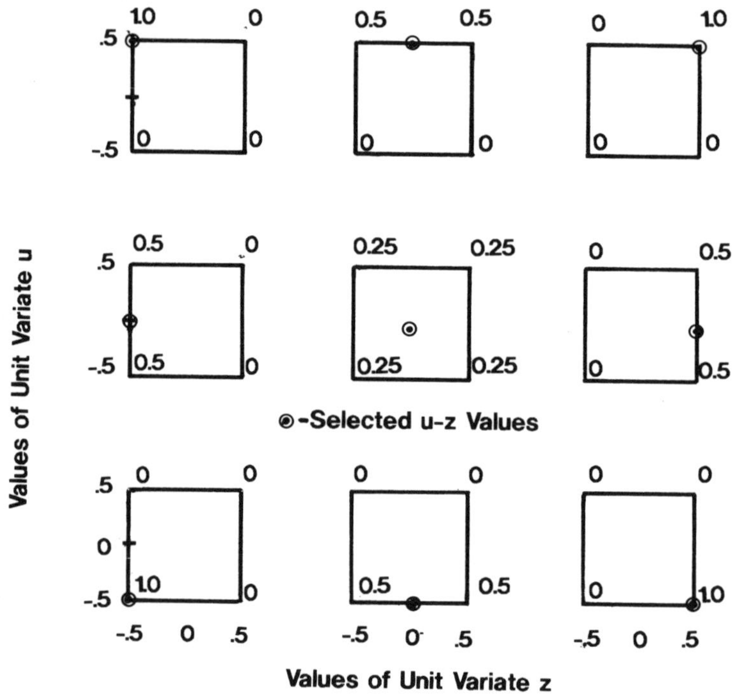

Fig. 7-9. **Sliding Polynomial Coefficients C(u)C(z) at Selected u-z Values in the Two-Point System**

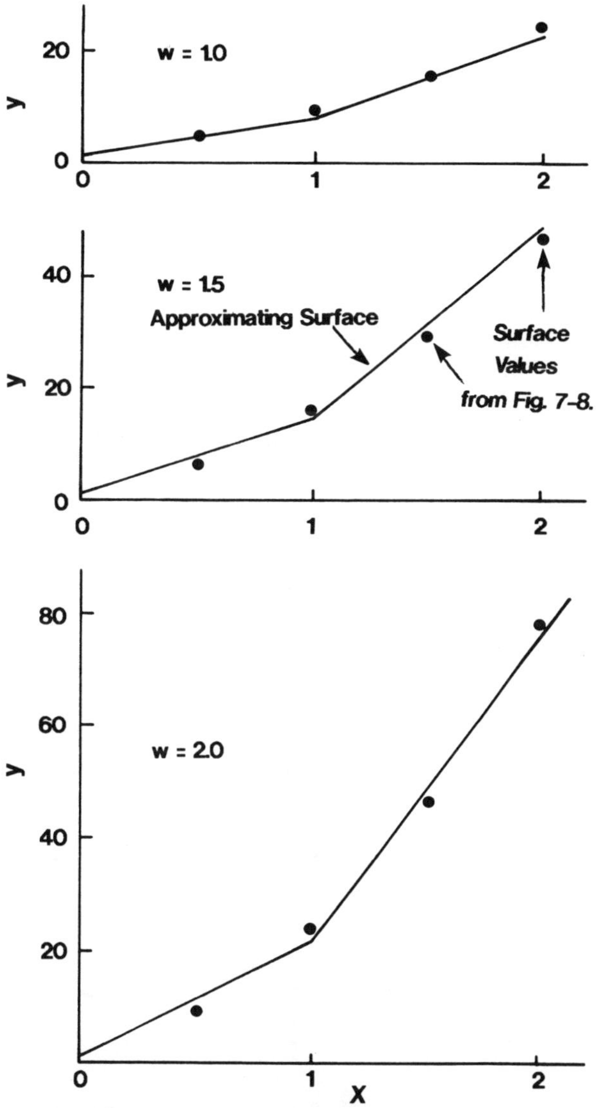

Fig. 7-11. Cross Sections Through the Approximating Surface of Example 7-9.

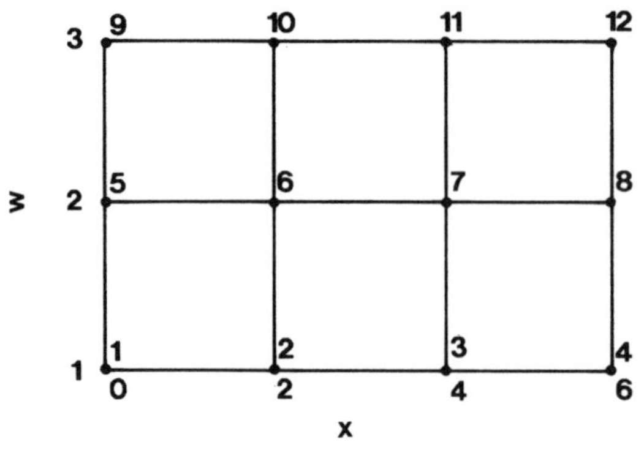

Fig. 7-12. **Numbering of Nodes for Example 7-10**

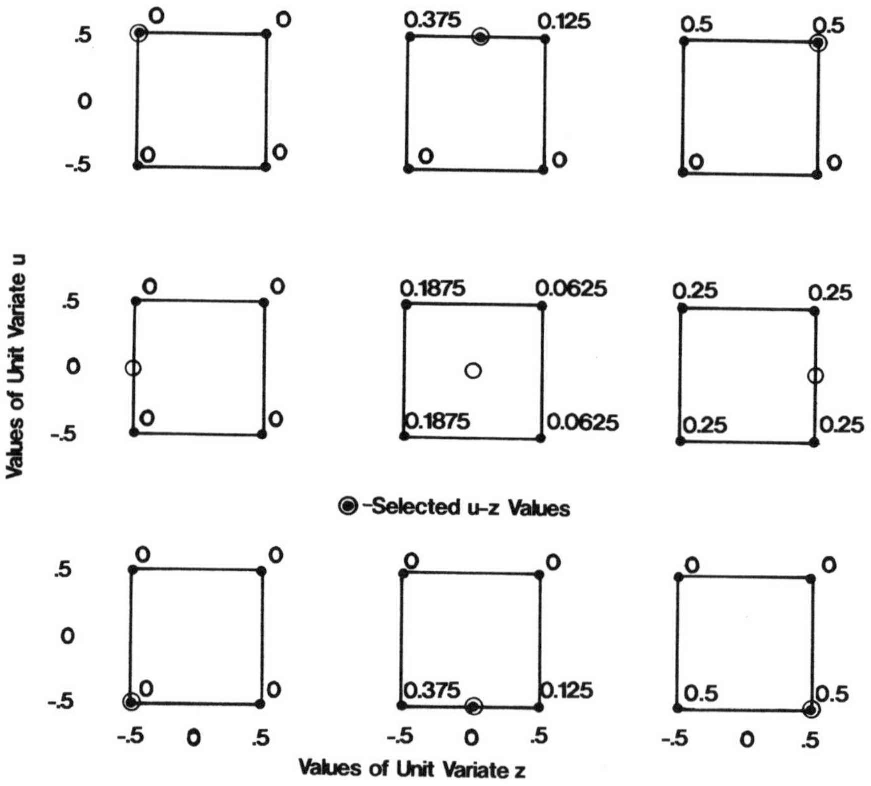

Fig. 7-13. **Sliding Polynomial Coefficients** $c(u)\int_0^z c(z)dz$ **at Selected u-z values in the Two-Point System**

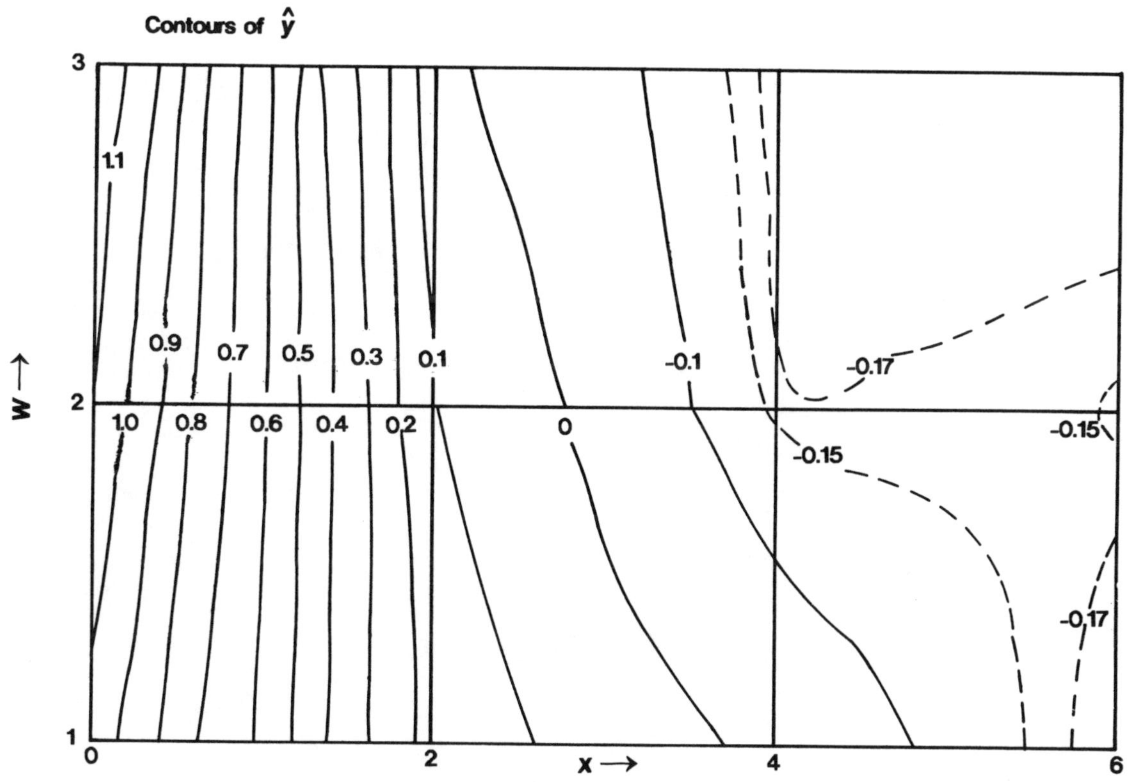

Fig. 7-14. Contours of the Approximating Surface of Example 7-10.

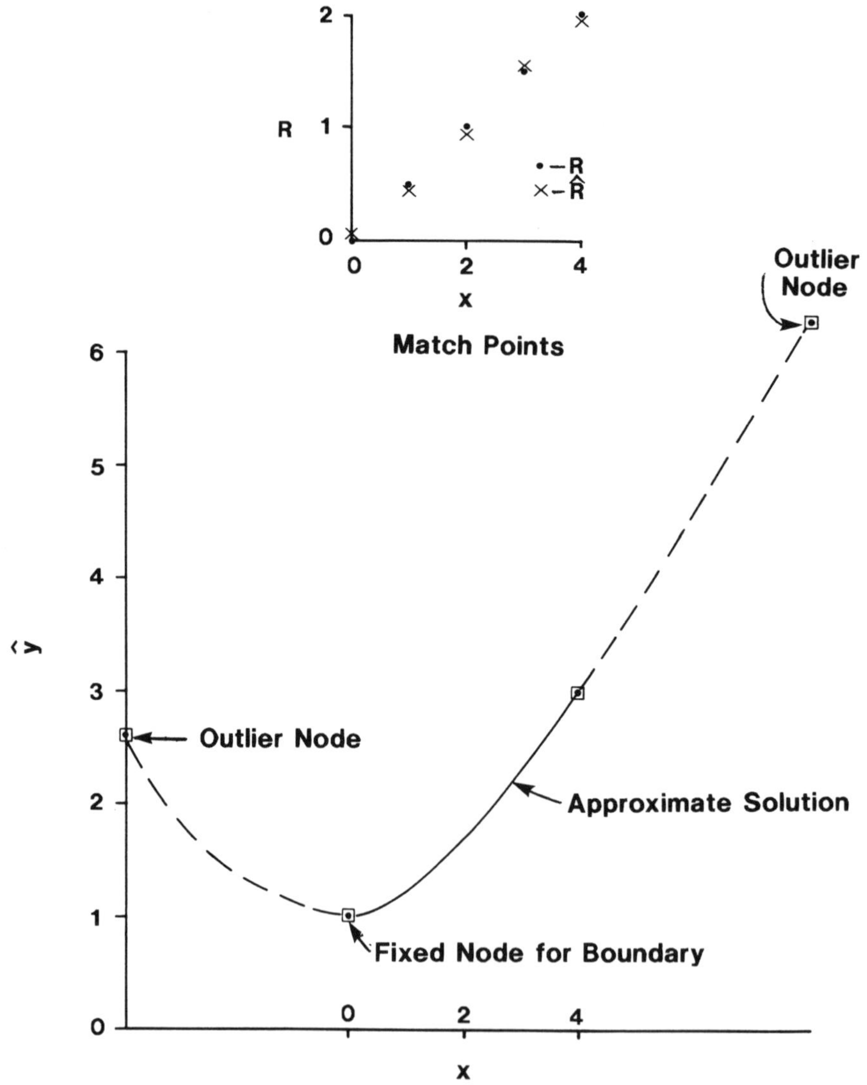

Fig. 7-15. Solution of Equation 7-21

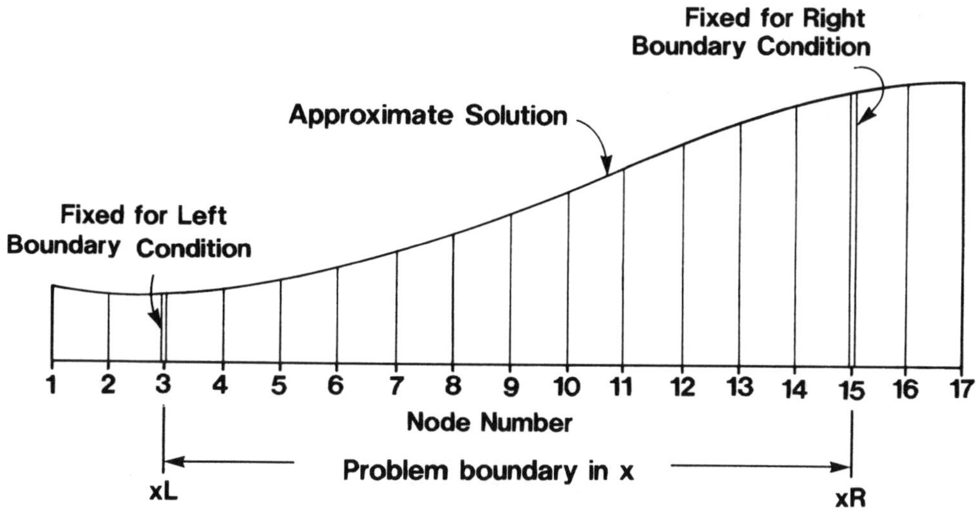

Fig. 7-16. Problem Boundaries for Examples

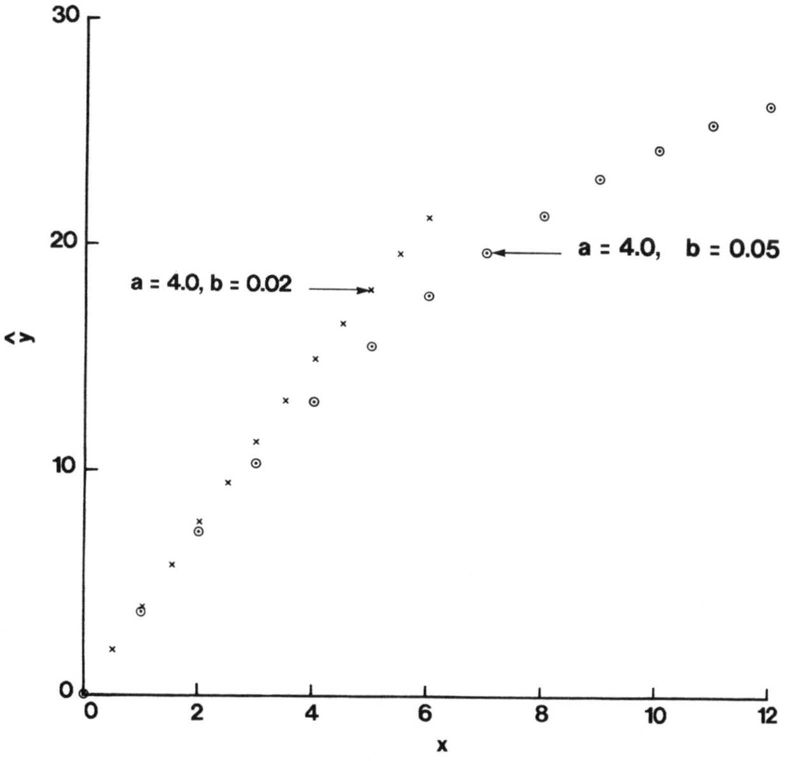

Fig. 7-17. Solutions of Equation 7-22

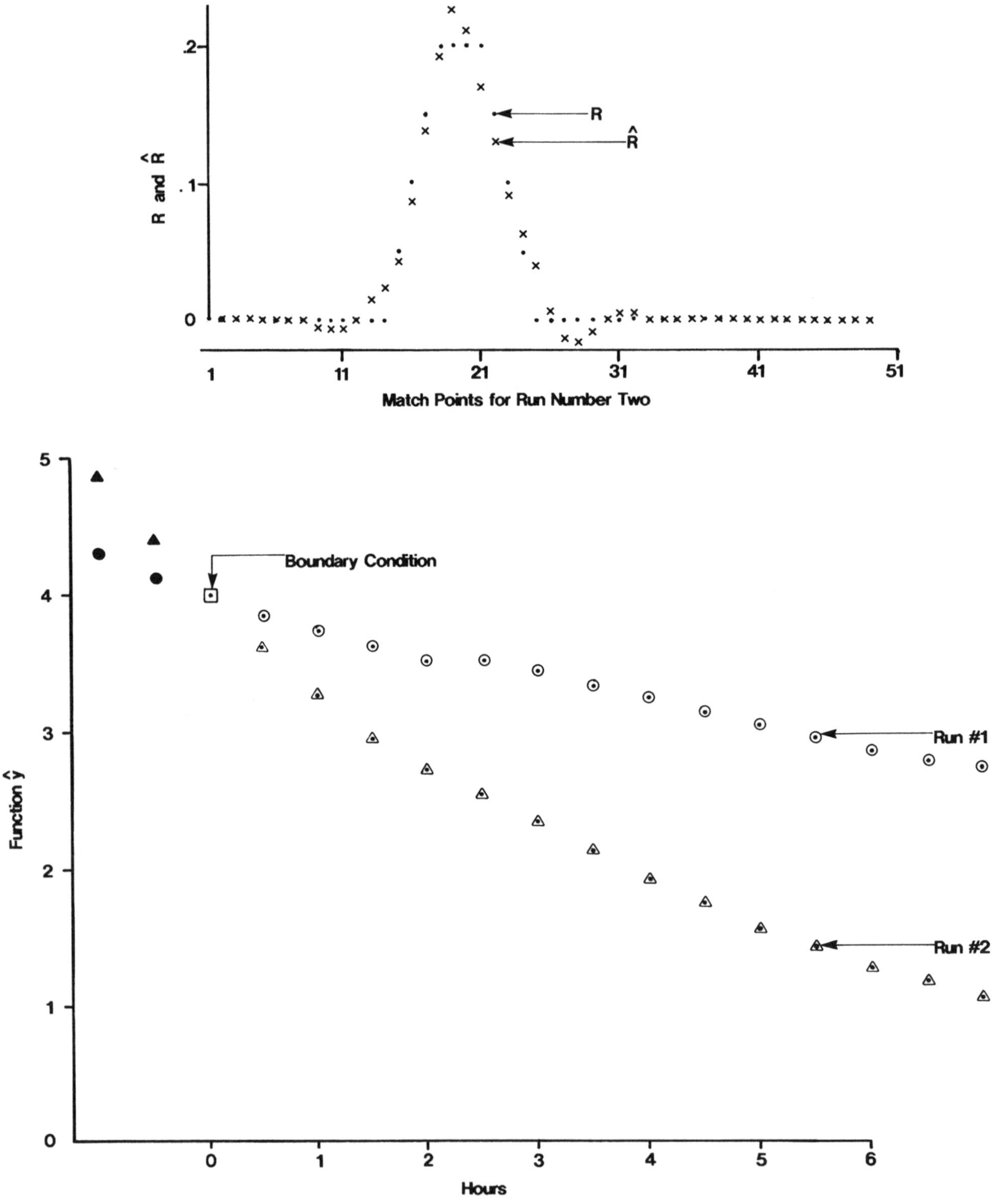

Fig. 7-18. Solutions One and Two of Equation 7-28

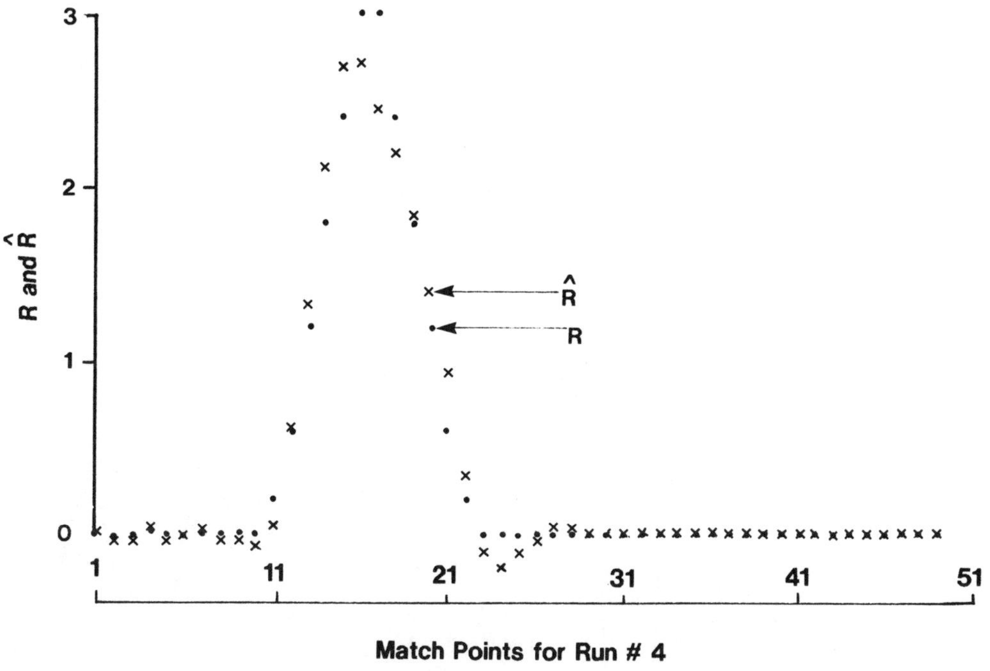

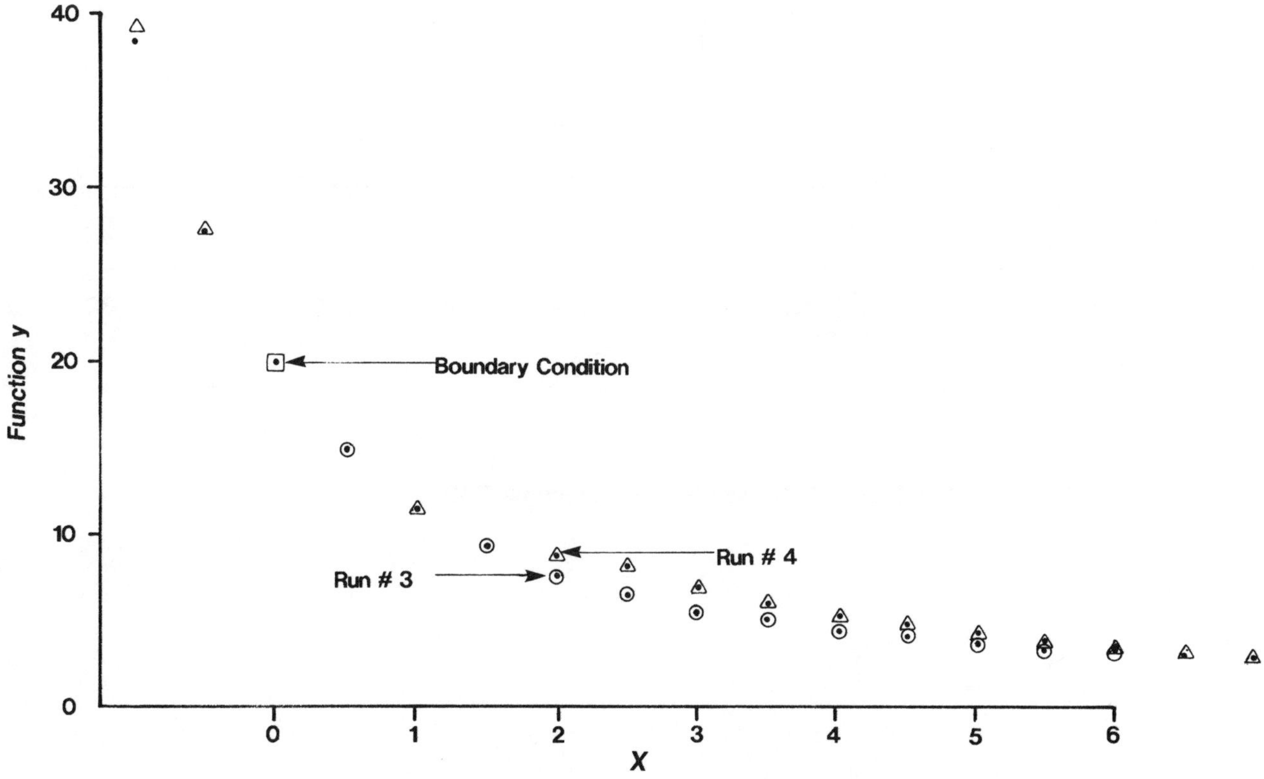

Fig. 7-19. Solutions Three and Four of Equation 7-28

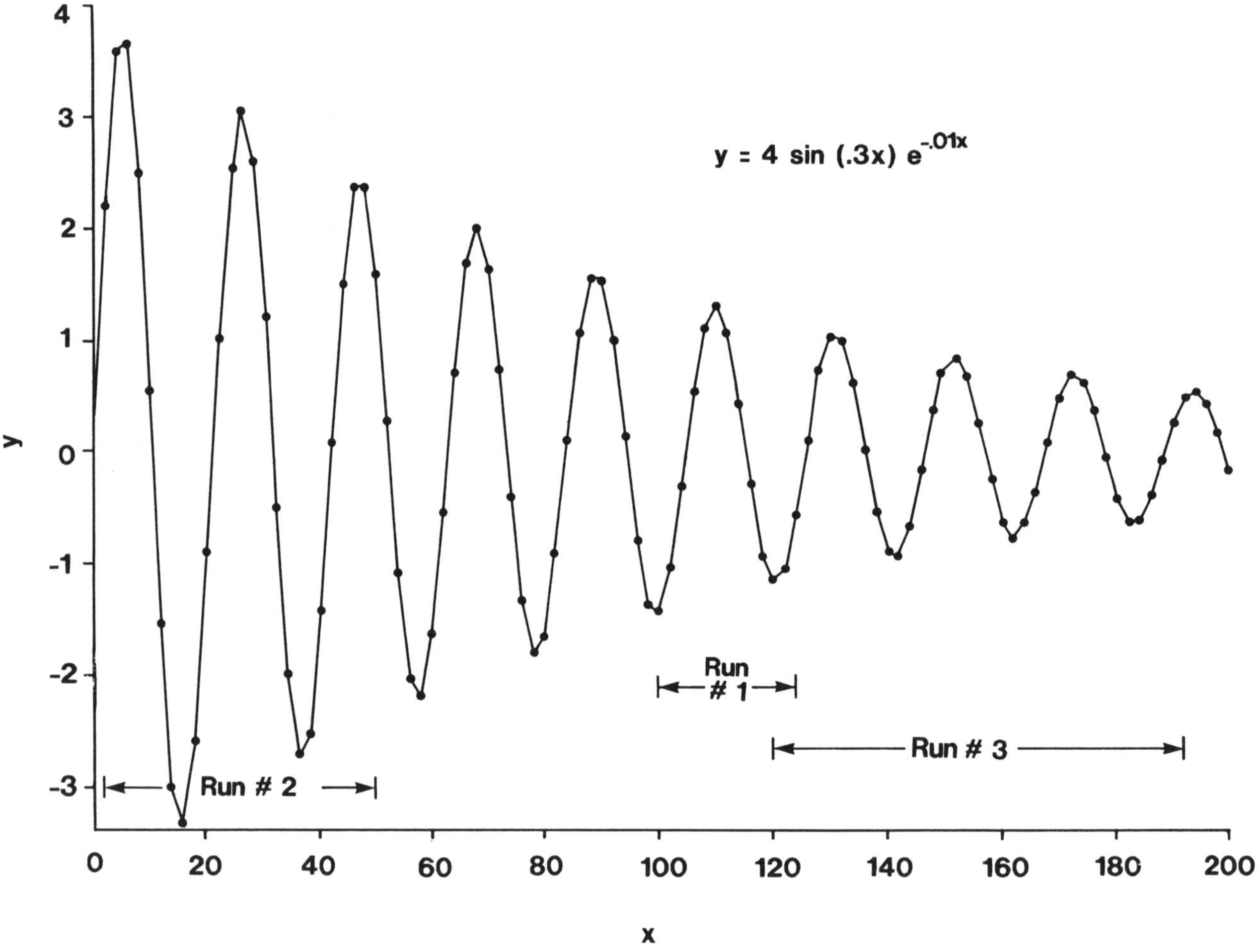

Fig. 7-20. Test Function for Example 7.13.

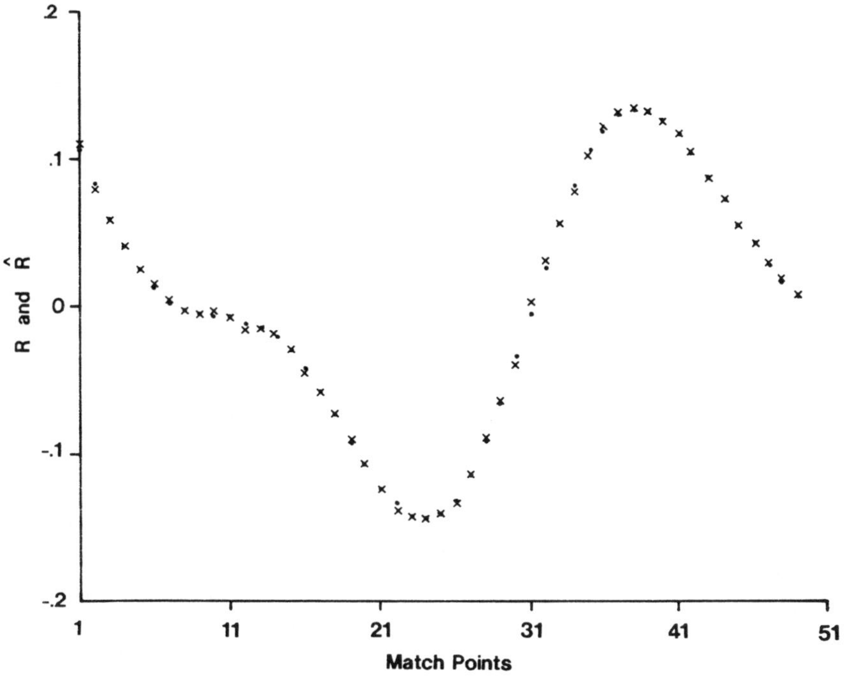

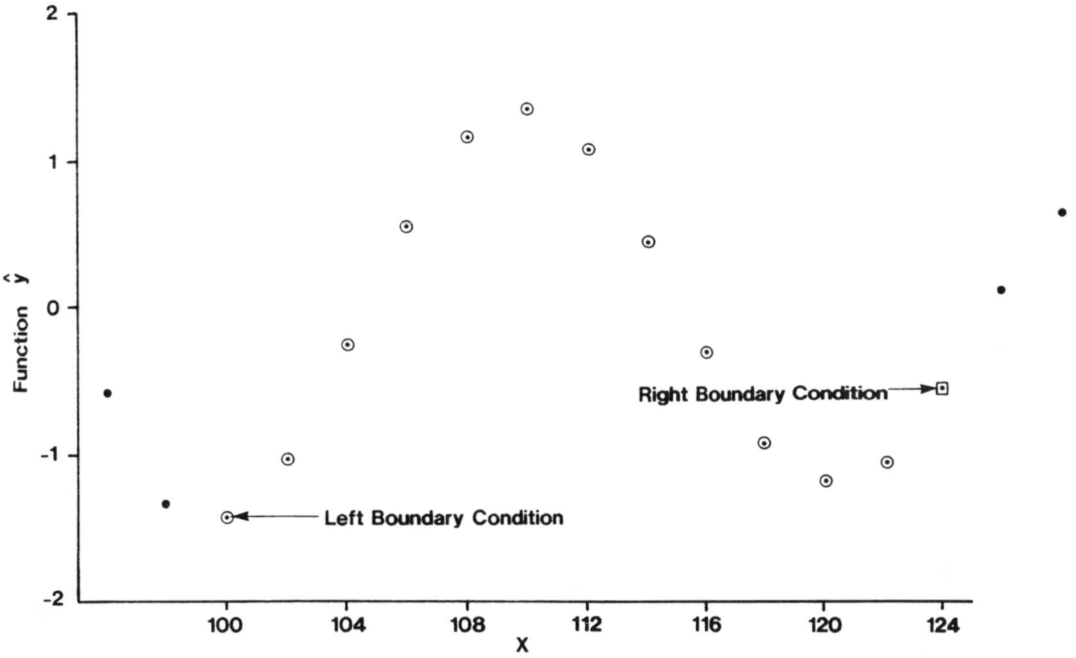

Fig. 7-21. Run Number One of Example 7-13.

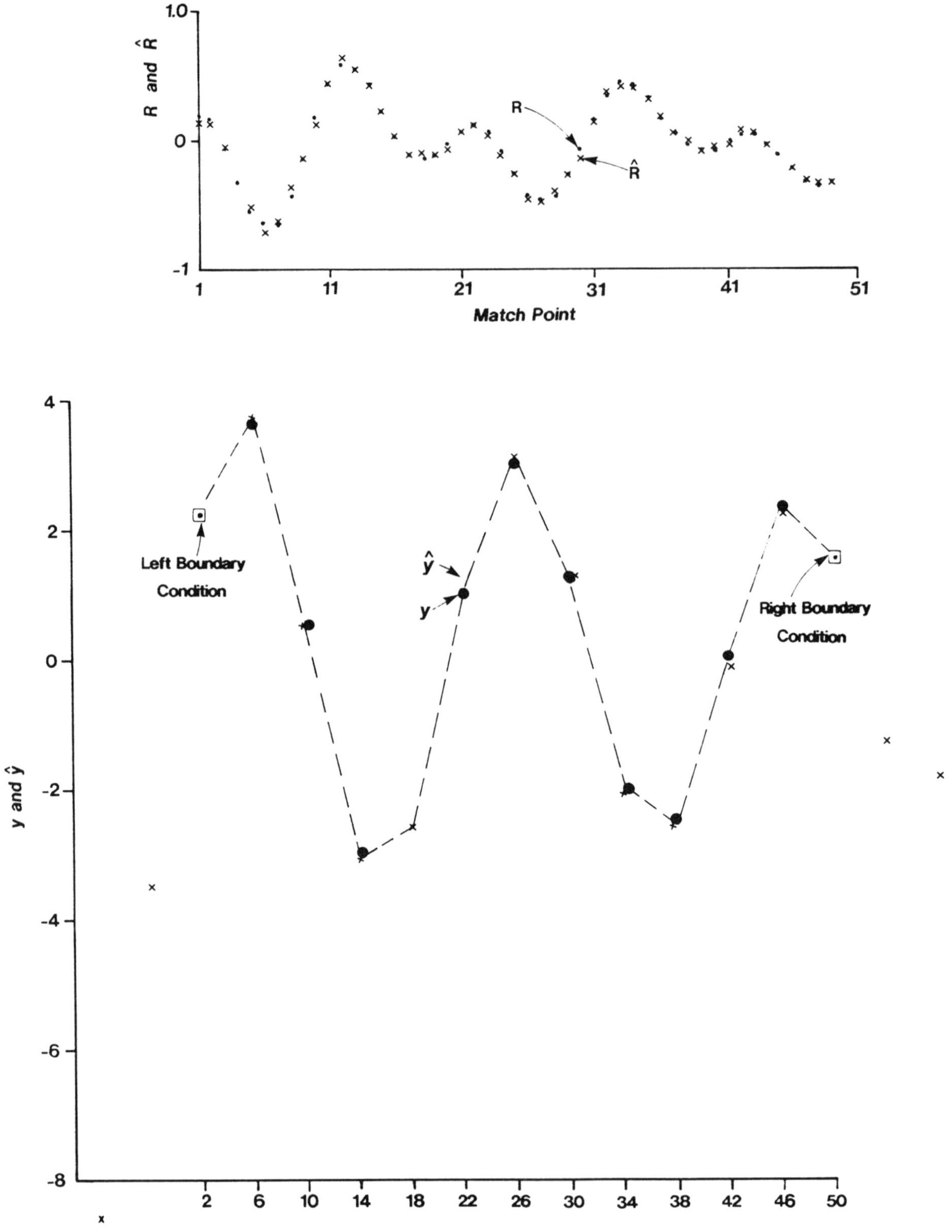

Fig. 7-22. Run Number Two of Example 7-13

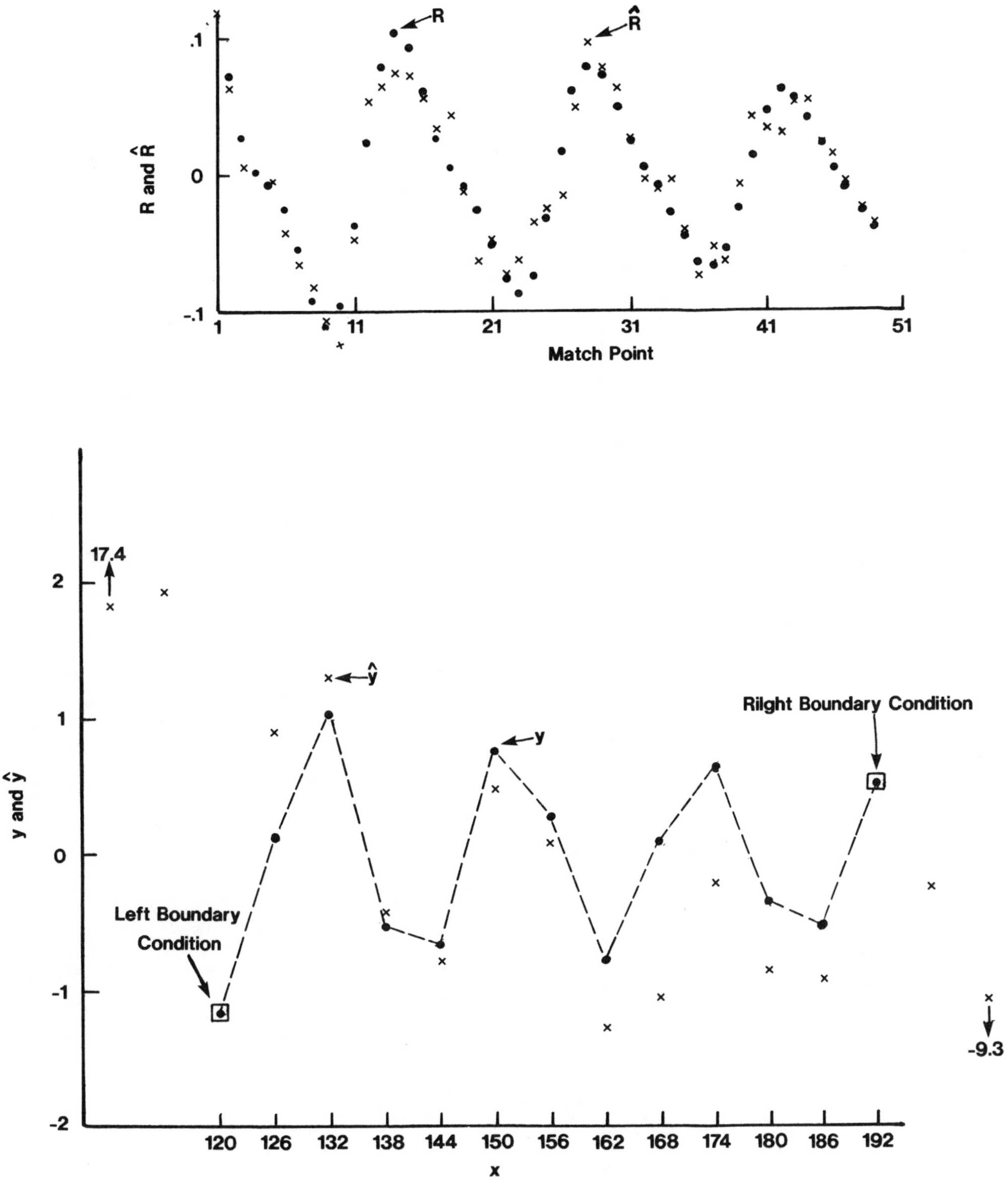

Fig. 7-23. Run Number Three of Example 7-13.

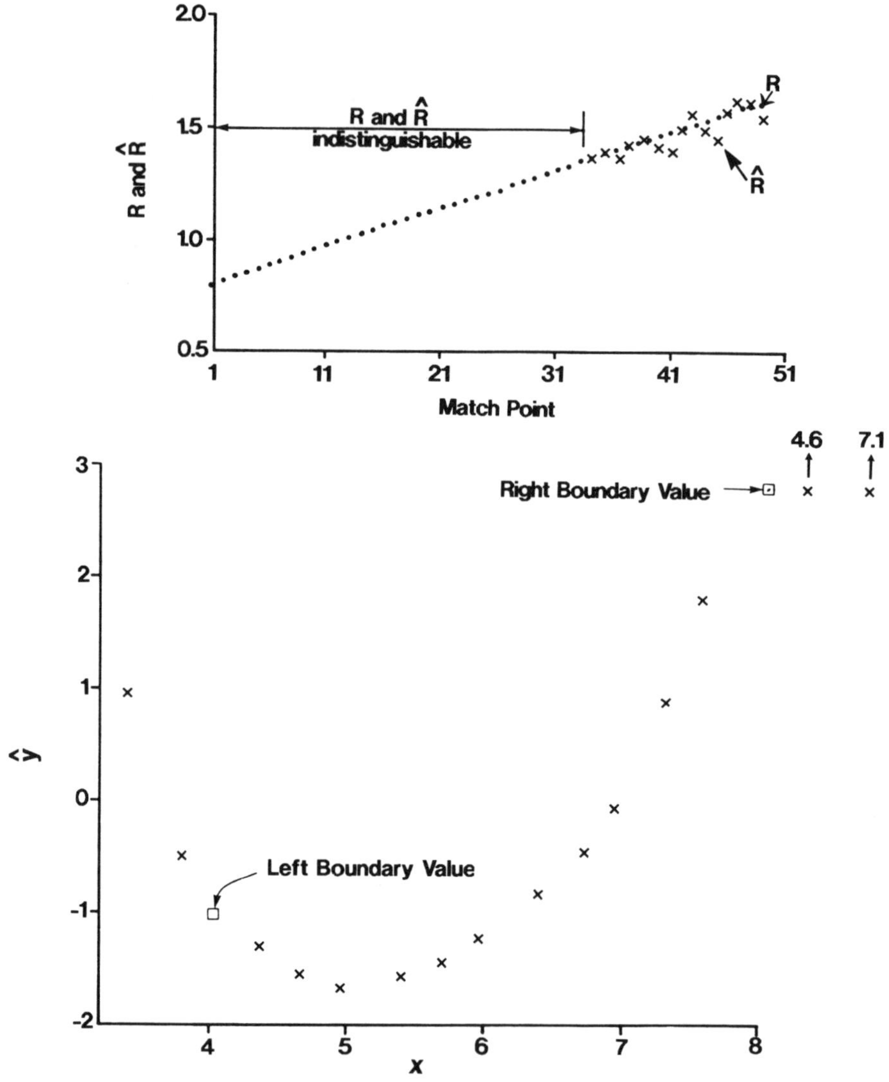

Fig. 7-24. Run Number One of Example 7-14

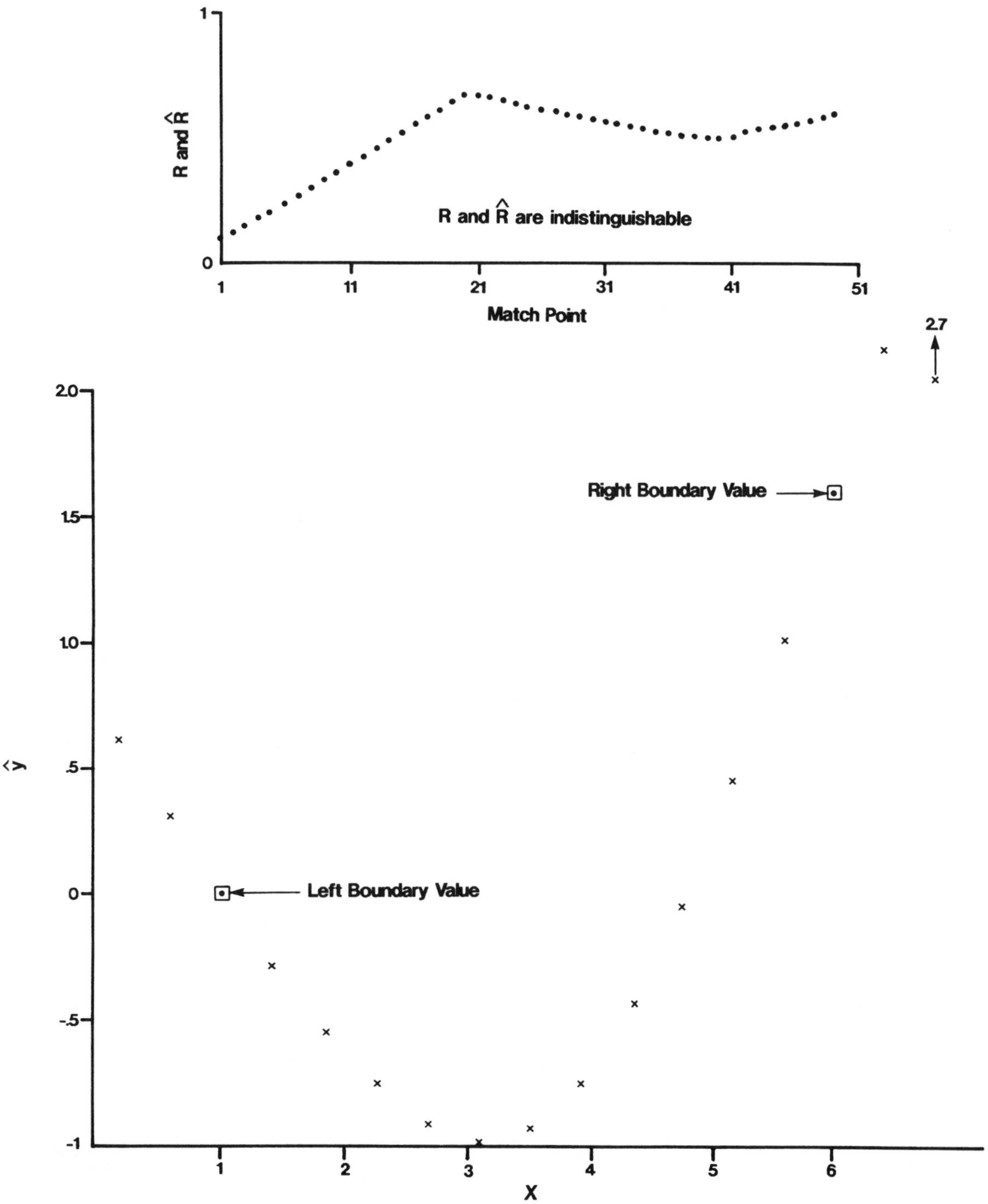

Fig. 7-25. Run Number Two of Example 7-14

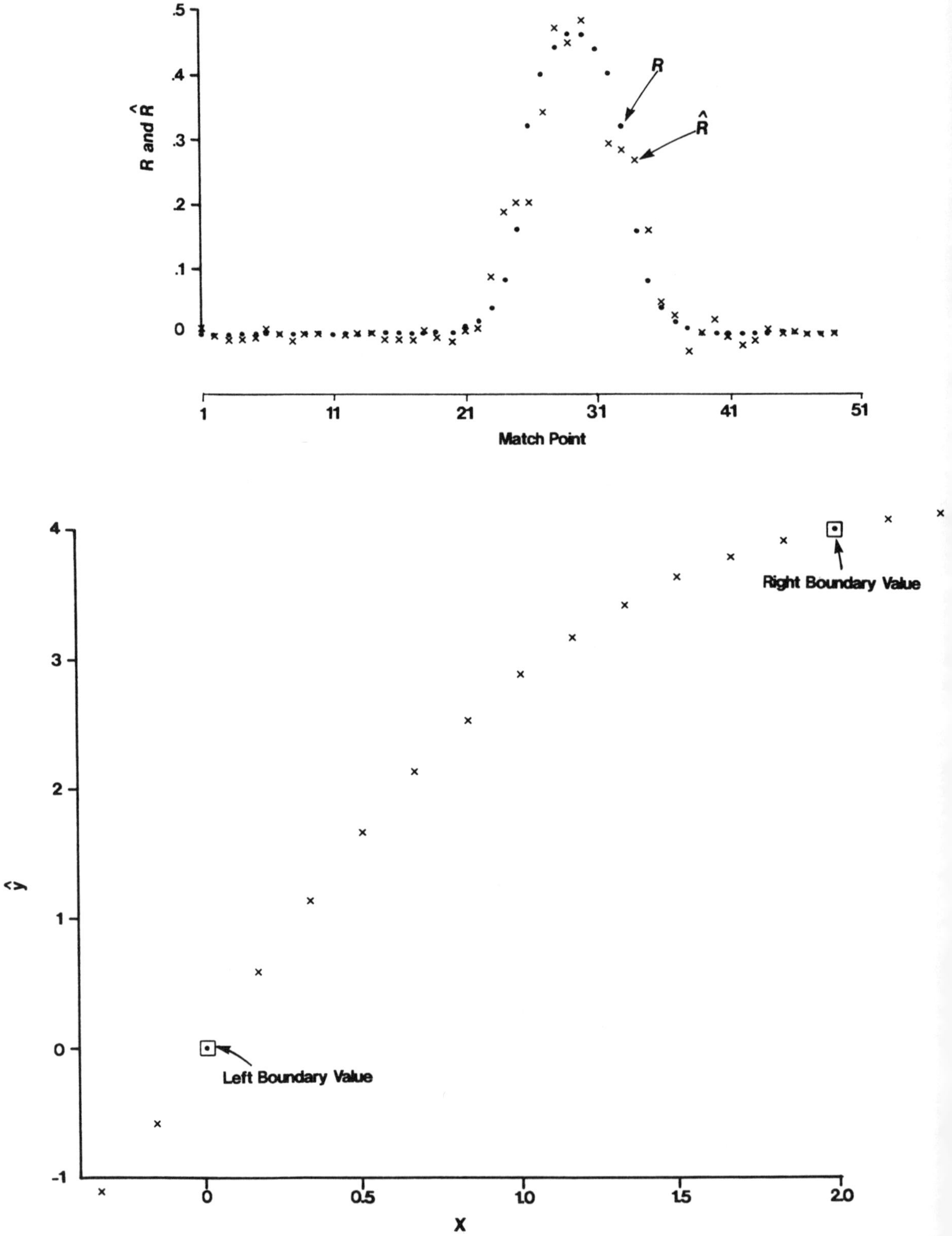

Fig. 7-26. Run Number Three of Example 7-14.

CHAPTER 8

Numerical Approximation Of Functions By Finite Elements

8.1 INTRODUCTION

Solution of differential equations by the method of finite elements has increased greatly in the last decade. Texts such as Segerlind (1976), Pinder and Gray (1977), and Cheng (1978) presented background literature and overviews of the topic easily understood by those beginning work in this field. Applications to specific problems can be found in technical journals of many disciplines.

Finite element solution is based on a span-to-span principle of optimal correspondence between a function and its approximation. We will see later that weighting functions are integrated across the entire span between nodes. In Chapter 7 we used a point-to-point principle of correspondence to evaluate an approximating function. Least squares produced an optimal correspondence based on the quarter- or half-points of the spans. Obviously, if the number of points were increased greatly the two methods should converge in the sense of producing the same approximation. Least squares produces a mean point difference of zero between the function and its approximation. Finite elements produces a zero difference between the integrals of a function and an approximation.

The advantage of the least-squares method is its simplicity and its similarity to data smoothing. Its precision might be increased in some problems by using more than quarter- or half-points in the solution. The advantage of the finite element method is its rigorous definition of correspondence between the function and its approximation. A disadvantage is the cumbersome requirement for calculation of numerous span integrals prior to problem solution. In further comparison, it may be noted that the least-squares method is more flexible. Discrete variations were made match-point by match-point in Chapter 7. In Chapter 8 we wll find that discrete variation is possible only span by span.

Our study will be limited to presentation of a combination of finite elements and sliding polynomials. Specifically, we will use the four-point system. The linear, or two-point, system is well documented in the literature. The six-point system is too cumbersome for routine usage, requiring a large numbers of integrals. We will rely on the three references cited above for rigorous background. This chapter will be limited to a description of the use of sliding polynomials as the approximating and weighting functions in the Galerkin method for weighted residuals in the finite element method. We hypothesize that the mathematical continuities forming the core of the piecewise sliding polynomials will allow relaxation of the number of elements in problem formulation, thus leading to simplicities and economies in the computational implementation.

Presentation will be descriptive, rather than narrowly rigorous. Following the establishment of the methodology to combine sliding polynomials with finite elements, examples with some typical functional and differential forms will be presented. Both one- and two-dimensional problems will be discussed. However, because of the large number of coefficient integrals required, the two-dimensional problems will be limited to a linear stepwise procedure. It will be seen that a large class of problems can be solved by this method.

8.2 BASIC METHODOLOGY

Pinder and Gray (1977, p. 55) state the method of weighted residuals as,

$$<L\hat{y} - f(x), w_i> = 0 \tag{8-1}$$

where L is an operation on some interpolation or trial function $\hat{y}$; $\hat{y}$ is a linear combination of suitable functions: $\hat{y} = \Sigma d_j \phi_j(x)$; f(x) is known function; w_i is a weighting function; $<v,w>$ denotes an inner product defined as $_B$ vwdB; and d_j is a set of linear weights on the basic functions $\Phi_j(x)$. In brief, the residuals between the approximating operation, $L\hat{y}$, and the known function, f(x), weighted by w_i and integrated across B, are required to be zero.

In the Galerkin method of weighted residuals, the interpolation or basic functions, $\phi(x)$, are used for the weighting functions w_i. The specific functions used herein as the basis and weighting will be the four-point sliding polynomial.

For convenience the four-point sliding polynomial interpolation function can be expressed as:

$$\hat{y} = C_1Y_1 + C_2Y_2 + C_3Y_3 + C_4Y_4 \tag{8-2}$$

In Eq. 8-2 the Y_i are equally-spaced nodal values and the $C_i(z)$ are known functions of x as presented earlier with a transformation to the unit variate z. Interpolation with Eq. 8-2 is limited to the inner range between nodal values Y_2 and Y_3. In combining sliding polynomials and finite elements, this inner range becomes one element of the finite-element method and is also the domain B in the definition of the inner product of Eq. 8-1. The $C_i(z)$ become the $\Phi_i(x)$ of Eq. 8-1 and the values of the equally-spaced nodes Y_i become the d_i of Equation 8-1.

In this chapter we will use two designations for the independent variate. When the variate is local to one element or to one range of the sliding polynomial, we will use z. When the variate is global, covering the full range of a problem, we will use x. The transform from z to x will be linear, as z shifts, normally, from left to right across x. From this transformation it is understood that we may write either f(x) or f(z) in weighted residual formulation. Nodes are spaced uniformly in both z and x scales; however, this requirement can be relaxed later.

Consider the Y_i of Equation 8-2 to be placed in local z-scale with Y_1 at z = -1.5, Y_2 at z = -0.5, Y_3 at z = 0.5, and Y_4 at z = 1.5. In this local scale it was shown in Chapter 1 that the $C_i(z)$ of Eq. 8-2 are functions of z given in Eqs. 8-3, with the domain B of Eq. 8-1 for this scale being $-0.5 \leq z \leq 0.5$:

$$C_1(z) = -\frac{1}{16} + \frac{1}{8}z + \frac{1}{4}z^2 - \frac{1}{2}z^3 \tag{8-3a}$$

$$C_2(z) = \frac{9}{16} - \frac{11}{8}z - \frac{1}{4}z^2 + \frac{3}{2}z^3 \tag{8-3b}$$

$$C_3(z) = \frac{9}{16} + \frac{11}{8}z - \frac{1}{4}z^2 - \frac{3}{2}z^3 \tag{8-3c}$$

$$C_4(z) = -\frac{1}{16} - \frac{1}{8}z + \frac{1}{4}z^2 + \frac{1}{2}z^3 \tag{8-3d}$$

All the ingredients of a sliding polynomial formulation of the finite-element method have now been presented. Further understanding can be developed with a simple example. Consider the trivial exercise of representing a straight line. Express the line as:

$$y = a_1 + b_1 z \tag{8-4}$$

Now approximate y with $\hat{y}$ as in Eq. 8-2. $L\hat{y}$ is simply $\hat{y}$ and f(z) is $a_1 + b_1 z$. Making the appropriate substitutions in Eq. 8-1 and using the $C_i(z)$ for w_i yields

$$\langle L\hat{y} - f(z)_1 w_i \rangle = \int_B C_i(z)\ (\hat{y} - a_1 - b_1 z)\ dB = 0 \tag{8-5}$$

From this product of vector $C_i(z)$ times the scalar $(\hat{y} - a_1 - b_1 z)$, we obtain the four equations, with the notation on the $C_i(z)$ simplified to C_i

$$\int_{-\frac{1}{2}}^{\frac{1}{2}} C_1\ (C_1Y_1 + C_2Y_2 + C_3Y_3 + C_4Y_4 - a_1 - b_1 z)\ dz = 0 \tag{8-6a}$$

$$\int_{-\frac{1}{2}}^{\frac{1}{2}} C_2\ (C_1Y_1 + C_2Y_2 + C_3Y_3 + C_4Y_4 - a_1 - b_1 z)\ dz = 0 \tag{8-6b}$$

$$\int_{-\frac{1}{2}}^{\frac{1}{2}} C_3\ (C_1Y_1 + C_2Y_2 + C_3Y_3 + C_4Y_4 - a_1 - b_1 z)\ dz = 0 \tag{8-6c}$$

$$\int_{-\frac{1}{2}}^{\frac{1}{2}} C_4\ (C_1Y_1 + C_2Y_2 + C_3Y_3 + C_4Y_4 - a_1 - b_1 z)\ dz = 0 \tag{8-6d}$$

In general, all terms in Eqs. 8-6 are known if the function $a_1 + b_1 z$ is specified. The exceptions are the nodal values Y_i which are unknown. Now if each of the integrals in Eq. 8-6 is evaluated, four simultaneous linear equations that can be solved for the unknowns Y_i will result.

Before proceeding with any particular solution of Eq. 8-6, we will now consider some details of the terms in these equations. The terms are of three types:

$$\int_{-\frac{1}{2}}^{\frac{1}{2}} C_j C_i Y_i \, dz = Y_i \int_{-\frac{1}{2}}^{\frac{1}{2}} C_j C_i \, dz \tag{8-7a}$$

$$\int_{-\frac{1}{2}}^{\frac{1}{2}} C_j \quad a dz = a \int_{-\frac{1}{2}}^{\frac{1}{2}} C_j \, dz \tag{8-7b}$$

$$\int_{-\frac{1}{2}}^{\frac{1}{2}} C_j \quad b z dz = b \int_{-\frac{1}{2}}^{\frac{1}{2}} C_j \, z \, dz \tag{8-7c}$$

All such integrals on the right-hand side of Eqs. 8-7 can be evaluated in advance of any particular problem formulation. In addition, when we later formulate differential operations, we will need terms of the form

$$\int_{-\frac{1}{2}}^{\frac{1}{2}} C_j \frac{d\hat{y}}{dz} = \int_{-\frac{1}{2}}^{\frac{1}{2}} C_j \left(\frac{dC_1}{dz} Y_1 + \frac{dC_2}{dz} Y_2 + \frac{dC_3}{dz} Y_3 + \frac{dC_4}{dz} Y_4 \right) dz \tag{8-8a}$$

$$\int_{-\frac{1}{2}}^{\frac{1}{2}} C_j \frac{d^2\hat{y}}{dz^2} = \int_{-\frac{1}{2}}^{\frac{1}{2}} C_j \left(\frac{d^2C_1}{dz^2} Y_1 + \frac{d^2C_2}{dz^2} Y_2 + \frac{d^2C_3}{dz^2} Y_3 + \frac{d^2C_4}{dz^2} Y_4 \right) dz \tag{8-8b}$$

The integrals in the terms such as $Y_1 \int_{-\frac{1}{2}}^{\frac{1}{2}} C_j \frac{d^2C_1}{dz^2}$ can also be evaluated in advance of problem formulation.

Table A-8.1 in Appendix A gives numerical values of certain integrals that occur frequently in problem formulation. Returning to Eq. 8-6, we can use the integrals in Table 8-1 to write directly the four linear equations:

$$0.0023807Y_1-0.0279762Y_2-0.0178572Y_3+0.0017857Y_4=-0.0416667a_1+0.0041667b_1 \tag{8-9a}$$

$$-0.0279762Y_1+0.4047619Y_2+0.1827381Y_3-0.0178572Y_4= 0.5416667a_1-0.0958333b_1 \tag{8-9b}$$

$$-0.0178572Y_1+0.1827381Y_2+0.4047619Y_3-0.0279762Y_4= 0.5416667a_1+0.0958333b_1 \tag{8-9c}$$

$$0.0017857Y_1-0.0178572Y_2-0.0279762Y_3+0.0023807Y_4=-0.0416667a_1-0.0041667b_1 \tag{8-9d}$$

Hypothetically, one now solves Eqs. 8-9 for the four unknown nodes Y_i. Practically, for this trivial exercise in solution formulation, Eqs. 8-9 express an over-determined situation with four unknowns dependent on only two parameters, a_1 and b_1. In the section "Imposing Boundary or Other Conditions" we will show how to select some subset of the four unknowns for solution.

8.2.1 Linking of Elements

Equations 8-9 cover only one element of the finite-element method. In this element we are approximating the simple linear function, Equation 8-4. Coverage of the complete solution domain of a particular problem will normally mean developing sets of equations, similar to Eqs. 8-9, which will represent adjacent elements. Then the equations for all elements must be combined and solved simultaneously for the expanded set of nodes Y_i.

Suppose, for example, we wish to add an element with linear function $y = a_2 + b_2x$ to the first element. The two elements are joined to each other so that nodes Y_1, Y_2, Y_3, and Y_4 define the approximating function for the first element, and nodes Y_2, Y_3, Y_4, and Y_5 define the approximating function for the second element. Finite-element equations for the second element are:

$$0.0023807Y_2-0.0279762Y_3-0.0178572Y_4+0.0017857Y_5=-0.0416667a_2+0.0041667b_2 \quad (8\text{-}10a)$$

$$-0.0279762Y_2+0.4047619Y_3+0.1827381Y_4-0.0178572Y_5= 0.5416667a_2-0.0958333b_2 \quad (8\text{-}10b)$$

$$-0.0178572Y_2+0.1827381Y_3+0.4047619Y_4-0.0279762Y_5= 0.5416667a_2+0.0958333b_2 \quad (8\text{-}10c)$$

$$0.0017857Y_2-0.0178572Y_3-0.0279762Y_4+0.0023807Y_5=-0.0416667a_2-0.0041667b_2 \quad (8\text{-}10d)$$

Since nodes Y_2, Y_3, and Y_4 are common to both elements, we must solve some set of simultaneous equations covering both elements.

The method of coupling Eqs. 8-9 and 8-10 is explained in Fig. 8-1. This schematic shows a problem domain broken into five elements. Each element must be covered by a set of four equations such as Eqs. 8-9 or 8-10. We can number the four separate equations in Eq. 8-9 as 1-1, 1-2, 1-3, and 1-4, with the first digit meaning the element number. Equations 8-10, covering a second element, could be numbered 2-1, 2-2, 2-3, and 2-4. Each such set of four equations is so placed in the domain of ordered nodes Y_i that its central interval corresponds to the element covered. The compositing of equations is now self-evident. One simply adds all equations relevant to the same node. For example, equation 1-1 stands alone, but equations 1-2 and 2-1, at node Y_2 must be added. Since the four-point system of sliding polynomials is being used, as many as four equations can be added at any one node. We can anticipate a resulting set of simultaneous equations whose coefficient matrix will be a band matrix seven elements wide.

8.2.2 Imposing Boundary or Other Conditions

We return now to the set of Equations 8-9. We wish to reduce the number of unknown nodes to two in order to match the number of parameters, a_1 and b_1, which determine f(z) for this problem. Such reduction is imposed by making some of the unknown nodes functions of other nodes. The choice of the functions imposes geometrical conditions that become part of the solution. Since we want to approximate a straight line, we can impose the obvious conditions:

$$Y_2 = \frac{2Y_1 + Y_4}{3} \tag{8-11a}$$

$$Y_3 = \frac{Y_1 + 2Y_4}{3} \tag{8-11b}$$

These values can be substituted into Eqs. 8-9, reducing the unknowns to Y_1 and Y_4. At the same time we must reduce the number of equations to two. Equations 1-2 and 1-3 (Fig. 8-1) are dropped since we eliminated the associated nodes Y_2 and Y_3.

Substituting Eqs. 8-11 into 1-1 and 1-4 of Eqs. 8-9, we get the following reduced set of equations:

$$-0.0222225Y_1 - 0.0194445Y_4 = -0.0416667a_1 + 0.0041667b_1 \tag{8-12a}$$

$$-0.0194445Y_1 - 0.0222225Y_4 = -0.0416667a_1 - 0.0041667b_1 \tag{8-12b}$$

These equations yield the following solution, allowing for truncation error:

$$Y_1 = a_1 - 1.5\ b_1 \tag{8-13a}$$

$$Y_4 = a_1 + 1.5\ b_1 \tag{8-13b}$$

Substitution into Eqs. 8-11 yields:

$$Y_2 = a_1 - 0.5\ b_1 \tag{8-14a}$$

$$Y_3 = a_1 + 0.5\ b_1 \tag{8-14b}$$

While such a solution is trivial, it does provide first confirmation of the methodology for combining sliding polynomials with finite elements under imposed problem geometry.

The general procedure for imposing conditions either at the boundaries or at internal nodes can be stated as follows. Establish the conditions that need to be imposed on the solution. This can only come from analysis of the particular problem. Normally it means setting a slope, or fixing a nodal value. Certain nodes are thus eliminated as unknowns and are no longer free in solution. Equations based on those nodes are dropped, maintaining a one-on-one relationship of unknowns to equations.

Further explanation of the coupling of finite elements with sliding polynomials is given in the following case examples of specific functions.

8.2.3 Example 8-1: Piecewise Linear Representation

Equations 8-9 and 8-10 covered two adjacent elements with f(z) specified as a straight line. The combination of Eqs. 8-9 and 8-10, which are needed to cover both elements simultaneously, would yield the five equations (1-1), (1-2) + (2-1), (1-3) + (2-2), (1-4) + (2-3), and (2-4). If $a_1 = 2$, $b_1 = 4$, $a_2 = 5$, $b_2 = 2$ and if we remember that the local scale of z ranges from -1.5 to 1.5, which shifts with the elements, then the two linear pieces are common at their common node. Count a_1, b_1, and b_2 as parameters with a_2 redundant. Then we need to reduce the combination of Eqs. 8-9 and 8-10 from five to three.

Conditions at the left boundary must be assumed so that $Y_2 = 0$ and the slope at Y_2 is b_1. By fixing Y_2 we eliminated Eqs. 1-2 and 2-1. The slope of the four-point sliding polynomials at Y_2 can be shown to be ½ $(Y_3 - Y_1)$. Setting this equal to $b_1 = 4$, the value for Y_1 becomes $(8 - Y_3)$. We substitute this for Y_1 and eliminate Eq.1-1. Our three reduced equations are (1-3) + (2-2), (1-4) + (2-3), and (2-4), as modified. This calculation is shown in Table 8-1, Part a.

In Table 8-1, Part b and Part c, are shown the construction of the simultaneous equations to extend the piecewise linear representation across three and four elements. These solutions are retabulated for ready comparison in Table 8-2. Interpolations with the four-point sliding polynomial were carried out at 1/5 of the element intervals. Also included in Table 8-2 is the trivial exercise of fitting a straight line to one element with the same left boundary conditions.

The four-element solutions are displayed in Fig. 8-2. In general, the sliding polynomial approximation tends to smooth across the angles of the given piecewise linear function. Thus, we have achieved a smoothing operation whereby a piecewise function with a continuous first derivative has replaced a piecewise linear function that has a discontinuous first derivative.

In summary, the approximation of a piecewise linear function has served to test the sliding polynomial representation. The method for maintaining one-on-one correspondence between free unknowns and the number of equations has been illustrated. While this elementary case study has served these purposes in initial presentation of methodology, practical applications are probably limited.

8.2.4 Example 8-2: Piecewise Constant Gradient

For this example we assume the simple differential $dy/dx = b_i$. We represent y with $\hat{y}$ and dy with $d\hat{y}$. f(x) is b_i. A Galerkin formulation for one element is given in Eq. 8-15. Since x does not occur in f(x), x and z can be used interchangeably, provided b_i is the slope in z scale.

$$\int_B \left[C_1\left(\frac{d\hat{y}}{dz} - b_i\right) + C_2\left(\frac{d\hat{y}}{dz} - b_i\right) + C_3\left(\frac{d\hat{y}}{dz} - b_i\right) + C_4\left(\frac{d\hat{y}}{dz} - b_i\right)\right] = 0 \quad (8\text{-}15)$$

Now $\frac{d\hat{y}}{dz} = \frac{dC_1}{dz} Y_1 + \frac{dC_2}{dz} Y_2 + \frac{dC_3}{dz} Y_3 + \frac{dC_4}{dz} Y_4$ and the four linear equations in Eq. 8-15 can be written as:

$$\int_{-\frac{1}{2}}^{\frac{1}{2}} C_1 \left(\frac{dC_1}{dz} Y_1 + \frac{dC_2}{dz} Y_2 + \frac{dC_3}{dz} Y_3 + \frac{dC_4}{dz} Y_4 - b_i \right) dz = 0 \quad (8\text{-}16a)$$

$$\int_{-\frac{1}{2}}^{\frac{1}{2}} C_2 \left(\frac{dC_1}{dz} Y_1 + \frac{dC_2}{dz} Y_2 + \frac{dC_3}{dz} Y_3 + \frac{dC_4}{dz} Y_4 - b_i \right) dz = 0 \quad (8\text{-}16b)$$

$$\int_{-\frac{1}{2}}^{\frac{1}{2}} C_3 \left(\frac{dC_1}{dz} Y_1 + \frac{dC_2}{dz} Y_2 + \frac{dC_3}{dz} Y_3 + \frac{dC_4}{dz} Y_4 - b_i \right) dz = 0 \quad (8\text{-}16c)$$

$$\int_{-\frac{1}{2}}^{\frac{1}{2}} C_4 \left(\frac{dC_1}{dz} Y_1 + \frac{dC_2}{dz} Y_2 + \frac{dC_3}{dz} Y_3 + \frac{dC_4}{dz} Y_4 - b_i \right) dz = 0 \quad (8\text{-}16d)$$

The integrals can be copied directly from Table A-8.1 in Appendix A, yielding for element 1:

1-1: $0Y_1 + 0.0458333Y_2 - 0.05Y_3 + 0.0041667Y_4 + 0.0416667b_1 = 0$ (8-17a)

1-2: $-0.0458333Y_1 - 0.5Y_2 + 0.5958333Y_3 - 0.05Y_4 - 0.5416667b_1 = 0$ (8-17b)

1-3: $0.05Y_1 - 0.5958333Y_2 + 0.5Y_3 + 0.045833Y_4 - 0.5416667b_1 = 0$ (8-17c)

1-4: $-0.0041667Y_1 + 0.05Y_2 - 0.0458333Y_3 + 0Y_4 + 0.0416667b_1 = 0$ (8-17d)

We need to allow for one constant of integration. Set one condition by fixing Y_1 at some value $\underline{\underline{Y}}_1$, and compensate by dropping 1-1 from Eq. 8-17. Then a reduced set is given by:

$$-0.5Y_2 + 0.5958333Y_3 - 0.05Y_4 = 0.5416667b_1 + 0.0458333\underline{\underline{Y}}_1 \quad (8\text{-}18a)$$

$$-0.5958333Y_2 + 0.5Y_3 + 0.0458333Y_4 = 0.5416667b_1 - 0.05\underline{\underline{Y}}_1 \qquad (8\text{-}18b)$$

$$0.05Y_2 - 0.0458333Y_3 = -0.041667b_1 + 0.0041667\underline{\underline{Y}}_1 \qquad (8\text{-}18c)$$

The solution for Eqs. 8-18 is given in

$$Y_2 = \underline{\underline{Y}}_1 + b_1 \qquad (8\text{-}19a)$$

$$Y_3 = \underline{\underline{Y}}_1 + 2b_1 \qquad (8\text{-}19b)$$

$$Y_4 = \underline{\underline{Y}}_1 + 3b_1 \qquad (8\text{-}19c)$$

Equations 8-19 confirm a straight line obviously given by the known integral of our governing differential equation.

We now extend this case study across two elements with $b_1 = 2$ and $b_2 = 1$. Also set $\underline{\underline{Y}}_1 = 100$. For the second element we can write directly

$$\text{2-1:} \quad 0Y_2 + 0.0458333Y_3 - 0.05Y_4 + 0.0041667Y_5 + 0.0416667b_2 = 0 \qquad (8\text{-}20a)$$

$$\text{2-2:} \quad -0.0458333Y_2 - 0.5Y_3 + 0.5958333Y_4 - 0.05Y_5 - 0.5416667b_2 = 0 \qquad (8\text{-}20b)$$

$$\text{2-3:} \quad 0.05Y_2 - 0.5958333Y_3 + 0.5Y_4 + 0.0458333Y_5 - 0.5416667b_2 = 0 \qquad (8\text{-}20c)$$

$$\text{2-4:} \quad -0.0041667Y_2 + 0.05Y_3 - 0.0458333Y_4 + 0Y_5 + 0.0416667b_2 = 0 \qquad (8\text{-}20d)$$

We now form the composite set of equations (1-2) + (2-1), (1-3) + (2-2), (1-4) + (2-3), and (2-4) given by

$$-0.5Y_2 + 0.641667Y_3 - 0.1Y_4 + 0.0041667Y_5 = 5.6249967 \qquad (8\text{-}21a)$$

$$- 0.6416667Y_2 + 0.641667Y_4 - 0.05Y_5 = -3.3750 \tag{8-21b}$$

$$0.1Y_2 - 0.6416667Y_3 + 0.5Y_4 + 0.0458333Y_5 = 0.8750033 \tag{8-21c}$$

$$- 0.0041667Y_2 + 0.05Y_3 - 0.0458333Y_4 = -0.0416667 \tag{8-21d}$$

These equations have the solution

$$Y_2 = 102.2072 \tag{8-22a}$$

$$Y_3 = 104.0167 \tag{8-22b}$$

$$Y_4 = 105.2072 \tag{8-22c}$$

$$Y_5 = 106.0000 \tag{8-22d}$$

This solution is plotted in Fig. 8-3. The smooth curvature was developed by sliding polynomial interpolation between Y_2 and Y_4.

A three-element solution is also plotted in Fig. 8-3. The same imposed condition of $Y_1 = 100$ was used as in the two-element example. The values $b_1 = 2$, $b_2 = 0$, $b_3 = -0.5$ were used for the piecewise gradients through the three elements. The solution is achieved by first writing a set of four equations, as in Eqs. 8-20, but based on nodes Y_3 through Y_6. Then a composite set of equations (1-2) + (2-1), (1-3) + (2-2) + (3-1), (1−4) + (2-3) + (3-2), (2-4) + (3-3), and (3-4) are formed and solved.

8.2.5 Example 8-3: Exponential Approximation

Assume the following differential equation for this study, with the equation extended over several elements

$$\frac{dy}{dz} = ky + a \tag{8-23}$$

Approximate y with $\hat{y}$ and dy with $d\hat{y}$. Set Y_2 at the left boundary of the first element to $\underline{\underline{Y}}_2 = 10$, and set k = -0.5 and a = -0.5. The Galerkin formulation is given by

$$\int_B \left[C_1\left(\frac{d\hat{y}}{dz}-k\hat{y}-a\right) + C_2\left(\frac{d\hat{y}}{dz}-k\hat{y}-a\right) + C_3\left(\frac{d\hat{y}}{dz}-k\hat{y}-a\right) + C_4\left(\frac{d\hat{y}}{dz}-k\hat{y}-a\right) \right] = 0 \qquad (8\text{-}24)$$

The first linear equation is derived by expanding the first term

$$\text{1-1:} \quad \int_B C_1 \left(\frac{d\hat{y}}{dz} - k\hat{y} - a\right) = \int_{-\frac{1}{2}}^{\frac{1}{2}} C_1 \left(\begin{array}{l} \frac{dC_1}{dz} Y_1 + \frac{dC_2}{dz} Y_2 + \frac{dC_3}{dz} Y_3 + \frac{dC_4}{dz} Y_4 \\ - k \left[C_1Y_1 + C_2Y_2 + C_3Y_3 + C_4Y_4 \right] - a \end{array} \right) dz$$

$$= 0Y_1 + 0.0458333\underline{\underline{Y}}_2 - 0.05Y_3 + 0.0041667Y_4$$

$$+ \frac{1}{2} \Big[0.0023807Y_1 - 0.0279762\underline{\underline{Y}}_2 - 0.0178572Y_3$$

$$+ 0.0017857Y_4 - 0.0416667 \Big] = 0 \qquad (8\text{-}25)$$

Equation 8-25 for element 1-1 is skipped since Y_2 was set to $\underline{\underline{Y}}_2$ and thus removed from free solution. The complete sets of equations for one, two, and three elements are shown in Table 8-3. The solutions are retabulated with fifth-point interpolations in Table 8-4.

The differential Eq. 8-23 has the closed solution given by

$$y = \left(y_1 + \frac{a}{k}\right) e^{kz} - \frac{a}{k} \qquad (8\text{-}26)$$

where y_1 is the value at the left boundary (corresponds to position of Y_2 in the approximation). Values of this function are also shown in Table 8-5. The values of the closed solution and the one-, two-, or three-element approximations by finite-element sliding polynomials agree very closely. We should also note that the approximation in element 1 does not change much when the second and third elements are added. This case study, therefore, adds a level of confidence concerning the stability of the finite-element sliding-polynomial method.

8.2.6 Example 8-4: Polynomial Approximation

In this study we will investigate the effect of a changing number of elements used within a fixed problem dimension. The following parabolic function will be assumed, and we will study the effect of approximating it with sliding polynomials:

$$y = a + bx + cx^2 \tag{8-27}$$

Assume that our problem lies in the range $1 \leq x \leq 2$ of the global variate x, and that a = 2, b = -2, and c = 1. Note that $(dy/dx)_{x=1}$ is zero, and that the curve of Eq. 8-27 passes through x = 2, y = 2. These two conditions will be imposed as boundary conditions in the approximating formulation that follows.

We now take the position that we do not know that Eq. 8-27 is the solution to the problem, and that we only know the boundary conditions and the governing differential equation

$$\frac{d^2y}{dx^2} = 2c \tag{8-28}$$

If we represent the global range $1 \leq x \leq 2$ with a varying number of elements and use sliding polynomials as the approximating function, a varying scale change from the global x to local z is necessary. The scale relationship can be expressed as

$$z = h + jx \tag{8-29}$$

Then x^2 in Eq. 8-27 is $\left(\frac{z-h}{j}\right)^2$ and our governing differential equation in local variate z is

$$\frac{d^2y}{dz^2} = \frac{2c}{j^2} = c' \tag{8-30}$$

Now j is 1, 2, 3, and 4 as we use one, two, three or four elements to cover the problem space $1 \leq x \leq 2$, and c' takes values 2, 1/2, 2/9, and 1/8.

The nodes of the approximating function are numbered as Y_1, Y_2, Y_3, ... Y_n from left to right. Regardless of the number of elements, Y_2 is at the left boundary (x = 1), and

Y_1 is to the left of this boundary. Now the slope of the sliding polynomials can be shown to be 1/2 (Y_3 - Y_1) at the left boundary. Since this is zero in our problem, then Y_1 = Y_3. We can eliminate Y_1 from free solution and drop the linear equation corresponding to this node. For the second condition we need to set the node that falls at the right boundary equal to 2. The actual node falling at the right boundary varies with the number of elements we use.

The relationships between the global and local scales and the placement of nodes are shown in Fig. 8-4. All element equations are given in Table 8-5. The results are shown in Table 8-6. A closed solution, simply an evaluation of Eq. 8-27, is also shown.

Note in Table 8-6 that all solutions exactly match the closed solution. This is to be expected since the four-point sliding polynomial is a special cubic and can take on a parabolic shape. This study, however, has helped verify the integral values in Table A-2.1, has helped validate the method for maintaining one-to-one correspondence between unknown nodes and linear equations, and has demonstrated the required scale changes from global to local variates in order to maintain continuity of imposed conditions. Lastly, we have demonstrated the feasibility of direct calculation of the second derivatives of the approximating sliding polynomials in contrast to the somewhat awkward integration by parts (Pinder and Gray, p. 80; Cheng, p. 215).

8.2.7 Example 8-5: Approximation for $\frac{d^2y}{dz^2} = ky + a$

We will derive an approximate solution for this equation using four elements. Nodes will run from Y_1 to Y_7. Y_2 at the left boundary and Y_6 at the right boundary will be fixed.

If y is approximated by ŷ, then the Galerkin formula is given by

$$\int_{-\frac{1}{2}}^{\frac{1}{2}} \left[C_1\left(\frac{d^2\hat{y}}{dz^2} - k\hat{y} - a\right) + C_2\left(\frac{d^2\hat{y}}{dz^2} - k\hat{y} - a\right) + C_3\left(\frac{d^2\hat{y}}{dz^2} - k\hat{y} - a\right) + C_4\left(\frac{d^2\hat{y}}{dz^2} - k\hat{y} - a\right)\right] dz = 0$$

(8-31)

The linear equations for the four elements are given in Table 8-7. For this example k was set to 2 and a to 1. The linear equations were solved twice; once with boundaries $Y_2 = 10$, $Y_6 = 50$, and again with boundaries $Y_2 = 70$, $Y_6 = 60$. The two approximating solutions are shown in Figure 8-5.

8.2.8 Example 8-6: Approximation for $\frac{d^2y}{dx^2} + x\frac{dy}{dx} + ay + b = 0$

The governing equation for this case has the variable coefficient x in the second term. In order to use the integrals in the z-scale in Table A-2.1 in the Appendix, it will be necessary to transform the governing equation to the z-scale.

Let the transformation from z to x be the linear

$$x = h_i + k_i\, z \tag{8-32}$$

The coefficients h_i and k_i could vary for each element i. However, for uniform node spacing k_i will be constant. From Eq. 8-32 we get $dx = k\,dz$ and $dx^2 = k^2\,dz^2$. Then the governing differential equation can be transformed to Eq. 8-33 and further transformed to Eq. 8-34:

$$\frac{d^2y}{k^2dz^2} + (h_i + kz)\frac{dy}{kdz} + ay + b = 0 \tag{8-33}$$

$$\frac{1}{k^2}\frac{d^2y}{dz^2} + \left(\frac{h_i}{k}\right)\frac{dy}{dz} + z\frac{dy}{dz} + ay + b = 0 \tag{8-34}$$

The first element equation using a Galerkin formulation is given by

$$\frac{1}{k^2}\int C_1\frac{d^2\hat{y}}{dz^2} + \frac{h_i}{k}\int C_1\frac{d\hat{y}}{dz} + \int C_1\, z\frac{d\hat{y}}{dz} + a\int C_1\,\hat{y} + b\int C_1 = 0 \tag{8-35}$$

The second, third, and fourth equations have similar formulations. All of the integrals in Eq. 8-35 can be found in Table A-8.1 in the Appendix. We should note that all terms except the second term in Eq. 8-35 and the additional three equations will be constant across all elements. Therefore, these terms can be computed as a subtotal. Then the second term can be calculated using $\frac{h_i}{k}$ as calculated for each element.

For a specific numerical example assume the range of the problem to be $0 \leq x \leq 3.0$. This range can be separated into six elements, thus placing node Y_2 at x = 0, node Y_3 at x = 0.5, and so on. Y_1 will be beyond the left boundary at x = -0.5. Y_9 will be beyond the right boundary at x = 3.5. Then with k = 0.5, $\frac{h_i}{k}$ will have values of 0.5, 1.5, and so on to 5.5 for element number 6. Values of a = 3 and b = 5 were used. For boundary conditions, set $\underline{Y}_1$ = 10, and $\frac{dy}{dx}_{x=0}$. Then $\frac{dy}{dz}_{x=0}$ will be 0.5.

Construction of the element equations is exemplified in Table 8-9. For brevity some detail steps are omitted. The coefficient matrix of the resulting set of simultaneous linear equations can be inverted using any standard computer library program. The nodal values Y_i are then calculated from a matrix-vector product. The solution is displayed in Fig. 8-6.

8.3 TWO-DIMENSIONAL PROBLEMS

Examples 8-1 through 8-6 are based on one independent variable. The solutions are line functions of the dependent variable versus the independent variable. In two-dimensional problems we have two independent variables.

It is extremely cumbersome to expand the four-point sliding polynomial finite element structure to two dimensions. A unit surface in two dimensions requires 16 nodes. The number of finite-element integral coefficients would be 16^2 or 256, for each form such as $c_j\ dc_i/dz$. Fortunately, many solutions to problems can be satisfactorily approximated by a simpler structure.

In Chapter 1 the concept of function-of-functions was discussed as the basis of multi-dimensional sliding polynomials. This concept was used in Chapter 5 for smoothing data in two dimensions and in Chapter 7 for solving differential equations in two-dimensions by least squares. We will use one particular form of function-of-functions illustrated in Part a of Fig. 8-7. A unit surface is defined on a four-point system in dimension z and on a two-point system in dimension u.

Equations 8-3 show the four-point coefficients used in the following form to approximate y(z):

$$y(z) \approx \hat{y}(z) = \sum_{i=1}^{4} C_i(z)\ Y_i \qquad (8\text{-}36)$$

We have seen previously that the simple two-point system has coefficients given by

$$c_1(u) = \frac{1}{2} - u$$
$$c_2(u) = \frac{1}{2} + u \tag{8-37}$$

Then the four-point by two-point unit surface in Part a of Figure 8-7 can be written directly as

$$y(u,z) \approx \hat{y}(u,z) = \sum_{j=1}^{2} c_j(u) \sum_{i=1}^{4} c_i(z)\ Y_{j,i} \tag{8-38}$$

Equation 8-38 can be re-written as

$$\hat{y}(u,z) = \sum_{i=1}^{4} c_1(u)\ c_i(z)\ Y_i + \sum_{i=1}^{4} c_2(u)\ c_i(z)\ Y_{i+4} \tag{8-39}$$

Now if the nodes Y_i, i = 1 to 4, define a boundary or initial set of nodes then only the nodes Y_{i+4}, i = 1 to 4, need to be evaluated. Given such values for Y_{i+4}, then a second set Y_{i+8} could be evaluated. Thus the solution steps forward in linear increments in dimension u.

A more practical scope of problem is illustrated in Part b of Figure 8-7. Here five unit surfaces have been linked in the x-dimension (or z-dimension). Such linking follows the method outlined in the section "Linking of Elements" (Section 8.2.1). The vector of nodes $Y_{1,i}$ i = 1 to 8, would be known from boundary conditions and the vector $Y_{2,i}$ would be evaluated. From this vector $Y_{3,i}$ would be evaluated, and so on. The stepping here is shown as occurring in dimension t, or time. Many practical two-dimensional problems are made up of functions such as in Examples 8-1 through 8-6 that change in time.

Finite element solution requires integrals of the composite coefficients for each of the eight nodes in Figure 8-7, Part a. Therefore, only 8^2 or 64 finite-element integrals are required, versus the 256 required for a full four-by-four-node unit surface. The algebraic form of the composite coefficients is given by the eight $c_j(u)\ c_i(z)$ products in Eq. 8-39. For example, this product for node Y_2 is given by

$$c_2(u)\ c_2(z) = \left(\frac{1}{2} - u\right)\left(-\frac{1}{16} + \frac{1}{8}z + \frac{1}{4}z^2 - \frac{1}{2}z^3\right) \tag{8-40}$$

Then the finite-element integral is the double integral

$$\int_{-\frac{1}{2}}^{\frac{1}{2}} \int_{-\frac{1}{2}}^{\frac{1}{2}} C_2(u)\ C_2(z)\ dzdu = \int_{-\frac{1}{2}}^{\frac{1}{2}} \int_{-\frac{1}{2}}^{\frac{1}{2}} \left(\frac{1}{2} - u\right) \left(-\frac{1}{16} + \frac{1}{8}z + \frac{1}{4}z^2 - \frac{1}{2}z^3\right) dzdu \quad (8\text{-}41)$$

The other seven finite-element integral can be expressed similarly. Also, forms such as $C_j(u)\ dC_i(z)/dz$ must be formulated and integrated. Several functional forms of the finite-element integrals have been computed and are shown in Table A-2.1 in Appendix A. Many two-dimensional stepwise problems can be solved with these listed coefficients.

The coefficient integrals in Table A-2.1 were calculated by numeric integration with sliding polynomials as discussed in Chapter 4.

8.3.1 Example 8-7: The Dispersion Equation

The dispersion equation is the partial differential equation

$$\frac{\partial C}{\partial t} = D\,\frac{\partial^2 C}{\partial x^2} + q \quad (8\text{-}42)$$

This equation expresses the concentration of some material, C, in a fluid as turbulent mixing gradually disperses the material throughout a large body of the fluid. The term q expresses input of additional material as dispersion takes place. The solution of this equation when q is zero is equivalent to the Gaussian probability density function, or the normal curve (Weiss, 1983).

In Eq. 8-42 the equation in (m, l, t) units is

$$\frac{m/l^3}{t} = D\,\frac{m/l^3}{l^2} + \frac{m/l^3}{t}$$

and D must have units of l^2/t. In real units we could write

$$\frac{\text{gms/meter}^3}{\text{hrs}} = \frac{\text{meters}^2}{\text{hrs}} \cdot \frac{\text{gms/meter}^3}{\text{meter}^2} + \frac{\text{gms/meter}^3}{\text{hrs}}$$

The last term can be converted to a more meaningful expression of input as follows. Consider that the solution of Eq. 8-42 is for a fluid body of unit cross-section. Then the units of q become gms/meter/hr.

Program B-8.1 was prepared to solve Eq. 8-42 using finite elements with sliding polynomial coefficients. The program yields the changing shape of the concentration curve through a series of linear time steps. Some special features were incorporated to accommodate the specific boundary conditions. We assume that the solution of Eq. 8-42 covers an infinite body of fluid with the concentration equal to zero at the left and right boundaries of minus and plus infinity. Also, the slope of the concentration curve will be zero at these infinite boundaries. The infinite boundaries must be changed to real boundaries through a variate transformation of the distance scale.

A transformed variable, v, can be constructed:

$$v_2 = -4 + 4e^{bx} \ , \ x \le 0 \qquad (8\text{-}43a)$$

$$v_1 = 4 - 4e^{-bx}, \ x \ge 0 \qquad (8\text{-}43b)$$

The parameter, b, can be calculated by making the transformation function v pass through some point (v, x). For this example the point (1, 50) as used, yielding a value of b of 0.0057536. Table 8-10 gives corresponding values of v and x calculated from Eq. 8-43. The full transformation of the distance scale is from x to the intermediate variable, v, and then to the local unit variable, z. Now dx/dz = dx/dv • dv/dz. If we place the sliding polynomial nodes at the unit values of v in Table 8-9, then dv/dz is unity. From Eq. 8-43 we can calculate $dv_1/dx = 4be^{-bx}$ for positive x and $dv_2/dx = 4be^{b(-x)}$ for negative x. These values are identical and we can write $dx^2 = dv^2/(4be^{-bx})^2$ for any value of x.

Equation 8-42 in local variable z is given by

$$\frac{\partial C}{\partial t} = (4be^{-bx})^2 \ D \ \frac{\partial^2 C}{\partial z^2} + q \qquad (8\text{-}44)$$

We will use unit time steps; therefore, the transform to a local unit variable in the time scale is unity.

With an approximation we can eliminate the variable x in Eq. 8-44 as follows. Evaluate the expression in parenthesis at the nodes as shown in Table 8-10. As an approximation, the averages of $(4be^{-bx})^2$ by spans can be used in the solution. This is simply a step-function of the effect of scale change from x to v.

Program B-8.1 was prepared to cover eight spans of the sliding polynomials in solution of Eq. 8-42. Nodes are placed at the plus and minus unit values of v in Table 8-9. Outlier nodes are positioned at v = -5 and v = 5. Nodes at v = $\pm$ 5 and v = $\pm$ 4 are used for boundary conditions. The solution of the concentration equation is given by the seven nodes from v = -3 to v = 3.

Program B-8.1 was run for three different conditions of initial values of concentrations and configuration of input of additional materials. The three solutions are shown in Figs. 8-8, 8-9, and 8-10.

Figure 8-8 shows a solution for an initially symmetrical concentration curve with a high central value. Additional material is input in a span to the left and a span to the right of center. The dispersion curve at first reflects the initial high central value but over continued time the inputs cause high concentrations to the left and right of center. Additionally, the material disperses, left and right, causing a concentration increase beyond 400 meters.

Figure 8-9 displays a solution for dispersion with no additional inputs. The initial concentration is high for the two central spans. The solution shows the high central values gradually diminishing as the material spreads outward. The abrupt high flat block of central values cannot be exactly represented by the sliding polynomials and a "bounce" occurs at the center. This bounce diminishes with time as the initial discrete jump between nodes at 120 meters and 50 meters is smoothed by dispersion.

In Figure 8-10 all initial values show zero concentration. Material is input in the two spans immediately to the left of the center. The solution is initially asymmetric, but tends toward symmetry as time progresses. The later curves show the typical normal-curve shape after six rounds of iteration.

8.3.2 Example 8-8: Simple Continuity of Mass

In this two-dimensional example we will analyze the profile of the concentration of some material carried in a stream. The profile will change with time as the material is carried along, and may also change by the input of additional material.

Figure 8-11 shows a differential cube fixed at some point in the stream. The volume of this cube is a $A\Delta x$ where A is the crossectional area of the stream and Δx is the length of the cube. An amount of material, m_1, may be added directly to the cube. At the same time an amount m_2 may be carried into the cube by the stream, and an amount m_3 may be carried out. Then the change of material is

$$\Delta m = m_1 + m_2 - m_3 \qquad (8\text{-}45)$$

If we express this as a concentration of material in the cube we write

$$\Delta m/A\Delta x = m_1/A\Delta x + (m_2 - m_3)/A\Delta x \qquad (8\text{-}46)$$

The change in concentration per unit of time is

$$(\Delta m/\Delta x)/\Delta t = (m_1/\Delta x)/\Delta t + ((m_2 - m_3)/\Delta x)/\Delta t \qquad (8\text{-}47)$$

In Eq. 8-47 the cross-sectional area, A, has been set to unity. If there is a uniform gradient of m from m_2 to m_3 we can write

$$\Delta C/\Delta t = M/\Delta t + (C_2 - C_3)/\Delta x \cdot \Delta x/\Delta t \qquad (8\text{-}48)$$

In the limit this becomes

$$\partial C/\partial t = M/\Delta t + \partial C/\partial x \cdot v \qquad (8\text{-}49)$$

In Eq. 8-49 C is the concentration in, say, grams per kilometer along the stream, v is the velocity of the stream in kilometers per hour, $M/\Delta t$ represents the material added directly to some reach of the stream, expressed as grams per kilometer per hour.

Program B-8.2 in the Appendix was prepared to numerically integrate Eq. 8-49 using finite elements with sliding polynomials. The program as written covers ten spans along the stream and five time steps. Input material $M/\Delta t$ can vary by span and by time steps. Initial values of the nodes, (an initial concentration profile) are input to the program. Boundary values of the concentration at the head of the study reach for each time step are also input. Problem specifics of stream velocity, length of each span, and length of the time step are additional program inputs.

Three runs of the computer program are presented. For Run #1 the input information is given in Table 8-11. Direct input for this run is limited to the first span at the head of the ten-span reach. The direct input is triangular in shape with the largest amount in the center of the five time steps. Results for Run #1 are shown in Fig. 8-12. The concentration profiles rise steeply with direct input of material in the first spans. As input falls off in time steps 4 and 5 the maximum concentration moves downstream. The input for Run #2 is shown in Table 8-12. For this run the direct input is spread across several spans and occurs in three separate time steps. The concentration profiles are shown in Fig. 8-13. These profiles show mainly the direct input of material, but also show the translation down stream.

Information for Run #3 is shown in Table 8-13. For this run material was input to all 10 spans for the first two time steps. The maximum amount of material is input at mid-reach. Twice as much material is input in time step 2 as in time step 1. For Runs #1 and #2 the head-of-reach boundary node was zero, meaning no advection of material by the stream into the reach. For this run the boundary node gradually increases. The concentration profiles are shown in Fig. 8-14. These profiles show that most material has been carried out of the reach by the end of the fifth time step. Concentrations gradually increase with advected material into the head of the reach.

8.3.3 Example 8-9: The Biological Oxygen Demand (BOD) Equation

When organic material that degrades enters a stream oxygen is used in the degradation process and the level of oxygen in the water is reduced. The partial differential equation governing BOD is given by

$$\frac{\partial L}{\partial t} = -U \frac{\partial L}{\partial x} + E \frac{\partial^2 L}{\partial x^2} - (K_1 + K_3)L + a \qquad (8\text{-}50)$$

In this equation L is the BOD concentration, U and E are parameters for a particular site, k_1 is the reaction rate coefficient, and k_3 is the rate coefficient for BOD removal by sedimentation and adsorption. "a" represents the addition of BOD along the reach by sewers and outfalls. Basically, $-(K_1+K_3)$ governs the rate of removal of BOD and "a" accounts for addition of BOD by discrete values by span. The differential terms of the equation account for the changing shape of the BOD concentration with time.

Program B-8.3 was prepared for numerical solution of Eq. 8-50. The program is similar in structure to Program B-8.2. Program B-8.3 was run three times using values shown in Table 8-14. Results are plotted in Figs. 8-15, 8-16, and 8-17. Initial values and head-of-reach boundary conditions are shown in the figures.

Figure 8-15 displays the BOD concentration when BOD loadings are added to a stream with zero initial demand. Material is added only during time steps 1 and 2. This material disperses and moves downstream. The inflow at the head of the reach has zero BOD throughout the time analyzed.

In Fig. 8-16 the stream also has a zero initial BOD. A constant boundary value of 2.0 mg/1 indicates organic material being carried in from upstream. The material gradually creates a BOD through the entire reach. The abrupt change from zero to 2.0 mg/1 for the boundary causes some polynomial bounce in the solution at the head of the reach.

The complex curves in Fig. 8-17 are the result of non-zero initial values, changing head-of-reach boundary values, and the changing span inputs shown in Table 8-15. The wave of initial values passes downstream. The span inputs create a new wave of BOD which disperses and moves downstream. Simultaneously, an increasing BOD is created by the inflow at the head of the reach.

EXERCISES

8-1. Check the value of $\int_{-0.5}^{0.5} C_1(z)dz$ in Table A-2.1 of the Appendix (-0.041667).

8-2. Check the value of $\int_{-0.5}^{0.5} C_1(z)d^2C_2(z)/dz^2$ in Table A-2.1 of the Appendix (−0.0333333).

8-3. Using Example 8-3 as a guide prepare a detailed listing of the steps needed to solve Eq. 8-23.

8-4. Solve the differential equation dy/dx = a(x) using the finite element method. The solution will cover three spans. Use the values for a(x) of 1.0, 2.0, 1.0 as a step function for the three spans. Choose your own transform function z = a + bx. Set a boundary value at the left edge of the three spans.

8-5. How many finite element integrals would be needed for the term $C_j(z)\ C_i(z)$ in the six-point system for one-dimensional sliding polynomials?

8-6. How many finite element integrals would be needed for the term $C_j(z)\ C_i(z)$ in the six-point system for two-dimensional sliding polynomials?

8-7. Compute the finite element integrals needed to solve the equation $dy/dx = c + dx$ using the two-point system.

8-8. Using the integral coefficients of Problem 8-7 solve the equation over three spans with the following values of c and d.

Span:	1	2	3
c:	1	2	1
d:	0.5	0.25	0.5

8-9. Use Program B-8.1.

8-10. Use Program B-8.2.

8-11. Use Program B-8.3.

Table 8-1. Calculations for Piece-wise Linear Representation

PART A. TWO ELEMENTS

Equation							
1-3	$-.0178572\ Y_1$	$+.1827381\ Y_2$	$+.4047619\ Y_3$	$-.0279762\ Y_4$	=	$.5416667\ a_1$	$+.0958333\ b_1$
1-4	$.0017857\ Y_1$	$-.0178572\ Y_2$	$-.0279762\ Y_3$	$+.0023807\ Y_4$	=	$-.0416667\ a_1$	$-.0041667\ b_1$
2-2	$-.0279762\ Y_2$	$+.4047619\ Y_3$	$+.1827381\ Y_4$	$-.0178572\ Y_5$	=	$.5416667\ a_2$	$-.0958333\ b_2$
2-3	$-.0178572\ Y_2$	$+.1827381\ Y_3$	$+.4047619\ Y_4$	$-.0279762\ Y_5$	=	$.5416667\ a_2$	$+.0958333\ b_2$
2-4	$.0017857\ Y_2$	$-.0178572\ Y_3$	$-.0279762\ Y_4$	$+.0023807\ Y_5$	=	$-.0416667\ a_2$	$-.0041667\ b_2$

Set $a_1 = 2$, $b_1 = 4$, $a_2 = 5$, $b_2 = 2$, $Y_1 = Y_3 - 2b$, and combine

I: (1-3)+(2-2)	$.7916667\ Y_3$	$+.1547619\ Y_4$	$-.0178572\ Y_5$	=	3.8404755
II: (1-4)+(2-3)	$.1565476\ Y_3$	$+.4071426\ Y_4$	$-.0279762\ Y_5$	=	2.81428558
III: (2-4)	$-.0178572\ Y_3$	$-.0279762\ Y_4$	$+.0023807\ Y_5$	=	$-.2166667$

Solution

$Y_5 = 9.2244$ $\quad Y_2 = 0$

$Y_4 = 6.0561$ $\quad Y_1 = -4.1247$

$Y_3 = 3.8753$

PART B. THREE ELEMENTS

3-4	$.0023807\ Y_3$	$-.0279762\ Y_4$	$-.0178572\ Y_5$	$+.0017857\ Y_6$	=	$-.0416667\ a_3$	$+.0041667\ b_3$
3-5	$-.0279762\ Y_3$	$+.4047619\ Y_4$	$+.1827381\ Y_5$	$-.0178572\ Y_6$	=	$.5416667\ a_3$	$-.0958333\ b_3$
3-6	$-.0178572\ Y_3$	$+.1827381\ Y_4$	$+.4047619\ Y_5$	$-.0279762\ Y_6$	=	$.5416667\ a_3$	$+.0958333\ b_3$
3-7	$.0017857\ Y_3$	$-.0178572\ Y_4$	$-.0279762\ Y_5$	$+.0023807\ Y_6$	=	$-.0416667\ a_3$	$-.0041667\ b_3$

Set $a_3 = 6.25$, $b_3 = 0.5$, and combine

PART B. THREE ELEMENTS
(Continued)

I': I + (3-1)	.7940474 Y_3	+.1267857 Y_4	-.0357144 Y_5	+.0017857 Y_6	= 3.5821422
II': II + (3-2)	.1285714 Y_3	+.8119045 Y_4	+.1547619 Y_5	-.0178572 Y_6	= 6.1517856
III': III + (3-3)	-.0357144 Y_3	+.1547619 Y_4	+.4071426 Y_5	-.0279762 Y_6	= 3.2166667
IV': (3-4)	.0017857 Y_3	-.0178572 Y_4	-.0279762 Y_5	+.0023807 Y_6	= -.2625

Solution

Y_6 = 8.0483 Y_2 = 0
Y_5 = 6.5497 Y_1 = -4.1538
Y_4 = 5.8964
Y_3 = 3.8462

PART C. FOUR ELEMENTS

4-1	.0023807 Y_4	-.0279762 Y_5	-.0178572 Y_6	+.0017857 Y_7	= -.0416667 a_4 +.0041667 b_4
4-2	-.0279762 Y_4	+.4047619 Y_5	+.1827381 Y_6	-.0178572 Y_7	= .5416667 a_4 -.0958333 b_4
4-3	-.0178572 Y_4	+.1827381 Y_5	+.4047619 Y_6	-.0279762 Y_7	= .5416667 a_4 +.0958333 b_4
4-4	.0017857 Y_4	-.0178572 Y_5	-.0279762 Y_6	+.0023807 Y_7	= -.0416667 a_4 -.0041667 b_4

Set a_4 = 6.625, b_4 = 0.25', and combine

I": I'	.7940474 Y_3	+.1267857 Y_4	-.0357144 Y_5	+.0017857 Y_6		= 3.5821422
II": II' + (4-1)	.1285714 Y_3	+.8142852 Y_4	+.1267857 Y_5	-.0357144 Y_6	+.0017857 Y_7	= 5.8767856
III": III' + (4-2)	-.0357144 Y_3	+.1267857 Y_4	+.8119045 Y_5	+.1547619 Y_6	-.0178572 Y_7	= 6.7812503
IV": IV' + (4-3)	.0017857 Y_3	-.0357144 Y_4	+.1547619 Y_5	+.4071426 Y_6	-.0279762 Y_7	= 3.35
V": (4-4)		.0017857 Y_4	-.0178572 Y_5	-.0278762 Y_6	+.0023807 Y_7	=-.2770833

Solution

Y_7 = 7.2419 Y_4 = 5.8825 Y_2 = 0
Y_6 = 6.7644 Y_3 = 3.8479 Y_1 = -4.1521
Y_5 = 6.4728

Table 8-2. Piece-Wise Linear Representation

	Number of Elements			
NODE	1	2	3	4
Y_1	-3.9997	-4.1247	-4.1538	-4.1521
Y_2	0	0	0	0
.2		0.8182	0.8177	0.8181
.4		1.6494	1.6468	1.6481
.6		2.4592	2.4518	2.4539
.8		3.2127	3.1968	3.1992
Y_3	4.0003	3.8753	3.8462	3.8479
.2		4.4041	4.3935	4.3940
.4		4.8222	4.8626	4.8616
.6		5.1940	5.2631	5.2596
.8		5.5839	5.6045	5.5970
Y_4	8.0040	6.0561	5.8964	5.8825
.2			6.1029	6.0978
.4			6.2177	6.2369
.6			6.2946	6.3275
.8			6.3873	6.3970
Y_5		9.2244	6.5497	6.4728
.2				6.5473
.4				6.6020
.6				6.6487
.8				6.6990
Y_6			8.0483	6.7644
Y_7				7.2419

Table 8-3. Exponential Approximation

PART A. ELEMENT ONE

1-1:	$0\ Y_1$	$+.0458333\ \underline{\underline{Y}}_2$	$-.05\ Y_2$	$+.0041667\ Y_2$	$+\frac{1}{2}[.0023807\ Y_1$	$-.0279762\ \underline{\underline{Y}}_2$	$-.0178572\ Y_3$	$+.0017857\ Y_4$	$-.0416667] = 0$
1-3:	$.05\ Y_1$	$-.5958333\ \underline{\underline{Y}}_2$	$+.5\ Y_3$	$+.0458333\ Y_4$	$+\frac{1}{2}[-.0178572\ Y_2$	$+.1827381\ \underline{\underline{Y}}_2$	$+.4047619\ Y_3$	$-.0279762\ Y_4$	$+.5416667] = 0$
1-4:	$-.0041667\ Y_0$	$+.05\ \underline{\underline{Y}}_1$	$-.0458333\ Y_2$	$+0\ Y_3$	$+\frac{1}{2}[.0017857\ Y_1$	$-.0178572\ \underline{\underline{Y}}_2$	$-.0279762\ Y_3$	$+.0023807\ Y_4$	$-.0416667] = 0$

Solution with $Y_2 = 10$: $Y_1 = 16.6521$, $Y_3 = 5.6722$, $:Y_4 = 3.3234$

PART B. ELEMENTS ONE AND TWO

2-2:	$-.0458333\ \underline{\underline{Y}}_2$	$-.5\ Y_3$	$+.5958333\ Y_3$	$-.05\ Y_5$	$+\frac{1}{2}[-.0279762\ \underline{\underline{Y}}_2$	$+.4047619\ Y_3$	$+.1827381\ Y_4$	$-.0178572\ Y_5$	$+.5416667] = 0$
2-3:	$.05\ \underline{\underline{Y}}_2$	$-.5958333\ Y_3$	$+.5\ Y_4$	$+.0458333\ Y_5$	$+\frac{1}{2}[-.0178572\ \underline{\underline{Y}}_2$	$+.1827381\ Y_3$	$+.4047619\ Y_4$	$-.0279762\ Y_5$	$+.5416667] = 0$
2-4:	$-.0041667\ \underline{\underline{Y}}_2$	$+.05\ Y_3$	$-.0458333\ Y_4$	$+\ 0\ Y_5$	$+\frac{1}{2}[.0017857\ \underline{\underline{Y}}_2$	$-.0178572\ Y_3$	$-.0279762\ Y_4$	$+.0023807\ Y_5$	$-.0416667] = 0$

Composite: (1-1), (1-3) + (2-2), (1-4) + (2-3), (2-4)

Solution with $\underline{\underline{Y}}_2 = 10$: $Y_1 = 17.7125$, $Y_3 = 5.6684$, $Y_4 = 3.0279$, $Y_5 = 1.6822$

PART C. ELEMENTS ONE, TWO, AND THREE

3-1:	$.0\ Y_3$	$+.0458333\ Y_4$	$+.05\ Y_5$	$+.0041667\ Y_6$	$+\frac{1}{2}[.0023807\ Y_3$	$-.0279762\ Y_4$	$-.0178572\ Y_5$	$+.0017857\ Y_6$	$-.0416667] = 0$
3-2:	$-.0458333\ Y_3$	$-.5\ Y_4$	$+.5958333\ Y_5$	$-.05\ Y_6$	$+\frac{1}{2}[-.0279762\ Y_3$	$+.4047619\ Y_4$	$+.1827381\ Y_5$	$-.0178572\ Y_6$	$+.5416667] = 0$
3-3:	$.05\ Y_3$	$-.5958333\ Y_4$	$+.5\ Y_5$	$+.0458333\ Y_6$	$+\frac{1}{2}[-.0178572\ Y_3$	$+.1827381\ Y_4$	$+.4047619\ Y_5$	$-.0279762\ Y_6$	$+.5416667] = 0$
3-4:	$-.0041667\ Y_3$	$+.05\ Y_4$	$-.0458333\ Y_5$	$+\ 0\ Y_6$	$+\frac{1}{2}[.0017857\ Y_3$	$-.0178572\ Y_4$	$-.0279762\ Y_5$	$+.0023807\ Y_6$	$-.0416667] = 0$

Composite: (1-1), (1-3) + (2-2) + (3-1), (1-4) + (2-3) + (3-2), (2-4) + (3-3), (3-4)

Solution with $\underline{\underline{Y}}_2 = 10$: $Y_1 = 17.0771$, $Y_3 = 5.6575$, $Y_4 = 3.0520$, $Y_5 = 1.4500$, $Y_6 = 0.6285$

Table 8-4.

Comparisons of Exponential Approximation

Global x	SLIDING POLYNOMIAL Number of Elements 1	2	3	Closed Solution
0	16.6521	17.7125	17.0771	
1	10	10	10	10
.2	8.9540	8.8902	8.9287	8.9532
.4	8.0065	7.9427	7.9827	8.0060
.6	7.1492	7.1169	7.1382	7.1490
.8	6.3739	6.3723	6.3711	6.3735
2	5.6722	5.6684	5.6575	5.6718
.2		5.0117	5.0092	5.0369
.4		4.4291	4.4421	4.4624
.6		3.9110	3.9386	3.9426
.8		3.4477	3.4811	3.4723
3	3.3234	3.0297	3.0520	3.0467
.2			2.6549	2.6616
.4			2.3015	2.3131
.6			1.9864	1.9978
.8			1.7044	1.7126
4		1.6822	1.4500	1.4544
5			0.6285	

Table 8-5.

Polynomial Approximation - Element Equations

ONE ELEMENT

1-1:	skip			
1-2:	-1.1333333 Y_2	-.0166667 Y_4	+(.5583333 +.5916667) $\underline{\underline{Y}}_3$	-C'(.5416667) = 0
1-3:	skip			
1-4:	-.0166667 Y_2	-.0333333 Y_4	+(.0583333 -.0083333) $\underline{\underline{Y}}_3$	-C'(-.0416667) = 0

TWO ELEMENTS

1-1:	skip				
1-2:	-1.1333333 Y_2	+(.5583333	+.5916667) Y_3	-.0166667 $\underline{\underline{Y}}_4$	-C'(.5416667) = 0
1-3:	.5916667 Y_2	-(1.1333333	+.0166667) Y_3	+.5583333 $\underline{\underline{Y}}_4$	-C'(.5416667) = 0
1-4:	skip				
2-1:	- .0333333 Y_2	+.0583333 Y_3	-.0083333 Y_5	-.0166667 $\underline{\underline{Y}}_4$	-C'(-.6416667) = 0
2-2:	.5583333 Y_2	-1.1333333 Y_3	-.0166667 Y_5	+.5916667 $\underline{\underline{Y}}_4$	-C'(.5416667) = 0
2-3:	skip				
2-4:	-.0083333 Y_2	-.0166667 Y_3	-.0333333 Y_5	+.0583333 $\underline{\underline{Y}}_4$	-C'(-.0416667) = 0

Composite: (1-2)+(2-1), (1-3)+(2-2), (2-4)

THREE ELEMENTS

1-1:	skip				
1-2:	-1.1333333 Y_2	+(.5583333	+.5916667) Y_3	-.0166667 Y_4	-C'(.5416667) = 0
1-3:	.5916666 Y_2	-(1.1333333	+.0166667) Y_3	+.5583333 Y_4	-C'(.5416667) = 0
1-4:	-.0166667 Y_2	+(.0583333	-.0083333) Y_3	-.0333333 Y_4	-C'(-.0416667) = 0
2-1:	-.0333333 Y_2	+.0583333 Y_3	-.0166667 Y_4	-.0083333 $\underline{\underline{Y}}_5$	-C'(-.0416667) = 0
2-2:	.5583333 Y_2	-1.1333333 Y_3	+.5916667 Y_4	-.0166667 $\underline{\underline{Y}}_5$	-C'(.5416666) = 0
2-3:	-.0166667 Y_2	+.5916667 Y_3	-1.13333333 Y_4	+.5583333 $\underline{\underline{Y}}_5$	-C'(.5416667) = 0
2-4:	skip				

```
3-1:                  -.0333333 Y3    +.0583333 Y4  -.0083333 Y6   -.016666y Y5   -C'(-.0416667) = 0
3-2:                   .5583333 Y3   -1.1333333 Y4  -.0166667 Y6   +.5916667 Y5   -C'( .5416667)  = 0
3-3:                  skip
3-4:                  -.0083333 Y3   - .0166667 Y4  -.0333333 Y6   +/-593333 Y5   -C'(-.0416667) = 0
                      Composite: (1-2)+(2-1), (1-3)+(2-2)+(3-1), (1-4)+(2-3)+(3-2), (3-4)
```

FOUR ELEMENTS

```
1-1:                  Same as Three Elements
1-2:                  Same as Three Elements
1-3:                  Same as Three Elements
1-4:                  Same as Three Elements

2-1:      -.0333333 Y2  +.0583333 Y3  -.0166667 Y4  -.0083333 Y5                              -C'(-.0416667) = 0
2-2:       .5583333 Y2 -1.1333333 Y3  +.5916667 Y4  -.0166667 Y5                              -C'( .5416667)  = 0
2-3:      -.0166667 Y2  +.5916667 Y3 -1.1333333 Y4  +.5583333 Y5                              -C'( .5416667)  = 0
2-4:      -.0083333 Y2  -.0166667 Y3  +.0583333 Y4  -.0333333 Y5                              -C'(-.0416667) = 0

3-1:                    -.0333333 Y3  +.0583333 Y4  -.0166667 Y5               -.0083333 Y6  -C'(-.0416667)  = 0
3-2:                     .5583333 Y3 -1.1333333 Y4  +.5916667 Y5               -.0166667 Y6  -C'( .5416667)  = 0
3-3:                    -.0166667 Y3  +.5916667 Y4 -1.1333333 Y5               +.5583333 Y6  -C'( .5416667)  = 0
3-4:                    skip

4-1:                                  -.0333333 Y4  +.0583333 Y5  -.083333 Y7  -.0166667 Y6  -C'(-.0416667)  = 0
4-2:                                   .5583333 Y4 -1.1333333 Y5  -.016667 Y7  +.5916667 Y6  -C'( .5416667)  = 0
4-3:                    skip
4-4:                                  -.0083333 Y4  -.0166667 Y5  -.0333333 Y7 +.0583333 Y6  -C'(-.0416667) = 0
                        Composite: (1-2)+(2-1), (1-3)+(2-2)+(3-1), (1-4)+(2-3)+(3-2)+(4-1), (2-4)+(3-3)+(4-2), (4-4)
```

Table 8-6.

Comparison of Polynomial Approximations

Global x	Number of Elements: 1	2	3	4	Closed Solution
0	2.0				
.5		1.25			
.667			1.1111		
.75				1.0625	
1.0	1.0	1.0	1.0	1.0	1.0
1.25	1.0625*			1.0625	1.0625
1.3333	1.1111*		1.1111		1.1111
1.5	1.25*	1.25		1.25	1.25
1.6667	1.4444*		1.4444		1.4444
1.75	1.5625*			1.5625	1.5625
2.0	2.0	2.0	2.0	2.0	2.0
2.25				2.5625	
2.3333			2.7778		
2.5		3.25			
3.0	5.0				

* Interpolated with sliding polynomial

Table 8-7. Element Equations for Example 8-5

1-0:	$-.0333333\ Y_0$	$+.0583333\ \underline{\underline{Y_1}}$	$-.0166667\ Y_2$	$-.0083333\ Y_3$	$-.0023807k\ Y_0$	$+.0279762k\ \underline{\underline{Y_1}}$	$+.0178572k\ Y_2$	$-.0017857k\ Y_3$	$+.0416667\ a = 0$
1-1:	skip								
1-2:	$-.0166667\ Y_0$	$+.5916667\ \underline{\underline{Y_1}}$	$-1.1333333\ Y_2$	$+.5583333\ Y_3$	$+.0178572k\ Y_0$	$-.1827381k\ \underline{\underline{Y_1}}$	$-.4047619k\ Y_2$	$+.0279762k\ Y_3$	$-.5416667\ a = 0$
1-3:	$-.0083333\ Y_0$	$-.0166667\ \underline{\underline{Y_1}}$	$+.0583333\ Y_2$	$-.0333333\ Y_3$	$-.0017857k\ Y_0$	$+.0178572k\ \underline{\underline{Y_1}}$	$+.0279762k\ Y_2$	$-.0023807k\ Y_3$	$+.0416667\ a = 0$
2-0:	skip								
2-1:	$.5583333\ \underline{\underline{Y_1}}$	$-1.1333333\ Y_2$	$+.5916667\ Y_3$	$-.01666672k\ Y_1$	$+.0279762k\ \underline{\underline{Y_1}}$	$-.4047619k\ Y_2$	$-.1827381k\ Y_3$	$+.0178572k\ Y_4$	$-.5416667\ a = 0$
2-2:	$-.0166667\ \underline{\underline{Y_1}}$	$+.5916667\ Y_2$	$-1.1333333\ Y_3$	$+.5583333\ Y_4$	$+.0178572k\ \underline{\underline{Y_1}}$	$-.1827381k\ Y_2$	$-.4047619k\ Y_3$	$+.0279762k\ Y_4$	$-.5416667\ a = 0$
2-3:	$-.0083333\ \underline{\underline{Y_1}}$	$-.0166667\ Y_2$	$+.0583333\ Y_3$	$-.0333333\ Y_4$	$-.0017857k\ \underline{\underline{Y_1}}$	$+.0178572k\ Y_2$	$+.0279762k\ Y_3$	$-.0023807k\ Y_4$	$+.0416667\ a = 0$
3-0:	$-.0333333\ Y_2$	$+.0583333\ Y_3$	$-.0166667\ Y_4$	$-.0083333\ \underline{\underline{Y_5}}$	$-.0023807k\ Y_2$	$+.0279762k\ Y_3$	$+.0178572k\ Y_4$	$-.00178527k\ \underline{\underline{Y_5}}$	$+.0416667\ a = 0$
3-1:	$.5583333\ Y_2$	$-1.1333333\ Y_3$	$+.5916667\ Y_4$	$-.0166667\ \underline{\underline{Y_5}}$	$+.0279762k\ Y_2$	$-.4047619k\ Y_3$	$-.1827381k\ Y_4$	$+.0178572k\ \underline{\underline{Y_5}}$	$-.5416667\ a = 0$
3-2:	$-.0166667\ Y_2$	$+.5916667\ Y_3$	$-1.1333333\ Y_4$	$+.5583333\ \underline{\underline{Y_5}}$	$+.0178572k\ Y_2$	$-.1827381k\ Y_3$	$-.4047619k\ Y_4$	$+.0279762k\ \underline{\underline{Y_5}}$	$-.5416667\ a = 0$
3-3:	skip								
4-0:	$-.0333333\ Y_3$	$+.0583333\ Y_4$	$-.0166667\ \underline{\underline{Y_5}}$	$-.0083333\ Y_6$	$-.0023807k\ Y_3$	$+.0279762k\ Y_4$	$+.0178572k\ \underline{\underline{Y_5}}$	$-.0017857k\ Y_6$	$+.0416667\ a = 0$
4-1:	$.5583333\ Y_3$	$-1.1333333\ Y_4$	$+.5916667\ \underline{\underline{Y_5}}$	$-.0166667\ Y_6$	$+.0279762k\ Y_3$	$-.4047619k\ Y_4$	$-.1827381k\ \underline{\underline{Y_5}}$	$+.0178572k\ Y_6$	$-.5416667\ a = 0$
4-2:	skip								
4-3:	$-.0083333\ Y_3$	$-.0166667\ Y_4$	$+.0583333\ Y_5$	$-.0333333\ Y_6$	$-.0017857k\ Y_3$	$+.0178572k\ Y_4$	$+.0279762k\ \underline{\underline{Y_5}}$	$-.0023807k\ Y_6$	$+.0416667\ a = 0$

Composite: (1-0), (1-2)+(2-1)+(3-0), (1-3)+(2-2)+(3-1)+(4-0), (2-3)+(3-2)+(4-1), (4-3)

Solutions: $Y_1 = 10$, $Y_5 = 50$: $Y_0 = 31.5956$, $Y_2 = 2.6243$, $Y_3 = 2.5943$, $Y_4 = 11.8291$, $Y_6 = 156.1998$

$Y_1 = 70$, $Y_5 = 60$: $Y_0 = 217.5713$, $Y_2 = 17.8066$, $Y_3 = 7.7725$, $Y_4 = 15.3428$, $Y_6 = 186.3389$

Table 8-8. Element Equations for Example 8-6

Equation	Y_0	Y_1	Y_2	Y_3	Y_4	Y_5	Y_6	Y_7	Y_8	b
		ELEMENT TWO								
2-0'		-.1273817	.1467261	-.1154764	-.0288691					
2-1'		2.1800595	-3.2714286	2.8335566	-.1071431					
		-.0687500	-.75	.8937500	-.075					
2-1		2.1113095	-4.0214286	3.7273066	-.1821431					-2.7083333
2-2'		-.1071431	2.8335566	-3.2714286	2.1800595					
		.075	-.89375	.75	.0687500					
2-2		-.0321431	1.9408066	-2.5214286	2.2488095					-2.7083333
2-3'		-.0288691	-.1154764	.1467261	-.1273817					
		-.0062501	.075	-.0687500	0					
2-3		-.0351192	-.0404764	.0779761	-.1273817					.208333

Equations for Elements Three through Six constructed similarly.

Skip 1-0, 1-1, 2-0. In 1-2, 1-3, 2-1, 2-2, 2-3 use $Y_1 = 10$ and $Y_0 = Y_2 - 1$

Composite		Y_2	Y_3	Y_4	Y_5	Y_6	Y_7	Y_8		b
(1-2)+(2-1)+(3-0)		-7.2523819	6.1915922	-.4226195	-.0184523	0.0	0.0	0.0	=	-41.34151
(1-3)+(2-2)+(3-1)+(4-0)		4.0991397	-7.2976206	6.8790921	-.5226195	-.0142856	0.0	0.0	=	6.19524
(2-3)+(3-2)+(4-1)+(5-0)		-.0226198	3.4415923	-7.2976206	7.566592	-.6226195	-.0101189	0.0	=	5.35119
(3-3)+(4-2)+(5-1)+(6-0)		-.0392859	.0773805	2.7540924	-7.2976206	8.2540919	-.7226195	-.0059522	=	5.00
(4-3)+(5-2)+(6-1)		0.0	-.0434526	.1773805	2.0665925	-7.1702389	8.4963733	-.3821431	=	5.208333
(5-3)+(6-2)		0.0	0.0	-.0476193	.2773805	-.5030504	-.6488103	2.4321427	=	2.50
(6-3)		0.0	0.0	0.0	-.0517860	.1595236	-.1053571	-.1273817	=	-.208333
Solution:		1.4425	-5.3951	-5.8662	-2.4374	0.6349	1.9190	1.8343		

ELEMENT TWO

2-0'	-.1273817	.1467261	-.1154764	-.0288691	
2-1'	2.1800595	-3.2714286	2.8335566	-.1071431	
	-.0687500	-.75	.8937500	-.075	
2-1	2.1113095	-4.0214286	3.7273066	-.1821431	-2.7083333
2-2'	-.1071431	2.8335566	-3.2714286	2.1800595	
	.075	-.89375	.75	.0687500	
2-2	-.0321431	1.9408066	-2.5214286	2.2488095	-2.7083333
2-3'	-.0288691	-.1154764	.1467261	-.1273817	
	-.0062501	.075	-.0687500	0	
2-3	-.0351192	-.0404764	.0779761	-.1273817	-.208333

Equations for Elements Three through Six constructed similarly.

Skip 1-0, 1-1, 2-0. In 1-2, 1-3, 2-1, 2-2, 2-3 use $Y_1 = 10$ and $Y_0 = Y_2 - 1$

Composite

(1-2)+(2-1)+(3-0)	-7.2523819	6.1915922	-.4226195	-.0184523	0.0	0.0	0.0	=-41.34151
(1-3)+(2-2)+(3-1)+(4-0)	4.0991397	-7.2976206	6.8790921	-.5226195	-.0142856	0.0	0.0	= 6.19524
(2-3)+(3-2)+(4-1)+(5-0)	-.0226198	3.4415923	-7.2976206	7.566592	-.6226195	-.0101189	0.0	= 5.35119
(3-3)+(4-2)+(5-1)+(6-0)	-.0392859	.0773805	2.7540924	-7.2976206	8.2540919	-.7226195	-.0059522	= 5.00
(4-3)+(5-2)+(6-1)	0.0	-.0434526	.1773805	2.0665925	-7.1702389	8.4963733	-.3821431	= 5.208333
(5-3)+(6-2)	0.0	0.0	-.0476193	.2773805	-.5030504	-.6488103	2.4321427	= 2.50
(6-3)	0.0	0.0	0.0	-.0517860	.1595236	-.1053571	-.1273817	= -.208333
Solution:	1.4425	-5.3951	-5.8662	-2.4374	0.6349	1.9190	1.8343	

Table 8-9. Corresponding Values Of x And v For Example 8 − 7

±v	±x
0	0
1	50
2	120.5
3	240.9
3.999	1441.5
4	∞

Table 8-10. Scale Change From x To v For Example 8-7

Nodal Value (x)	$(4b\ e^{-bx})$	Average for Span
0	.000529670	
		.0004138
50	.000297939	
		.0002152
120.5	.000132373	
		.0000828
240.9	.000033120	
		.0000166
1441.5	0	

Table 8-11. Input Information For Run #1 Of Program B-8.2

Direct Input of Material in Gms/Kilometer/Hour

	Time Step				
Span	1	2	3	4	5
1	10	20	30	20	10
2	0	0	0	0	0
3	0	0	0	0	0
4	0	0	0	0	0
5	0	0	0	0	0
6	0	0	0	0	0
7	0	0	0	0	0
8	0	0	0	0	0
9	0	0	0	0	0
10	0	0	0	0	0

Initial nodes: All zero

Head of reach boundary nodes: All zero

Stream velocity: 1.5 kilometers per hour

Span length: 1 kilometer

Time step: 1 hour

Table 8-12. Input Information For Run #2 Of Program B-8.2

Direct Input of Material in Gms/Kilometer/Hour

	Time Step				
Span	1	2	3	4	5
1	5	0	3	0	0
2	4	0	2	0	0
3	3	0	1	0	0
4	2	0	0	1	0
5	1	0	0	2	0
6	0	0	0	3	0
7	0	0	0	2	0
8	0	0	0	1	0
9	0	0	0	0	0
10	0	0	0	0	0

Initial nodes: All zero

Head of reach boundary nodes: All zero

Stream velocity: 1 kilometer per hour

Span length: 0.5 kilometer

Time step: 0.5 hour

Table 8-13. Input Information For Run #3 Of Program B-8.2

Direct Input of Material in Gms/Kilometer/Hour

	Time Step				
Span	1	2	3	4	5
1	0.05	0.1	0	0	0
2	0.1	0.2	0	0	0
3	0.2	0.4	0	0	0
4	0.35	0.7	0	0	0
5	0.55	1.1	0	0	0
6	0.55	1.1	0	0	0
7	0.35	0.7	0	0	0
8	0.2	0.4	0	0	0
9	0.1	0.2	0	0	0
10	0.05	0.1	0	0	0

Initial nodes: All zero

Head of reach boundary nodes:

Time Step:	1	2	3	4	5
Node:	0	0.2	0.4	0.6	0.8

Stream velocity: 1.5 kilometer per hour

Span length: 0.5 kilometer

Time step: 0.25 hour

Table 8-14. Input Information For Example Program 8-9

	Run 1	Run 2	Run 3
U-Kilometer/hr	3	5	2
E-Kilometer^2/hr	4	5	2
$(K_1 + K_3)$-1/hrs	0.6	0.4	0.2
a-(mg/l)/hrs	*	**	***

* 3 units in Span 2 for time steps 1 and 2. All other values were zero.

** All values were zero

***

	Time Step				
Span	1	2	3	4	5
1	0	0	0	0	0
2	0	0.5	1.0	0.5	0
3	0	1.0	2.0	1.0	0
4	0	0.5	1.0	0.5	0
5	0	0	0	0	0
6	0	0	0	0	0
7	0	0	0	0	0
8	0	0	0	0	0
9	0	0	0	0	0
10	0	0	0	0	0

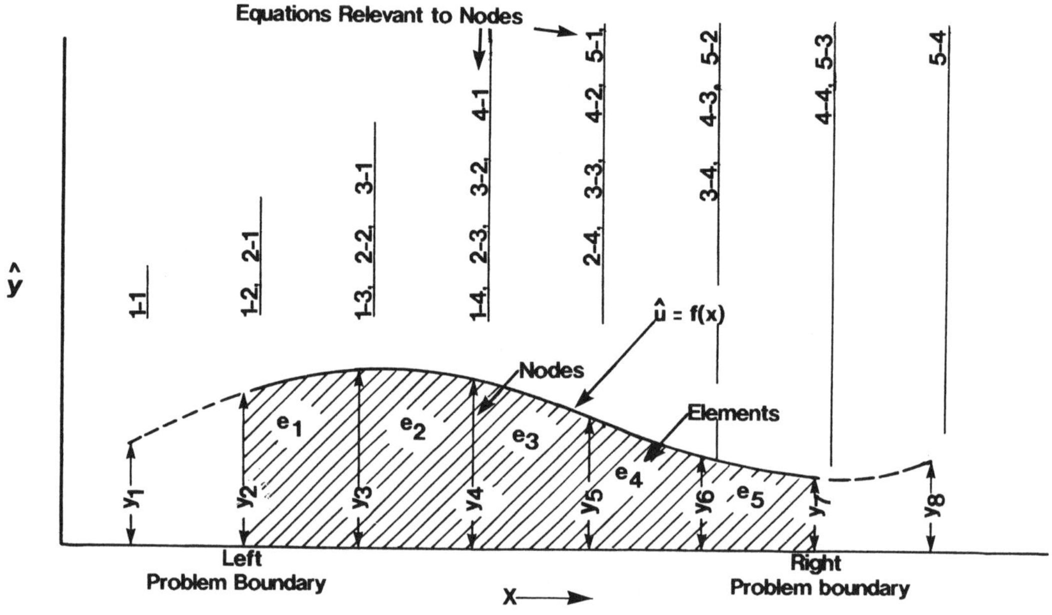

Fig. 8-1. Schematic for Compositing Equations

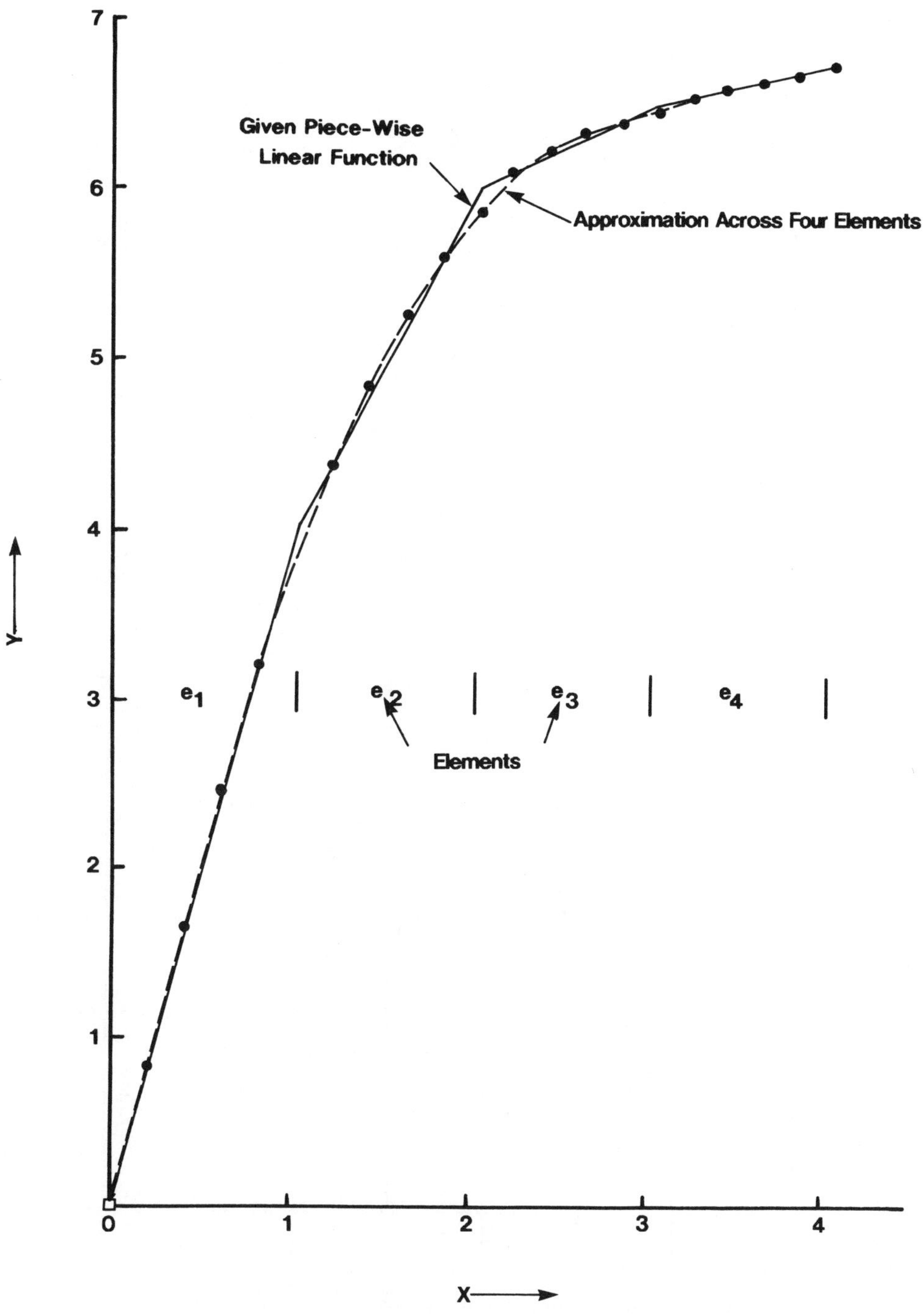

Fig. 8-2. Piece-wise Linear Approximation

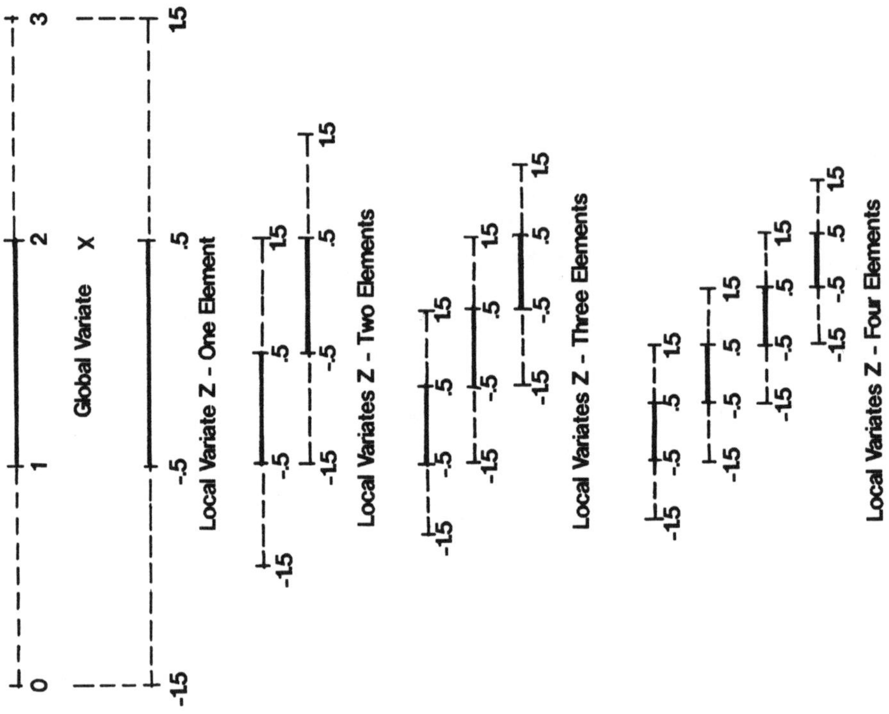

Fig. 8-4. **Scales for Polynomial Approximation**

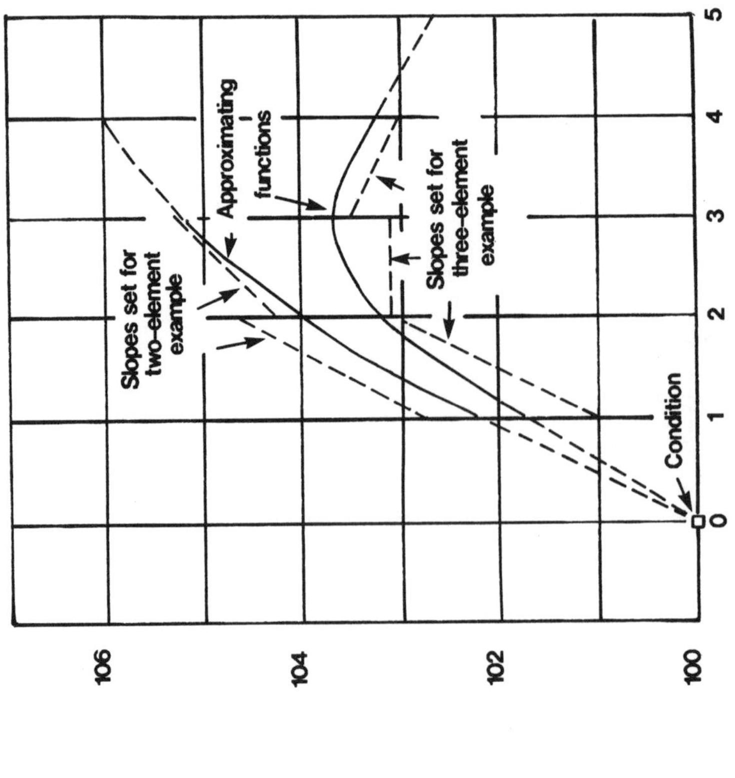

Fig. 8-3. **Piece-wise Constant Gradients**

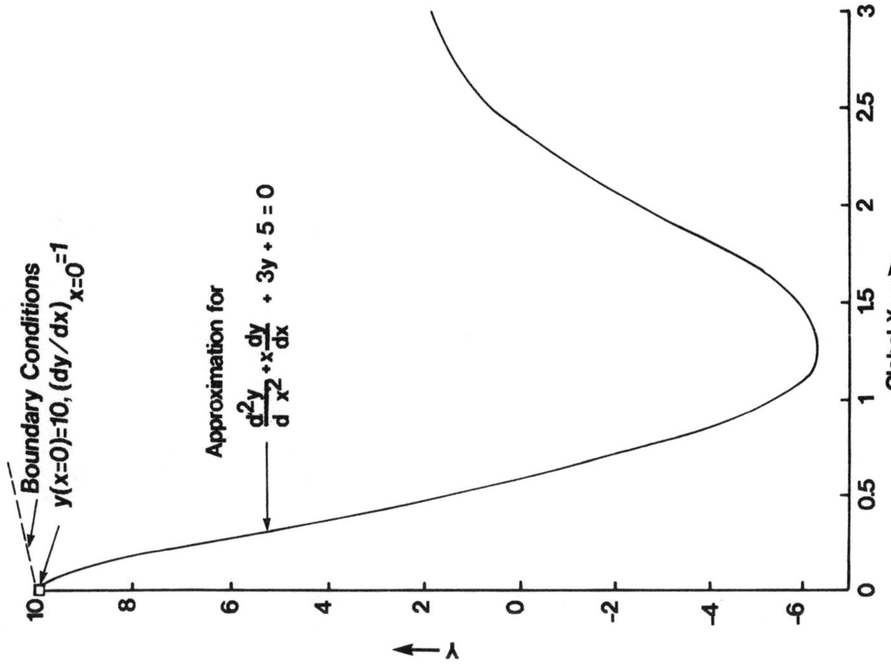

Fig. 8-6. **Solution for Example 8-6**

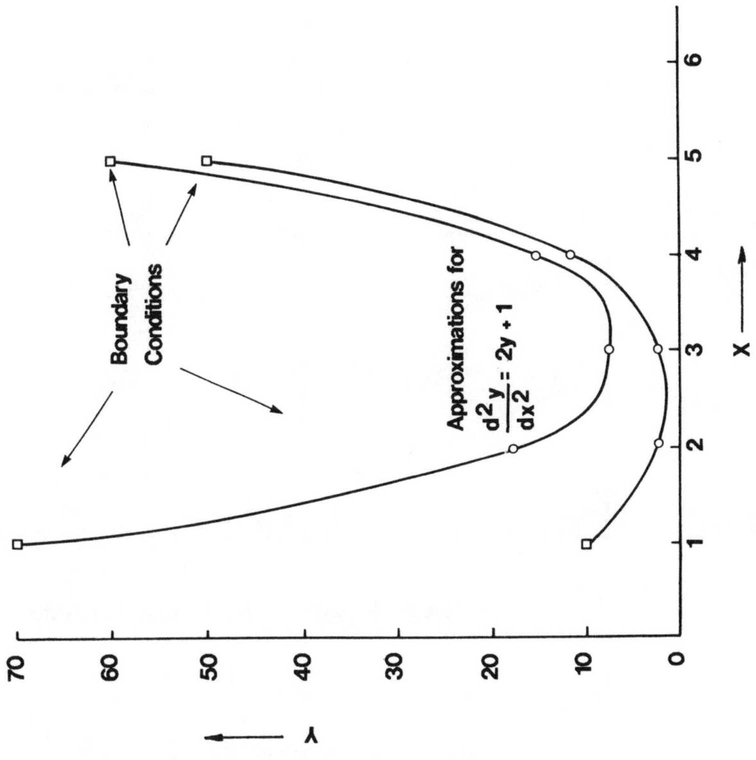

Fig. 8-5. **Solutions for Example 8-5**

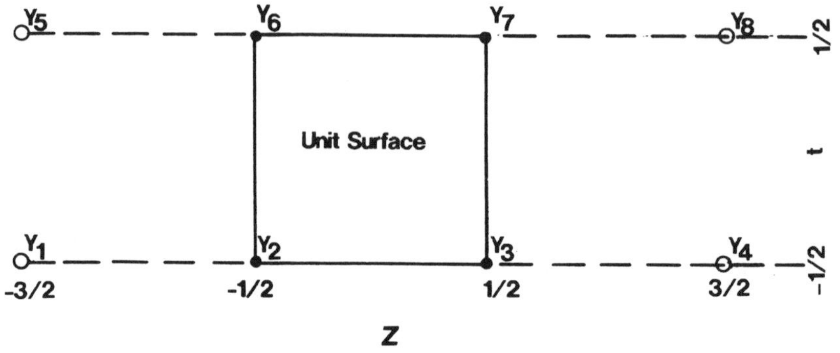

Part a. Definition of Unit Surface

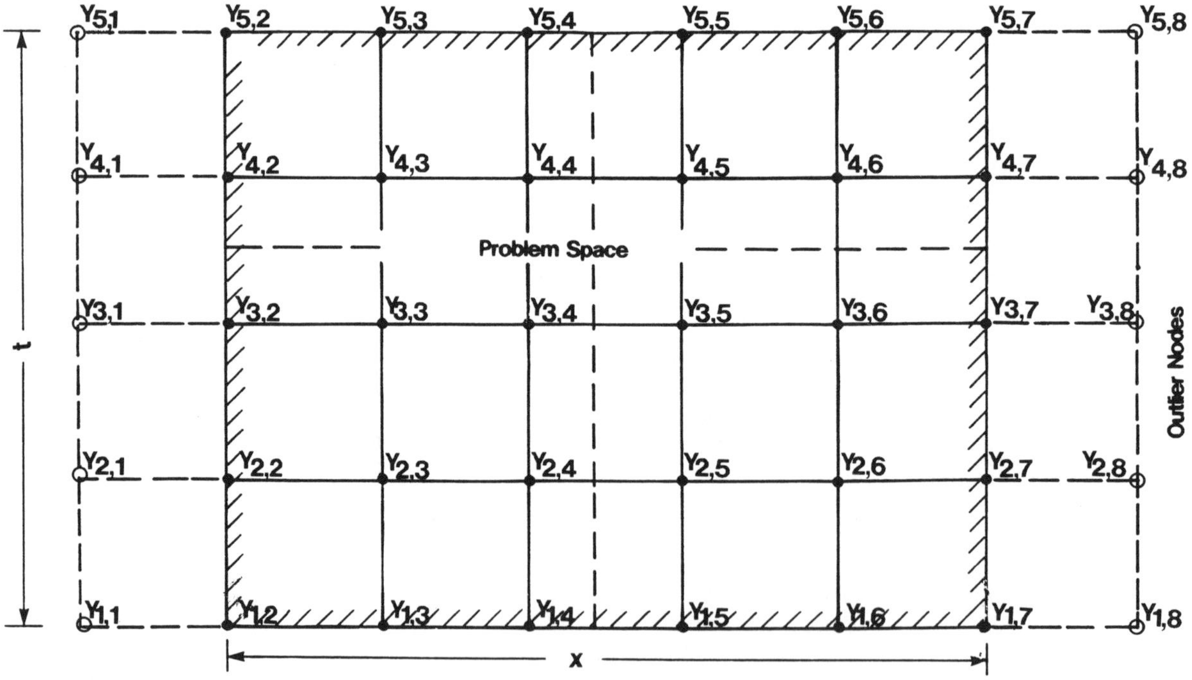

Part b. Schematic for Problem Solution

Fig. 8-7. **Step-wise Finite Elements**

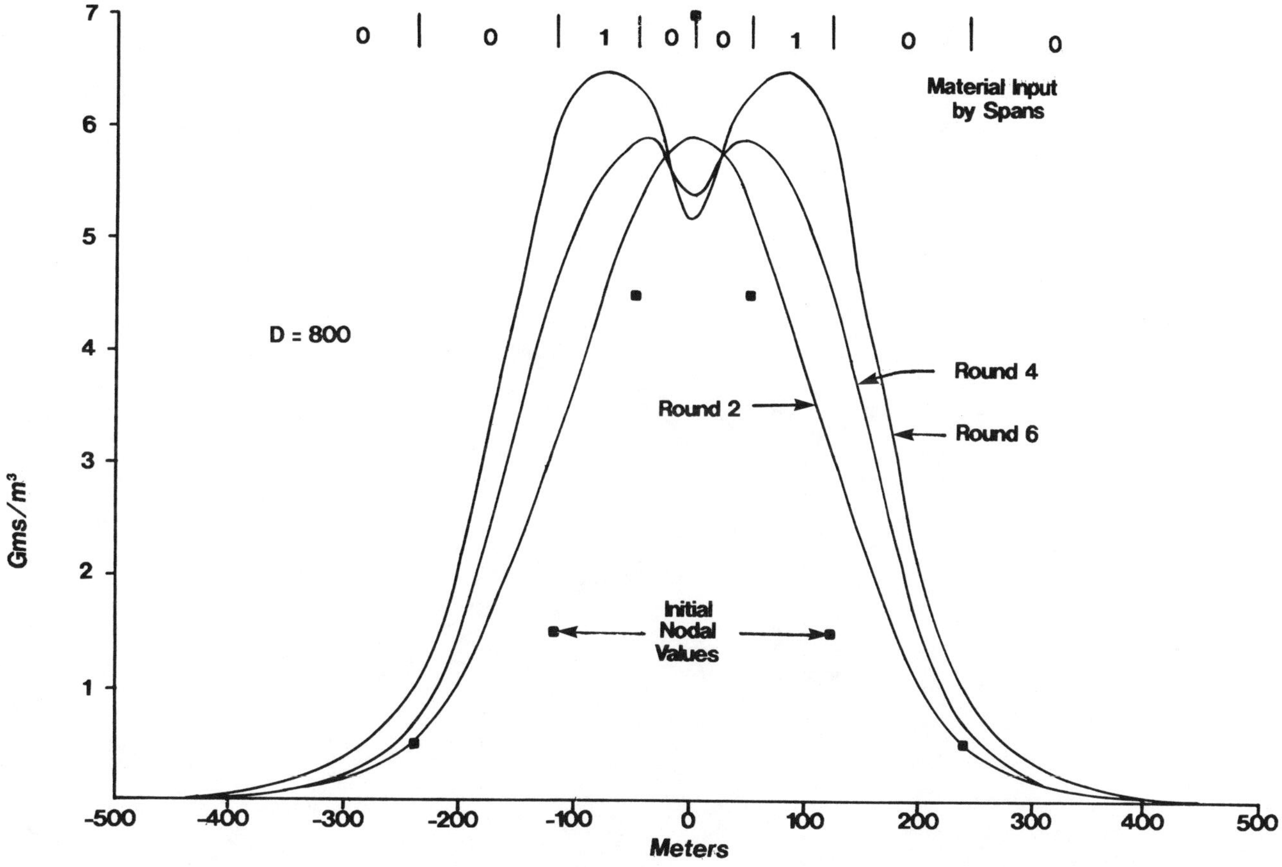

Fig. 8-8. Dispersion with Divided Inputs

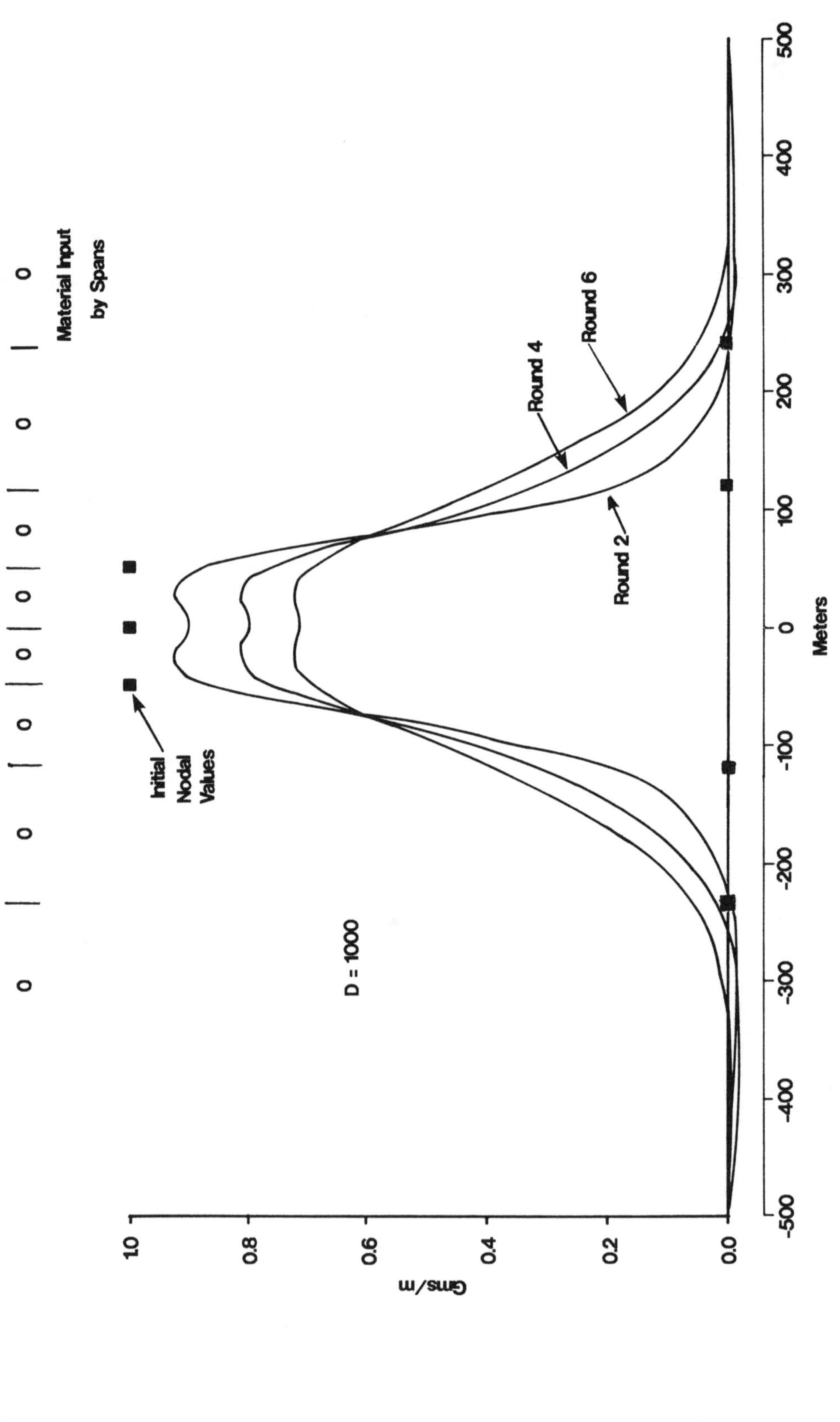

Fig. 8-9. Dispersion With No Additional Inputs

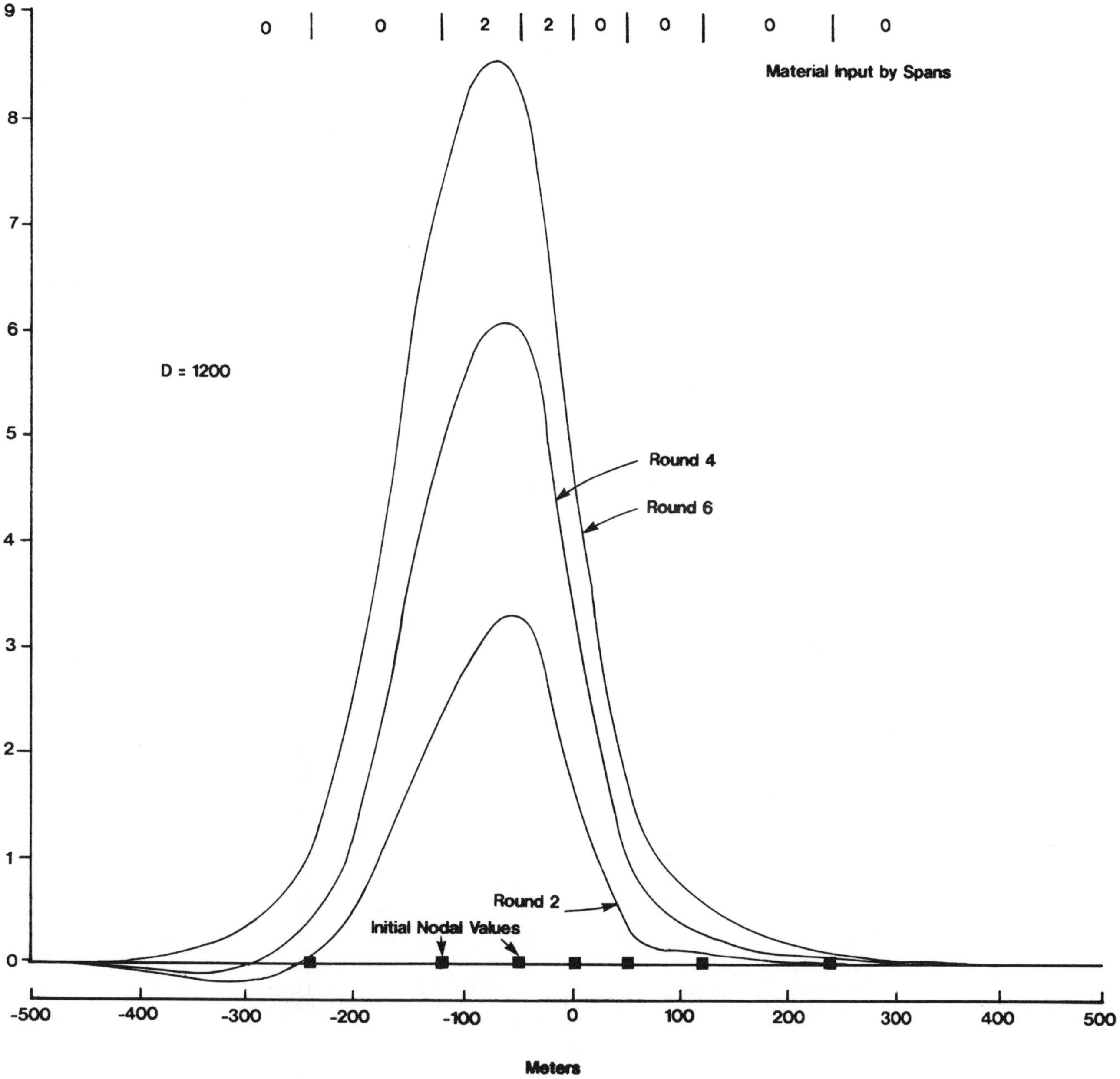

Fig. 8-10. Dispersion With No Initial Concentration

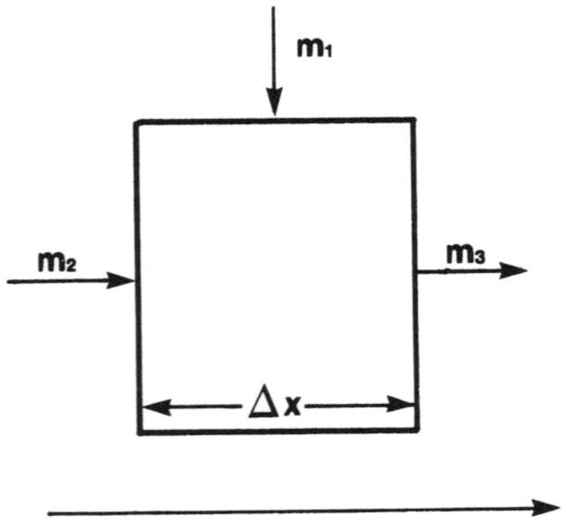

Fig. 8-11. **Unit Volume for Example 8.7.**

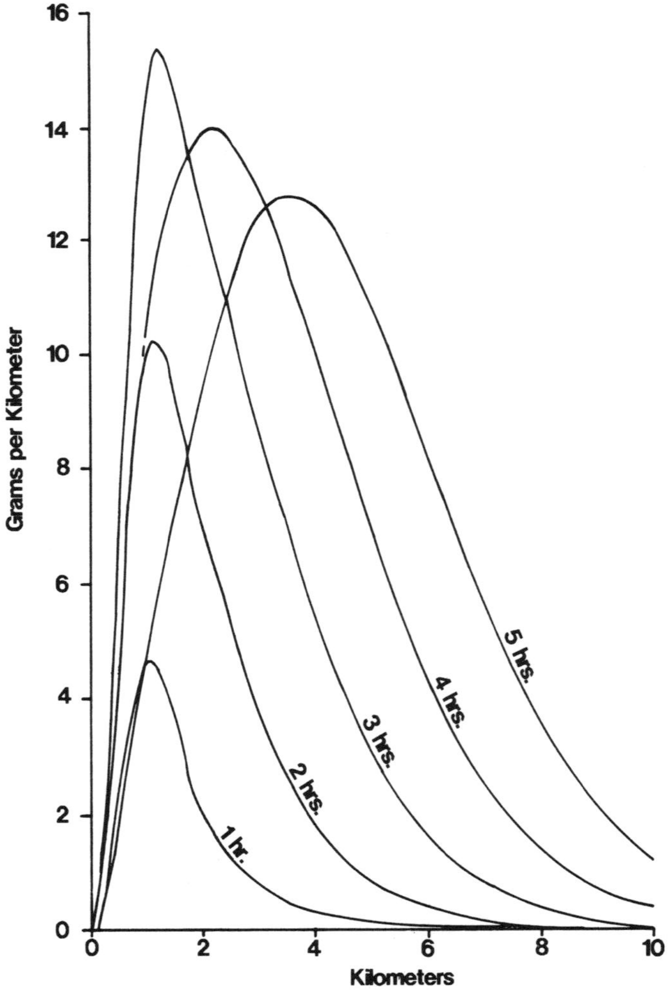

Fig. 8-12. **Concentration Profiles for Run #1 of Program B-8.2**

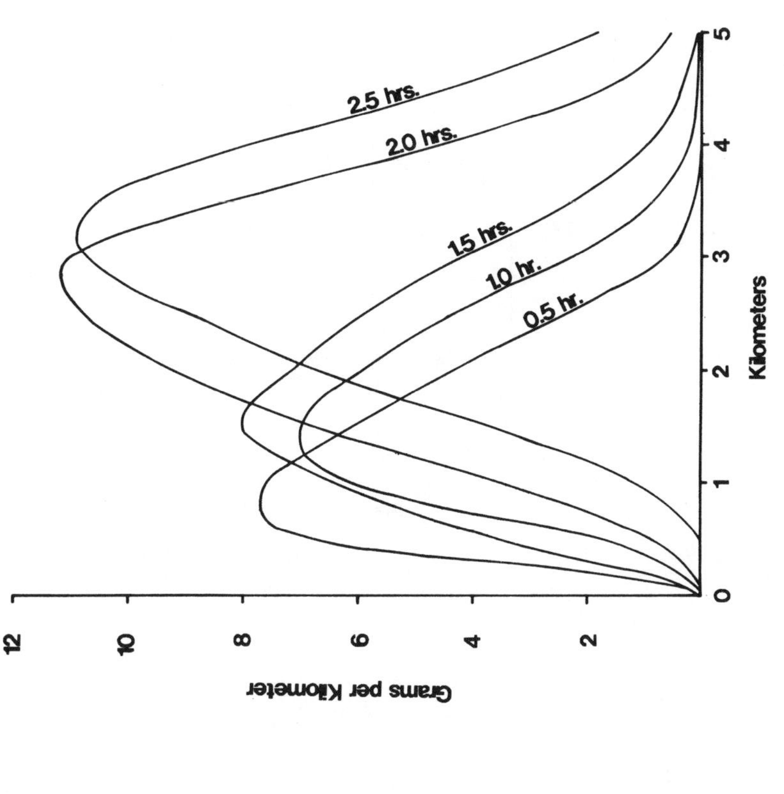

Fig. 8-13. **Concentration Profiles for Run #2 of Program B-8.2**

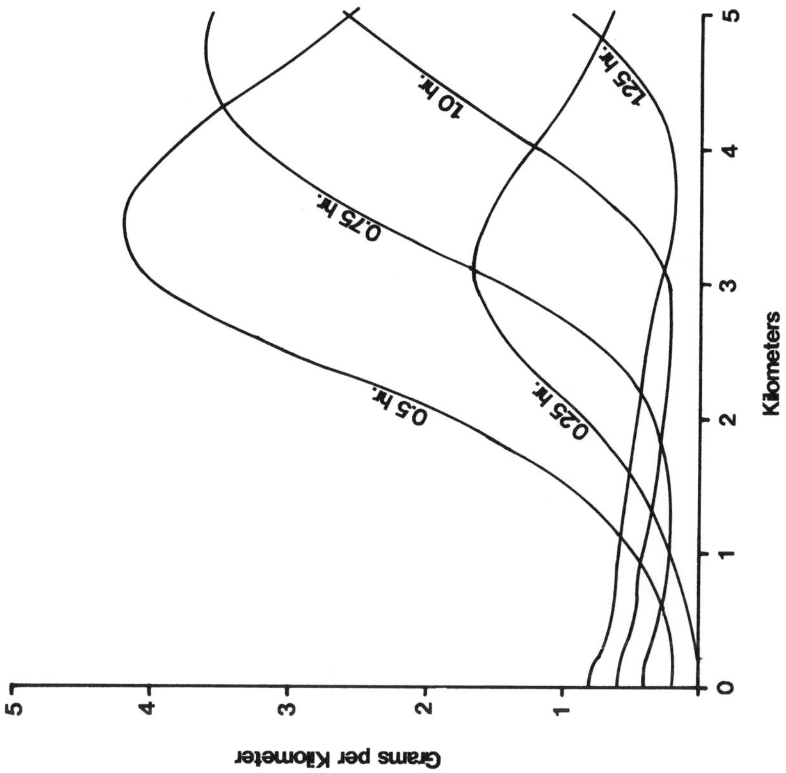

Fig. 8-14. **Concentration Profiles for Run #3 of Program B-8.2**

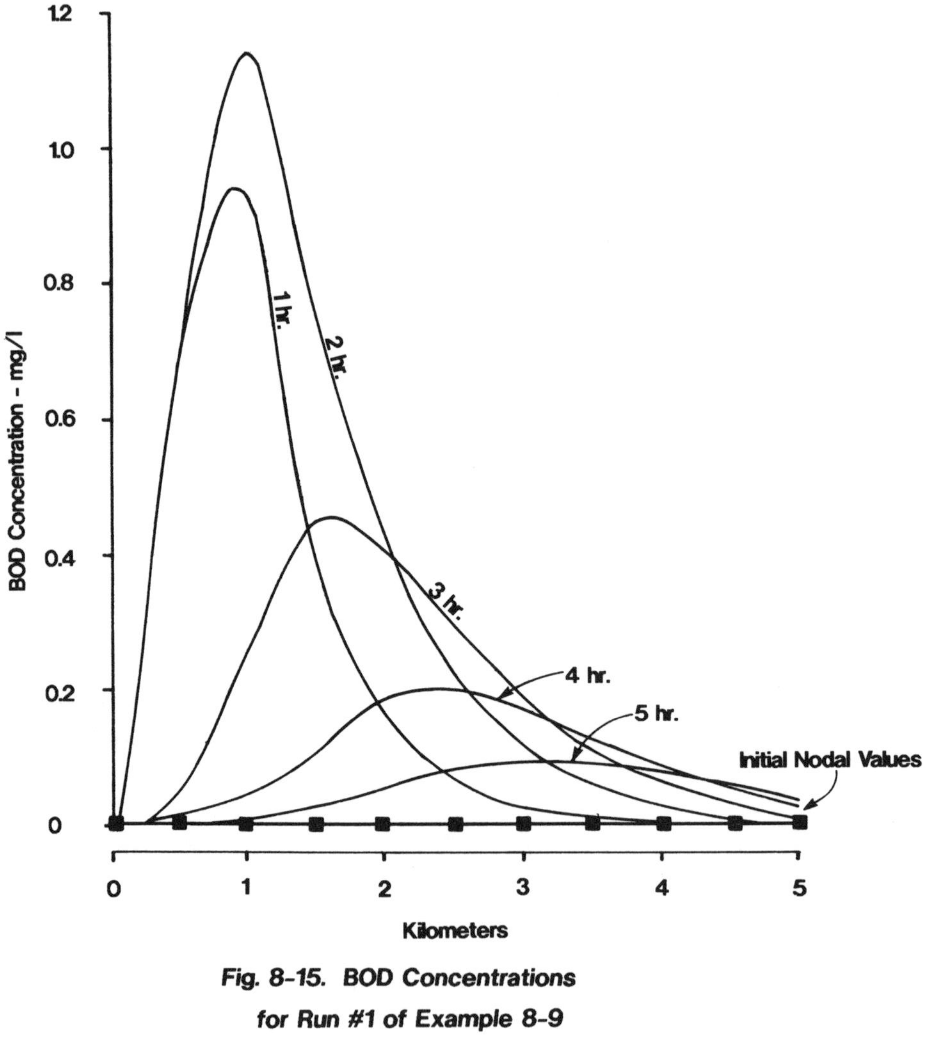

Fig. 8-15. BOD Concentrations for Run #1 of Example 8-9

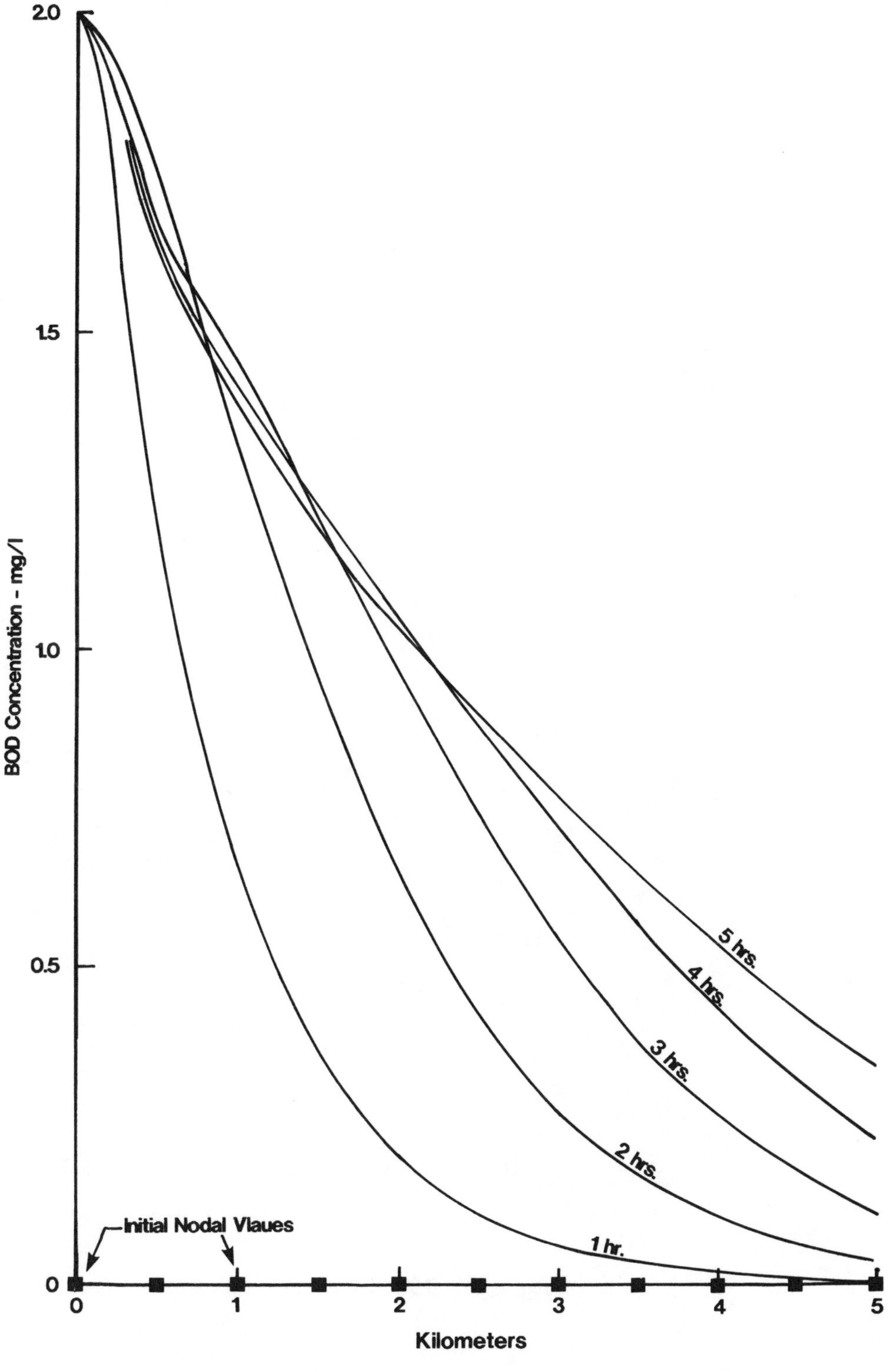

Fig. 8-16. BOD Concentrations for Run 2 of Example 8-9

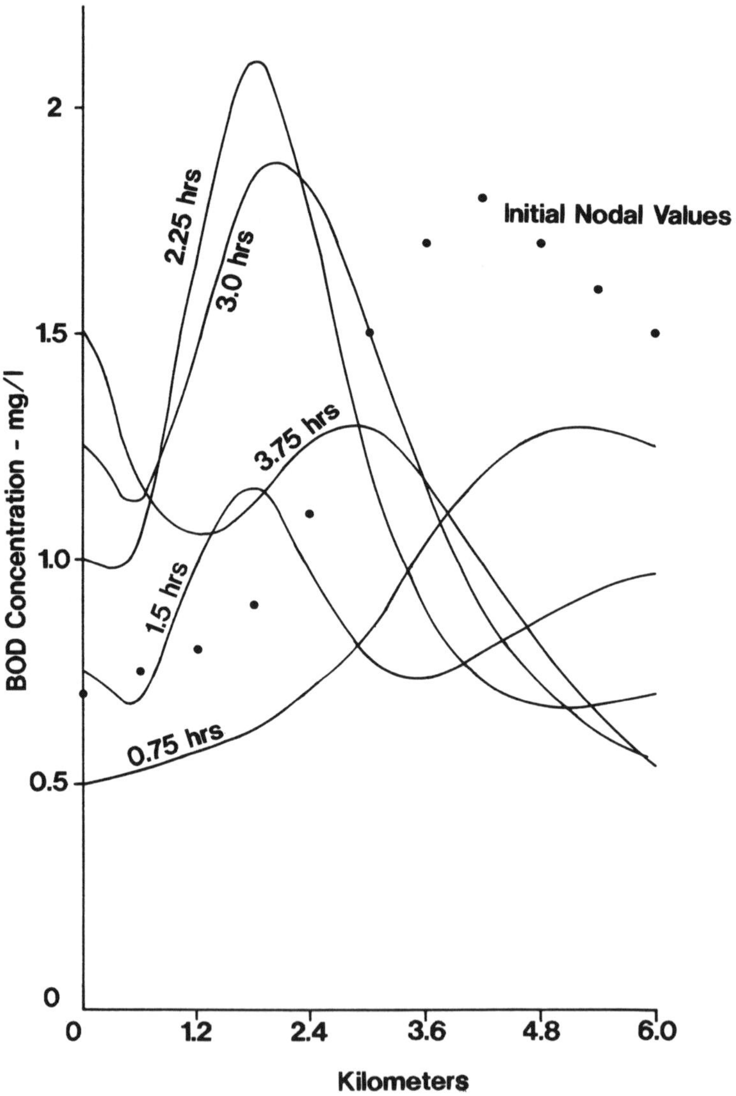

Fig. 8-17. BOD Concentrations for Run 3 of Example 8.9.

REFERENCES

Cheng, R.T. 1978. "Modeling of Hydraulic Systems by Finite-Element Methods," in Advances in Hydroscience, Vol. II, Academic Press, pp. 208-279.

Cover, K.A. and T.E. Unny, 1986. "Application of Computer Intensive Statistics to Parameter Uncertainty in Streamflow Synthesis," Water Res. Bul., 22(2) pp. 495-507.

Harper, L.A. and W.M. Snyder, 1987. "Testing for Significant Difference Between Functions from Non-Standard Experimental Data Sets," Trans. ASAE, 30(2) pp. 411-414.

Kinard, D.J. and H.H. Tippins, 1982. "Weather Station Summary for the Georgia Experiment Station 1926-1980 Research Report 406," College of Agricultural Experiment Stations, Univ. of GA., 51p.

Koch, R.W. and G.M. Smillie, 1986. "Bias in Hydrologic Prediction Using Log-transformed Regression Models," Water Res. Bul., 22(5), pp. 717-723.

Natrella, M.G., 1963. Experimental Statistics, Handbook 91, Nat'l Bureau of Standards, 528 p.

Nielsen, K.L., 1957. "Methods in Numerical Analysis," The MacMillan Co. 382 p.

NOAA, 1972-1976. "Climatological Data for Georgia, U.S. Dept. of Commerce, (Monthly and Annual).

Pinder, G.F. and W.G. Gray, 1977. Finite Element Simulation in Surface and Subsurface Hydrology, Academic Press, 295 p.

Rango, A. and Martinec, J., "Snow Accumulation Derived from Modified Depletion Curves of Snow Coverage," Hydrological Aspects of Alpine and High Mountain Areas, IAHS Publ. No. 138, pp 83-90, July, 1982.

Segerlind, L.J., 1976. Applied Finite Element Analysis, Wiley and Sons, 422 p.

Snyder, W.M., 1961. "Continuous Parabolic Interpolation," Proc. ASCE 87(HY4), pp. 99-111.

Snyder, W.M., 1962. "Closure to 'Continuous Parabolic Interpolation,'" Proc. ASCE 88(HY4), pp. 265-274.

Snyder, W.M., 1967. "Extended Continuous Interpolation," Proc. ASCE 93(HY5), pp. 261-280.

Snyder, W.M., 1975. "Continuous Seasonal Probability of Extreme Rainfall Events," Hydrological Sciences Bull. XX, 26 pp 275-283.

Snyder, W.M., 1976a. "Interpolation and Smoothing of Experimental Data with Sliding Polynomials," USDA - Agric. Res. Service, ARS-S-83, 34p.

Snyder, W.M., 1976b. "Time Series Data Analysis and Synthesis for Research Watersheds," USDA - Agric. Res. Service, ARS-S-112, 33p.

Snyder, W.M., 1978. "Running Details for Fully Computerized Smooth Data and Gradients Using Sliding Polynomials," Southeast Watershed Research Center, USDA-ARS Athens, GA, 38p.

Snyder, W.M., 1979. "Moving Average Calculation Using Sliding Polynomial Coefficients," Southeast Watershed Research Center, USDA-ARS, Athens, GA, 13p.

Snyder, W.M., 1980. "Smoothed Data and Gradients Using Sliding Polynomials with Optional Controls," Water Res. Bul. 16(1) pp. 22-30.

Snyder, W.M. and A.W. Thomas, 1983. "Return-Period Analysis with Sliding Polynomials," Trans. ASAE., 26(6), pp. 1732-1737.

Snyder, W.M. and A.W. Thomas, 1986. "Stochastic Impacts on Farming, II: Seasonally Continuous Probability Distributions," Trans. ASAE, 29(4), pp. 1017-1025.

Snyder, W.M. and A.W. Thomas, 1987a. "Stochastic Impacts on Farming, IV: A Form-free Bivariate Distribution to Generate Inputs to Agricultural Models," Trans ASAE, 30(4), pp. 946-952.

Snyder, W.M. and A.W. Thomas, 1987b. "Patterns of Watershed Monthly Runoff," Water Res. Bul., 23(6), pp. 1133-1140.

Snyder, W.M. and L.A. Harper, 1986. "Testing for Differences between Integrals of Functions Defined by Limited Agronomic Data," Trans. ASAE, 29(2), pp. 597-600.

Snyder, W.M., R.R. Bruce, L.A. Harper, and A.W. Thomas, 1984. "Two-Dimensional Sliding Polynomials," College of Agriculture Experiment Stations, Univ. of Georgia Research Bulletin 320, 57p.

Staff, Southeast Watershed Research Center, 1977. "The Hydrology and Hydrology of Ahoskie Creek Watershed, North Carolina: Data and Analysis," USDA Tech. Bul. 1563, 163pp.

Thomas, A.W. and W.M. Snyder, 1984a. "Return Period Analysis of Minimum Events Using Sliding Polynomials," Trans. ASAE, 27(2), pp. 464-469.

Thomas, A.W. and W.M. Snyder, 1984b. "Testing the Climatic Representativeness of Short-period Records through Simulation," Trans. ASAE, 27(4), pp. 1027-1033.

Thomas, A.W. and W.M. Snyder, 1986a. "Stochastic Impacts on Farming I: Transformation of Agriculturally Relevant Variates for Probability Analysis," Trans. ASAE, 29(1), pp. 128-134.

Thomas, A.W. and W.M. Snyder, 1986b. "Stochastic Impacts on Farming III: Simulation of Seasonal Variation of Climatic Risk," Trans. ASAE, 29(4), pp. 1026-1031.

Thomas, A.W., W.M. Snyder, and R.R. Bruce, 1977. "Smoothing, Interpolation, and Gradients from Limited Data," Agronomy Journal Vol., 69 (Sept.-Oct), pp. 747-750.

Weiss, G.H., 1983. "Random Walks and their Applications," Amer. Scientist, 71(1), pp. 65-71

Appendix A - Coefficient Tables

TABLE A-1.1 Coefficients for four-point interpolation

Z	$C_1(Z)$	$C_2(Z)$	$C_3(Z)$	$C_4(Z)$
-.50	0.00000000	1.00000000	.00000000	0.00000000
-.49	-.00490050	.99975150	.00519850	-.00004950
-.48	-.00960400	.99901200	.01078800	-.00019600
-.47	-.01411350	.99779050	.01675950	-.00043650
-.46	-.01843200	.99609600	.02310400	-.00076800
-.45	-.02256250	.99393750	.02981250	-.00118750
-.44	-.02650800	.99132400	.03687600	-.00169200
-.43	-.03027150	.98826450	.04428550	-.00227850
-.42	-.03385600	.98476800	.05203200	-.00294400
-.41	-.03726450	.98084350	.06010650	-.00368550
-.40	-.04050000	.97650000	.06850000	-.00450000
-.39	-.04356550	.97174650	.07720350	-.00538450
-.38	-.04646400	.96659200	.08620800	-.00633600
-.37	-.04919850	.96104550	.09550450	-.00735150
-.36	-.05177200	.95511600	.10508400	-.00842800
-.35	-.05418750	.94881250	.11493750	-.00956250
-.34	-.05644800	.94214400	.12505600	-.01075200
-.33	-.05855650	.93511950	.13543050	-.01190350
-.32	-.06051600	.92774800	.14605200	-.01328400
-.31	-.06232950	.92003850	.15691150	-.01462050
-.30	-.06400000	.91200000	.16800000	-.01600000
-.29	-.06553050	.90364150	.17930850	-.01741950
-.28	-.06692400	.89497200	.19082800	-.01887600
-.27	-.06818350	.88600050	.20254950	-.02036650
-.26	-.06931200	.87673600	.21446400	-.02188800
-.25	-.07031250	.86718750	.22656250	-.02343750
-.24	-.07118800	.85736400	.23883600	-.02501200
-.23	-.07194150	.84727450	.25127550	-.02660850
-.22	-.07257600	.83692800	.26387200	-.02822400
-.21	-.07309450	.82633350	.27661650	-.02985550
-.20	-.07350000	.81550000	.28950000	-.03150000
-.19	-.07379550	.80443650	.30251350	-.03315450
-.18	-.07398400	.79315200	.31564800	-.03481600
-.17	-.07406850	.78165550	.32889450	-.03648150
-.16	-.07405200	.76995600	.34224400	-.03814800

TABLE A-1.1 Coefficients for four-point interpolation - Continued

Z	$C_1(Z)$	$C_2(Z)$	$C_3(Z)$	$C_4(Z)$
-.15	-.07393750	.75806250	.35568750	-.03981250
-.14	-.07372800	.74598400	.36921600	-.04147200
-.13	-.07342650	.73372950	.38282050	-.04312350
-.12	-.07303600	.72130800	.39649200	-.04476400
-.11	-.07255950	.70872850	.41022150	-.04639050
-.10	-.07200000	.69600000	.42400000	-.04800000
-.09	-.07136050	.68313150	.43781850	-.04958950
-.08	-.07064400	.67013200	.45166800	-.05115600
-.07	-.06985350	.65701050	.46553950	-.05269650
-.06	-.06899200	.64377600	.47942400	-.05420800
-.05	-.06806250	.63043750	.49331250	-.05568750
-.04	-.06706800	.61700400	.50719600	-.05713200
-.03	-.06601150	.60348450	.52106550	-.05853850
-.02	-.06489600	.58988800	.53491200	-.05990400
-.01	-.06372450	.57622350	.54872650	-.06122550
.00	-.06250000	.56250000	.56250000	-.06250000
.01	-.06122550	.54872650	.57622350	-.06372450
.02	-.05990400	.53491200	.58988800	-.06489600
.03	-.05853850	.52106550	.60348450	-.06601150
.04	-.05713200	.50719600	.61700400	-.06706800
.05	-.05568750	.49331250	.63043750	-.06806250
.06	-.05420800	.47942400	.64377600	-.06899200
.07	-.05269650	.46553950	.65701050	-.06985350
.08	-.05115600	.45166800	.67013200	-.07064400
.09	-.04958950	.43781850	.68313150	-.07136050
.10	-.04800000	.42400000	.69600000	-.07200000
.11	-.04639050	.41022150	.70872850	-.07255950
.12	-.04476400	.39649200	.72130800	-.07303600
.13	-.04312350	.38282050	.73372950	-.07342650
.14	-.04147200	.36921600	.74598400	-.07372800
.15	-.03981250	.35568750	.75806250	-.07393750
.16	-.03814800	.34224400	.76995600	-.07405200
.17	-.03648150	.32889450	.78165550	-.07406850
.18	-.03481600	.31564800	.79315200	-.07398400
.19	-.03315450	.30251350	.80443650	-.07379550

TABLE A-1.1 Coefficients for four-point interpolation - Continued

Z	$C_1(Z)$	$C_2(Z)$	$C_3(Z)$	$C_4(Z)$
.20	-.03150000	.28950000	.81550000	-.07350000
.21	-.02985550	.27661650	.82633350	-.07309450
.22	-.02822400	.26387200	.83692800	-.07257600
.23	-.02660850	.25127550	.84727450	-.07194150
.24	-.02501200	.23883600	.85736400	-.07118800
.25	-.02343750	.22656250	.86718750	-.07031250
.26	-.02188800	.21446400	.87673600	-.06931200
.27	-.02036650	.20254950	.88600050	-.06818350
.28	-.01887600	.19082800	.89497200	-.06692400
.29	-.01741950	.17930850	.90364150	-.06553050
.30	-.01600000	.16800000	.91200000	-.0400000
.31	-.01462050	.15691150	.92003850	-.06232950
.32	-.01328400	.14605200	.92774800	-.06051600
.33	-.01199350	.13543050	.93511950	-.05855650
.34	-.01075200	.12505600	.94214400	-.05644800
.35	-.00956250	.11493750	.94881250	-.05418750
.36	-.00842800	.10508400	.95511600	-.05177200
.37	-.00735150	.09550450	.96104550	-.04919850
.38	-.00633600	.08620800	.96659200	-.04646400
.39	-.00538450	.07720350	.97174650	-.04356550
.40	-.00450000	.06850000	.97650000	-.04050000
.41	-.00368550	.06010650	.98084350	-.03726450
.42	-.00294400	.05203200	.98476800	-.03385600
.43	-.00227850	.04428550	.98826450	-.03027150
.44	-.00169200	.03687600	.99132400	-.02650800
.45	-.00118750	.02981250	.99393750	-.02256250
.46	-.00076800	.02310400	.99609600	-.01843200
.47	-.00043650	.01675950	.99779050	-.01411350
.48	-.00019600	.01078800	.99901200	-.00960400
.49	-.00004950	.00519850	.99975150	-.00490050
.50	0.00000000	-.00000000	1.00000000	0.00000000

TABLE A-2.1 Coefficients for six-point interpolation

Z	$C_1(Z)$	$C_2(Z)$	$C_3(Z)$	$C_4(Z)$	$C_5(Z)$	$C_6(Z)$
-.50	0.00000000	0.00000000	1.00000000	0.00000000	0.00000000	0.00000000
-.49	.00082880	-.00659840	.99987214	.00673603	-.00083885	.00000029
-.48	.00164709	-.01305409	.99947750	.01362118	-.00169393	.00000225
-.47	.00245281	-.01935826	.99880045	.02067012	-.00257259	.00000748
-.46	.00324403	-.02550272	.99782656	.02789632	-.00348160	.00001741
-.45	.00401895	-.03147988	.99654258	.03531211	-.00442715	.00003340
-.44	.00477586	-.03728275	.99493642	.04292866	-.00541487	.00005668
-.43	.00551320	-.04290490	.99299713	.05075603	-.00644985	.00008839
-.42	.00622950	-.04834048	.99071488	.05880320	-.00753664	.00012954
-.41	.00692343	-.05358418	.98808090	.06707807	-.00867927	.00018105
-.40	.00759375	-.05863125	.98508750	.07558750	-.00988125	.00024375
-.39	.00823933	-.06347744	.98172802	.08433735	-.01114561	.00031836
-.38	.00885914	-.06811904	.97799680	.09333248	-.01247488	.00040550
-.37	.00945226	-.07255283	.97388918	.10257678	-.01387113	.00050572
-.36	.01001788	-.07677607	.96940146	.11207322	-.01533595	.00061946
-.35	.01055527	-.08078652	.96453086	.12182383	-.01687051	.00074707
-.34	.01106381	-.08458240	.95927552	.13182976	-.01847552	.00088883
-.33	.01154295	-.08816237	.95363447	.14209130	-.02015128	.00104493
-.32	.01199225	-.09152253	.94760758	.15260790	-.02189769	.00121549
-.31	.01241136	-.09467142	.94119558	.16337819	-.02371423	.00140052
-.30	.01280000	-.09760000	.93440000	.17440000	-.02560000	.00160000
-.29	.01315798	-.10031161	.92722315	.18567042	-.02755374	.00181381
-.28	.01348519	-.10280699	.91966810	.19718578	-.02957383	.00204175
-.27	.01378159	-.10508727	.91173866	.20894170	-.03165829	.00228359
-.26	.01404723	-.10715392	.90343936	.22093312	-.03380480	.00253901

TABLE A-2.1 Coefficients for six-point interpolation - Continued

Z	$C_1(Z)$	$C_2(Z)$	$C_3(Z)$	$C_4(Z)$	$C_5(Z)$	$C_6(Z)$
-.25	.01428223	-.10900879	.89477539	.23315430	-.03601074	.00280762
-.24	.01448676	-.11065405	.88575262	.24559886	-.03827317	.00308898
-.23	.01466108	-.11209221	.87637755	.25825982	-.04058884	.00338261
-.22	.01480550	-.11332608	.86665728	.27112960	-.04295424	.00368794
-.21	.01492041	-.11435879	.85659951	.28420005	-.04536556	.00400437
-.20	.01500625	-.11519375	.84621250	.29746250	-.04781875	.00433125
-.19	.01506351	-.11583465	.83550503	.31090774	-.05030950	.00466788
-.18	.01509274	-.11628544	.82448640	.32452608	-.05283328	.00501350
-.17	.01509454	-.11655033	.81316640	.33830737	-.05538532	.00536734
-.16	.01506958	-.11663377	.80155526	.35224102	-.05796065	.00572856
-.15	.01501855	-.11654043	.78966367	.36631602	-.06055410	.00609629
-.14	.01494221	-.11627520	.77750272	.38052096	-.06316032	.00646963
-.13	.01484133	-.11584317	.76508388	.39484409	-.06577378	.00684765
-.12	.01471675	-.11524963	.75241898	.40927330	-.06838879	.00722939
-.11	.01456934	-.11450003	.73952019	.42379617	-.07099952	.00761384
-.10	.01440000	-.11360000	.72640000	.43840000	-.07360000	.00800000
-.09	.01420966	-.11255531	.71307116	.45307181	-.07618414	.00838682
-.08	.01399929	-.11137189	.69954670	.46779838	-.07874573	.00877325
-.07	.01376987	-.11005577	.68583988	.48256629	-.08127848	.00915821
-.06	.01352243	-.10861312	.67196416	.49736192	-.08377600	.00954061
-.05	.01325801	-.10705020	.65793320	.51217148	-.08623184	.00991934
-.04	.01297766	-.10537335	.64376082	.52698106	-.08863947	.01029328
-.03	.01268246	-.10358901	.62946096	.54177661	-.09099234	.01066132
-.02	.01237350	-.10170368	.61504768	.55654400	-.09328384	.01102234
-.01	.01205190	-.09972390	.60053513	.57126904	-.09550735	.01137519

TABLE A-2.1 Coefficients for six-point interpolation - Continued

Z	$C_1(Z)$	$C_2(Z)$	$C_3(Z)$	$C_4(Z)$	$C_5(Z)$	$C_6(Z)$
.00	.01171875	-.09765625	.58593750	.58593750	-.09765625	.01171875
.01	.01137519	-.09550735	.57126904	.60053513	-.09972390	.01205190
.02	.01102234	-.09328384	.55654400	.61504768	-.10170368	.01237350
.03	.01066132	-.09099234	.54177661	.62946096	-.10358901	.01268246
.04	.01029328	-.08863947	.52698106	.64376082	-.10537335	.01297766
.05	.00991934	-.08623184	.51217148	.65793320	-.10705020	.01325801
.06	.00954061	-.08377600	.49736192	.67196416	-.10861312	.01352243
.07	.00915821	-.08127848	.48256629	.68583988	-.11005577	.01376987
.08	.00877325	-.07874573	.46779838	.69954670	-.11137189	.01399929
.09	.00838682	-.07618414	.45307181	.71307116	-.11255531	.01420966
.10	.00800000	-.07360000	.43840000	.72640000	-.11360000	.01440000
.11	.00761384	-.07099952	.42379617	.73952019	-.11450003	.01456934
.12	.00722939	-.06838879	.40927330	.75241898	-.11524963	.01471675
.13	.00684765	-.06577378	.39484409	.76508388	-.11584317	.01484133
.14	.00646963	-.06316032	.38052096	.77750272	-.11627520	.01494221
.15	.00609629	-.06055420	.36631602	.78966367	-.11654043	.01501855
.16	.00572856	-.05796065	.35224102	.80155526	-.11663377	.01506958
.17	.00536734	-.05538532	.33830737	.81316640	-.11655033	.01509454
.18	.00501350	-.05283328	.32452608	.82448640	-.11628544	.01509274
.19	.00466788	-.05030950	.31090774	.83550503	-.11583465	.01506351
.20	.00433125	-.04781875	.29746250	.84621250	-.11519375	.01500625
.21	.00400437	-.04536556	.28420005	.85659951	-.11435879	.01492041
.22	.00368794	-.04295424	.27112960	.86665728	-.11332608	.01480550
.23	.00338261	-.04058884	.25825982	.87637755	-.11209221	.01466108
.24	.00308898	-.03827317	.24559886	.88575262	-.11065405	.01448676

TABLE A-2.1 Coefficients for six-point interpolation - Continued

Z	$C_1(Z)$	$C_2(Z)$	$C_3(Z)$	$C_4(Z)$	$C_5(Z)$	$C_6(Z)$
.25	.00280762	-.03601074	.23315430	.89477539	-.10900879	.01428223
.26	.00253901	-.03380480	.22093312	.90343936	-.10715392	.01404723
.27	.00228359	-.03165829	.20894170	.91173866	-.10508727	.01378159
.28	.00204175	-.02957383	.19718578	.91966810	-.10280699	.01348519
.29	.00181381	-.02755374	.18567042	.92722315	-.10031161	.01315798
.30	.00160000	-.02560000	.17440000	.93440000	-.09760000	.01280000
.31	.00140052	-.02371423	.16337819	.94119558	-.09467142	.01241136
.32	.00121549	-.02189769	.15260790	.94760758	-.09152553	.01199225
.33	.00104493	-.02015128	.14209130	.95363447	-.08816237	.01154295
.34	.00088883	-.01847552	.13182976	.95927552	-.08458240	.01106381
.35	.00074707	-.01687051	.12182383	.96453086	-.08078652	.01055527
.36	.00061946	-.01533595	.11207322	.96940146	-.07677607	.01001788
.37	.00050572	-.01387113	.10257678	.97388918	-.07255283	.00945226
.38	.00040550	-.01247488	.09333248	.97799680	-.06811904	.00885914
.39	.00031836	-.01114561	.08433735	.98172802	-.06347744	.00823933
.40	.00024375	-.00988125	.07558750	.98508750	-.05863125	.00759375
.41	.00018105	-.00867927	.06707807	.98808090	-.05358418	.00692343
.42	.00012954	-.00753664	.05880320	.99071488	-.04834048	.00622950
.43	.00008839	-.00644985	.05075603	.99299713	-.04290490	.00551320
.44	.00005668	-.00541487	.04292866	.99493642	-.03728275	.00477586
.45	.00003340	-.00442715	.03531211	.99654258	-.03147988	.00401895
.46	.00001741	-.00348160	.02789632	.99782656	-.02550272	.00324403
.47	.00000748	-.00257259	.02067012	.99880045	-.01935826	.00245281
.48	.00000225	-.00169393	.01362118	.99947750	-.01305409	.00164709
.49	.00000029	-.00083885	.00673603	.99987214	-.00659840	.00082880
.50	0.00000000	0.00000000	0.00000000	1.00000000	0.00000000	0.00000000

TABLE A-3.1 Coefficients $dC_i(z)/dz$ for six-point differentiation

Z	$C_1(Z)$	$C_2(Z)$	$C_3(Z)$	$C_4(Z)$	$C_5(Z)$	$C_6(Z)$
-.50	.08333333	-.66666667	0.00000000	.66666667	-.08333333	0.00000000
-.49	.08238966	-.65285645	-.02585410	.68080444	-.08456905	.00008551
-.48	.08123383	-.63813450	-.05333367	.69646967	-.08656950	.00033417
-.47	.07987849	-.62256295	-.08231644	.71354210	-.08927555	.00073434
-.46	.07833600	-.60620267	-.11268267	.73190400	-.09262933	.00127467
-.45	.07661849	-.58911328	-.14431510	.75144010	-.09657422	.00194401
-.44	.07473783	-.57135317	-.17709900	.77203767	-.10105483	.00273150
-.43	.07270566	-.55297945	-.21092210	.79358644	-.10601705	.00362651
-.42	.07053333	-.53404800	-.24567467	.81597867	-.11140800	.00461867
-.41	.06823199	-.51461345	-.28124944	.83910910	-.11717605	.00569784
-.40	.06581250	-.49472917	-.31754167	.86287500	-.12327083	.00685417
-.39	.06328549	-.47444728	-.35444910	.88717610	-.12964322	.00807801
-.38	.06066133	-.45381867	-.39187200	.91191467	-.13624533	.00936000
-.37	.05795016	-.43289295	-.42971310	.93699544	-.14303055	.01069101
-.36	.05516183	-.41171850	-.46787767	.96232567	-.14995350	.01206217
-.35	.05230599	-.39034245	-.50627344	.98781510	-.15697005	.01346484
-.34	.04939200	-.36881067	-.54481067	1.01337600	-.16403733	.01489067
-.33	.04642899	-.34716778	-.58340210	1.03892310	-.17111372	.01633151
-.32	.04342583	-.32545717	-.62196300	1.06437367	-.17815883	.01777950
-.31	.04039116	-.30372095	-.66041110	1.08964744	-.18513355	.01922701
-.30	.03733333	-.28200000	-.69866667	1.11466667	-.19200000	.02066667
-.29	.03426049	-.26033395	-.73665244	1.13935610	-.19872155	.02209134
-.28	.03118050	-.23876117	-.77429367	1.16364300	-.20526283	.02349417
-.27	.02810099	-.21731878	-.81151810	1.18745710	-.21158972	.02486851
-.26	.02502933	-.19604267	-.84825600	1.21073067	-.21766933	.02620800

TABLE A-3.1 Coefficients for six-point differentiation - Continued

Z	$C_1(Z)$	$C_2(Z)$	$C_3(Z)$	$C_4(Z)$	$C_5(Z)$	$C_6(Z)$
-.25	.02197266	-.17496745	-.88444010	1.23339844	-.22347005	.02750651
-.24	.01893783	-.15412650	-.92000567	1.25539767	-.22896150	.02875817
-.23	.01593149	-.13355195	-.95489044	1.27666810	-.23411455	.02995734
-.22	.01296000	-.11327467	-.98903467	1.29715200	-.23890133	.03109867
-.21	.01002949	-.09332428	-1.02238110	1.31679410	-.24329522	.03217701
-.20	.00714583	-.07372917	-1.05487500	1.33554167	-.24727083	.03318750
-.19	.00431466	-.05451645	-1.08646410	1.35334444	-.25080405	.03412551
-.18	.00154133	-.03571200	-1.11709867	1.37015467	-.25387200	.03498667
-.17	-.00116901	-.01734045	-1.14673144	1.38592710	-.25645305	.03576684
-.16	-.00381150	.00057483	-1.17531767	1.40061900	-.25852683	.03646217
-.15	-.00638151	.01801172	-1.20281510	1.41419010	-.26007422	.03706901
-.14	-.00887467	.03494933	-1.22918400	1.42660267	-.26107733	.03758400
-.13	-.01128684	.05136805	-1.25438710	1.43782144	-.26151955	.03800401
-.12	-.01361417	.06724950	-1.27838967	1.44781367	-.26138550	.03832617
-.11	-.01585301	.08257655	-1.30115944	1.45654910	-.26066105	.03854784
-.10	-.01800000	.09733333	-1.32266667	1.46400000	-.25933333	.03866667
-.09	-.02005201	.11150522	-1.34288410	1.47014110	-.25739072	.03868051
-.08	-.02200617	.12507883	-1.36178700	1.47494967	-.25482283	.03858750
-.07	-.02385984	.13804205	-1.37935310	1.47840544	-.25162055	.03838601
-.06	-.02561067	.15038400	-1.39556267	1.48049067	-.24777600	.03807467
-.05	-.02725651	.16209505	-1.41039844	1.48119010	-.24328255	.03765234
-.04	-.02879550	.17316683	-1.42384567	1.48049100	-.23813483	.03711817
-.03	-.03022601	.18359222	-1.43589210	1.47838310	-.23232872	.03647151
-.02	-.03154667	.19336533	-1.44652800	1.47485867	-.22586133	.03571200
-.01	-.03275634	.20248155	-1.45574610	1.46991244	-.21873105	.03483951

TABLE A-3.1 Coefficients for six-point differentiation - Continued

Z	$C_1(Z)$	$C_2(Z)$	$C_3(Z)$	$C_4(Z)$	$C_5(Z)$	$C_6(Z)$
.00	-.03385417	.21093750	-1.46354167	1.46354167	-.21093750	.03385417
.01	-.03483951	.21873105	-1.46991244	1.45574610	-.20248155	.03275634
.02	-.03571200	.22586133	-1.47485867	1.44652860	-.19336533	.03154667
.03	-.03647151	.23232872	-1.47838310	1.43589260	-.18359222	.03022601
.04	-.03711817	.23813483	-1.48049100	1.42384557	-.17316683	.02879550
.05	-.03765234	.24328255	-1.48119010	1.41039844	-.16209505	.02725651
.06	-.03807467	.24777600	-1.48049067	1.39556267	-.15038400	.02561067
.07	-.03838601	.25162055	-1.47840544	1.37935310	-.13804205	.02385984
.08	-.03858750	.25482283	-1.47494967	1.36178700	-.12507883	.02200617
.09	-.03868051	.25739072	-1.47014110	1.34288410	-.11150522	.02005201
.10	-.03866667	.25933333	-1.46400000	1.32266667	-.09733333	.01800000
.11	-.03854784	.26066105	-1.45654910	1.30115944	-.08257655	.01585301
.12	-.03832617	.26138550	-1.44781367	1.27838967	-.06724950	.01361417
.13	-.03800401	.26151955	-1.43782144	1.25438710	-.05136805	.01128684
.14	-.03758400	.26107733	-1.42660267	1.22918400	-.03494933	.00887467
.15	-.03706901	.26007422	-1.41419010	1.20281510	-.01801172	.00638151
.16	-.03646217	.25852683	-1.40061900	1.17531767	-.00057483	.00381150
.17	-.03576684	.25645305	-1.38592710	1.14673144	.01734045	.00116901
.18	-.03498667	.25387200	-1.37015467	1.11709867	.03571200	-.00154133
.19	-.03412551	.25080405	-1.35334444	1.08646410	.05451645	-.00431466
.20	-.03318750	.24727083	-1.33554167	1.05487500	.07372917	-.00714583
.21	-.03217701	.24329522	-1.31679410	1.02238110	.09332428	-.01002949
.22	-.03109867	.23890133	-1.29715200	.98903467	.11327467	-.01296000
.23	-.02995734	.23411455	-1.27666810	.95489044	.13355195	-.01593149
.24	-.02875817	.22896150	-1.25539767	.92000567	.15412650	-.01893783

TABLE A-3.1 Coefficients for six-point differentiation - Continued

Z	$C_1(Z)$	$C_2(Z)$	$C_3(Z)$	$C_4(Z)$	$C_5(Z)$	$C_6(Z)$
.25	-.02750651	.22347005	-1.23339844	.88444010	.17496745	-.02197266
.26	-.02620800	.21766933	-1.21073067	.84825600	.19604267	-.02502933
.27	-.02486851	.21158972	-1.18745710	.81151810	.21731878	-.02810099
.28	-.02349417	.20526283	-1.16364300	.77429367	.23876117	-.03118050
.29	-.02209134	.19872155	-1.13935610	.73665244	.26033395	-.03426049
.30	-.02066667	.19200000	-1.11466667	.69866667	.28200000	-.03733333
.31	-.01922701	.18513355	-1.08964744	.66041110	.30372095	-.04039116
.32	-.01777950	.17815883	-1.06437367	.62196300	.32545717	-.04342583
.33	-.01633151	.17111372	-1.03892310	.58340210	.34716778	-.04642899
.34	-.01489067	.16403733	-1.01337600	.54481067	.36881067	-.04939200
.35	-.01346484	.15697005	-.98781510	.50627344	.39034245	-.05230599
.36	-.01206217	.14995350	-.96232567	.46787767	.41171850	-.05516183
.37	-.01069101	.14303055	-.93699544	.42971310	.43289295	-.05795016
.38	-.00936000	.13624533	-.91191467	.39187200	.45381867	-.06066133
.39	-.00807801	.12964322	-.88717610	.35444910	.47444728	-.06328549
.40	-.00685417	.12327083	-.86287500	.31754167	.49472917	-.06581250
.41	-.00569784	.11717605	-.83910910	.28124944	.51461345	-.06823199
.42	-.00461867	.11140800	-.81597867	.24567467	.53404800	-.07053333
.43	-.00362651	.10601705	-.79358644	.21092210	.55297945	-.07270566
.44	-.00273150	.10105483	-.77203767	.17709900	.57135317	-.07473783
.45	-.00194401	.09657422	-.75144010	.14431510	.58911328	-.07661849
.46	-.00127467	.09262933	-.73190400	.11268267	.60620267	-.07833600
.47	-.00073434	.08927555	-.71354210	.08231644	.62256295	-.07987849
.48	-.00033417	.08656950	-.69646967	.05333367	.63813450	-.08123383
.49	-.00008551	.08456905	-.68080444	.02585410	.65285645	-.08238966
.50	0.00000000	.08333333	-.66666667	0.00000000	.66666667	-.08333333

TABLE A-4.1 Coefficients for four-point integration

Z	$C_1(Z)$	$C_2(Z)$	$C_3(Z)$	$C_4(Z)$
-0.50	-.00000000	-.00000000	0.00000000	.00000000
-0.49	-.00002467	.00999917	.00002566	-.00000017
-0.48	-.00009735	.01999339	.00010527	-.00000131
-0.47	-.00021610	.02997780	.00024270	-.00000440
-0.46	-.00037899	.03994763	.00044171	-.00001035
-0.45	-.00058411	.04989818	.00070599	-.00002005
-0.44	-.00082962	.05982486	.00103914	-.00003438
-0.43	-.00111367	.06972317	.00144466	-.00005417
-0.42	-.00143445	.07958869	.00192597	-.00008021
-0.41	-.00179020	.08941710	.00248640	-.00011330
-0.40	-.00217917	.09920417	.00312917	-.00015417
-0.39	-.00259963	.10894574	.00385743	-.00020353
-0.38	-.00304992	.11863776	.00467424	-.00026208
-0.37	-.00352837	.12827627	.00558256	-.00033047
-0.36	-.00403335	.13785739	.00658527	-.00040931
-0.35	-.00456328	.14737734	.00768516	-.00049922
-0.34	-.00511659	.15683243	.00888491	-.00060075
-0.33	-.00569173	.16621904	.01018713	-.00071443
-0.32	-.00628722	.17553366	.01159434	-.00084078
-0.31	-.00690157	.18477287	.01310896	-.00098027
-0.30	-.00753333	.19393333	.01473333	-.00113333
-0.29	-.00818110	.20301180	.01646970	-.00130040
-0.28	-.00884349	.21200513	.01832021	-.00148185
-0.27	-.00951913	.22091024	.02028693	-.00167803
-0.26	-.01020672	.22972416	.02237184	-.00188928
-0.25	-.01090495	.23844401	.02457682	-.00211589
-0.24	-.01161255	.24706699	.02690367	-.00235811
-0.23	-.01232830	.25559040	.02935410	-.00261620
-0.22	-.01305099	.26401163	.03192971	-.00289035
-0.21	-.01377943	.27232814	.03463203	-.00318073
-0.20	-.01451250	.28053750	.03746250	-.00348750
-0.19	-.01524907	.28863737	.04042246	-.00381077
-0.18	-.01598805	.29662549	.04351317	-.00415061
-0.17	-.01672840	.30449970	.04673580	-.00450710
-0.16	-.01746909	.31225793	.05009141	-.00488025

TABLE A-4.1 Coefficients for four-point integration - Continued

Z	$C_1(Z)$	$C_2(Z)$	$C_3(Z)$	$C_4(Z)$
-0.15	-.01820911	.31989818	.05358099	-.00527005
-0.14	-.01894752	.32741856	.05720544	-.00567648
-0.13	-.01968337	.33481727	.06096556	-.00609947
-0.12	-.02041575	.34209259	.06486207	-.00653891
-0.11	-.02114380	.34924290	.06889560	-.00699470
-0.10	-.02186667	.35626667	.07306667	-.00746667
-0.09	-.02258353	.36316244	.07737573	-.00795463
-0.08	-.02329362	.36992886	.08182314	-.00845838
-0.07	-.02399617	.37656467	.08640916	-.00897767
-0.06	-.02469045	.38306869	.09113397	-.00951221
-0.05	-.02537578	.38943984	.09599766	-.01006172
-0.04	-.02605149	.39567713	.10100021	-.01062585
-0.03	-.02671693	.40177964	.10614153	-.01120423
-0.02	-.02737152	.40774656	.11142144	-.01179648
-0.01	-.02801467	.41357717	.11683966	-.01240217
0.00	-.02864583	.41927083	.12239583	-.01302083
0.01	-.02926450	.42482700	.12808950	-.01365200
0.02	-.02987019	.43024523	.13392011	-.01429515
0.03	-.03046243	.43552514	.13988703	-.01494973
0.04	-.03104082	.44066646	.14598954	-.01561518
0.05	-.03160495	.44566901	.15222682	-.01629089
0.06	-.03215445	.45053269	.15859797	-.01697621
0.07	-.03268900	.45525750	.16510200	-.01767050
0.08	-.03320829	.45984353	.17173781	-.01837305
0.09	-.03371203	.46429094	.17850423	-.01908313
0.10	-.03420000	.46860000	.18540000	-.01980000
0.11	-.03467197	.47277107	.19242376	-.02052287
0.12	-.03512775	.47680459	.19957407	-.02125091
0.13	-.03556720	.48070110	.20684940	-.02198330
0.14	-.03599019	.48446123	.21424811	-.02271915
0.15	-.03639661	.48808568	.22176849	-.02345755
0.16	-.03678642	.49157526	.22940874	-.02419758
0.17	-.03715957	.49493087	.23716696	-.02493827
0.18	-.03751605	.49815349	.24504117	-.02567861
0.19	-.03785590	.50124420	.25302930	-.02641760

TABLE A-4.1 Coefficients for four-point integration - Continued

Z	$C_1(Z)$	$C_2(Z)$	$C_3(Z)$	$C_4(Z)$
0.20	-.03817917	.50420417	.26112917	-.02715417
0.21	-.03848593	.50703464	.26933853	-.02788723
0.22	-.03877632	.50973696	.27765504	-.02861568
0.23	-.03905047	.51231257	.28607626	-.02933837
0.24	-.03930855	.51476299	.29459967	-.03005411
0.25	-.03955078	.51708984	.30322266	-.03076172
0.26	-.03977739	.51929483	.31194251	-.03145995
0.27	-.03998863	.52137974	.32075643	-.03214753
0.28	-.04018482	.52334646	.32966154	-.03282318
0.29	-.04036627	.52519697	.33865486	-.03348557
0.30	-.04053333	.52693333	.34773333	-.03413333
0.31	-.04068640	.52855770	.35689380	-.03476510
0.32	-.04082589	.53007233	.36613301	-.03537945
0.33	-.04095223	.53147954	.37544763	-.03597493
0.34	-.04106592	.53278176	.38483424	-.03655008
0.35	-.04116745	.53398151	.39428932	-.03710339
0.36	-.04125735	.53508139	.40380927	-.03763331
0.37	-.04133620	.53608410	.41339040	-.03813830
0.38	-.04140459	.53699243	.42302891	-.03861675
0.39	-.04146313	.53780924	.43272093	-.03906703
0.40	-.04151250	.53853750	.44246250	-.03948750
0.41	-.04155337	.53918027	.45224956	-.03987647
0.42	-.04158645	.53974069	.46207797	-.04023221
0.43	-.04161250	.54022200	.47194350	-.04055300
0.44	-.04163229	.54062753	.48184181	-.04083705
0.45	-.04164661	.54096068	.49176849	-.04108255
0.46	-.04165632	.54122496	.50171904	-.04128768
0.47	-.04166227	.54142397	.51168886	-.04145057
0.48	-.04166535	.54156139	.52167327	-.04156931
0.49	-.04166650	.54164100	.53166750	-.04164200
0.50	-.04166667	.54166667	.54166667	-.04166667

TABLE A-8.1 Values of One-dimensional Finite-Element Integrals

	$\int c_1$	$\int c_2$	$\int c_3$	$\int c_4$
	-.0416667	.5416667	.5416667	-.0416667

	$\int c_1 z$	$\int c_2 z$	$\int c_3 z$	$\int c_4 z$
	.0041667	-.0958333	.0958333	-.0041667

	$\int c_1 z^2$	$\int c_2 z^2$	$\int c_3 z^2$	$\int c_4 z^2$
	-.0020833	.0437500	.0437500	-.0020833

	$\int c_1 z^3$	$\int c_2 z^3$	$\int c_3 z^3$	$\int c_4 z^3$
	.0004464	-.0138393	.0138393	-.0004464

j	$\int c_j c_1$	$\int c_j c_2$	$\int c_j c_3$	$\int c_j c_4$
1	.0023807	-.0279762	-.0178572	.0017857
2	-.0279762	.4047619	.1827381	-.0178572
3	-.0178572	.1827381	.4047619	-.0279762
4	.0017857	-.0178572	-.0279762	.0023807

j	$\int c_j\ dc_1/dz$	$\int c_j\ dc_2/dz$	$\int c_j\ dc_3/dz$	$\int c_j\ dc_4/dz$
1	.0	.0458333	-.05	.0041667
2	-.0458333	-.5	.5958333	-.05
3	.05	-.5958333	.5	.0458333
4	-.0041667	.05	-.0458333	.0

j	$\int c_j d^2c_1/dz^2$	$\int c_j d^2c_2/dz^2$	$\int c_j d^2c_3/dz^2$	$\int c_j d^2c_4/dz^2$
1	-.0333333	.0583333	-.0166667	-.0083333
2	.5583333	-1.1333333	.5916667	-.0166667
3	-.0166667	.5916667	-1.1333333	.5583333
4	-.0083333	-.0166667	.0583333	-.0333333

j	$\int c_j c_1 z$	$\int c_j c_2 z$	$\int c_j c_3 z$	$\int c_j c_4 z$
1	-.0002976	.0047619	-.0002976	.0
2	.0047619	-.1008929	.0	.0002976
3	-.0002976	.0	.1008929	-.0047619
4	.0	.0002976	-.0047619	.0002976

j	$\int c_j \cdot dc_1/dz \cdot z$	$\int c_j \cdot dc_2/dz \cdot z$	$\int c_j \cdot dc_3/dz \cdot z$	$\int c_j \cdot dc_4/dz \cdot z$
1	-.0011905	-.0026786	.0047619	-.0008929
2	.0306548	.0476190	-.0913690	.0130952
3	.0130952	-.0913690	.0476190	.0306548
4	-.0008929	.0047619	-.0026786	-.0011905

TABLE A-8.2 Values of Two-dimensional Finite-Element Integrals

i		j = 1	2	3	4	5	6	7	8
1	$\iint c_j(u,z)$	-0.02083330	+0.27083333	+0.27083333	-0.02083333	-0.02083333	+0.27083333	+0.27083333	-0.02083333
2	$\iint c_j(u,z)\,z$	+0.00208391	-0.04791839	+0.04791839	-0.00208391	+0.00208391	-0.04791839	+0.04791839	-0.00208391
3	$\iint c_j(u,z)\,t$	+0.00354167	-0.04604167	-0.04604167	+0.00354167	-0.00354167	+0.04604167	+0.04604167	-0.00354167
4	$\iint c_j(u,z)\,tz$	-0.00035426	+0.00814613	-0.00814613	+0.00035426	+0.00035426	-0.00814613	+0.00814613	-0.00035426
5	$\iint c_j(u,z)\,z^2$	-0.00104195	+0.02187529	+0.02187529	-0.00104195	-0.00104195	+0.02187529	+0.02187529	-0.00104195
6	$\iint c_j(u,z)\,t^2$	-0.00177083	+0.02302083	+0.02302083	-0.00177083	-0.00177083	+0.02302083	+0.02302083	-0.00177083

TABLE A-8.2 Values of Two-dimensional Finite-Element Integrals - Continued

j	i = 1	2	3	4	5	6	7	8
				$\iint c_j c_i$				
1	.00079743	-.00937193	-.00598214	.00059831	.00039276	-.00461602	-.00294643	.00029469
2	-.00937193	.13559619	.06121621	-.00598214	-.00461602	.06678618	.03015127	-.00294643
3	-.00598214	.06121621	.13559619	-.00937193	-.00294643	.03015127	.06678618	-.00461602
4	.00059831	-.00598214	-.00937193	.00079743	.00029469	-.00294643	-.00461602	.00039276
5	.00039276	-.00461602	-.00294643	.00029469	.00079743	-.00937193	-.00598214	.00059831
6	-.00461602	.06678618	.03015127	-.00294643	-.00937193	.13559619	.06121621	-.00598214
7	-.00294643	.03015127	.06679618	-.00461602	-.00598214	.06121621	.13559619	-.00937193
8	.00029469	-.00294643	-.00461602	.00039276	.00059831	-.00598214	-.00937193	.00079743
				$\iint c_j\, \partial c_i/\partial u$				
1	-.00119019	.0139879	.0089286	-.00089299	.0011902	-.01398789	-.00892859	.000893
2	.0139879	.20238239	-.09136749	.0089286	-.01398789	.20238239	.0913675	-.00892859
3	.0089286	-.09136749	-.20238239	.0139879	-.00892859	.0913675	.20238239	-.01398789
4	-.00089299	.0089286	.0139879	-.00119019	.000893	-.00892859	-.01398789	.0011902
5	-.00119019	.0139879	.0089286	-.00089299	.0011902	-.01398789	-.00892859	.000893
6	.0139879	-.20238239	-.09136749	.0089286	-.01398789	.2023824	.0913675	-.00892859
7	.0089286	-.09136749	-.20238239	.0139879	-.00892859	.0913675	.2023824	-.01398789
8	.00089299	.0089286	.0139879	-.00119019	.000893	-.00892859	-.01398789	.0011902
				$\iint c_j\, \partial c_i/\partial z$				
1	0	.01535311	-.01674933	.00139574	0	.00756198	-.00824967	.00068745
2	-.01535503	-.16749856	.19960388	-.01675029	-.00756293	-.08249929	.09831236	-.00825014
3	.01675029	.19960388	.16749856	.01535503	.00825014	-.09831236	.08249929	.00756293
4	-.00139574	.01674933	-.01535311	0	-.00068745	.00824967	-.00756198	0
5	0	.00756198	-.00824967	.00068745	0	.01535311	-.01674933	.00139574
6	-.00756293	-.08249929	.09831236	-.00825014	-.01535503	-.16749856	.19960388	-.01675029
7	.00825014	-.09831236	.08249929	.00756293	.01675029	-.19960388	.16749856	.01535503
8	-.00068745	.00824967	-.00756198	0	-.00139574	.01674933	-.01535311	0
				$\iint c_j \partial^2 c_i \partial z^2$				
1	-.01116782	.01954512	-.00558679	-.00279052	-.00550057	.0096267	-.0027517	-.00137443
2	.18704512	-.37967703	.1982187	-.00558679	.0921267	-.1870051	.0976301	-.0027517
3	-.00558679	.1982187	-.37967703	.18704512	-.0027517	.0976301	-.1870051	.0921267
4	-.00279052	-.00558679	.01954512	-.01116782	-.00137443	-.0027517	.0096267	-.00550057
5	-.00550057	.0096267	-.0027517	-.00137443	-.01116782	.01954512	-.00558679	-.00279052
6	.0921267	-.1870051	.0976301	-.0027517	.18704512	-.37967703	.1982187	-.00558679
7	-.0027517	.0076301	-.1870051	.0921367	-.00558679	.1982187	-.37967703	.18704512
8	-.00137443	-.0027517	.0096267	-.00550057	-.00279052	-.00558679	.01954512	-.01116782

Appendix B - Program Listings

Program B-2.1

Four-Point Sliding Polynomial Interpolation

```
SOURCE
PRECISION 12
OPTION BASE 1
PROCS=1
REAL ARRAY(20): Ynode
REAL ARRAY(4): Co
REAL: Xl,Xr,Dx,Xc,Z
INTEGER: I,J,L,N
INTEGER: Nx,Nspx,L1
STRING: Title[60]
REAL ARRAY(11): Xin,Yin

PROCEDURE: Fourpoint
  REAL ARG: Z
END PROCEDURE

PROCEDURE: Fourpoint
  EXTERNAL: Co
    10 Co(1)=(((-8*Z+4)*Z+2)*Z-1)/16
    20 Co(2)=(((24*Z-4)*Z-22)*Z+9)/16
    30 Co(3)=(((-24*Z-4)*Z+22)*Z+9)/16
    40 Co(4)=(((8*Z+4)*Z-2)*Z-1)/16
END PROCEDURE

'MAIN Program:


  10 REM - Interpolate a curve through a given set of nodes.
  20 REM - Uses the 4-point sliding polynomial,calculating 10 interpolations
  30 REM - per span. Total number of nodes to input includes one outlier node
  40 REM - to left and right. Input boundary values xl and xr do not include
  50 REM - outlier positions,but instead are the real boundaries
  60 REM - of the problem space.
  70 OPEN "prn" FOR OUTPUT AS #3
  80 INPUT "problem title";Title
  90 PRINT#3 Title
 100 INPUT "value of xl,which is at node 2 from left";Xl
 110 INPUT " value of xr,which is at node 2 from right";Xr
 120 INPUT "total number of nodes";Nx
 130 Nspx=Nx-3
 140 Dx=(Xr-Xl)/Nspx
 150 PRINT "input the line of nodes"
```

```
160 FOR I = 1 TO Nx
170   PRINT I;"   ";:INPUT Ynode (I)
180 NEXT
190 PRINT#3 "                    the interpolated curve  - x over y"
200 FOR J= 1 TO Nspx
210   Xc=Xl+(J-1)*Dx
220   IF J=1 THEN L1 = 1:Z=-0.5 ELSE L1 = 2:Z=-0.4
230   CLEAR (Yin)
240   PRINT#3
250   FOR L= L1 TO 11
260     Xin(L) =Xc+(L-1)*Dx/10
270     Fourpoint Z
280     FOR N= 1 TO 4
290       Yin(L)=Yin(L)+Co(N)*Ynode(J+N-1)
300     NEXT
310     Z=Z+0.1
320   NEXT
330   IF L1=1 THEN Z=-0.5 ELSE Z=-0.4
340   PRINT#3
350   PRINT#3 "  span number  ";J
360   FOR L=L1 TO 11
370     PRINT#3 USING "##########";L;
380   NEXT
390   PRINT#3
400   FOR L=L1 TO 11
410     PRINT#3 USING "++######.###";Xin(L);
420   NEXT
430   PRINT#3
440   FOR L=L1 TO 11
450     PRINT#3 USING"++######.###";Yin(L);
460   NEXT
470   PRINT#3
480 NEXT
490 PRINT#3
500 CLOSE#3
510 STOP

ENDFILE
```

Program B-2.2

Six-Point Sliding Polynomial Interpolation

```
SOURCE
PRECISION 12
OPTION BASE 1
PROCS=1
REAL ARRAY(20): Ynode
REAL ARRAY(11): Yint,Xint
REAL ARRAY(6): Co
REAL: Xl,Xr,Dx,Xc,Z
INTEGER: I,J,L
```

```
INTEGER: N,Nx,Nspx,L1
STRING: Title[60]
PROCEDURE: Sixpoint
  REAL ARG: Z
END PROCEDURE

PROCEDURE: Sixpoint
  EXTERNAL: Co
    10 Co(1)=(((((-160*Z+16)*Z+144)*Z-40)*Z-26)*Z+9)/768
    20 Co(2)=(((((800*Z-48)*Z-848)*Z+312)*Z+162)*Z-75)/768
    30 Co(3)=(((((-1600*Z+32)*Z+1824)*Z-272)*Z-1124)*Z+450)/768
    40 Co(4)=(((((1600*Z+32)*Z-1824)*Z-272)*Z+1124)*Z+450)/768
    50 Co(5)=(((((-800*Z-48)*Z+848)*Z+312)*Z-162)*Z-75)/768
    60 Co(6)=(((((160*Z+16)*Z-144)*Z-40)*Z+26)*Z+9)/768
END PROCEDURE

'MAIN Program:


  10 REM - Interpolate a curve through a given set of nodes.
  20 REM - Uses the 6-point sliding polynomial,calculating 10 interpolations
  30 REM - per span. Total number of nodes to input includes two outlier nodes
  40 REM - to left and right. Input boundary values xl and xr do not include
  50 REM - outlier positions,but instead are the real boundaries
  60 REM - of the problem space.
  70 OPEN "prn" FOR OUTPUT AS #3
  80 INPUT "problem title";Title
  90 PRINT#3 Title
 100 INPUT "value of xl,which is at node 3 from left";Xl
 110 INPUT " value of xr,which is at node 3 from right";Xr
 120 INPUT "total number of nodes";Nx
 130 Nspx=Nx-5
 140 Dx=(Xr-Xl)/Nspx
 150 PRINT "input the line of nodes"
 160 FOR I = 1 TO Nx
 170   PRINT I;"  ";:INPUT Ynode (I)
 180 NEXT
 190 PRINT#3 "                    the interpolated curve  - x over y"
 200 FOR J= 1 TO Nspx
 210   Xc=Xl+(J-1)*Dx
 220   IF J=1 THEN L1 = 1:Z=-0.5 ELSE L1 = 2:Z=-0.4
 230   CLEAR (Yint)
 240   PRINT#3
 250   FOR L= L1 TO 11
 260     Xint(L) =Xc+(L-1)*Dx/10
 270     Sixpoint Z
 280     FOR N= 1 TO 6
 290       Yint(L)=Yint(L)+Co(N)*Ynode(J+N-1)
 300     NEXT
 310     Z=Z+0.1
 320   NEXT
 330   IF L1=1 THEN Z=-0.5 ELSE Z=-0.4
 340   PRINT#3
 350   PRINT#3 "  span number  ";J
 360   FOR L=L1 TO 11
 370     PRINT#3 USING "##########";L;
```

```
380  NEXT
390  PRINT#3
400  FOR L=L1 TO 11
410    PRINT#3 USING "++######.###";Xint(L);
420  NEXT
430  PRINT#3
440  FOR L=L1 TO 11
450    PRINT#3 USING"++######.###";Yint(L);
460  NEXT
470  PRINT#3
480 NEXT
490 PRINT#3
500 CLOSE#3
510 STOP

ENDFILE
```

Program B-2.3

Reverse Interpolation

```
SOURCE
PRECISION 12
OPTION BASE 1
PROCS=1
REAL ARRAY(4): Co,Ynode
REAL: Xl,Xm,Xr,Yl,Ym,Yr,Y
REAL: Z,Xlp,Xrp
INTEGER: I,J,Is
REAL: Xls,Xrs,Ymb

PROCEDURE: Fourpoint
  REAL ARG: Z
END PROCEDURE

PROCEDURE: Fourpoint
  EXTERNAL: Co
    10 Co(1)=(((-8*Z+4)*Z+2)*Z-1)/16
    20 Co(2)=(((24*Z-4)*Z-22)*Z+9)/16
    30 Co(3)=(((-24*Z-4)*Z+22)*Z+9)/16
    40 Co(4)=(((8*Z+4)*Z-2)*Z-1)/16
END PROCEDURE

'MAIN Program:

  10 REM - Perform reverse interpolation by interval halving.
  20 REM - Continues to ask for points for reverse interpolation.
  30 REM - On query input other than "1" to stop.
  40 PRINT " input four nodes"
  50 OPEN "prn" FOR OUTPUT AS #3
  60 PRINT#3 "        the input nodes"
  70 FOR I = 1 TO 4
  80   INPUT Ynode(I)
```

```
 90  PRINT#3 I;" ";Ynode(I);"     ";
100 NEXT I
110 PRINT#3
120 INPUT "left boundary in x";Xls
130 INPUT "right boundary in x"; Xrs
140 INPUT "one to go;any other to stop";Is
150 IF Is=1 THEN GOTO 160 ELSE GOTO 430
160 INPUT "reverse interpolation value";Y
170 PRINT#3
180 PRINT#3 "reverse interpolation at y=  ";Y
190 Xlp=Xls:Xl=Xls
200 Xrp=Xrs:Xr=Xrs
210 J=1
220 Ymb=9999999
230 Yl=Ynode(2)
240 Yr=Ynode(3)
250 PRINT#3"    xl        xm        xr            yl         ym          yr"
260 Xm=(Xl+Xr)/2
270 Z=(-(Xrp+Xlp)/2+Xm)/(Xrp-Xlp)
280 PRINT#3 "z";Z
290 Fourpoint Z
300 Ym=0
310 FOR I = 1 TO 4
320   Ym=Ym+Co(I)*Ynode(I)
330 NEXT I
340 PRINT#3 USING "++###.######";Xl;Xm;Xr;
350 PRINT #3 "     ";
360 PRINT#3 USING "++####.#####";Yl;Ym;Yr
370 IF Ym>Yl THEN GOTO 380 ELSE GOTO 400
380 IF Y<Ym THEN Xr=Xm:Yr=Ym ELSE Xl=Xm:Yl=Ym
390 GOTO 420
400 IF Y<Ym THEN Xl=Xm:Yl=Ym ELSE Xr=Xm:Yr=Ym
410 IF ABS(Ymb-Ym)<.00001 THEN GOTO 140 ELSE Ymb=Ym
420 IF J=15 THEN GOTO 140 ELSE J=J+1: GOTO 260
430 PRINT#3
440 CLOSE#3
450 STOP

ENDFILE
```

Program B-2.4

Two-Dimensional Six-Point Sliding Polynomial Interpolation - Contouring

```
SOURCE
PRECISION 12
OPTION BASE 1
PROCS=1
REAL ARRAY(20,20): Ynode
REAL ARRAY(11,11): Yint
REAL ARRAY(6): Co,Co1,Co2
REAL: X1l,X1r,X2b,X2t,Dx1,Dx2
REAL: Xc1,Xc2,Z1,Z2,Z
INTEGER: I,J,K,L
```

```
INTEGER: M,N,Nx1,Nx2,Nspx1,Nspx2,L1,K1
STRING: Title[60]
REAL ARRAY(11): X1int,X2int

PROCEDURE: Sixpoint
  REAL ARG: Z
END PROCEDURE

PROCEDURE: Sixpoint
  EXTERNAL: Co
    10 Co(1)=(((((-160*Z+16)*Z+144)*Z-40)*Z-26)*Z+9)/768
    20 Co(2)=(((((800*Z-48)*Z-848)*Z+312)*Z+162)*Z-75)/768
    30 Co(3)=(((((-1600*Z+32)*Z+1824)*Z-272)*Z-1124)*Z+450)/768
    40 Co(4)=(((((1600*Z+32)*Z-1824)*Z-272)*Z+1124)*Z+450)/768
    50 Co(5)=(((((-800*Z-48)*Z+848)*Z+312)*Z-162)*Z-75)/768
    60 Co(6)=(((((160*Z+16)*Z-144)*Z-40)*Z+26)*Z+9)/768
END PROCEDURE

'MAIN Program:


  10 REM - Contour a two-dimensional surface by interpolation on a given set of
  20 REM - nodes. Uses the sixpoint sliding polynomial.
  30 REM - Total number of nodes to input includes two outlier nodes left and
  40 REM - right and top and bottom. Input boundary values of x1 and x2 do not
  50 REM - include outlier positions, but instead are the real boundaries of
  60 REM - the problem space.
  70 OPEN "prn" FOR OUTPUT AS #3
  80 INPUT "problem title";Title
  90 PRINT#3 Title
 100 INPUT "value of x1 at node 3 from left";X1l
 110 INPUT "value of x1 at node 3rd from right";X1r
 120 INPUT "value of x2 at node 3 from bottom";X2b
 130 INPUT "value of x2 at node 3rd from top";X2t
 140 INPUT "total number of nodes in x1 scale";Nx1
 150 INPUT "total number of nodes in x2 scale";Nx2
 160 Nspx1=Nx1-5
 170 Nspx2=Nx2-5
 180 Dx1=(X1r-X1l)/Nspx1
 190 Dx2=(X2t-X2b)/Nspx2
 200 PRINT "input the grid of nodes by row-column format"
 210 FOR J= 1 TO Nx2
 220   PRINT "row" ;J
 230   FOR I= 1 TO Nx1
 240     PRINT I;" ";:INPUT Ynode(J,I)
 250   NEXT
 260 NEXT
 270 PRINT#3 "    the input grid of nodes"
 280 FOR J = 1 TO Nx2
 290   PRINT#3 "  row  ";J
 300   FOR I = 1 TO Nx1
 310     PRINT#3 I;" ";Ynode(J,I); "   ";
 320   NEXT
 330   PRINT#3
 340 NEXT
 350 PRINT#3:PRINT#3
```

```
360 FOR J= 1 TO Nspx2
370   Xc2=X2b+(J-1)*Dx2
380   IF J=1 THEN L1 = 1:Z2=-0.5 ELSE L1 = 2:Z2=-0.4
390   FOR I= 1 TO Nspx1
400     Xc1 =X1l+(I-1)*Dx1
410     IF I = 1 THEN K1 = 1:Z1=-0.5 ELSE K1 = 2:Z1=-0.4
420     CLEAR (Yint)
430     PRINT#3
440     PRINT#3 "surface with lower left coordinates x2= ";Xc2;"  x1= ";Xc1
450     PRINT#3 "  values are x1 over y by x2 rows"
460     FOR L= L1 TO 11
470       X2int(L)=Xc2+(L-1)*Dx2/10
480       Z=Z2
490       Sixpoint Z
500       FOR M = 1 TO 6
510         Co2(M)= Co(M)
520       NEXT
530       FOR K= K1 TO 11
540         X1int(K)=Xc1+(K-1)*Dx1/10
550         Z=Z1
560         Sixpoint Z
570         FOR M = 1 TO 6
580           Co1(M)=Co(M)
590         NEXT
600         FOR M= 1 TO 6
610           FOR N= 1 TO 6
620             Yint(L,K)=Yint(L,K)+Co2(M)*Co1(N)*Ynode(J+M-1,I+N-1)
630           NEXT
640         NEXT
650         Z1=Z1+0.1
660       NEXT
670       Z2=Z2+0.1
680       IF K1=1 THEN Z1=-0.5 ELSE Z1=-0.4
690     NEXT
700     IF L1=1 THEN Z2=-0.5 ELSE Z2=-0.4
710     PRINT#3
720     PRINT#3"                    the surface"
730     FOR L=L1 TO 11
740       PRINT#3 " row ";L;"   x2  =";X2int(L)
750       FOR K= K1 TO 11
760         PRINT#3 USING "#######";K;
770       NEXT
780       PRINT#3
790       FOR K= K1 TO 11
800         PRINT#3 USING "++###.###";X1int(K);
810       NEXT
820       PRINT#3
830       FOR K = K1 TO 11
840         PRINT#3 USING "++###.###";Yint(L,K);
850       NEXT
860       PRINT#3
870       PRINT#3
880     NEXT
890     PRINT#3
900   NEXT
910   PRINT#3
```

```
920 NEXT
930 PRINT#3
940 CLOSE#3
950 STOP
```

ENDFILE

Program B-2.5

Three-Dimensional Four-Point Sliding Polynomial Interpolation - Contouring

```
SOURCE
PRECISION 12
OPTION BASE 1
PROCS=1
REAL ARRAY(6,6,6): Y
REAL ARRAY(9,9): Yi
REAL ARRAY(4): Cc,Cv,Cw,Cx
REAL: Xl,Xr,Wl,Wr,Vl
REAL: Vr,Dx,Dw,Dv,Vi,Wi,Xi,Vz
REAL: Wz,Xz
INTEGER: Nni,Nnj,Nnk,Nsi,Nsj,Nsk,K
INTEGER: J,I,L,Kk,Jj,Ii,Mk,Mj
INTEGER: Mi,Kkk,Jjj,Iii
REAL: Wz1,Vz1,Xz1

PROCEDURE: Coefs4
  REAL ARG: Zz
END PROCEDURE

PROCEDURE: Coefs4
  EXTERNAL: Cc
    10 Cc(1)=(((-8*Zz+4)*Zz+2)*Zz-1)/16
    20 Cc(2)=(((24*Zz-4)*Zz-22)*Zz+9)/16
    30 Cc(3)=(((-24*Zz-4)*Zz+22)*Zz+9)/16
    40 Cc(4)=(((8*Zz+4)*Zz-2)*Zz-1)/16
END PROCEDURE

'MAIN Program:

  10 REM - Triple interpolation  using
  20 REM - four-point sliding polynomials in the three dimensions.
  30 REM - Limited to three elements each dimension, however
  40 REM - this could be expanded.
  50 REM - k indexes in variable domain v
  60 REM - j indexes in variable domain w
  70 REM - i indexes in variable domain x
  80 INPUT "number of nodes in x-scale";Nni
  90 Nsi=Nni-3
 100 INPUT "number of nodes in w-scale";Nnj
 110 Nsj=Nnj-3
 120 INPUT "number of nodes in v-scale";Nnk
```

```
130 Nsk=Nnk-3
135 OPEN "prn" FOR OUTPUT AS #3
136 PRINT #3 "                    triple interpolation"
140 PRINT "input the left/right or top/bottom positions of nodes in each"
150 PRINT "dimension. note that these are not the outlier nodes, but"
160 PRINT "are the nodes at the problem boundaries."
170 INPUT "left and right in x-scale xl,xr";Xl,Xr
180 INPUT "left and right in w-scale wl,wr";Wl,Wr
190 INPUT "left and right in v-scale vl,vr";Vl,Vr
200 Dx=(Xr-Xl)/Nsi
210 Dw=(Wr-Wl)/Nsj
220 Dv=(Vr-Vl)/Nsk
230 PRINT "input the cube of nodes on which interpolation will be based."
240 PRINT "indexing is y(k,j,i)"
245 PRINT #3 "          input nodes"
250 FOR K= 1 TO Nnk
255   PRINT #3 "*  *  *    k  is ";K;" *    *    *"
260   FOR J= 1 TO Nnj
265     PRINT #3 "   j is ";J
270     PRINT "k ";K;" j ";J
280     FOR I = 1 TO Nni
290       PRINT "i ";I;
300       INPUT Y(K,J,I)
305       PRINT #3 I;" ";Y(K,J,I);"   ";
310     NEXT
315     PRINT #3
320   NEXT
330 NEXT
340 REM - Set k-j-i elemental cube then interpolate at eighth-points
350 REM - in x and w and at third-points in v in the cube.
360 FOR K= 1 TO Nsk
370   IF K=1 THEN GOTO 410
380   Kk=2
390   Vi=Vl+(K-1)*Dv+Dv/3
395   Vz =-0.5+1/3
400   GOTO 440
410   Kk=1
420   Vi=Vl
430   Vz=-0.5
440   FOR J= 1 TO Nsj
450     IF J= 1 THEN GOTO 490
460     Jj=2
470     Wi=Wl+(J-1)*Dw+Dw/8
475     Wz=-0.5+1/8
480     GOTO 520
490     Jj=1
500     Wi=Wl
510     Wz=-0.5
520     FOR I= 1 TO Nsi
530       IF I= 1 THEN GOTO 570
540       Ii=2
550       Xi=Xl+(I-1)*Dx+Dv/8
555       Xz= -0.5+1/8
560       GOTO 605
570       Ii=1
580       Xi=Xl
```

```
 590       Xz=-0.5
 600       REM - Interpolate within the cube
 605       Vz1=Vz
 610       FOR Kkk= Kk  TO 4
 615         CLEAR (Yi)
 620         Coefs4 Vz1
 630         FOR L= 1 TO 4
 640           Cv(L)=Cc(L)
 650         NEXT
 655         Wz1=Wz
 660         FOR Jjj= Jj TO 9
 670           Coefs4 Wz1
 680           FOR L= 1 TO 4
 690             Cw(L)=Cc(L)
 700           NEXT
 705           Xz1=Xz
 710           FOR Iii= Ii TO 9
 720             Coefs4 Xz1
 730             FOR L= 1 TO 4
 740               Cx(L)=Cc(L)
 750             NEXT
 760             FOR Mk= 1 TO 4
 770               FOR Mj= 1 TO 4
 780                 FOR Mi = 1 TO 4
 790                   Yi(Jjj,Iii)=Yi(Jjj,Iii)+Cv(Mk)*Cw(Mj)*Cx(Mi)*Y(K-1+Mk,J-1+Mj,I-1+Mi)
 800                 NEXT
 810               NEXT
 820             NEXT
 830             Xz1=Xz1+1/8
 840           NEXT
 850           Wz1=Wz1+1/8
 860         NEXT
 870         PRINT#3:PRINT#3
 880         PRINT#3 "element k,j,i  ";K;"  ";J;"  ";I
 890         PRINT #3 " for level  ";Kkk
 895         FOR Jjj = Jj TO 9
 900           PRINT #3 "row  ";Jjj
 910           FOR Iii = Ii TO 9
 920             PRINT#3 USING "####";Iii;
 930             PRINT#3 USING "--###.###";Yi(Jjj,Iii);
 940           NEXT
 950           PRINT #3
 960         NEXT
 970         Vz1=Vz1+1/3
 980       NEXT
 990       PRINT#3:PRINT#3
1000     NEXT
1010   NEXT
1020 NEXT
1030 PRINT#3
1040 CLOSE #3
1050 END
```

ENDFILE

Program B-3.1

Six-Point Sliding Polynomial Differentiation

```
SOURCE
PRECISION 12
OPTION BASE 1
PROCS=3
REAL ARRAY(6): Co
REAL ARRAY(6): D1co,D2co
REAL ARRAY(20): Ynode
REAL ARRAY(11): Xint,Yint,Ypint,Yppint
REAL: Xl,Xr,Dx,Xc
REAL: Z
INTEGER: I,J,L,N,Nx,Nspx,L1
STRING: Title[60]

PROCEDURE: Sixpoint
  REAL ARG: Z
END PROCEDURE

PROCEDURE: D1sixpt
  REAL ARG: Z
END PROCEDURE

PROCEDURE: D2sixpt
  REAL ARG: Z
END PROCEDURE

PROCEDURE: Sixpoint
  EXTERNAL: Co
    10 Co(1)=(((((-160*Z+16)*Z+144)*Z-40)*Z-26)*Z+9)/768
    20 Co(2)=(((((800*Z-48)*Z-848)*Z+312)*Z+162)*Z-75)/768
    30 Co(3)=(((((-1600*Z+32)*Z+1824)*Z-272)*Z-1124)*Z+450)/768
    40 Co(4)=(((((1600*Z+32)*Z-1824)*Z-272)*Z+1124)*Z+450)/768
    50 Co(5)=(((((-800*Z-48)*Z+848)*Z+312)*Z-162)*Z-75)/768
    60 Co(6)=(((((160*Z+16)*Z-144)*Z-40)*Z+26)*Z+9)/768
END PROCEDURE

PROCEDURE: D1sixpt
  EXTERNAL: D1co
    10 D1co(1)=((((-400*Z+32)*Z+216)*Z-40)*Z-13)/384
    20 D1co(2)=((((2000*Z-96)*Z-1272)*Z+312)*Z+81)/384
    30 D1co(3)=((((-4000*Z+64)*Z+2736)*Z-272)*Z-562)/384
    40 D1co(4)=((((4000*Z+64)*Z-2736)*Z-272)*Z+562)/384
    50 D1co(5)=((((-2000*Z-96)*Z+1272)*Z+312)*Z-81)/384
    60 D1co(6)=((((400*Z+32)*Z-216)*Z-40)*Z+13)/384
END PROCEDURE

PROCEDURE: D2sixpt
  EXTERNAL: D2co
    10 D2co(1)=(((-200*Z+12)*Z+54)*Z-5)/48
    20 D2co(2)=(((1000*Z-36)*Z-318)*Z+39)/48
```

```
    30 D2co(3)=(((-2000*Z+24)*Z+684)*Z-34)/48
    40 D2co(4)=(((2000*Z+24)*Z-684)*Z-34)/48
    50 D2co(5)=(((-1000*Z-36)*Z+318)*Z+39)/48
    60 D2co(6)=(((200*Z+12)*Z-54)*Z-5)/48
END PROCEDURE

'MAIN Program:

 10 REM - Interpolate a curve through a given set of nodes.
 20 REM - Uses the six-point sliding polynomial,calculating 10 interpolations
 30 REM - per span. Total number of nodes to input includes two outlier nodes
 40 REM - to left and right. Input boundary values xl and xr do not include
 50 REM - outlier positions,but instead are the real boundaries
 55 REM - of the problem space.
 60 OPEN "prn" FOR OUTPUT AS #3
 70 INPUT "problem title";Title
 80 PRINT#3 Title
 90 INPUT "value of xl,which is at node 3 from left";Xl
100 INPUT " value of xr,which is at node 3 from right";Xr
110 INPUT "total number of nodes";Nx
120 Nspx=Nx-5
130 Dx=(Xr-Xl)/Nspx
140 PRINT "input the line of nodes"
150 FOR I = 1 TO Nx
160   PRINT I;"   ";:INPUT Ynode (I)
170 NEXT
180 PRINT#3 "                    the interpolated curve  - x over y"
190 FOR J= 1 TO Nspx
200   Xc=Xl+(J-1)*Dx
210   IF J=1 THEN L1 = 1:Z=-0.5 ELSE L1 = 2:Z=-0.4
220   CLEAR (Yint,Ypint,Yppint)
230   PRINT#3
240   FOR L= L1 TO 11
250     Xint(L) =Xc+(L-1)*Dx/10
260     Sixpoint Z
270     D1sixpt Z
280     D2sixpt Z
290     FOR N= 1 TO 6
300       Yint(L)=Yint(L)+Co(N)*Ynode(J+N-1)
310       Ypint(L)=Ypint(L)+D1co(N)*Ynode(J+N-1)
320       Yppint(L)=Yppint(L)+D2co(N)*Ynode(J+N-1)
330     NEXT
340     Ypint(L)=Ypint(L)/Dx
350     Yppint(L)=Yppint(L)/Dx/Dx
360     Z=Z+0.1
370   NEXT
380   IF L1=1 THEN Z=-0.5 ELSE Z=-0.4
390   PRINT#3
400   PRINT#3 "   span number  ";J
410   FOR L=L1 TO 11
420     PRINT#3 USING "##########";L;
430   NEXT
440   PRINT#3
450   FOR L=L1 TO 11
460     PRINT#3 USING "++######.###";Xint(L);
```

```
470  NEXT
480  PRINT#3
490  FOR L=L1 TO 11
500    PRINT#3 USING"++######.###";Yint(L);
510  NEXT
520  PRINT#3
530  FOR L=L1 TO 11
540    PRINT#3 USING"++######.###";Ypint(L);
550  NEXT
560  PRINT#3
570  FOR L= L1 TO 11
580    PRINT#3 USING"++######.###";Yppint(L);
590  NEXT
600  PRINT#3
610 NEXT
620 PRINT#3
630 CLOSE#3
640 STOP

ENDFILE
```

Program B-3.2

Two-Dimensional Six-Point Sliding Polynomial Differentiation

```
SOURCE
PRECISION 12
OPTION BASE 1
PROCS=3
REAL ARRAY(20,20): Ynode
REAL ARRAY(11,11): Yint,Ypint,Yppint
REAL ARRAY(6): Co,Dco,D2co,Co1,Dco1,D2co1
REAL ARRAY(6): Co2,Dco2,D2co2
REAL: X1l,X1r,X2b,X2t,Dx1,Dx2
REAL: Xc1,Xc2,Z1,Z2,Z
INTEGER: I,J,K,L
INTEGER: M,N,Nx1,Nx2,Nspx1,Nspx2
STRING: Title[60]
INTEGER: L1,K1
REAL ARRAY(11,11): Ypint2,Yppint2
REAL ARRAY(11,11): Yptot,Ypptot,Ypasp
REAL: Pi
PROCEDURE: Sixpoint
  REAL ARG: Z
END PROCEDURE

PROCEDURE: Dsixpoint
  REAL ARG: Z
END PROCEDURE

PROCEDURE: D2sixpoint
  REAL ARG: Z
END PROCEDURE
```

```
PROCEDURE: Sixpoint
  EXTERNAL: Co
    10 Co(1)=(((((-160*Z+16)*Z+144)*Z-40)*Z-26)*Z+9)/768
    20 Co(2)=(((((800*Z-48)*Z-848)*Z+312)*Z+162)*Z-75)/768
    30 Co(3)=(((((-1600*Z+32)*Z+1824)*Z-272)*Z-1124)*Z+450)/768
    40 Co(4)=(((((1600*Z+32)*Z-1824)*Z-272)*Z+1124)*Z+450)/768
    50 Co(5)=(((((-800*Z-48)*Z+848)*Z+312)*Z-162)*Z-75)/768
    60 Co(6)=(((((160*Z+16)*Z-144)*Z-40)*Z+26)*Z+9)/768
END PROCEDURE

PROCEDURE: Dsixpoint
  EXTERNAL: Dco
    10 Dco(1)=((((-400*Z+32)*Z+216)*Z-40)*Z-13)/384
    20 Dco(2)=((((2000*Z-96)*Z-1272)*Z+312)*Z+81)/384
    30 Dco(3)=((((-4000*Z+64)*Z+2736)*Z-272)*Z-562)/384

    50 Dco(5)=((((-2000*Z-96)*Z+1272)*Z+312)*Z-81)/384
    60 Dco(6)=((((400*Z+32)*Z-216)*Z-40)*Z+13)/384
END PROCEDURE

PROCEDURE: D2sixpoint
  EXTERNAL: D2co
    10 D2co(1)=(((-200*Z+12)*Z+54)*Z-5)/48
    20 D2co(2)=(((1000*Z-36)*Z-318)*Z+39)/48
    30 D2co(3)=(((-2000*Z+24)*Z+684)*Z-34)/48
    40 D2co(4)=(((2000*Z+24)*Z-684)*Z-34)/48
    50 D2co(5)=(((-1000*Z-36)*Z+318)*Z+39)/48
    60 D2co(6)=(((200*Z+12)*Z-54)*Z-5)/48
END PROCEDURE

'MAIN Program:


  10 REM - Contour a two-dimensional surface by interpolation on a given set of
  20 REM - nodes. Uses the sixpoint sliding polynomial. Includes differentials.
  30 REM - Total number of nodes to input includes two outlier nodes left and
  40 REM - right and top and bottom. Input boundary values of x1 and x2 do not
  50 REM - include outlier positions, but instead are the real boundaries of
  60 REM - the problem space.
  61 REM - Enter nodes left to right of grid by columns.
  62 REM - Enter nodes bottom to top of grid by rows.
  70 OPEN "prn" FOR OUTPUT AS #3
  80 INPUT "problem title";Title
  90 PRINT#3 Title
 100 INPUT "value of x1 at node 3 from left";X1l
 110 INPUT "value of x1 at node 3rd from right";X1r
 120 INPUT "value of x2 at node 3 from bottom";X2b
 130 INPUT "value of x2 at node 3rd from top";X2t
 140 INPUT "total number of nodes in x1 scale";Nx1
 150 INPUT "total number of nodes in x2 scale";Nx2
 160 Nspx1=Nx1-5
 170 Nspx2=Nx2-5
 180 Dx1=(X1r-X1l)/Nspx1
 190 Dx2=(X2t-X2b)/Nspx2
 200 Pi=3.14159265
```

```
210 PRINT "input the grid of nodes by row-column format"
220 FOR J= 1 TO Nx2
230   PRINT "row" ;J
240   FOR I= 1 TO Nx1
250     PRINT I;" ";:INPUT Ynode(J,I)
260   NEXT
270 NEXT
280 FOR J= 1 TO Nspx2
290   Xc2=X2b+(J-1)*Dx2
300   IF J=1 THEN L1 = 1:Z2=-0.5 ELSE L1 = 2:Z2=-0.4
310   FOR I= 1 TO Nspx1
320     Xc1 =X1l+(I-1)*Dx1
330     IF I = 1 THEN K1 = 1:Z1=-0.5 ELSE K1 = 2:Z1=-0.4
340     CLEAR (Yint,Ypint,Yppint)
350     CLEAR (Ypint2,Yppint2,Yptot,Ypptot,Ypasp)
360     PRINT#3
370     PRINT#3 "surface with lower left coordinates x2= ";Xc2;" x1= ";Xc1
380     FOR L= L1 TO 11
390       Z=Z2
400       Sixpoint Z
410       Dsixpoint Z
420       D2sixpoint Z
430       FOR M = 1 TO 6
440         Co2(M)= Co(M)
450         Dco2(M)=Dco(M)/Dx2
460         D2co2(M)=D2co(M)/Dx2/Dx2
470       NEXT
480       FOR K= K1 TO 11
490         Z=Z1
500         Sixpoint Z
510         Dsixpoint Z
520         D2sixpoint Z
530         FOR M = 1 TO 6
540           Co1(M)=Co(M)
550           Dco1(M)=Dco(M)/Dx1
560           D2co1(M)=D2co(M)/Dx1/Dx1
570         NEXT
580         FOR M= 1 TO 6
590           FOR N= 1 TO 6
600             Yint(L,K)=Yint(L,K)+Co2(M)*Co1(N)*Ynode(J+M-1,I+N-1)
610             Ypint(L,K)=Ypint(L,K)+Co2(M)*Dco1(N)*Ynode(J+M-1,I+N-1)
620             Yppint(L,K)=Yppint(L,K)+Co2(M)*D2co1(N)*Ynode(J+M-1,I+N-1)
630             Ypint2(L,K)=Ypint2(L,K)+Dco2(M)*Co1(N)*Ynode(J+M-1,I+N-1)
640             Yppint2(L,K)=Yppint2(L,K)+D2co2(M)*Co1(N)*Ynode(J+M-1,I+N-1)
650           NEXT
660         NEXT
670         Z1=Z1+0.1
680         Yptot(L,K)=SQR(Ypint(L,K)*Ypint(L,K)+Ypint2(L,K)*Ypint2(L,K))
690         Ypptot(L,K)=SQR(Yppint(L,K)*Yppint(L,K)+Yppint2(L,K)*Yppint2(L,K))
700         Ypasp(L,K)=ATN(ABS(Ypint(L,K)/Ypint2(L,K)))
710         IF SGN(Ypint(L,K))=-1 AND SGN(Ypint2(L,K))=1 THEN Ypasp(L,K)=Pi-Ypasp(L,K):GOTO
740
720         IF SGN(Ypint(L,K))=1 AND SGN(Ypint2(L,K))=1 THEN Ypasp(L,K)=Pi+Ypasp(L,K):GOTO
740
730         IF SGN(Ypint(L,K))=1 AND SGN(Ypint2(L,K))=-1 THEN Ypasp(L,K)=2*Pi-Ypasp(L,K)
740       NEXT
```

```
750    Z2=Z2+0.1
760    IF K1=1 THEN Z1=-0.5 ELSE Z1=-0.4
770   NEXT
780   IF L1=1 THEN Z2=-0.5 ELSE Z2=-0.4
790   PRINT#3
800   PRINT#3"              The surface"
810   FOR L=L1 TO 11
820     PRINT#3 " Row ";L
830     FOR K= K1 TO 11
840       PRINT#3 USING "###";K;
850       PRINT#3 USING "++###.###";Yint(L,K);
860     NEXT
870     PRINT#3
880   NEXT
890   PRINT#3
900   PRINT#3"              The first partial derivative dy/dx1"
910   FOR L= L1 TO 11
920     PRINT#3 "  Row ";L
930     FOR K= K1 TO 11
940       PRINT#3 USING "###";K;
950       PRINT#3 USING"++###.###";Ypint(L,K);
960     NEXT
970     PRINT#3
980   NEXT
990   PRINT#3
1000  PRINT#3"              The second partial derivative d2y/dx12"
1010  FOR L = L1 TO 11
1020    PRINT#3 " row ";L
1030    FOR K = K1 TO 11
1040      PRINT#3 USING "###";K;
1050      PRINT#3 USING "++###.###";Yppint(L,K);
1060    NEXT
1070    PRINT#3
1080  NEXT
1090  PRINT#3
1100  PRINT#3"                 The first partial derivative dy/dx2"
1110  FOR L= L1 TO 11
1120    PRINT#3"  row ";L
1130    FOR K=K1 TO 11
1140      PRINT#3 USING "###";K;
1150      PRINT#3 USING "++###.###";Ypint2(L,K);
1160    NEXT
1170    PRINT#3
1180  NEXT
1190  PRINT#3
1200  PRINT#3  "              The second partial derivative d2y/dx22"
1210  FOR L=L1 TO 11
1220    PRINT#3 "  row ";L
1230    FOR K=K1 TO 11
1240      PRINT#3 USING "###";K;
1250      PRINT#3 USING "++###.###";Yppint2(L,K);
1260    NEXT
1270    PRINT#3
1280  NEXT
1290  PRINT#3
1300  PRINT#3"              The vector first derivative"
```

```
1310     FOR L=L1 TO 11
1320       PRINT#3 " row ";L
1330       FOR K=K1 TO 11
1340         PRINT#3 USING "###";K;
1350         PRINT#3 USING "++###.###";Yptot(L,K);
1360       NEXT
1370       PRINT#3
1380     NEXT
1390     PRINT#3
1400     PRINT#3 "          Aspect angle of vector first derivative"
1410     FOR L= L1 TO 11
1420       PRINT#3 " row ";L
1430       FOR K= K1 TO 11
1440         PRINT#3 USING "###";K;
1450         PRINT#3 USING "++###.###";Ypasp(L,K);
1460       NEXT
1470       PRINT#3
1480     NEXT
1490     PRINT#3
1500     PRINT#3 "            The vector second derivative"
1510     FOR L= L1 TO 11
1520       PRINT#3 " row ";L
1530       FOR K= K1 TO 11
1540         PRINT#3 USING "###";K;
1550         PRINT#3 USING "++###.###";Ypptot(L,K);
1560       NEXT
1570       PRINT#3
1580     NEXT
1590   NEXT
1600   PRINT#3
1610 NEXT
1620 PRINT#3
1630 CLOSE#3
1640 STOP

ENDFILE
```

Program B-4.1

Four-Point Sliding Polynomial Integration

```
SOURCE
PRECISION 12
OPTION BASE 1
PROCS=2
REAL ARRAY(15): Node
REAL ARRAY(10): Intv,Tintv,Sx
REAL ARRAY(4): Coef,Icoe
STRING: Title[50]
REAL: Xl,Xr,X,Barea,Dx
REAL: Z,Dxi
INTEGER: Nn,Ns,I,J,K
```

```
PROCEDURE: Ccoe4
  REAL ARG: Z
END PROCEDURE

PROCEDURE: Icoe4
  REAL ARG: Z
END PROCEDURE

PROCEDURE: Ccoe4
  EXTERNAL: Coef
    10 Coef(1)=(((-8*Z+4)*Z+2)*Z-1)/16
    20 Coef(2)=(((24*Z-4)*Z-22)*Z+9)/16
    30 Coef(3)=(((-24*Z-4)*Z+22)*Z+9)/16
    40 Coef(4)=(((8*Z+4)*Z-2)*Z-1)/16
END PROCEDURE

PROCEDURE: Icoe4
  EXTERNAL: Icoe
    10 Icoe(1)=((((-48*Z+32)*Z+24)*Z-24)*Z-11)/384
    20 Icoe(2)=((((144*Z-32)*Z-264)*Z+216)*Z+161)/384
    30 Icoe(3)=((((-144*Z-32)*Z+264)*Z+216)*Z+47)/384
    40 Icoe(4)=((((48*Z+32)*Z-24)*Z-24)*Z-5)/384
END PROCEDURE

'MAIN Program:


  10 REM - Run an interpolation curve thru a set of nodes.
  20 REM - Use the four-point sliding polynomial;
  30 REM - also compute the area under the curve.
  40 REM - Tthe number of nodes is nn. Node number 1 and node number nn are
  50 REM - outlier nodes. The boundaries of the problem are node number 2 and
  60 REM - node number nn-1. The left boundary of x is xl at node number 2. The
  70 REM - right boundary of x is xr at node number nn-1.
  80 OPEN "prn" FOR OUTPUT AS #3
  90 INPUT "problem title  ";Title
 100 PRINT#3 Title
 110 INPUT "number of nodes  ";Nn
 120 PRINT "input the ";Nn;"  nodes"
 130 FOR I = 1 TO Nn
 140   PRINT I;"  ";:INPUT Node(I)
 150 NEXT
 160 PRINT#3 "The input nodes"
 170 K=0
 180 FOR I = 1 TO Nn
 190   K=K+1
 200   PRINT#3 I;"  ";
 210   PRINT#3 USING "--####.##";Node(I);
 220   IF K=7 THEN GOTO 250
 230   PRINT#3"   ";
 240   GOTO 260
 250   K=0:PRINT#3
 260 NEXT
 270 PRINT#3
 280 REM - Number of spans is ns
 290 Ns=Nn-3
```

```
300 INPUT "left boundary of x  ";Xl
310 INPUT "right boundary of x  ";Xr
320 PRINT#3 "Left boundary of x ";Xl;"  Right boundary of x ";Xr
330 PRINT#3
340 REM - x-increment for each span is dx
350 Dx=(Xr-Xl)/Ns
360 Dxi=Dx/10
370 REM - Interpolate 10 values on the curve for each span. Note: there is an
380 REM - additional value at xl which is the value of node number 2 which
390 REM - is not printed. Interpolated function values are intv(_).
395 REM - Interpolated integrals are tintv(_).
396 Barea=0.0
400 X=Xl
410 PRINT#3 "     Values of x over interpolated y over integral of y
420 FOR I = 1 TO Ns
430   PRINT#3 "          Span number   ";I
440   CLEAR (Intv,Tintv)
450   Z=-0.5
460   FOR J= 1 TO 10
470     Z=Z+.10
480     X=X+Dxi
490     Sx(J)=X
500     Ccoe4 Z
505     Icoe4 Z
510     FOR K= 1 TO 4
520       Intv(J)=Intv(J)+Coef(K)*Node(I+K-1)
525       Tintv(J)=Tintv(J)+Icoe(K)*Node(I+K-1)
530     NEXT
535     Tintv(J)=Tintv(J)*Dx+Barea
540   NEXT
550   Barea=Tintv(10)
570   FOR J= 1 TO 10
580     PRINT#3 USING "--####.###";Sx(J);
590   NEXT
600   PRINT#3
610   FOR J = 1 TO 10
620     PRINT#3 USING "--####.###";Intv(J);
630   NEXT
640   PRINT#3
650   FOR J= 1 TO 10
660     PRINT#3 USING "--#####.##";Tintv(J);
670   NEXT
680   PRINT#3:PRINT#3
690 NEXT
700 PRINT#3
710 CLOSE#3
720 STOP
```

ENDFILE

Program B-4.2

Four-Point Sliding Polynomial Integration and Center of Gravity

```
SOURCE
PRECISION 12
OPTION BASE 1
PROCS=3
REAL ARRAY(15): Node
REAL ARRAY(10): Intv,Tintv,Sx
REAL ARRAY(4): Coef,Icoe
STRING: Title[50]
REAL: Xl,Xr,X,Barea,Dx
REAL: Z,Dxi
INTEGER: Nn,Ns,I,J,K
REAL ARRAY(10): Xintv,Cofg
REAL ARRAY(4): Xicoe
REAL: A,Bxa

PROCEDURE: Ccoe4
  REAL ARG: Z
END PROCEDURE

PROCEDURE: Icoe4
  REAL ARG: Z
END PROCEDURE

PROCEDURE: Xicoe4
  REAL ARG: Z
END PROCEDURE

PROCEDURE: Ccoe4
  EXTERNAL: Coef
    10 Coef(1)=(((-8*Z+4)*Z+2)*Z-1)/16
    20 Coef(2)=(((24*Z-4)*Z-22)*Z+9)/16
    30 Coef(3)=(((-24*Z-4)*Z+22)*Z+9)/16
    40 Coef(4)=(((8*Z+4)*Z-2)*Z-1)/16
END PROCEDURE
PROCEDURE: Icoe4
  EXTERNAL: Icoe
    10 Icoe(1)=((((-48*Z+32)*Z+24)*Z-24)*Z-11)/384
    20 Icoe(2)=((((144*Z-32)*Z-264)*Z+216)*Z+161)/384
    30 Icoe(3)=((((-144*Z-32)*Z+264)*Z+216)*Z+47)/384
    40 Icoe(4)=((((48*Z+32)*Z-24)*Z-24)*Z-5)/384
END PROCEDURE

PROCEDURE: Xicoe4
  EXTERNAL: Xicoe,Dx,A
                                                                 1     0
Xicoe(1)=(((((-8*Dx/5*Z-(8*A-4*Dx)/4)*Z+(4*A+2*Dx)/3)*Z+(2*A-Dx)/2)*Z-A)*Z+(-110*A+23*Dx)/
240)/16
                                                                 2     0
Xicoe(2)=(((((24*Dx/5*Z+(24*A-4*Dx)/4)*Z-(4*A+22*Dx)/3)*Z-(22*A-9*Dx)/2)*Z+9*A)*Z+(1610*A-4
39*Dx)/240)/16
```

```
                                                                                  3         0
Xicoe(3)=(((((-24*Dx/5*Z-(24*A+4*Dx)/4)*Z-(4*A-22*Dx)/3)*Z+(22*A+9*Dx)/2)*Z+9*A)*Z+(470*A-
71*Dx)/240)/16
                                                                                  4         0
Xicoe(4)=(((((8*Dx/5*Z+(8*A+4*Dx)/4)*Z+(4*A-2*Dx)/3)*Z-(2*A+Dx)/2)*Z-A)*Z+(-50*A+7*Dx)/24
0)/16
END PROCEDURE

'MAIN Program:


 10 REM - Run an interpolation curve thru a set of nodes.
 20 REM - Use the four-point sliding polynomial;
 30 REM - also compute the center of gravity of the area under the curve.
 40 REM - The number of nodes is nn. Node number 1 and node number nn are
 50 REM - outlier nodes. the boundaries of the problem are node number 2 and
 60 REM - node number nn-1. The left boundary of x is xl at node number 2. The
 70 REM - right boundary of x is xr at node number nn-1.
 80 OPEN "prn" FOR OUTPUT AS #3
 90 INPUT "problem title ";Title
100 PRINT#3 Title
110 INPUT "number of nodes ";Nn
120 PRINT "input the ";Nn;" nodes"
130 FOR I = 1 TO Nn
140   PRINT I;" ";:INPUT Node(I)
150 NEXT
160 PRINT#3 "The input nodes"
170 K=0
180 FOR I = 1 TO Nn
190   K=K+1
200   PRINT#3 I;" ";
210   PRINT#3 USING "--####.##";Node(I);
220   IF K=7 THEN GOTO 250
230   PRINT#3" ";
240   GOTO 260
250   K=0:PRINT#3
260 NEXT
270 PRINT#3
280 REM - Number of spans is ns
290 Ns=Nn-3
300 INPUT "left boundary of x ";Xl
310 INPUT "right boundary of x ";Xr
320 PRINT#3 "Left boundary of x ";Xl;" Right boundary of x ";Xr
330 PRINT#3
340 REM - x-increment for each span is dx
350 Dx=(Xr-Xl)/Ns
360 Dxi=Dx/10
370 REM - Interpolate 10 values on the curve for each span. note: there is an
380 REM - additional value at xl which is the value of node number 2 which
390 REM - is not printed. Interpolated function values are intv(_).
400 REM - Interpolated integrals are tintv(_).First moment is xintv(_).
410 Barea=0.0 :Bxa=0.0
420 X=Xl
430 PRINT#3 "    Values of x over interpolated y over integral of y over center of gravity"
440 FOR I = 1 TO Ns
450   PRINT#3 "          Span number   ";I
```

```
460  CLEAR (Intv,Tintv,Xintv)
470  Z=-0.5
480  A=Xl+(I-0.5)*Dx
490  FOR J= 1 TO 10
500    Z=Z+.10
510    X=X+Dxi
520    Sx(J)=X
530    Ccoe4 Z
540    Icoe4 Z
550    Xicoe4 Z
560    FOR K= 1 TO 4
570      Intv(J)=Intv(J)+Coef(K)*Node(I+K-1)
580      Tintv(J)=Tintv(J)+Icoe(K)*Node(I+K-1)
590      Xintv(J)=Xintv(J)+Xicoe(K)*Node(I+K-1)
600    NEXT
610    Tintv(J)=Tintv(J)*Dx+Barea
620    Xintv(J)=Xintv(J)*Dx+Bxa
630    Cofg(J)=Xintv(J)/Tintv(J)
640  NEXT
650  Barea=Tintv(10):Bxa=Xintv(10)
660  FOR J= 1 TO 10
670    PRINT#3 USING "--####.###";Sx(J);
680  NEXT
690  PRINT#3
700  FOR J = 1 TO 10
710    PRINT#3 USING "--####.###";Intv(J);
720  NEXT
730  PRINT#3
740  FOR J= 1 TO 10
750    PRINT#3 USING "--#####.##";Tintv(J);
760  NEXT
770  PRINT#3
780  FOR J = 1 TO 10
790    PRINT#3 USING "--####.###";Cofg(J);
800  NEXT
810  PRINT#3:PRINT#3
820 NEXT
830 PRINT#3
840 CLOSE#3
850 STOP

ENDFILE
```

Program B-4.3

Two-Dimensional Four-Point Sliding Polynomial Integration

```
SOURCE
PRECISION 12
OPTION BASE 1
PROCS=0
REAL ARRAY(25,25): Ynode,Voly
```

```
STRING: Title[60]
INTEGER: I,J,K,L
REAL: X1l,X1r,X2b,X2t
INTEGER: N1,N2,Nspx1,Nspx2
REAL ARRAY(4): Coe
REAL: Bvol1
REAL ARRAY(25): Bvol2
REAL: Dx1,Dx2

'MAIN Program:


 10 REM - Compute the volume under a two-dimensional,four-point surface.
 20 REM - Volumes are computed by whole units of the four-point sliding
 30 REM - polynomial. Total nodes to input include one outlier node left,
 40 REM - one outlier node  right, one on top, and one on bottom.
 50 REM - Input boundaries xl and xr do not include the outlier positions.
 60 REM - They are the real boundaries of the problem space.
 70 REM - Enter nodes left to right of grid by columns.
 80 REM - Enter nodes bottom to top of grid by rows.
 90 OPEN "prn" FOR OUTPUT AS#3
100 INPUT "problem title";Title
110 PRINT#3 Title
120 INPUT "value of x1 at node 2 from left";X1l
130 INPUT "value of x1 at node 2nd from right";X1r
140 INPUT "value of x2 at node 2 from bottom";X2b
150 INPUT "value of x2 at node 2nd from top";X2t
160 INPUT "total number of nodes in x1 scale";N1
170 INPUT "total number of nodes in x2 scale";N2
180 Nspx1=N1-3
190 Nspx2=N2-3
200 Dx1=(X1r-X1l)/Nspx1
210 Dx2=(X2t-X2b)/Nspx2
220 PRINT "input the grid of nodes by row-column format"
230 FOR J = 1 TO N2
240   PRINT " row  ";J
250   FOR I = 1 TO N1
260     PRINT I;"  ";:INPUT Ynode(J,I)
270   NEXT
280 NEXT
290 Coe(1)=-0.041666667
300 Coe(2)= 0.541666667
310 Coe(4)= Coe(1):Coe(3)=Coe(2)
320 CLEAR (Voly)
330 FOR J = 1 TO N2
340   FOR I = 1 TO N1
350     FOR K= 1 TO 4
360       FOR L= 1 TO 4
370         Voly(J,I)=Voly(J,I)+Coe(K)*Coe(L)*Ynode(J+K-1,I+L-1)
380       NEXT
390     NEXT
400     Voly(J,I)=Voly(J,I)*Dx1*Dx2
410   NEXT
420 NEXT
430 CLEAR (Bvol2)
440 FOR J= 1 TO Nspx2
```

```
 450   Bvol1 = 0.0
 460   FOR I = 1 TO Nspx1
 470     Voly(J,I)=Voly(J,I)+Bvol1
 480     Bvol1=Voly(J,I)
 490     Voly(J,I)=Voly(J,I)+Bvol2(I)
 500     Bvol2(I)=Voly(J,I)
 510   NEXT
 520 NEXT
 530 PRINT#3 " Total integral"
 540 FOR J= 1 TO Nspx2
 550   PRINT#3 " Row  ";J
 560   K=1
 570   FOR I = 1 TO Nspx1
 580     PRINT#3 USING "####";I;
 590     PRINT#3 USING "####.##";Voly(J,I);
 600     IF K=8 THEN PRINT#3:K=1: GOTO 620
 610     K=K+1
 620   NEXT
 630   PRINT#3
 640 NEXT
 650 PRINT#3
 660 CLOSE#3
 670 STOP

ENDFILE
```

Program B-5.1

Simple Regression

```
SOURCE
PRECISION 12
OPTION BASE 1
PROCS=1
REAL ARRAY(30,30): A,B
REAL ARRAY(30): X,Yp,Cd
INTEGER: Ni,Nd,I1,J1,Id
INTEGER: Ix,M,Ij,I
REAL: Qi,Qd,Sy,Ssr,Sse
REAL: Sd2

PROCEDURE: Matrixinv
END PROCEDURE

PROCEDURE: Matrixinv
  EXTERNAL: A,B,Yp,Ni
  REAL: S,T
  INTEGER: I,J,K,L
     5 CLEAR (B)
    10 FOR J = 1 TO Ni
    20   B(J,J)=1.
```

```
   30 NEXT
   40 FOR J = 1 TO Ni
   45   K=0
   50   FOR I = J TO Ni
   55     L=I
   60     IF A(I,J) <>0 THEN EXIT 1 LEVELS
   80     K=K+1
   90   NEXT
   94   IF K<>Ni-J+1 THEN GOTO 100
   95   PRINT "singular matrix"
   96   STOP
  100   FOR K = 1 TO Ni
  110     S=A(J,K)
  120     A(J,K)=A(L,K)
  130     A(L,K)=S
  140     S=B(J,K)
  150     B(J,K)=B(L,K)
  160     B(L,K)=S
  170   NEXT
  180   T=1/A(J,J)
  190   FOR K = 1 TO Ni
  200     A(J,K)=T*A(J,K)
  210     B(J,K)=T*B(J,K)
  220   NEXT
  230   FOR L = 1 TO Ni
  240     IF L=J THEN GOTO 300
  250     T=-A(L,J)
  260     FOR K=1 TO Ni
  270       A(L,K)=A(L,K)+T*A(J,K)
  280       B(L,K)=B(L,K)+T*B(J,K)
  290     NEXT
  300   NEXT
  310 NEXT
  320 LPRINT "matrix inverse"
  330 FOR I= 1 TO Ni
  340   FOR J= 1 TO Ni
  350     PRINT#2 USING "--##.######";B(I,J);
  360   NEXT
  370   LPRINT
  380 NEXT
  390 LPRINT "solution of equations"
  400 FOR I= 1 TO Ni
  410   Yp(I)=0.0
  420   FOR J= 1 TO Ni
  430     Yp(I)=Yp(I)+B(I,J)*A(Ni+1,J)
  440   NEXT
  450   LPRINT I;"  ";Yp(I);"    ";
  460 NEXT
END PROCEDURE

'MAIN Program:


  10 REM - Perform simple least-squares fit
  14 INPUT "number of independent variates";Ni
  18 INPUT "number of observations";Nd
```

```
 20 OPEN "prn" FOR OUTPUT AS #2
 30 PRINT #2 " Input data for least-squares solution - 'y' is on the right"
 40 FOR I1 = 1 TO 30
 50   FOR J1 = 1 TO 30
 60     A(I1,J1)=0.0
 70   NEXT
 80 NEXT
 90 Qi=Ni+1
100 Qd= Nd
110 REM - Input data  - - y comes last
120 Sy=0.0
130 FOR Id = 1 TO Qd
140   PRINT "row";Id
150   FOR Ix = 1 TO Qi
160     PRINT Ix;
170     INPUT X(Ix)
180   NEXT
190   FOR M = 1 TO Qi
200     PRINT #2 USING "--####.##"; X(M);
210   NEXT
215   PRINT #2
220   Sy=Sy+X(Ni+1)
230   FOR Ix= 1 TO Qi
240     FOR Ij= 1 TO Qi
250       A(Ix,Ij)=A(Ix,Ij)+X(Ix)*X(Ij)
260     NEXT
270   NEXT
280 NEXT
290 PRINT #2 "Sums of squares matrix"
300 FOR Ix = 1 TO Qi
310   FOR Ij = 1 TO Qi
320     PRINT #2 USING "--####.####"; A(Ix,Ij);
330   NEXT
340   PRINT #2
350 NEXT
351 Matrixinv
352 Ssr=0.0
360 FOR I=1 TO Ni
370   Ssr=Ssr+Yp(I)*A(Ni+1,I)
380 NEXT
390 Sse=A(Ni+1,Ni+1)-Ssr
400 Sd2=Sse/(Nd-Ni)
405 PRINT #2
410 PRINT #2 "Residual variance";Sd2
420 PRINT #2 "Standard deviation of coefficients"
430 FOR I = 1 TO Ni
440   Cd(I)=SQR(B(I,I)*Sd2)
450   PRINT #2 I;Cd(I);"   ";
460 NEXT
470 PRINT #2
480 CLOSE #2
490 STOP
```

ENDFILE

Program B-5.2

Two-Point Sliding Polynomial Smoothing with Error of Prediction

```
SOURCE
PRECISION 12
OPTION BASE 1
PROCS=1
STRING: Title[60]
REAL ARRAY(100,20): C
REAL ARRAY(100): X,Y,E
REAL ARRAY(20,20): Sc
REAL ARRAY(22): Yc
REAL ARRAY(20): Xn,Ce,C2,C1,Ee,Ve
INTEGER: P,N,Ip,I,J,M,Nq,Pe
INTEGER: Pc
REAL: Xl,Xr,Dx,K,Jc,Dp,Z,Xp
REAL: Nc,Xb,Xs
REAL ARRAY(5): W
REAL: Xi,Yi,Sy,Sw,Yp,Si
REAL: Area,Sp,Sn

PROCEDURE: Simeqns
  INTEGER ARG: N
END PROCEDURE

PROCEDURE: Simeqns
  EXTERNAL: Sc
  REAL: X,Y
  INTEGER: J,I,K,M,Ic
   10 REM - solve set of n equations
   20 FOR J = 1 TO N
   30  REM - statements 480 to 520 find the first equation with a
   40  REM - non-zero coefficient for the current column.
   50  Ic=0:M=1
   60  FOR I = J TO N
   70   IF Sc(I,J)=0 THEN GOTO 80 ELSE GOTO 100
   80   Ic=Ic+1
   90   GOTO 130
  100   ON M GOTO 110,130
  110   K=I
  120   M=2
  130  NEXT
  140  IF Ic=N THEN GOTO 160
  150  GOTO 180
  160  PRINT "no unique solution": STOP
  170  REM  statements 540 to 580 move that equation up to current row.
  180  I=K
  190  FOR K= 1 TO N+1
  200   X=Sc(J,K)
  210   Sc(J,K)=Sc(I,K)
  220   Sc(I,K)=X
  230  NEXT
  240  REM - statements 610 to 640 get a 1 coefficient in the first non-zero
```

```
    250  REM - column of the current row
    260  Y=1/Sc(J,J)
    270  FOR K= 1 TO N+1
    280    Sc(J,K)=Y*Sc(J,K)
    290  NEXT
    300  REM - statements 670 to 730 subtract the current equation from
    310  REM - th other rows,
    320  FOR I = 1 TO N
    330    IF I = J THEN GOTO 380
    340    Y=-Sc(I,J)
    350    FOR K = 1 TO N+1
    360      Sc(I,K)=Sc(I,K)+Y*Sc(J,K)
    370    NEXT
    380  NEXT
    390  REM - this process is repeated for all equations.
    400 NEXT
    410 PRINT #6 " solution of equations"
    420 M = 1
    430 FOR I = 1 TO N
    440   M= M+1
    450   PRINT #6 USING "#####"; I;
    460   PRINT#6 USING "--#####.####";Sc(I,N+1);
    470   IF M=>6 GOTO 500
    480   PRINT #6 "   ";
    490   GOTO 510
    500   PRINT #6: M= 1
    510 NEXT
    520 PRINT #6
END PROCEDURE

'MAIN Program:

   10 REM - Smooth data by fitting two-point sliding polynomial.
   20 REM - Includes calculation of continuous curve of standard deviation.
   30 OPEN "prn" FOR OUTPUT AS #6
   40 PRINT#6 "Two-point sliding polynomial smoothing"
   50 PRINT#6
   60 INPUT "problem title";Title
   70 PRINT#6 Title
   80 INPUT "number of data points";P
   90 INPUT "number of nodes";N
  100 INPUT "x-position of node at left side of data";Xl
  110 INPUT "x-position of node at right side of data";Xr
  120 REM - Input the data by x-y pairs
  130 FOR Ip = 1 TO P
  140   PRINT Ip;:INPUT X(Ip),Y(Ip)
  150 NEXT Ip
  160 Dx=(Xr-Xl)/(N-1)
  170 Xn(1)=Xl:Xn(N)=Xr
  180 FOR I = 2 TO N-1
  190   Xn(I)=Xl+(I-1)*Dx
  200 NEXT I
  210 PRINT#6 "     x-location of nodes"
  220 K=1
  230 FOR I = 1 TO N
```

```
240  PRINT#6 USING "#####";I;
250  PRINT#6 USING "--####.####";Xn(I);
260  IF K= 8 THEN PRINT#6:K=1:GOTO 280
270  K=K+1
280 NEXT I
290 PRINT#6
300 PRINT#6 "    The n coefficients for each data point"
310 CLEAR (C)
320 FOR K = 1 TO P
330  Jc=1
340  IF X(K)<=(Xl+Jc*Dx) THEN GOTO 360
350  Jc=Jc+1: GOTO 340
360  Dp=(X(K)-Xl-(Jc-1)*Dx)/Dx
370  Z=-0.5+Dp
380  C(K,Jc)=0.5-Z
390  C(K,Jc+1)=0.5+Z
400  PRINT#6 USING "###";K;
410  Ip=1
420  FOR M= 1 TO N
430    PRINT#6 USING "--##.#####";C(K,M);
440    IF Ip=10 THEN PRINT#6:PRINT#6 "   ";Ip=1:GOTO 460
450    Ip=Ip+1
460  NEXT M
470  PRINT#6
480 NEXT K
490 FOR K=  1 TO P
500  C(K,N+1)=Y(K)
510 NEXT K
520 CLEAR (Sc)
530 REM - Compute sums of squares matrix
540 FOR K = 1 TO P
550  FOR I = 1 TO N+1
560    FOR J= I TO N+1
570      Sc(I,J)=Sc(I,J)+C(K,I)*C(K,J)
580    NEXT J
590  NEXT I
600 NEXT K
610 REM - Fill in complete matrix
620 FOR I = 1 TO N
630  FOR J = I TO N
640    Sc(J,I)=Sc(I,J)
650  NEXT J
660 NEXT I
670 PRINT#6 "          Sums of squares - - coefficients and y"
680 FOR I = 1 TO N+1
690  K=1
700  FOR J=1 TO N+1
710    PRINT#6 USING "####.######";Sc(I,J);
720    IF K= 10 THEN K=1:PRINT#6: GOTO 740
730    K=K+1
740  NEXT J
750  PRINT#6
760 NEXT I
770 Simeqns N
780 FOR I= 1 TO N
790  Yc(I)=Sc(I,N+1)
```

```
800 NEXT I
810 PRINT#6 "                Residual errors"
820 K=0
830 CLEAR (Ce,C2,C1)
840 FOR I = 1 TO P
850   Yp=0
860   K=K+1
870   FOR J = 1 TO N
880     Yp=Yp+C(I,J)*Yc(J)
890   NEXT J
900   E(I)=Y(I)-Yp
910   PRINT#6 USING "#####";I;
920   PRINT#6 USING "--###.####";E(I);
930   IF K<>8 GOTO 950
940   PRINT#6: K=0
950   E(I)=E(I)*E(I)
960   FOR J=1 TO N
970     Ce(J)=Ce(J)+E(I)*C(I,J)*C(I,J)
980     C2(J)=C2(J)+C(I,J)*C(I,J)
990     C1(J)=C1(J)+C(I,J)
1000   NEXT J
1010 NEXT I
1020 PRINT#6
1030 FOR J= 1 TO N
1040   Ee(J)=Ce(J)/C2(J)
1050   Ve(J)=Ee(J)*P/(P-N)/C1(J)
1060 NEXT J
1070 PRINT#6 "Node     c2*e2         e2           c            var.    std.dev."
1080 FOR J= 1 TO N
1090   PRINT#6 USING "###";J;
1100   PRINT#6 USING "--####.######";Ce(J);C2(J);C1(J);Ve(J);SQR(Ve(J))
1110 NEXT J
1120 Xp=Xl
1130 Nq=N-1
1140 Nc=0
1150 PRINT#6 "         The smooth curve and its localized standard deviation"
1160 PRINT#6 "    Span number;  x;  y;  sd of y"
1170 Nc=Nc+1
1180 Xb=1+(Nc-1)*3
1190 Xs=Xb+2
1200 IF Xs>Nq THEN Xs=Nq+Xb-1
1210 Pe= INT(Xs-Xb+1)
1220 Z=-0.5
1230 FOR J= 1 TO 20
1240   Z=Z+0.05
1250   Xp=Xp+Dx/20
1260   W(1)=0.5-Z
1270   W(2)=0.5+Z
1280   Pc=1
1290   FOR Xi=Xb TO Xs
1300     I=INT(Xi)
1310     Yi=W(1)*Yc(I)+W(2)*Yc(I+1)
1320     Sy=0:Sw=0
1330     FOR K = 1 TO 2
1340       Sy=Sy+W(K)*W(K)*Ve(I+K-1)
1350       Sw=Sw+W(K)*W(K)
```

```
1360      NEXT K
1370      Si=SQR(Sy/Sw)
1380      PRINT#6 USING "#####";I;
1390      PRINT#6 USING "--####.####";Xp+(Xi-Nc)*Dx;Yi;Si;
1400      IF Pc=Pe THEN GOTO 1440
1410      Pc=Pc+1
1420      PRINT#6 "   ";
1430      GOTO 1460
1440      Pc=1
1450      PRINT#6
1460    NEXT I
1470 NEXT J
1480 PRINT#6
1490 IF Nq>3 THEN GOTO 1510
1500 GOTO 1530
1510 Nq=Nq-3
1520 GOTO 1170
1530 PRINT#6 "    Area by spans"
1540 FOR I = 1 TO N-1
1550    Area=(Yc(I)+Yc(I+1))*Dx/2
1560    PRINT#6 USING "###";I;
1570    PRINT#6 USING "--######.####";Arca;
1580    IF Area<0 THEN GOTO 1610
1590    Sp=Sp+Area
1600    GOTO 1620
1610    Sn=Sn+Area
1620 NEXT I
1630 PRINT#6
1640 PRINT#6 " Total positive area  ";Sp
1650 PRINT#6 " Total negative area  ";Sn
1660 PRINT#6
1670 CLOSE #6
1680 STOP

ENDFILE
```

Program B-5.3

Four-Point Sliding Polynomial Smoothing with Error of Prediction

```
SOURCE
PRECISION 12
OPTION BASE 1
PROCS=5
REAL ARRAY(100,20): C
REAL ARRAY(20,20): Sc
REAL ARRAY(100): X,Y
REAL ARRAY(20): Xn
INTEGER: P,N
REAL: Xl,Xr
REAL ARRAY(22): Yc
REAL ARRAY(20): Ve
```

```
REAL: Dx
STRING: T[16]
REAL: Ip,I,J

PROCEDURE: Coefs
END PROCEDURE

PROCEDURE: Sims
  INTEGER ARG: N
END PROCEDURE

PROCEDURE: Fullset
END PROCEDURE

PROCEDURE: Interp
END PROCEDURE

PROCEDURE: Locunc
  INTEGER ARG: N,P
END PROCEDURE

PROCEDURE: Coefs
  EXTERNAL: N,Xl,Xr,C,Xn,P
  REAL: Dx,Dp,Z
  INTEGER: I,K,Jc
  EXTERNAL: X
  REAL: M,J
  EXTERNAL: Y,Sc
   930 CLEAR (C)
   940 Dx=(Xr-Xl)/(N-1)
   950 Xn(1)=Xl
   955 Xn(N)=Xr
   960 FOR I = 2 TO N-1
   970   Xn(I)=Xl+(I-1)*Dx
   975 NEXT
   980 PRINT #6 "       x-location of nodes"
   990 FOR I = 1 TO N
  1000   PRINT #6 I;" ";Xn(I);"   ";
  1005 NEXT
  1010 PRINT #6
  1020 PRINT #6 "   n+2 coefficients for each data point"
  1030 FOR K = 1 TO P
  1035   PRINT #6 " data point number  ";K
  1040   Jc = 1
  1050   IF X(K)<=(Xl+Jc*Dx) THEN GOTO 1070
  1060   Jc=Jc+1
  1065   GOTO 1050
  1070   Dp=(X(K)-Xl-(Jc-1)*Dx)/Dx
  1080   Z=-0.5+Dp
  1090   C(K,Jc)=(((-8*Z+4)*Z+2)*Z-1)/16
  1100   C(K,Jc+1)=(((24*Z-4)*Z-22)*Z+9)/16
  1110   C(K,Jc+2)=(((-24*Z-4)*Z+22)*Z+9)/16
  1120   C(K,Jc+3)=(((8*Z+4)*Z-2)*Z-1)/16
  1130   FOR M = 1 TO N+2
  1140     PRINT #6 M;INT(C(K,M)*100000+.5)/100000;"  ";
  1150   NEXT
```

```
  1155  PRINT #6
  1160 NEXT
  1170 PRINT #6
  1180 FOR K = 1 TO P
  1190  C(K,4)=C(K,4)+C(K,1)
  1200  C(K,3)=C(K,3)-3*C(K,1)
  1210  C(K,2)=C(K,2)+3*C(K,1)
  1220  C(K,N-1)=C(K,N-1)+C(K,N+2)
  1230  C(K,N)=C(K,N)-3*C(K,N+2)
  1240  C(K,N+1)=C(K,N+1)+3*C(K,N+2)
  1250 NEXT
  1260 FOR K = 1 TO P
  1270  FOR J = 1 TO N
  1280   C(K,J)=C(K,J+1)
  1290  NEXT
  1300  C(K,N+1)=Y(K)
  1310 NEXT
  1320 FOR K = 1 TO P
  1330  FOR I = 1 TO N+1
  1340   FOR J = 1 TO N+1
  1350    Sc(I,J)=Sc(I,J)+C(K,I)*C(K,J)
  1360   NEXT
  1365  NEXT
  1370 NEXT
END PROCEDURE

PROCEDURE: Sims
 EXTERNAL: Sc
 REAL: X,Y
 INTEGER: J,I,K,M,Ic
  440 REM - solve set of n equations
  450 FOR J = 1 TO N
  460  REM - statements 480 to 520 find the first equation with a
  470  REM - non-zero coefficient for the current column.
  475  Ic=0:M=1
  480  FOR I = J TO N
  490   IF Sc(I,J)=0 THEN GOTO 491 ELSE GOTO 493
  491   Ic=Ic+1
  492   GOTO 500
  493   ON M GOTO 494,500
  494   K=I
  495   M=2
  500  NEXT
  505  IF Ic=N THEN GOTO 520
  510  GOTO 535
  520  PRINT "no unique solution": STOP
  530  REM  statements 540 to 580 move that equation up to current row.
  535  I=K
  540  FOR K= 1 TO N+1
  550   X=Sc(J,K)
  560   Sc(J,K)=Sc(I,K)
  570   Sc(I,K)=X
  580  NEXT
  590  REM - statements 610 to 640 get a 1 coefficient in the first non-zero
  600  REM - column of the current row
  610  Y=1/Sc(J,J)
```

```
620  FOR K= 1 TO N+1
630    Sc(J,K)=Y*Sc(J,K)
640  NEXT
650  REM - statements 670 to 730 subtract the current equation from
660  REM - th other rows,
670  FOR I = 1 TO N
680    IF I = J THEN GOTO 730
690    Y=-Sc(I,J)
700    FOR K = 1 TO N+1
710      Sc(I,K)=Sc(I,K)+Y*Sc(J,K)
720    NEXT
730  NEXT
740  REM - thia process is repeated for all equations.
750 NEXT
760 PRINT #6 " solution of equations"
770 M = 1
780 FOR I = 1 TO N
785  M= M+1
786  PRINT #6 I;" ";Sc(I,N+1);
800  IF M=>6 GOTO 820
805  PRINT #6 "  ";
810  GOTO 830
820  PRINT #6: M= 1
830 NEXT
840 PRINT #6
END PROCEDURE

PROCEDURE: Fullset
  EXTERNAL: Yc,Sc,N
  INTEGER: I
   1390 Yc(1) = 3*Sc(1,N+1)-3*Sc(2,N+1)+Sc(3,N+1)
   1400 Yc(N+2) = 3*Sc(N,N+1)-3*Sc(N-1,N+1)+Sc(N-2,N+1)
   1410 FOR I = 2 TO N+1
   1420   Yc(I) = Sc(I-1,N+1)
   1425 NEXT
END PROCEDURE
PROCEDURE: Interp
  EXTERNAL: Xl,N,Dx,Yc,Ve
  REAL: Xp,Nc,Xb,Xs
  REAL: Z,Xi,Yi,Sy,Sw,Si
  INTEGER: Nq,Pe,J
  INTEGER: Pc,I,K
  REAL ARRAY(4): W,Wm
    10 Xp=Xl
    20 Nq=N-1
    30 Nc=0
    40 PRINT #6:PRINT #6
    50 PRINT #6 "   the smooth curve and its localized standard deviation"
    60 PRINT #6 "    x;      y;     y+sd;    y-sd"
    70 Nc=Nc+1
    80 Xb=1+(Nc-1)*4
    90 Xs=Xb+3
   100 IF Xs>Nq THEN Xs=Nq+Xb-1
   110 Pe = INT(Xs-Xb+1)
   120 Z=-0.5
   130 FOR J = 1 TO 20
```

```
140  Z= Z+ 0.05
150  Xp=Xp+Dx/20
160  W(1)=(((-8*Z+4)*Z+2)*Z-1)/16
170  W(2)=(((24*Z-4)*Z-22)*Z+9)/16
180  W(3)=(((-24*Z-4)*Z+22)*Z+9)/16
190  W(4)=(((8*Z+4)*Z-2)*Z-1)/16
200  Pc=1
210  FOR Xi = Xb TO Xs
220    I=INT(Xi)
230    Yi=W(1)*Yc(I)+W(2)*Yc(I+1)+W(3)*Yc(I+2)+W(4)*Yc(I+3)
240    Sy=0
250    Sw=0
260    IF I = 1 THEN GOTO 320
270    IF I = N-1 THEN GOTO 390
280    FOR K = 1 TO 4
290      Sy=Sy+W(K)*W(K)*Ve(I+K-2)
300    NEXT
310    GOTO 450
320    Wm(1)=W(2)+3*W(1)
330    Wm(2)=W(3)-3*W(1)
340    Wm(3)=W(4)+W(1)
350    FOR K = 1 TO 3
360      Sy=Sy+Wm(K)*Wm(K)*Ve(K)
370    NEXT
380    GOTO 450
390    Wm(1)=W(1)+W(4)
400    Wm(2)=W(2)-3*W(4)
410    Wm(3)=W(3)+3*W(4)
420    FOR K = 1 TO 3
430      Sy=Sy+Wm(K)*Wm(K)*Ve(N+K-3)
440    NEXT
450    Si=SQR(Sy)
460    PRINT#6 USING "--####.##";Xp+(Xi-Nc)*Dx;
470    PRINT#6 USING "--##.###";Yi;Yi+Si;Yi-Si;
480    IF Pc=Pe GOTO 510
490    Pc=Pc+1
500    GOTO 530
510    Pc=1
520    PRINT #6
530  NEXT
540 NEXT
550 PRINT #6
560 IF Nq>4 GOTO 580
570 GOTO 600
580 Nq=Nq-4
590 GOTO 70
600 PRINT #6
END PROCEDURE

PROCEDURE: Locunc
  EXTERNAL: C,Yc,Y,Ve
  REAL ARRAY(100): E
  REAL ARRAY(20): Ce,C2,C1,Ee
  REAL: Yp
  INTEGER: I,K
  INTEGER: J
```

```
REAL: Rt
 2099 FOR I = 1 TO 20
 2100   Ce(I)=0.0
 2110   C2(I)=0.0
 2120   C1(I)= 0.0
 2121   Ee(I)=0.0
 2122 NEXT
 2123 PRINT #6
 2125 PRINT #6 "         residual errors"
 2126 K=0
 2130 FOR I = 1 TO P
 2134   Yp=0.0
 2138   K=K+1
 2150   FOR J =  1 TO N
 2160     Yp = Yp + C(I,J)*Yc(J+1)
 2170   NEXT
 2180   E(I)=Y(I)-Yp
 2190   PRINT #6 I;" ";INT(E(I)*10000+.5)/10000;" ";
 2200   IF K<>8 THEN GOTO 2220
 2210   PRINT #6
 2220   E(I)=E(I)*E(I)
 2230   FOR J = 1 TO N
 2240     Ce(J)=Ce(J)+E(I)*C(I,J)*C(I,J)
 2250     C2(J)=C2(J)+C(I,J)*C(I,J)
 2260     C1(J)=C1(J)+C(I,J)
 2270   NEXT
 2280 NEXT
 2290 PRINT #6
 2300 FOR J = 1 TO N
 2310   Ee(J)=Ce(J)/C2(J)
 2320   Ve(J)=Ee(J)*P/(P-N)/C1(J)
 2330 NEXT
 2340 PRINT #6
 2345 PRINT #6
 2350 PRINT #6 "  node, sum c2*e2,sum c2,sum c,var, std. dev."
 2360 FOR J= 1 TO N
 2365   IF Ve(J)<0 THEN Rt=0.0 ELSE Rt=SQR(Ve(J))
 2370   PRINT #6 J;" ";Ce(J);" ";C2(J);" ";C1(J);"  ";Ve(J);" ";Rt
 2380 NEXT
END PROCEDURE
```

```
'MAIN Program:

  10 REM - Program to smooth data by fitting sliding polynomial arcs.
  20 REM - Uses four-point sliding polynomial.
  30 REM - Outlier nodes are evaluated by parabolic projection.
  40 REM - Includes calculation of continuous curve of std.dev. of predicition.
  50 OPEN "prn" FOR OUTPUT AS #6
  60 PRINT #6 "            Four-point sliding polynomial smoothing"
  70 INPUT "problem title";T
  80 PRINT #6 T
  90 PRINT #6:PRINT #6
 100 INPUT "number of data points";P
 110 INPUT " number of nodes not including the two outlier nodes";N
 120 INPUT "x=position of node at left side of data";Xl
 130 INPUT "x-position of node at right side of data";Xr
```

```
140 REM - Input the data x-y pairs
150 Ip=1
160 PRINT "input point";Ip;:INPUT X(Ip),Y(Ip)
170 IF Ip = P THEN GOTO 200
180 Ip= Ip+1: GOTO 160
190 REM - Use proc. COEFS to calculate sliding polynomial coefficients.
200 Coefs
210 PRINT #6" Sums of squares - - coefficients and 'y'"
220 FOR I = 1 TO N+1
230   PRINT #6 "Row  ";I
240   FOR J= 1 TO N+1
250     PRINT #6 J;"  ";Sc(I,J);"  ";
260   NEXT
270   PRINT #6
280 NEXT
290 Dx=(Xr-Xl)/(N-1)
300 REM - Fill in complete matrix.
310 FOR I = 1 TO N
320   FOR J = 1 TO N
330     Sc(J,I)=Sc(I,J)
340   NEXT
350 NEXT
360 REM - use proc. SIMS to solve simultaneous normal equations.
370 Sims N
380 REM - Proc. FULLSET lays in the full set of nodes.
390 Fullset
400 REM - Proc. LOCUNC calculates localized uncertainty.
410 Locunc N,P
420 REM - Proc. INTERP calculates the smooth curve through the nodes.
430 Interp
440 PRINT #6
450 CLOSE #6
460 END

ENDFILE
```

Program B-5.4

Smooth with Composite Cycle and Trend

```
SOURCE
PRECISION 12
OPTION BASE 1
PROCS=1
REAL ARRAY(120,10): C
REAL ARRAY(120): Y,Er,Yt
REAL ARRAY(8,8): Sc
REAL ARRAY(12): Ycp,No,Ce2,C2,C1,Nv
REAL ARRAY(12): Sd
STRING: T[100]
INTEGER: Nc,Nd,I,J,K,Ic,Id,N
```

```
REAL: B,A,P,W,Tw,Z,Pb,Ssc
REAL: Sx,Bb
INTEGER: Ip

PROCEDURE: Sims
  INTEGER ARG: N
END PROCEDURE

PROCEDURE: Sims
  EXTERNAL: Sc
  REAL: X,Y
  INTEGER: J,I,K,M,Ic
    10 REM - solve set of n equations
    20 FOR J = 1 TO N
    30   REM - statements 480 to 520 find the first equation with a
    40   REM - non-zero coefficient for the current column.
    50   Ic=0:M=1
    60   FOR I = J TO N
    70     IF Sc(I,J)=0 THEN GOTO 80 ELSE GOTO 100
    80     Ic=Ic+1
    90     GOTO 130
   100     ON M GOTO 110,130
   110     K=I
   120     M=2
   130   NEXT
   140   IF Ic=N THEN GOTO 160
   150   GOTO 180
   160   PRINT "no unique solution": STOP
   170   REM  statements 540 to 580 move that equation up to current row.
   180   I=K
   190   FOR K= 1 TO N+1
   200     X=Sc(J,K)
   210     Sc(J,K)=Sc(I,K)
   220     Sc(I,K)=X
   230   NEXT
   240   REM - statements 610 to 640 get a 1 coefficient in the first non-zero
   250   REM - column of the current row
   260   Y=1/Sc(J,J)
   270   FOR K= 1 TO N+1
   280     Sc(J,K)=Y*Sc(J,K)
   290   NEXT
   300   REM - statements 670 to 730 subtract the current equation from
   310   REM - th other rows,
   320   FOR I = 1 TO N
   330     IF I = J THEN GOTO 380
   340     Y=-Sc(I,J)
   350     FOR K = 1 TO N+1
   360       Sc(I,K)=Sc(I,K)+Y*Sc(J,K)
   370     NEXT
   380   NEXT
   390   REM - this process is repeated for all equations.
   400 NEXT
   410 PRINT #6 " solution of equations"
   420 M = 1
   430 FOR I = 1 TO N
   440   M= M+1
```

```
    450   PRINT #6 USING "#####"; I;
    460   PRINT#6 USING "--#####.####";Sc(I,N+1);
    470   IF M=>6 GOTO 500
    480   PRINT #6 "   ";
    490   GOTO 510
    500   PRINT #6: M= 1
    510 NEXT
    520 PRINT #6
END PROCEDURE
```

'MAIN Program:

```
 10 REM - Smooth data with composite function of cycle plus trend.
 20 REM - Phase variate "p" runs from 1 to 12 for monthly values.
 30 REM - Trend line runs from 1 to nc*12+1 where nc is number of cycles.
 40 REM - Wave is defined by nodes at odd values of "p".
 50 REM - Trend is defined by zero at 1 and one node at nc*12+1.
 60 OPEN "prn" FOR OUTPUT AS #6
 70 PRINT#6 "          Composite smoothing of linear trend plus wave"
 80 INPUT "problem title";T
 90 PRINT#6 "          ";T
100 INPUT "number of data  points";Nd
110 INPUT"input'1' to print sliding polynomial coefs; else any";Ip
120 Nc=Nd/12
130 REM - Input data from disk and calculate sliding polynomial coefficients.
140 OPEN "b:lawr" FOR INPUT AS #5
150 Bb=Nc
160 B=1/(12*Bb):A=-0.5-B
170 CLEAR(C)
180 FOR I = 1 TO Nd
190   P=((I-1) MOD 12) + 1
200   W =INT((I-1)/12)+1.0
210   INPUT #5 Y(I)
220   Tw=12*(W-1)+P:Z=A+B*Tw
230   C(I,1)=0.5+Z
240   FOR J = 1 TO 6
250     IF P<(2*J+1) THEN EXIT
260   NEXT J
270   Pb=2*J-1:Z=(-1.-Pb+P)/2
280   C(I,J+1)=(((-8*Z+4)*Z+2)*Z-1)/16
290   C(I,J+2)=(((24*Z-4)*Z-22)*Z+9)/16
300   C(I,J+3)=(((-24*Z-4)*Z+22)*Z+9)/16
310   C(I,J+4)=(((8*Z+4)*Z-2)*Z-1)/16
320 NEXT I
330 CLOSE #5
340 REM - Move and consolidate coefficients, add y for sums of squares.
350 FOR I = 1 TO Nd
360   C(I,3)=C(I,3)+C(I,9)
370   C(I,4)=C(I,4)+C(I,10)
380   C(I,8)=C(I,8)+C(I,2)
390 NEXT I
400 FOR I = 1 TO Nd
410   FOR J = 2 TO 7
420     C(I,J)=C(I,J+1)
430   NEXT J
```

```
440  C(I,8)=Y(I)
450 NEXT I
460 IF Ip<>1 THEN GOTO 560
470 PRINT#6:PRINT#6 "    The consolidated sliding polynomial coefficients."
480 FOR I = 1 TO Nd
490  PRINT#6 "   ";I
500  FOR J = 1 TO 8
510   PRINT#6 USING "--###.######";C(I,J);
520  NEXT J
530  PRINT#6
540 NEXT I
550 REM - Compute suns of squares matrix
560 CLEAR(Sc)
570 FOR K = 1 TO Nd
580  FOR I = 1 TO 8
590   FOR J = I TO 8
600    Sc(I,J)=Sc(I,J)+C(K,I)*C(K,J)
610   NEXT J
620  NEXT I
630 NEXT K
640 REM - square the matrix
650 FOR I = 1 TO 8
660  FOR J = I TO 8
670   Sc(J,I)=Sc(I,J)
680  NEXT J
690 NEXT I
700 PRINT#6:PRINT#6 "    The sums of squares matrix"
710 FOR I = 1 TO 8
720  FOR J = 1 TO 8
730   PRINT#6 USING "--####.#####";Sc(I,J);
740  NEXT J
750  PRINT#6
760 NEXT I
770 Sims 7
780 PRINT#6
790 REM - Set up the full vector of nodes:#1 isa trend,#2-#10 are wave
800 No(1)=Sc(1,8)
810 No(2)=Sc(7,8)
820 No(3)=Sc(2,8)
830 No(4)=Sc(3,8)
840 No(5)=Sc(4,8)
850 No(6)=Sc(5,8)
860 No(7)=Sc(6,8)
870 No(8)=Sc(7,8)
880 No(9)=Sc(2,8)
890 No(10)=Sc(3,8)
900 REM - Calculate wave and trend at 12th points of each cycle
910 REM - Compute wave for one cycle and save
920 FOR J = 1 TO 12
930  I=INT((J+1)/2)
940  IF J-2*INT(J/2)=0 THEN GOTO 970
950  Ycp(J)=No(I+2)
960  GOTO 980
970  Ycp(J)=-0.0625*(No(I+1)+No(I+4))+0.5625*(No(I+2)+No(I+3))
980 NEXT J
990 REM - add this wave to trend line
```

```
1000 I=0
1010 FOR Ic = 1 TO Nc
1020   FOR Id = 1 TO 12
1030     I=I+1
1040     Yt(I)=Ycp(Id)+(Id-1+12*(Ic-1))*No(1)/(12*Nc)
1050   NEXT Id
1060 NEXT Ic
1070 PRINT #6"     The basic cycle"
1080 FOR I = 1 TO 6
1090   PRINT#6 USING "####";I;
1100   PRINT#6 USING "--####.####";Ycp(I) ;
1110 NEXT I
1120 PRINT#6
1130 FOR I = 7 TO 12
1140   PRINT#6 USING "####";I;
1150   PRINT#6 USING "--####.####";Ycp(I);
1160 NEXT I
1170 PRINT#6
1180 REM - Print the full data range.
1190 PRINT#6 "     The full curve of cycles plus trend"
1200 FOR Ic = 1 TO Nc
1210   PRINT#6 "cycle number ";Ic
1220   FOR Id = 1 TO 12
1230     K = INT(Id+(Ic-1)*12)
1240     PRINT#6 USING "####";K;
1250     PRINT#6 USING "--####.####";Yt(K);
1260     IF Id = 6 THEN PRINT#6
1270   NEXT Id
1280   PRINT#6
1290 NEXT Ic
1300 REM - Compute residual errors
1310 CLEAR(Er)
1320 FOR Id = 1 TO Nd
1330   FOR N = 1 TO 7
1340     Er(Id)=Er(Id)+C(Id,N)*Sc(N,8)
1350   NEXT N
1360   Er(Id)=Y(Id)-Er(Id)
1370 NEXT Id
1380 PRINT#6:PRINT#6 "   The residual errors"
1390 K=0
1400 FOR Id = 1 TO Nd
1410   K=K+1
1420   PRINT#6 USING "####";Id;
1430   PRINT#6 USING "--###.#####";Er(Id);
1440   IF K = 8 THEN GOTO 1460
1450   GOTO 1470
1460   K=0:PRINT#6
1470 NEXT Id
1480 PRINT#6
1490 REM - Unitize sum of c's for each data point.
1500 FOR Id = 1 TO Nd
1510   Ssc=0.0
1520   FOR N = 1 TO 7
1530     Ssc=Ssc+C(Id,N)
1540   NEXT N
1550   FOR N = 1 TO 7
```

```
1560     C(Id,N)=C(Id,N)/Ssc
1570   NEXT N
1580 NEXT Id
1590 REM - Sum terms for nodal variance
1600 FOR N = 1 TO 7
1610   FOR Id = 1 TO Nd
1620     Sx=C(Id,N)*C(Id,N)
1630     Ce2(N)=Ce2(N)+Sx*Er(Id)*Er(Id)
1640     C2(N)=C2(N)+Sx
1650     C1(N)=C1(N)+C(Id,N)
1660   NEXT Id
1670   Nv(N)=Ce2(N)/C2(N)/C1(N)*(Nd-7)/Nd
1680   Sd(N)=SQR(Nv(N))
1690 NEXT N
1700 PRINT#6:PRINT#6"     Nodal variance terms"
1710 PRINT#6"      wtd err sqd   wt sqd     weight      variance   std deviation"
1720 FOR N = 1 TO 7
1730   PRINT#6 USING "####";N;
1740   PRINT#6 USING "--####.#####";Ce2(N);C2(N);C1(N);Nv(N);Sd(N)
1750 NEXT N
1760 PRINT#6
1770 CLOSE#6
1780 END

ENDFILE
```

Program B-5.5

Four-Point Sliding Polynomial Moving Average

```
SOURCE
PRECISION 12
OPTION BASE 1
PROCS=0
REAL ARRAY(200): Y,Ma
REAL ARRAY(5): Wt
STRING: T[200]
INTEGER: I,J,P
REAL: Sw

'MAIN Program:


  10 REM - Moving average using four=point sliding polynomial weights.
  20 REM - The averaging span is six data intervals; three either side.
  30 REM - Five data points enter into the calculation of the average.
  40 REM - Averages are computed at each data point except first and last two.
  50 OPEN "prn" FOR OUTPUT AS #6
  60 PRINT#6 "            Moving average calculation"
  70 INPUT "problem title";T
  80 PRINT#6 T
```

```
 90 REM - Set weights and compute sum of weights
100 Wt(1)=-0.0741
110 Wt(2)=0.3333
120 Wt(3)=1.0
130 Wt(4)=.3333
140 Wt(5)=-0.0741
150 Sw=0
160 FOR I = 1 TO 5
170   Sw=Sw+Wt(I)
180 NEXT I
190 REM - Input the data to be averaged
200 OPEN "b:lawr" FOR INPUT AS #5
210 FOR I = 1 TO 60
220   INPUT #5 Y(I)
230 NEXT I
240 PRINT#6 "Record to be averaged"
250 P=0
260 FOR J = 1 TO 60
270   P=P+1
280   PRINT#6 USING "####";J;
290   PRINT#6 USING "--###.#";Y(J);
300   IF P = 12 THEN P = 0:PRINT#6
310 NEXT J
320 CLOSE#5
330 REM - Compute 60-4=56 moving averages running from 3 to 58
340 FOR I = 3 TO 58
350   Ma(I)=0.0
360   FOR J = 1 TO 5
370     Ma(I)=Ma(I)+Wt(J)*Y(J+I-3)
380   NEXT J
390   Ma(I)=Ma(I)/Sw
400 NEXT I
410 P=0
420 PRINT#6:PRINT#6 " The calculated moving averages"
430 FOR I = 3 TO 58
440   P=P+1
450   PRINT#6 USING "####";I;
460   PRINT#6 USING "--####.####";Ma(I);
470   IF P = 8 THEN P=0:PRINT#6: GOTO 480
480 NEXT I
490 PRINT#6
500 CLOSE#6
510 END
```

ENDFILE

Program B-5.6

Six-Point Sliding Polynomial Smoothing With Optional Boundary Conditions

SOURCE
PRECISION 12

```
OPTION BASE 1
PROCS=3
STRING: Title[60]
REAL ARRAY(300): X,Y,V,Xint,Ypint
REAL ARRAY(300,16): W
REAL ARRAY(12): Sxy,Nodvar,Sumw,Sumwe
REAL ARRAY(12): Sum
REAL ARRAY(13,13): Ssx
REAL ARRAY(6): Co
INTEGER: Nd,Nn,Iprin,Ilb,Irb,Itran,Nnn,Nn4
INTEGER: Nn3,Nn2,Nn1,Ifun,I,J,K
REAL ARRAY(17,17): Sc
REAL: Dzplot,Xl
REAL: Xr,B,Xc,Sl,Sr,Fnn
REAL ARRAY(12): Xnode,Ynode
REAL: Dvdxl,Dvdxr
REAL: Vxr,Dzdv,Vxc,Vnode
INTEGER: Ip
REAL: Zf
INTEGER: Ii,Ir,Ic
REAL: Ssy
REAL ARRAY(6): D1co
REAL ARRAY(6): Dco
REAL: Zint
INTEGER: Mms,Mmi
REAL: Fz,Rmi
REAL ARRAY(300): Yintr
REAL: Yp,Ser2

PROCEDURE: Coe6
  REAL ARG: Z
END PROCEDURE

PROCEDURE: Simeqns
  INTEGER ARG: N
END PROCEDURE

PROCEDURE: Dcoe6
  REAL ARG: Z
END PROCEDURE

PROCEDURE: Coe6
  EXTERNAL: Co
    10 Co(1)=(((((-160*Z+16)*Z+144)*Z-40)*Z-26)*Z+9)/768
    20 Co(2)=(((((800*Z-48)*Z-848)*Z+312)*Z+162)*Z-75)/768
    30 Co(3)=(((((-1600*Z+32)*Z+1824)*Z-272)*Z-1124)*Z+450)/768
    40 Co(4)=(((((1600*Z+32)*Z-1824)*Z-272)*Z+1124)*Z+450)/768
    50 Co(5)=(((((-800*Z-48)*Z+848)*Z+312)*Z-162)*Z-75)/768
    60 Co(6)=(((((160*Z+16)*Z-144)*Z-40)*Z+26)*Z+9)/768
END PROCEDURE

PROCEDURE: Simeqns
  EXTERNAL: Sc
  REAL: X,Y
  INTEGER: J,I,K,M
    10 REM - solve set of n equations
```

```
   20 FOR J = 1 TO N
   30  REM - statements 480 to 520 find the first equation with a
   40  REM - non-zero coefficient for the current column.
   50  M=0
   60  FOR I = J TO N
   70   IF Sc(I,J)<>0 THEN GOTO 90
   80   GOTO 120
   90   M=1
  100   K=I
  110   EXIT
  120  NEXT
  130  IF M= 1 THEN GOTO 180
  160  PRINT "no unique solution": STOP
  170  REM  statements 540 to 580 move that equation up to current row.
  180  I=K
  190  FOR K= 1 TO N+1
  200   X=Sc(J,K)
  210   Sc(J,K)=Sc(I,K)
  220   Sc(I,K)=X
  230  NEXT
  240  REM - statements 610 to 640 get a 1 coefficient in the first non-zero
  250  REM - column of the current row
  260  Y=1/Sc(J,J)
  270  FOR K= 1 TO N+1
  280   Sc(J,K)=Y*Sc(J,K)
  290  NEXT
  300  REM - statements 670 to 730 subtract the current equation from
  310  REM - th other rows,
  320  FOR I = 1 TO N
  330   IF I = J THEN GOTO 380
  340   Y=-Sc(I,J)
  350   FOR K = 1 TO N+1
  360    Sc(I,K)=Sc(I,K)+Y*Sc(J,K)
  370   NEXT
  380  NEXT
  390  REM - this process is repeated for all equations.
  400 NEXT
  410 PRINT #3 " solution of equations"
  420 M = 1
  430 FOR I = 1 TO N
  440  M= M+1
  450  PRINT #3 USING "#####"; I;
  460  PRINT#3 USING "--#####.####";Sc(I,N+1);
  470  IF M=>6 GOTO 500
  480  PRINT #3 "   ";
  490  GOTO 510
  500  PRINT #3: M= 1
  510 NEXT
  520 PRINT #3
END PROCEDURE

PROCEDURE: Dcoe6
  EXTERNAL: Dco
   10 Dco(1)=((((-400*Z+32)*Z+216)*Z-40)*Z-13)/384
   20 Dco(2)=((((2000*Z-96)*Z-1272)*Z+312)*Z+81)/384
   30 Dco(3)=((((-4000*Z+64)*Z+2736)*Z-272)*Z-562)/384
```

```
    40 Dco(4)=((((4000*Z+64)*Z-2736)*Z-272)*Z+562)/384
    50 Dco(5)=((((-2000*Z-96)*Z+1272)*Z+312)*Z-81)/384
    60 Dco(6)=((((400*Z+32)*Z-216)*Z-40)*Z+13)/384
END PROCEDURE
```

'MAIN Program:

```
 10 REM - Smooth a data set of one independent variable and the dependent
 20 REM - variable. Up to 300 pairs. Left and right boundary conditions can
 30 REM - be set and internal options control nodal spacings. See remarks
 40 REM - at inputs.
 50 OPEN "prn" FOR OUTPUT AS #3
 60 INPUT "problem title"; Title
 70 PRINT #3 "      ";Title
 80 INPUT " number of data pairs  ";Nd
 90 INPUT " number of nodes to be evaluated  ";Nn
100 PRINT " input a '2' to avoid printing weighting coefficients"
110 PRINT " any other number will cause print":INPUT "iprin";Iprin
120 PRINT " input the points per span for plotting smoothing function"
130 PRINT " normally use 10":INPUT "dzplot";Dzplot
140 PRINT " input left and right boundary options:ilb and irb"
150 PRINT " option 1 -  set zero boundary slope by nodal symmetry"
160 PRINT " option 2 -  set any slope to be input as sl and/or sr"
170 PRINT " option 3 -  allow free slope as result of smoothing"
180 PRINT " left and right boundary conditions are set independently"
190 PRINT " boundary options do not apply for cylic transform, see later"
200 INPUT " option for left boundary,ilb  ";Ilb
210 INPUT " option for right boundary,ilb  ";Irb
220 PRINT " input the index for spacing of nodes -'itran'"
230 PRINT " option 1 -  uniform spacing of nodes"
240 PRINT " option 2 - right clustering of nodes"
250 PRINT " option 3 - left clustering of nodes"
260 PRINT " option 4 - central clustering of nodes"
270 PRINT " option 5 - cylic pattern of nodes with unifrom spacing"
280 INPUT " itran  "; Itran
290 INPUT " position of left boundary node in x-scale "; Xl
300 INPUT " position of right boundary node in x-scale ";Xr
310 PRINT " input the parameter 'b' for 'clustering' transform of nodal"
320 PRINT " spacing. this is an exponential parameter. use '3' for first"
330 PRINT " try for an x-scale in the range 0<x<10. use smaller values for"
340 PRINT " larger x's. may need several  tries."
350 PRINT " use a negative 'b' for transform option 3"
360 INPUT " b  ";B
370 PRINT " for transform option 4 only input the common point of the"
380 PRINT " two joined exponentials forming an ogee curve. this value is"
390 PRINT " an 'x' value in the range xl<xc<xr. use mid-value or slightly"
400 PRINT " left or right for first try."
410 INPUT " xc  ";Xc
420 PRINT " if you used ilb option 2 input the left slope dy/dx;otherwise any"
430 INPUT " sl  "; Sl
440 PRINT " if you used irb option 2 input the right slope dydx;otherwise any"
450 INPUT " sr ";Sr
460 PRINT #3 "                 Options used"
470 PRINT #3 "ilb= ";Ilb;"  irb= ";Irb;"  itran= ";Itran;"  xl= ";Xl;"  xr= ";Xr
480 PRINT #3 "b= ";B;"  xc= ";Xc;"  sl= ";Sl;"  sr= ";Sr
490 PRINT " input the x-y data pairs"
```

```
500 FOR I = 1 TO Nd
510   PRINT I:INPUT "x";X(I):INPUT "y";Y(I)
520 NEXT I
530 Fnn=Nn
540 Nnn=Nn-1
550 Nn4=Nn+4
560 Nn3=Nn+3
570 Nn2=Nn+2
580 Nn1=Nn+1
590 Ifun=1
600 Sr=-Sr
610 Xnode(1)=Xl
620 Xnode(Nn)=Xr
630 IF Itran = 1 THEN GOTO 680
640 IF Itran = 2 THEN GOTO 760
650 IF Itran = 3 THEN GOTO 1040
660 IF Itran = 5 THEN GOTO 960
670 GOTO 840
680 Dvdxl=(Fnn-1.)/(Xr-Xl)
690 Dvdxr=Dvdxl
700 Vxr=Fnn
710 Dzdv=1.
720 FOR I = 1 TO Nd
730   V(I)=Dvdxl*(X(I)-Xl)+1.
740 NEXT I
750 GOTO 1110
760 Vxr=(Xr-Xl+1.)^B
770 Dzdv=(Fnn-1.)/(Vxr-1.)
780 Dvdxl=B
790 Dvdxr=B*(Xr+1.-Xl)^(B-1.)
800 FOR I = 1 TO Nd
810   V(I)=(X(I)+1.-Xl)^B
820 NEXT I
830 GOTO 1110
840 Vxc=EXP(B*(Xc-Xl))
850 Vxr=Vxc*(2.-EXP(-B*(Xr-Xc)))
860 Dzdv=(Fnn-1.)/(Vxr-1.)
870 Dvdxl=B
880 Dvdxr=Vxc*B*EXP(-B*(Xr-Xc))
890 FOR I = 1 TO Nd
900   IF X(I)<=Xc THEN GOTO 930
910   V(I)=Vxc*(2.-EXP(-B*(X(I)-Xc)))
920   GOTO 940
930   V(I)=EXP(B*(X(I)-Xl))
940 NEXT I
950 GOTO 1110
960 Dvdxl=Fnn/(Xr-Xl)
970 Dvdxr=Dvdxl
980 Dzdv=1.
990 Vxr=1.+Fnn
1000 FOR I = 1 TO Nd
1010   V(I)=Dvdxl*(X(I)-Xl)+1.
1020 NEXT I
1030 GOTO 1110
1040 Vxr=2.-EXP(B*(Xr-Xl))
1050 Dzdv=(Fnn-1.)/(Vxr-1.)
```

```
1060 Dvdxl=-B
1070 Dvdxr=-B*EXP(B*(Xr-Xl))
1080 FOR I = 1 TO Nd
1090   V(I)=2.-EXP(B*(X(I)-Xl))
1100 NEXT I
1110 CLEAR(W)
1120 Vnode=1.
1130 FOR I= 2 TO Nnn
1140   Vnode=Vnode+1./Dzdv
1150   IF Itran = 1 THEN Xnode(I)=(Xr-Xl)/(Vxr-1.)*(Vnode-1.)+Xl
1160   IF Itran = 2 THEN Xnode(I)=EXP(LOG(Vnode)/B)+Xl-1.
1170   IF Itran = 3 THEN Xnode(I)= LOG(2.-Vnode)/B+Xl
1180   IF Itran = 4 AND Vnode < Vxc THEN Xnode(I)=LOG(Vnode)/B+Xl
1190   IF Itran=4 AND Vnode>=Vxc THEN Xnode(I)=(-LOG(2.*Vxc-Vnode)/B)+2.*Xc-Xl
1200 NEXT I
1210 IF Itran = 5 THEN GOTO 1220 ELSE GOTO 1280
1220 Xnode(Nn1)=Xr
1230 Vnode=1.
1240 FOR I = 2 TO Nn
1250   Vnode=Vnode+1./Dzdv
1260   Xnode(I)=(Vnode-1.)/Dvdxl+Xl
1270 NEXT I
1280 PRINT #3 "         Transformed data location"
1290 K=1
1300 FOR I = 1 TO Nd
1310   PRINT #3 USING "#####";I;
1320   PRINT #3 USING "--####.###";V(I);
1330   IF K=10 THEN PRINT #3:K=1: GOTO 1350
1340   K=K+1
1350 NEXT I
1360 PRINT #3:PRINT #3   " x-location of nodes"
1370 K=1
1380 FOR I = 1 TO Nn
1390   PRINT #3 USING "#####";I;
1400   PRINT #3 USING "--####.###";Xnode(I);
1410   IF K = 10 THEN PRINT #3:K=1: GOTO 1430
1420   K=K+1
1430 NEXT I
1440 PRINT #3
1450 Ip=0
1460 FOR I = 1 TO Nd
1470   K=1
1480   IF V(I)<=1.+K/Dzdv THEN GOTO 1510
1490   K=K+1
1500   GOTO 1480
1510   Zf=(V(I)-(1.+((K-1)/Dzdv)))*Dzdv-.5
1520   Coe6 Zf
1530   FOR J = 1 TO 6
1540     W(I,J+K-1)=Co(J)
1550   NEXT J
1560 NEXT I
1570 IF Iprin = 2 THEN GOTO 1700
1580 PRINT #3 " Coefficients for each data point"
1590 FOR J = 1 TO Nd
1600   PRINT #3 USING "####";J;
1610   Ii=1
```

```
1620  FOR I = 1 TO 5+K
1630    PRINT #3 USING "####";I;
1640    PRINT #3 USING "--##.######";W(J,I);
1650    IF Ii = 8 THEN Ii=1:PRINT #3:PRINT #3 "          ";:GOTO 1670
1660    Ii=Ii+1
1670  NEXT I
1680  PRINT#3
1690 NEXT J
1700 FOR I = 1 TO Nd
1710  IF V(I)<1.+(2./Dzdv) THEN GOTO 1730
1720  GOTO 1900
1730  IF Itran = 5 GOTO 1880
1740  IF Ilb = 1 THEN GOTO 1850
1750  IF Ilb = 2 THEN GOTO 1800
1760  W(I,3)=W(I,3)+2.*(W(I,1)+W(I,2))
1770  W(I,5)=W(I,5)-W(I,1)
1780  W(I,4)=W(I,4)-W(I,2)
1790  GOTO 1900
1800  W(I,3)=W(I,3)+2.*W(I,1)+.25*W(I,2)
1810  W(I,4)=W(I,4)+W(I,2)-W(I,1)
1820  W(I,5)=W(I,5)-0.25*W(I,2)
1830  Y(I)=Y(I)+1.5*Sl/Dvdxl/Dzdv*W(I,2)
1840  GOTO 1900
1850  W(I,4)=W(I,4)+W(I,2)
1860  W(I,5)=W(I,5)+W(I,1)
1870  GOTO 1900
1880  W(I,Nn+1)=W(I,Nn+1)+W(I,1)
1890  W(I,Nn+2)=W(I,Nn+2)+W(I,2)
1900  IF V(I)>1.+(Nn-3)/Dzdv THEN GOTO 1920
1910  GOTO 2100
1920  IF Itran = 5 THEN GOTO 2070
1930  IF Irb = 1 THEN GOTO 2040
1940  IF Irb = 2 THEN GOTO 1990
1950  W(I,Nn+2)=W(I,Nn+2)+2.*(W(I,Nn+4)+W(I,Nn+3))
1960  W(I,Nn+1)=W(I,Nn+1)-W(I,Nn+3)
1970  W(I,Nn)=W(I,Nn)-W(I,Nn+4)
1980  GOTO 2100
1990  W(I,Nn+2)=W(I,Nn+2)+2.*W(I,Nn+4)+0.25*W(I,Nn+3)
2000  W(I,Nn+1)=W(I,Nn+1)+W(I,Nn+3)-W(I,Nn+4)
2010  W(I,Nn)=W(I,Nn)-.25*W(I,Nn+3)
2020  Y(I)=Y(I)+1.5*Sr/Dvdxr/Dzdv*W(I,Nn+3)
2030  GOTO 2100
2040  W(I,Nn+1)=W(I,Nn+1)+W(I,Nn+3)
2050  W(I,Nn)=W(I,Nn)+W(I,Nn+4)
2060  GOTO 2100
2070  W(I,3)=W(I,3)+W(I,Nn+3)
2080  W(I,4)=W(I,4)+W(I,Nn+4)
2090  W(I,5)=W(I,5)+W(I,Nn+5)
2100 NEXT I
2110 IF Iprin = 2 THEN GOTO 2250
2120 PRINT #3 "    Reduced set of coefficients"
2130 FOR I = 1 TO Nd
2140  K=1
2150  PRINT #3 USING "####";I;
2160  FOR J = 3 TO Nn2
2170    PRINT #3 USING "###";J;
```

```
2180     PRINT #3 USING "--###.######";W(I,J);
2190     IF K= 8 THEN K=1:PRINT #3:PRINT #3 "    ";:GOTO 2210
2200     K=K+1
2210   NEXT J
2220   PRINT #3
2230 NEXT I
2240 REM - Generate matrix for least squares solution of nodes
2250 CLEAR (Ssx,Sxy)
2260 FOR Ir = 1 TO Nn
2270   FOR Ic = Ir TO Nn
2280     FOR I = 1 TO Nd
2290       Ssx(Ir,Ic)=Ssx(Ir,Ic)+W(I,Ir+2)*W(I,Ic+2)
2300     NEXT I
2310   NEXT Ic
2320   FOR I = 1 TO Nd
2330     Sxy(Ir)=Sxy(Ir)+W(I,Ir+2)*Y(I)
2340   NEXT I
2350 NEXT Ir
2360 Ssy = 0
2370 FOR I = 1 TO Nd
2380   Ssy=Ssy+Y(I)*Y(I)
2390 NEXT I
2400 FOR Ir = 1 TO Nn
2410   FOR Ic= Ir TO Nn
2420     Ssx(Ic,Ir)=Ssx(Ir,Ic)
2430   NEXT Ic
2440 NEXT Ir
2450 FOR Ir = 1 TO Nn
2460   FOR Ic = 1 TO Nn
2470     Sc(Ir,Ic)=Ssx(Ir,Ic)
2480   NEXT Ic
2490   Sc(Ir,Nn+1)=Sxy(Ir)
2500 NEXT Ir
2510 Simeqns Nn
2520 FOR I = 1 TO Nn
2530   Ynode(I+2)= Sc(I,Nn+1)
2540 NEXT I
2550 REM - Fill in outlier nodes for interpolating
2560 IF Itran = 5 THEN GOTO 2780
2570 IF Ilb = 1 THEN GOTO 2650
2580 IF Ilb = 2 THEN GOTO 2620
2590 Ynode(1)=2.*Ynode(3)-Ynode(5)
2600 Ynode(2)=2.*Ynode(3)-Ynode(4)
2610 GOTO 2670
2620 Ynode(1)=2.*Ynode(3)-Ynode(5)
2630 Ynode(2)=0.25*Ynode(3)+Ynode(4)-0.25*Ynode(5)-1.5*Sl/Dvdxl/Dzdv
2640 GOTO 2670
2650 Ynode(1)=Ynode(5)
2660 Ynode(2)=Ynode(4)
2670 IF Irb=1 THEN GOTO 2750
2680 IF Irb=2 THEN GOTO 2720
2690 Ynode(Nn+4)=2.*Ynode(Nn+2)-Ynode(Nn)
2700 Ynode(Nn+3)=2.*Ynode(Nn+2)-Ynode(Nn+1)
2710 GOTO 2840
2720 Ynode(Nn+4)=2.*Ynode(Nn+2)-Ynode(Nn)
2730 Ynode(Nn+3)=0.25*Ynode(Nn+2)+Ynode(Nn+1)-.25*Ynode(Nn)-1.5*Sr/Dvdxr/Dzdv
```

```
2740 GOTO 2840
2750 Ynode(Nn+4)=Ynode(Nn)
2760 Ynode(Nn+3)=Ynode(Nn+1)
2770 GOTO 2840
2780 Ynode(1)=Ynode(Nn+1)
2790 Ynode(2)=Ynode(Nn+2)
2800 Ynode(Nn+3)=Ynode(3)
2810 Ynode(Nn+4)=Ynode(4)
2820 Ynode(Nn+5)=Ynode(5)
2830 REM -  Interpolate function and derivative on full set of nodes
2840 Zint=1.
2850 CLEAR (Yintr,Ypint)
2860 I=1
2870 Mms=-1
2880 IF Itran = 1 OR Itran = 5 THEN Xint(I)=(Xr-Xl)/(Vxr-1.)*(Zint-1.)+Xl
2890 IF Itran = 2 THEN Xint(I)=EXP(LOG(Zint)/B)+Xl-1.
2900 IF Itran = 3 THEN Xint(I)=LOG(2.-Zint)/B+Xl
2910 IF Itran = 4 AND Zint<Vxc THEN Xint(I)=LOG(Zint)/B+Xl
2920 IF Itran = 4 AND Zint>=Vxc THEN  Xint(I)=(-LOG(2.*Vxc-Zint)/B)+2.*Xc-Xl
2930 Fz=(Zint-1.)*Dzdv
2940 Mmi=INT(Fz)
2950 Rmi=Mmi
2960 Zf=Fz-Rmi-.5
2970 Coe6 Zf
2980 Dcoe6 Zf
2990 FOR J = 1 TO 6
3000   Yintr(I)=Yintr(I)+Co(J)*Ynode(J+Mmi)
3010   Ypint(I)=Ypint(I)+Dco(J)*Ynode(J+Mmi)
3020 NEXT J
3030 IF Itran =1 OR Itran = 5 THEN Ypint(I)=Ypint(I)*Dvdxl
3040 IF Itran = 2 THEN Ypint(I)=Ypint(I)*B*Dzdv*(Xint(I)+1.-Xl)^(B-1)
3050 IF Itran = 3 THEN Ypint(I)=Ypint(I)*Dzdv*(-B)*EXP(B*(Xint(I)-Xl))
3060 IF Itran =4 AND Zint<Vxc THEN Ypint(I)=Ypint(I)*B*Dzdv*EXP(B*(Xint(I)-Xl))
3070 IF Itran=4 AND Zint>=Vxc THEN Ypint(I)=Ypint(I)*B*Vxc*Dzdv*EXP(-B*(Xint(I)-Xc))
3080 IF Zint>=Vxr THEN GOTO 3120
3090 Zint=Zint+1./Dzdv/Dzplot
3100 I=I+1
3110 GOTO 2870
3120 PRINT #3 "        x-y coordinates of nodes"
3130 K=1
3140 FOR J = 1 TO Nn
3150   PRINT #3 USING "####";J;
3160   PRINT #3 USING "--####.####";Xnode(J);Ynode(J+2);
3170   IF K=6 THEN K=1:PRINT #3: GOTO 3190
3180   K=K+1
3190 NEXT J
3200 PRINT #3:PRINT #3:PRINT #3 " x,y and dydx of interpolation function"
3210 K=1
3220 FOR J = 1 TO I
3230   PRINT #3 USING "####";J;
3240   PRINT #3 USING "--###.###";Xint(J);Yintr(J);
3250   PRINT #3 USING "--##.####";Ypint(J);
3260   IF K= 4 THEN K=1:PRINT #3: GOTO 3280
3270   K=K+1
3280 NEXT J
3290 PRINT #3
```

```
3300 REM - Comput residual errors and store in v(i)
3310 IF Ifun<>1 THEN GOTO 3700
3320 Ser2=0.
3330 FOR I = 1 TO Nd
3340   Yp =0.0
3350   FOR J= 3 TO Nn2
3360     Yp=Yp+Ynode(J)*W(I,J)
3370   NEXT J
3380   V(I)=Y(I)-Yp
3390   Ser2=Ser2+V(I)*V(I)
3400 NEXT I
3410 PRINT #3:PRINT #3 "              Residual errors"
3420 K=1
3430 FOR I = 1 TO Nd
3440   PRINT #3 USING "####";I;
3450   PRINT #3 USING "--###.####";V(I);
3460   IF K=9 THEN K=1:PRINT #3: GOTO 3480
3470   K=K+1
3480 NEXT I
3490 PRINT #3
3500 CLEAR(Sum,Sumwe,Sumw)
3510 FOR K = 1 TO Nn
3520   FOR I = 1 TO Nd
3530     Sum(K)=Sum(K)+W(I,K+2)
3540     Sumwe(K)=Sumwe(K)+V(I)*V(I)*W(I,K+2)*W(I,K+2)
3550     Sumw(K)=Sumw(K)+W(I,K+2)*W(I,K+2)
3560   NEXT I
3570   Nodvar(K)=Sumwe(K)/Sumw(K)*Nd/(Nd-Nn)/Sum(K)
3580 NEXT K
3590 PRINT #3
3600 PRINT #3" Node   wtd error    wt                    node"
3610 PRINT #3" number  squared   squared   wt   nodvar  std dev"
3620 FOR I = 1 TO Nn
3630   PRINT #3 USING "####";I;
3640   PRINT #3 USING "--#####.###";Sumwe(I);Sumw(I);Sum(I);Nodvar(I);SQR(Nodvar(I))
3650 NEXT I
3660 PRINT #3
3670 PRINT #3:PRINT #3 "   Correlation coefficient  ";
3680 PRINT #3 USING "###.####";SQR((Ssy-Ser2)/Ssy)
3690 PRINT #3
3700 IF Ifun = 3 THEN GOTO 3840
3710 Ifun = Ifun +1
3720 IF Ifun = 2  THEN GOTO 3740
3730 GOTO 3790
3740 FOR I = 1 TO Nn
3750   Ynode(I+2)=Ynode(I+2)+SQR(Nodvar(I))
3760 NEXT I
3770 PRINT #3 "      Basic function plus one standard deviation"
3780 GOTO 2560
3790 FOR I = 1 TO Nn
3800   Ynode(I+2)=Ynode(I+2)-2.*SQR(Nodvar(I))
3810 NEXT I
3820 PRINT #3 "      Basic function minus one standard deviation"
3830 GOTO 2560
```

```
3840 PRINT #3
3850 CLOSE #3
3860 STOP

ENDFILE
```

Program B-5.7

Two-Dimensional Two-Point Sliding Polynomial Smoothing

```
SOURCE
PRECISION 12
OPTION BASE 1
PROCS=1
REAL ARRAY(5): Xl
REAL ARRAY(6,6): C,Nod
REAL ARRAY(25,100): Wt
REAL ARRAY(100): Y,Erx
REAL ARRAY(25): Col0,Towe2,Tow2,Tow,Nodvar
STRING: T[200]
INTEGER: Nx,Nu
INTEGER: I,Nd,Ic,J,Wc,Jb,P,K
REAL ARRAY(26,26): Sc
REAL ARRAY(5): Ul
REAL: B,X,U,Xin,Uin
INTEGER: Sum
REAL: Colc,Predy,Txw
INTEGER: Skip,L
INTEGER: M
REAL ARRAY(6,6): Yc
INTEGER: Ji
REAL ARRAY(6): Uc,Xc
REAL: Zu
INTEGER: Ii
REAL: Zx,C1,C2,C3,C4

PROCEDURE: Sims
  INTEGER ARG: N
END PROCEDURE

PROCEDURE: Sims
  EXTERNAL: Sc
  REAL: X,Y
  INTEGER: J,I,K,M,Ic
    10 REM - solve set of n equations
    20 FOR J = 1 TO N
    30   REM - statements 480 to 520 find the first equation with a
    40   REM - non-zero coefficient for the current column.
    50   Ic=0:M=1
    60   FOR I = J TO N
    70     IF Sc(I,J)=0 THEN GOTO 80 ELSE GOTO 100
    80     Ic=Ic+1
    90     GOTO 130
   100     ON M GOTO 110,130
```

```
110     K=I
120     M=2
130   NEXT
140   IF Ic=N THEN GOTO 160
150   GOTO 180
160   PRINT "no unique solution": STOP
170   REM  statements 540 to 580 move that equation up to current row.
180   I=K
190   FOR K= 1 TO N+1
200     X=Sc(J,K)
210     Sc(J,K)=Sc(I,K)
220     Sc(I,K)=X
230   NEXT
240   REM - statements 610 to 640 get a 1 coefficient in the first non-zero
250   REM - column of the current row
260   Y=1/Sc(J,J)
270   FOR K= 1 TO N+1
280     Sc(J,K)=Y*Sc(J,K)
290   NEXT
300   REM - statements 670 to 730 subtract the current equation from
310   REM - th other rows,
320   FOR I = 1 TO N
330     IF I = J THEN GOTO 380
340     Y=-Sc(I,J)
350     FOR K = 1 TO N+1
360       Sc(I,K)=Sc(I,K)+Y*Sc(J,K)
370     NEXT
380   NEXT
390   REM - this process is repeated for all equations.
400 NEXT
410 PRINT #3 " solution of equations"
420 M = 1
430 FOR I = 1 TO N
440   M= M+1
450   PRINT #3 USING "#####"; I;
460   PRINT#3 USING "--#####.####";Sc(I,N+1);
470   IF M=>6 GOTO 500
480   PRINT #3 "   ";
490   GOTO 510
500   PRINT #3: M= 1
510 NEXT
520 PRINT #3
END PROCEDURE

'MAIN Program:

  10 REM - Fit a two-dimensional segmented linear surface. Each data point
  20 REM - weights to four nodes except for the trivial case where the data
  30 REM - point is at a node. Program will delete unweighted nodes, thereby
  40 REM - allowing for irregular surfaces.
  50 OPEN "prn" FOR OUTPUT AS #3
  60 PRINT#3 "       Two dimensional smoothing with linear segments"
  70 INPUT "problem title";T
  80 PRINT#3;T
  90 INPUT "number of nodes to use in x-scale";Nx
 100 PRINT "input x-locations of nx nodes;one at left and one at right boundary"
```

```
110 FOR I = 1 TO Nx
120   PRINT I;:INPUT Xl(I)
130 NEXT I
140 INPUT " number of nodes to use in u-scale";Nu
150 PRINT"input u=locations of nu nodes,one at bottom and one at top boundary"
160 FOR I = 1 TO Nu
170   PRINT I;:INPUT Ul(I)
180 NEXT I
190 PRINT#3 "Location of nodes in x-scale"
200 FOR I = 1 TO Nx
210   PRINT#3 I;Xl(I);
220 NEXT I
230 PRINT#3
240 PRINT#3 "Location of nodes in u-scale"
250 FOR I = 1 TO Nu
260   PRINT#3 I;Ul(I);
270 NEXT I
280 PRINT#3
290 INPUT "number of data points";Nd
300 INPUT "input a '1' to print reduced matrix, otherwise any";B
310 REM - Input data from disk, one point at a time;format x,u,y
320 OPEN "b:testdat" FOR INPUT AS #4
330 FOR Ic = 1 TO Nd
340   FOR I = 1 TO Nu
350     FOR J =1 TO Nx
360       C(I,J)=0.0
370     NEXT J
380   NEXT I
390   INPUT#4 X
400   INPUT#4 U
410   INPUT#4 Y(Ic)
420   J=1
430   IF X<=Xl(J+1) THEN Xin=(X-Xl(J))/(Xl(J+1)-Xl(J))-0.5 ELSE J=J+1:GOTO 430
440   I=1
450   IF U<=Ul(I+1) THEN Uin=(U-Ul(I))/(Ul(I+1)-Ul(I))-0.5 ELSE I=I+1:GOTO 450
460   C(I,J)=0.25-0.5*Uin-0.5*Xin+Xin*Uin
470   C(I,J+1)=0.25-0.5*Uin+0.5*Xin-Xin*Uin
480   C(I+1,J)=0.25+0.5*Uin-0.5*Xin -Xin*Uin
490   C(I+1,J+1)=0.25+0.5*Uin+0.5*Xin+Xin*Uin
500   Wc=0
510   FOR I = 1 TO Nu
520     FOR J= 1 TO Nx
530       Wc=Wc+1
540       Wt(Wc,Ic)=C(I,J)
550     NEXT J
560   NEXT I
570 NEXT Ic
580 CLOSE#4
590 PRINT "data loaded and coefficients computed"
600 Colc=0
610 Jb=1
620 FOR J = Jb TO Wc
630   Sum = 0
640   FOR I = 1 TO Nd
650     IF Wt(J,I)=0 THEN Sum=Sum+1
660   NEXT I
```

```
670   IF Sum=Nd THEN  EXIT
680 NEXT J
690 IF  Sum< >Nd THEN GOTO 800
700 FOR K = J TO Wc
710   FOR I = 1 TO Nd
720     Wt(K,I)=Wt(K+1,I)
730   NEXT I
740 NEXT K
750 Colc=Colc+1
760 Col0(J+Colc-1)=1
770 Wc=Wc-1
780 Jb=J
790 GOTO 620
800 IF B< >1 THEN GOTO 920
810 FOR J = 1 TO Nd
820   PRINT#3,J
830   P=0
840   FOR I = 1 TO Wc
850     P=P+1
860     PRINT#3 USING "####";I;
870     PRINT#3 USING "--##.#####";Wt(I,J);
880     IF P = 8 THEN P=0:PRINT#3
890   NEXT I
900   PRINT#3
910 NEXT J
920 PRINT#3:PRINT#3 "Number of zero columns is  ";Colc
930 PRINT#3 "The zero columns are 'ones' "
940 FOR K= 1 TO Nx*Nu
950   PRINT#3 K;Col0(K);"    ";
960 NEXT K
970 PRINT#3
980 REM - Add "y" to matrix and comput sums of squares"
990 CLEAR(Sc)
1000 FOR I = 1 TO Nd
1010   Wt(Wc+1,I)=Y(I)
1020   FOR J = 1 TO Wc+1
1030     FOR K = J TO Wc+1
1040       Sc(J,K)=Sc(J,K)+Wt(J,I)*Wt(K,I)
1050     NEXT K
1060   NEXT J
1070 NEXT I
1080 REM - Fill in the complete matrix
1090 FOR J = 1 TO Wc+1
1100   FOR K = J TO Wc+1
1110     Sc(K,J)=Sc(J,K)
1120   NEXT K
1130 NEXT J
1140 PRINT "sums of squares matrix calculate; call 'simultaneous equations'"
1150 Sims Wc
1160 REM - Compute residual errors
1170 FOR J = 1 TO Nd
1180   Predy = 0.0
1190   FOR I = 1 TO Wc
1200     Predy = Predy + Wt(I,J)*Sc(I,Wc+1)
1210   NEXT I
1220   Erx(J)=Y(J)-Predy
```

```
1230 NEXT J
1240 REM - Compute nodal variance
1250 FOR J = 1 TO Wc
1260   FOR K = 1 TO Nd
1270     Txw=Wt(J,K)*Wt(J,K)
1280     Towe2(J)=Towe2(J)+Txw*Erx(K)*Erx(K)
1290     Tow2(J)=Tow2(J)+Txw
1300     Tow(J)=Tow(J)+Wt(J,K)
1310   NEXT K
1320 NEXT J
1330 FOR K = 1 TO Wc
1340   Nodvar(K)=Towe2(K)/Tow2(K)/Tow(K)*(Nd-Wc)/Nd
1350 NEXT K
1360 P=0
1370 PRINT#3 "        Residual errors"
1380 FOR I = 1 TO Nd
1390   P= P+1
1400   PRINT#3 USING "####";I;
1410   PRINT#3 USING"--##.####";Erx(I);
1420   IF P = 10 THEN P=0:PRINT#3
1430 NEXT I
1440 PRINT#3 :PRINT#3 "      Nodal variance"
1450 PRINT#3"      wt err sqd. wt sqd.    weight    variance    std. dev."
1460 FOR J = 1 TO Wc
1470   PRINT#3 USING "####";J;
1480   PRINT#3 USING "--####.#####";Towe2(J);Tow2(J);Tow(J);Nodvar(J);SQR(Nodvar(J))
1490 NEXT J
1500 PRINT#3
1510 REM - Set up the full nu by nx grid of nodes. "9999999" means deleted node
1520 PRINT#3
1530 K=0
1540 Colc=0
1550 FOR J = 1 TO Nu
1560   FOR I = 1 TO Nx
1570     K=K+1
1580     IF Col0(K)=1 THEN Nod(J,I)=999999:Colc=Colc+1 ELSE Nod(J,I)=Sc(K-Colc,Wc+1)
1590   NEXT I
1600 NEXT J
1610 PRINT#3 "The full grid of nodes"
1620 FOR J = 1 TO Nu
1630   PRINT#3 "row  ";J
1640   FOR I = 1 TO Nx
1650     PRINT#3 USING "####";I;
1660     PRINT#3 USING "--###.####";Nod(J,I);
1670   NEXT I
1680   PRINT#3
1690 NEXT J
1700 PRINT#3
1710 REM - Compute values of surface at two-way fifth points of each
1720 REM - unit surface, for contouring.
1730 PRINT#3 "   Contours for each unit surface"
1740 Skip =0
1750 FOR J = 1 TO Nu-1
1760   FOR I = 1 TO Nx-1
1770     PRINT#3 " unit grid x = ";Xl(I);" u =  "; Ul(J)
1780     FOR L = 1 TO 2
```

```
1790       FOR M = 1 TO 2
1800         IF Nod(J+L-1,I+M-1)=999999 THEN Skip = 1
1810       NEXT M
1820     NEXT L
1830     IF Skip = 1 THEN PRINT#3"this unit grid was deleted":Skip=0:GOTO 2060
1840     REM - Store contouring values in yc(j,i)
1850     FOR Ji = 1  TO 6
1860       Uc(Ji)=Ul(J)+(Ji-1)*(Ul(J+1)-Ul(J))/5
1870       Zu=-0.5+0.2*(Ji-1)
1880       FOR Ii = 1 TO 6
1890         Xc(Ii)=Xl(I)+(Ii-1)*(Xl(I+1)-Xl(I))/5
1900         Zx=-0.5+0.2*(Ii-1)
1910         C1=0.25-0.5*Zu-0.5*Zx+Zu*Zx
1920         C2=0.25-0.5*Zu+0.5*Zx-Zu*Zx
1930         C3=0.25+0.5*Zu-0.5*Zx-Zu*Zx
1940         C4=0.25+0.5*Zu+0.5*Zx+Zu*Zx
1950         Yc(Ji,Ii)=C1*Nod(J,I)+C2*Nod(J,I+1)+C3*Nod(J+1,I)+C4*Nod(J+1,I+1)
1960       NEXT Ii
1970     NEXT Ji
1980     FOR Ji= 1 TO 6
1990       PRINT#3 " u-coordinate  ";
2000       PRINT#3 USING "--###.####";Uc(Ji)
2010       FOR Ii = 1 TO 6
2020         PRINT#3 USING "--###.####";Xc(Ii);Yc(Ji,Ii);
2030       NEXT Ii
2040       PRINT#3
2050     NEXT Ji
2060   NEXT I
2070 NEXT J
2080 PRINT#3
2090 CLOSE #3
2100 END
2110 STOP

ENDFILE
```

Program B-5.8

Two-Dimensional Four-Point Sliding Polynomial Smoothing

```
SOURCE
PRECISION 12
OPTION BASE 1
PROCS=1
REAL ARRAY(8,8): C2,Nod
REAL ARRAY(4): Cu,Cz
REAL ARRAY(64): Wt
REAL ARRAY(200): Y,Erx
REAL ARRAY(65,65): Sx
REAL ARRAY(65): Towe2,Tow2,Tow,Nodvar
INTEGER ARRAY(7,7): Ret
```

```
INTEGER ARRAY(65): Sum,Jb
STRING: T[200]
INTEGER: Nx,Nw,Nd,Ic,I,J,K,L
INTEGER: Wc,Colc,Ks,P,Ka
REAL: Xl,Xr,Wb,Wu
REAL: Dx,Dw,B,X,W,Tx,Tw,Z
REAL: U
REAL ARRAY(10,10): Sumc
REAL: Predy,Txw
REAL ARRAY(9,9): Yc
REAL ARRAY(9): Ww,Xx
REAL: Jc,Ip,Jp

PROCEDURE: Sims
  INTEGER ARG: N
END PROCEDURE

PROCEDURE: Sims
  REAL: X,Y
  INTEGER: J,I,K,M,Ic
  EXTERNAL: Sx
    10 REM - solve set of n equations
    20 FOR J = 1 TO N
    30  REM - statements 480 to 520 find the first equation with a
    40  REM - non-zero coefficient for the current column.
    50  Ic=0:M=1
    60  FOR I = J TO N
    70    IF Sx(I,J)=0 THEN GOTO 80 ELSE GOTO 100
    80    Ic=Ic+1
    90    GOTO 130
   100    ON M GOTO 110,130
   110    K=I
   120    M=2
   130   NEXT
   140   IF Ic=N THEN GOTO 160
   150   GOTO 180
   160   PRINT "no unique solution": STOP
   170   REM  statements 540 to 580 move that equation up to current row.
   180   I=K
   190   FOR K= 1 TO N+1
   200     X=Sx(J,K)
   210     Sx(J,K)=Sx(I,K)
   220     Sx(I,K)=X
   230   NEXT
   235   CLEAR(Ret,Colc)
   240   REM - statements 610 to 640 get a 1 coefficient in the first non-zero
   250   REM - column of the current row
   260   Y=1/Sx(J,J)
   270   FOR K= 1 TO N+1
   280     Sx(J,K)=Y*Sx(J,K)
   290   NEXT
   300   REM - statements 670 to 730 subtract the current equation from
   310   REM - th other rows,
   320   FOR I = 1 TO N
   330     IF I = J THEN GOTO 380
   340     Y=-Sx(I,J)
```

```
  350     FOR K = 1 TO N+1
  360       Sx(I,K)=Sx(I,K)+Y*Sx(J,K)
  370     NEXT
  380   NEXT
  390   REM - this process is repeated for all equations.
  400 NEXT
  410 PRINT #3 " solution of equations"
  420 M = 1
  430 FOR I = 1 TO N
  440   M= M+1
  450   PRINT #3 USING "#####"; I;
  460   PRINT#3 USING "--#####.####";Sx(I,N+1);
  470   IF M=>6 GOTO 500
  480   PRINT #3 "   ";
  490   GOTO 510
  500   PRINT #3: M= 1
  510 NEXT
  520 PRINT #3
END PROCEDURE
```

'MAIN Program:

```
 10 REM - Fit a two-dimensional four-point sliding polynomial surface.
 20 REM - Program will delete un-weoghted nodes thereby allowing for
 30 REM - an irregular shaped surface.
 40 REM - Outlier nodes are set by parabolic extrapolation.
 50 REM - Place data disk in drive 'a';place scratch disk in drive 'b'
 60 OPEN "prn" FOR OUTPUT AS #3
 70 PRINT#3: PRINT#3"                    Four-point surface smoothing"
 80 INPUT "problem title";T
 90 PRINT#3 T:PRINT#3
100 PRINT "input number of nodes horizontally and vertically,dont count outliers"
110 INPUT "number of nodes in x-scale  -horizontally";Nx
120 INPUT "number of nodes in w-scale - vertically";Nw
130 INPUT "x-location of left-most column of nodes";Xl
140 INPUT "x-location of right-most column of nodes";Xr
150 INPUT "w-location of bottom line of nodes";Wb
160 INPUT "w-location of upper line of nodes";Wu
170 Dx=(Xr-Xl)/(Nx-1)
180 Dw=(Wu-Wb)/(Nw-1)
190 INPUT "number of data points";Nd
200 INPUT"input a '1' to print reduced coefficient matrix,otherwiae any";B
210 REM - Input data from disk - one point at a time
220 OPEN "b:tempwts" FOR OUTPUT AS #5
230 OPEN "a:testdat" FOR INPUT AS #4
240 FOR Ic = 1 TO Nd
250   CLEAR(C2,Wt)
260   INPUT#4 X
270   INPUT#4 W
280   INPUT#4 Y(Ic)
290   Tx=Xl
300   REM - Find the unit surface and the unit variates z and u for the data point
310   REM - Ret( )=1 for all unit surfaces containing at least one data point
320   J=0
```

```
330  Tx=Tx+Dx
340  J=J+1
350  IF X<=Tx THEN GOTO 370
360  GOTO 330
370  Tw=Wb
380  I=0
390  Tw=Tw+Dw
400  I=I+1
410  IF W<=Tw THEN GOTO 430
420  GOTO 390
430  Z=0.5+(X-Tx)/Dx
440  U=0.5+(W-Tw)/Dw
450  Ret(I,J)=1
460  REM - Comput the 16 coefficients c2(-,-) for the point
470  Cz(1)=(((-8*Z+4)*Z+2)*Z-1)/16
480  Cu(1)=(((-8*U+4)*U+2)*U-1)/16
490  Cz(2)=(((24*Z-4)*Z-22)*Z+9)/16
500  Cu(2)=(((24*U-4)*U-22)*U+9)/16
510  Cz(3)=(((-24*Z-4)*Z+22)*Z+9)/16
520  Cu(3)=(((-24*U-4)*U+22)*U+9)/16
530  Cz(4)=(((8*Z+4)*Z-2)*Z-1)/16
540  Cu(4)=(((8*U+4)*U-2)*U-1)/16
550  FOR K = 1 TO 4
560    FOR L = 1 TO 4
570      C2(I+K-1,J+L-1)=Cu(K)*Cz(L)
580    NEXT L
590  NEXT K
600  REM - Delete the outlier nodes by parabolic extension
610  FOR I = 2 TO Nw+1
620    C2(I,2)=C2(I,2)+3*C2(I,1)
630    C2(I,3)=C2(I,3)-3*C2(I,1)
640    C2(I,4)=C2(I,4)+C2(I,1)
650    C2(I,Nx+1)=C2(I,Nx+1)+3*C2(I,Nx+2)
660    C2(I,Nx)=C2(I,Nx)-3*C2(I,Nx+2)
670    C2(I,Nx-1)=C2(I,Nx-1)+C2(I,Nx+2)
680  NEXT I
690  FOR J = 1 TO Nx+1
700    C2(2,J)=C2(2,J)+3*C2(1,J)
710    C2(3,J)=C2(3,J)-3*C2(1,J)
720    C2(4,J)=C2(4,J)+C2(1,J)
730    C2(Nw+1,J)=C2(Nw+1,J)+3*C2(Nw+2,J)
740    C2(Nw,J)=C2(Nw,J)-3*C2(Nw+2,J)
750    C2(Nw-1,J)=C2(Nw-1,J)+C2(Nw+2,J)
760  NEXT J
770  C2(2,2)=C2(2,2)+3*C2(1,1)
780  C2(3,3)=C2(3,3)-3*C2(1,1)
790  C2(4,4)=C2(4,4)+C2(1,1)
800  C2(2,Nx+1)=C2(2,Nx+1)+3*C2(1,Nx+2)
810  C2(3,Nx)=C2(3,Nx)-3*C2(1,Nx+2)
820  C2(4,Nx-1)=C2(4,Nx-1)+C2(1,Nx+2)
830  C2(Nw+1,2)=C2(Nw+1,2)+3*C2(Nw+2,1)
840  C2(Nw,3)=C2(Nw,3)-3*C2(Nw+2,1)
850  C2(Nw-1,4)=C2(Nw-1,4)+C2(Nw+2,1)
860  C2(Nw+1,Nx+1)=C2(Nw+1,Nx+1)+3*C2(Nw+2,Nx+2)
870  C2(Nw,Nx)=C2(Nw,Nx)-3*C2(Nw+2,Nx+2)
880  C2(Nw-1,Nx-1)=C2(Nw-1,Nx-1)+C2(Nw+2,Nx+2)
```

```
890  REM - Linearize the matrix
900  Wc=0
910  FOR I = 1 TO Nw
920    FOR J = 1 TO Nx
930      Wc=Wc+1
940      Wt(Wc)=C2(I+1,J+1)
950      Sumc(I,J)=Sumc(I,J)+C2(I+1,J+1)
960    NEXT J
970  NEXT I
980  FOR J = 1 TO Wc
990    PRINT#5 Wt(J)
1000   IF Wt(J)=0 THEN Sum(J)=Sum(J)+1
1010  NEXT J
1020 NEXT Ic
1030 CLOSE #4
1040 CLOSE #5
1050 PRINT "data loaded, coefficients computed and stored"
1060 REM - Read wts back in - delete wts for nodes with only zeroes
1070 REM - Save reduced wts and compute sums of squares
1080 FOR L = 1 TO Wc
1090  IF Sum(L)=Nd THEN Jb(L)=1:Colc=Colc+1
1100 NEXT L
1110 OPEN "b:tempwts" FOR INPUT AS #5
1120 OPEN "b:redwts" FOR OUTPUT AS #4
1130 CLEAR (Sx)
1140 FOR I = 1 TO Nd
1150  FOR J = 1 TO Wc
1160    INPUT#5 Wt(J)
1170  NEXT J
1180  Ks=0
1190  FOR K = 1 TO Wc - Colc
1200    Ks=Ks+Jb(K)
1210    Wt(K)=Wt(K+Ks)
1220    PRINT#4 Wt(K)
1230  NEXT K
1240  Wt(Wc-Colc+1)=Y(I)
1250  FOR J = 1 TO Wc-Colc+1
1260    FOR K = J TO Wc-Colc+1
1270      Sx(J,K)=Sx(J,K)+Wt(J)*Wt(K)
1280    NEXT K
1290  NEXT J
1300  IF B<> 1 THEN GOTO 1410
1310  IF I=1 THEN PRINT#3 "Reduced coefficient matrix"
1320  PRINT#3 I
1330  P=0
1340  FOR L = 1 TO Wc-Colc+1
1350    P=P+1
1360    PRINT#3 USING "####";L;
1370    PRINT#3 USING "--###.#####";Wt(L);
1380    IF P = 8 THEN P = 0:PRINT#3
1390  NEXT L
1400  PRINT#3
1410 NEXT I
1420 PRINT#3:PRINT#3 "Number of zero columns  ";Colc
1430 PRINT#3 "The zero columns are 'ones'"
1440 FOR K = 1 TO Nx*Nw
```

```
1450   PRINT#3 Jb(K);
1460 NEXT K
1470 PRINT#3
1480 REM - Fill in the complete matrix
1490 FOR J = 1 TO Wc-Colc+1
1500   FOR K = J TO Wc-Colc+1
1510     Sx(K,J)=Sx(J,K)
1520   NEXT K
1530 NEXT J
1540 CLOSE#5
1550 CLOSE#4
1560 PRINT "sums of squares matrix calculated; call proc. for sim. eqns."
1570 Sims (Wc-Colc)
1580 REM - Call reduced wts back in and compute residual errors
1590 OPEN "b:redwts" FOR INPUT AS #4
1600 KILL "b:tempwts"
1610 CLEAR(Towe2,Tow2,Tow)
1620 FOR J = 1 TO Nd
1630   FOR L = 1 TO Wc-Colc
1640     INPUT #4 Wt(L)
1650   NEXT L
1660   Predy =0.0
1670   FOR I = 1 TO Wc-Colc
1680     Predy=Predy+Wt(I)*Sx(I,Wc-Colc+1)
1690   NEXT I
1700   Erx(J)=Y(J)-Predy
1710   REM - Compute nodal variance
1720   FOR I= 1 TO Wc-Colc
1730     Txw=Wt(I)*Wt(I)
1740     Towe2(I)=Towe2(I)+Txw*Erx(J)*Erx(J)
1750     Tow2(I)=Tow2(I)+Txw
1760     Tow(I)=Tow(I)+Wt(I)
1770   NEXT I
1780 NEXT J
1790 FOR K = 1 TO Wc-Colc
1800     IF Tow(K)>0 THEN Nodvar(K)=Towe2(K)/Tow2(K)/Tow(K)*(Nd-Wc+Colc)/Nd ELSE
Nodvar(K)=999
1810 NEXT K
1820 P=0
1830 PRINT#3 "          Residual errors"
1840 FOR I = 1 TO Nd
1850   P=P+1
1860   PRINT #3 USING "####";I;
1870   PRINT #3 USING "--##.####";Erx(I);
1880   IF P = 8 THEN PRINT#3:P=0
1890 NEXT I
1900 PRINT#3:PRINT#3:PRINT#3"          Nodal variance"
1910 PRINT#3"Node  wtd err sqd   wt sqd  weight  variance   std dev"
1920 FOR J = 1 TO Wc-Colc
1930   PRINT#3 USING "####";J;
1940   PRINT#3 USING "--###.#####";Towe2(J);Tow2(J);Tow(J);Nodvar(J);
1950     IF Nodvar(J)=999 THEN PRINT#3 USING "#########";999 ELSE PRINT#3 USING
"--###.#####";SQR(Nodvar(J))
1960 NEXT J
1970 PRINT#3:PRINT#3
1980 CLOSE#4
```

```
1990 KILL "b:redwts"
2000 REM - Set up the full grid of nodes. 000000 means a deleted node.
2010 K=0
2020 Ka = 0
2030 FOR J = 1 TO Nw
2040   FOR I = 1 TO Nx
2050     K=K+1
2060     IF Jb(K)= 1 THEN Nod(J+1,I+1)=0:Ka=Ka+1 ELSE Nod(J+1,I+1)=Sx(K-Ka,Wc-Colc+1)
2070   NEXT I
2080 NEXT J
2090 REM - Fill in all boundary nodes
2100 REM - Left and right
2110 FOR I = 2 TO Nw+1
2120   Nod(I,1)=3*Nod(I,2)-3*Nod(I,3)+Nod(I,4)
2130   Nod(I,Nx+2)=3*Nod(I,Nx+1)-3*Nod(I,Nx)+Nod(I,Nx-1)
2140 NEXT I
2150 REM - Top and bottom
2160 FOR J = 2 TO Nx+1
2170   Nod(1,J)=3*Nod(2,J)-3*Nod(3,J)+Nod(4,J)
2180   Nod(Nw+2,J)=3*Nod(Nw+1,J)-3*Nod(Nw,J)+Nod(Nw-1,J)
2190 NEXT J
2200 REM - Four corners
2210 Nod(1,1)=3*Nod(2,2)-3*Nod(3,3)+Nod(4,4)
2220 Nod(Nw+2,1)=3*Nod(Nw+1,2)-3*Nod(Nw,3)+Nod(Nw-1,4)
2230 Nod(1,Nx+2)=3*Nod(2,Nx+1)-3*Nod(3,Nx)+Nod(4,Nx-1)
2240 Nod(Nw+2,Nx+2)=3*Nod(Nw+1,Nx+1)-3*Nod(Nw,Nx)+Nod(Nw-1,Nx-1)
2250 REM - Compute values of surfaces at two-way 8th points for contouring.
2260 PRINT#3:PRINT#3 "          contouring values for each retained unit surface"
2270 CLEAR(Yc)
2280 FOR I = 1 TO Nw-1
2290   W=Wb+(I-1)*Dw
2300   FOR J = 1 TO Nx-1
2310     X=Xl+(J-1)*Dx
2320     PRINT#3:PRINT#3"Lower left coordinates of unit surface:w= ";W;" x= ";X
2330     IF Ret(I,J)=1 THEN GOTO 2360
2340     PRINT#3"  This unit surface was deleted"
2350     GOTO 2740
2360     FOR Ic = 1 TO 9
2370       Ww(Ic)=W+(Ic-1)*Dw/8
2380       U=-0.5+(Ww(Ic)-W)/Dw
2390       Cu(1)=(((-8*U+4)*U+2)*U-1)/16
2400       Cu(2)=(((24*U-4)*U-22)*U+9)/16
2410       Cu(3)=(((-24*U-4)*U+22)*U+9)/16
2420       Cu(4)=(((8*U+4)*U-2)*U-1)/16
2430       FOR Jc= 1 TO 9
2440         Yc(Ic,Jc)=0.0
2450         Xx(Jc)=X+(Jc-1)*Dx/8
2460         Z=-0.5+(Xx(Jc)-X)/Dx
2470         Cz(1)=(((-8*Z+4)*Z+2)*Z-1)/16
2480         Cz(2)=(((24*Z-4)*Z-22)*Z+9)/16
2490         Cz(3)=(((-24*Z-4)*Z+22)*Z+9)/16
2500         Cz(4)=(((8*Z+4)*Z-2)*Z-1)/16
2510         FOR Ip = 1 TO 4
2520           FOR Jp = 1 TO 4
2530             Yc(Ic,Jc)=Yc(Ic,Jc)+Cu(Ip)*Cz(Jp)*Nod(I+Ip-1,J+Jp-1)
2540           NEXT Jp
```

```
2550        NEXT Ip
2560      NEXT Jc
2570    NEXT Ic
2580    PRINT#3 " Values of the surface at w = **  ** and x = ________"
2590    FOR Jc= 1 TO 9
2600      PRINT#3 USING "--####.####";Xx(Jc);
2610    NEXT Jc
2620    PRINT#3:PRINT#3 " ";
2630    FOR Jc= 1 TO 9
2640      PRINT#3 "  ------- ";
2650    NEXT Jc
2660    PRINT#3
2670    FOR Ic = 1 TO 9
2680      PRINT#3 "                 ** ";Ww(Ic);"  **"
2690      FOR Jc = 1 TO 9
2700        PRINT#3 USING "--####.####";Yc(Ic,Jc);
2710      NEXT Jc
2720      PRINT#3
2730    NEXT Ic
2740  NEXT J
2750 NEXT I
2760 PRINT#3
2770 CLOSE#3
2780 CLOSE#6
2790 END

ENDFILE
```

Program B-5.9

Monthly Cyclic Two-Dimensional Smoothing

```
SOURCE
PRECISION 12
OPTION BASE 1
PROCS=3
REAL ARRAY(500): Xm,Xp,Xr,Vp
REAL ARRAY(17,17): Sc
REAL ARRAY(42): W,Yfull
REAL ARRAY(8,8): Xpnode
REAL ARRAY(6,7): Cw
REAL ARRAY(16): Swro,Dw,Sumw,Sumwe
REAL ARRAY(16): Sumww,Nodvar
REAL ARRAY(20): Surf,Xpi
REAL ARRAY(8): Xmnode
REAL ARRAY(5): Cp,Cm,Coe
INTEGER: I,J,K,L
INTEGER: M,Nd,Nxp,Is,Ir,Ixm,Ivp,Inod
INTEGER: Im,Ik,Nnu,Iw,Nodset
STRING: Title[60]
REAL: A1p,A2p,A3p,B
```

```
REAL: Fxpc,Xpl,Ubjan,Ubmay,Ubsep,Vxpc,Vlim,Vlitr
REAL: A2,A1,Ang,Xpr,Xlit,Vlit,Vnode,Zp
REAL: Zm,Nxn,Xmi,Vpi,Dsurf,Surfer,Sige2,Sigy
REAL: Sigy2,Cor,Xk,Fm,Fivp,Fk
REAL ARRAY(6,7): Ynod

REAL FUNCTION: Seas
  REAL ARG: Ang
END FUNCTION

PROCEDURE: Coef
  REAL ARG: Z
END PROCEDURE

PROCEDURE: Simeqns
  INTEGER ARG: N
END PROCEDURE

REAL FUNCTION: Seas
  EXTERNAL: A1p,A2p,A3p
    10 RESULT=A1p+A2p*SIN(Ang)+A3p*COS(Ang)
END FUNCTION
PROCEDURE: Coef
  EXTERNAL: Coe
    10 Coe(1)=(((-8*Z+4)*Z+2)*Z-1)/16
    20 Coe(2)=(((24*Z-4)*Z-22)*Z+9)/16
    30 Coe(3)=(((-24*Z-4)*Z+22)*Z+9)/16
    40 Coe(4)=(((8*Z+4)*Z-2)*Z-1)/16
END PROCEDURE

PROCEDURE: Simeqns
  EXTERNAL: Sc
  REAL: X,Y
  INTEGER: J,I,K,M
    10 REM - solve set of n equations
    20 FOR J = 1 TO N
    30   REM - statements 480 to 520 find the first equation with a
    40   REM - non-zero coefficient for the current column.
    50   M=0
    60   FOR I = J TO N
    70     IF Sc(I,J)<>0 THEN GOTO 90
    80     GOTO 120
    90     M=1
   100     K=I
   110     EXIT
   120   NEXT
   130   IF M= 1 THEN GOTO 180
   160   PRINT "no unique solution": STOP
   170   REM  statements 540 to 580 move that equation up to current row.
   180   I=K
   190   FOR K= 1 TO N+1
   200     X=Sc(J,K)
   210     Sc(J,K)=Sc(I,K)
   220     Sc(I,K)=X
   230   NEXT
   240   REM - statements 610 to 640 get a 1 coefficient in the first non-zero
```

```
   250  REM - column of the current row
   260  Y=1/Sc(J,J)
   270  FOR K= 1 TO N+1
   280    Sc(J,K)=Y*Sc(J,K)
   290  NEXT
   300  REM - statements 670 to 730 subtract the current equation from
   310  REM - th other rows,
   320  FOR I = 1 TO N
   330    IF I = J THEN GOTO 380
   340    Y=-Sc(I,J)
   350    FOR K = 1 TO N+1
   360      Sc(I,K)=Sc(I,K)+Y*Sc(J,K)
   370    NEXT
   380  NEXT
   390  REM - this process is repeated for all equations.
   400 NEXT
   410 PRINT #3 " solution of equations"
   420 M = 1
   430 FOR I = 1 TO N
   440  M= M+1
   450  PRINT #3 USING "#####"; I;
   460  PRINT#3 USING "--#####.####";Sc(I,N+1);
   470  IF M=>6 GOTO 500
   480  PRINT #3 "   ";
   490  GOTO 510
   500  PRINT #3: M= 1
   510 NEXT
   520 PRINT #3
END PROCEDURE
```

'MAIN Program:

```
 10 REM - Smooth a record of monthly rainfall and runoff to yield a seasonallydelete 15
 20 REM - continuous surface of runoff in the rainfall-calendar month space.
 30 INPUT "problem title";Title
 40 OPEN "prn" FOR OUTPUT AS #3
 50 PRINT #3 "          Smoothed monthly rainfall-runoff surface"
 60 PRINT#3 Title
 70 PRINT " input parameter to control central cluster transform of precip"
 80 PRINT " use 0.6 initially. probable range 0.5 to 2.0?"
 90 INPUT " central cluster control";B
100 PRINT " input location of common point in two joined exponential functions"
110 PRINT " that produce central cluster transform. use 0.5 initially.
120 PRINT " probable range 0.3 to 0.7?"
130 INPUT " common point ";Fxpc
140 PRINT " input location of lowest row of nodes in precip scale"
150 PRINT " normally use 0.0"
160 INPUT " low node location";Xpl
170 Nxp=4
180 PRINT "input location of largest node in precip scale in january"
190 PRINT " use maximum monthly january rainfall plus about 0.25 inch."
200 INPUT " max. january node location";Ubjan
210 PRINT " same for may"
220 INPUT " max may node location";Ubmay
230 PRINT " same for september"
240 INPUT  "max september node location";Ubsep
```

```
250 REM - Parameters of sin-cos wave upper boundary thru ubjan,ubmay and ubsep.
260 A3p=(Ubjan-Ubmay)/1.732
270 A2p= (Ubmay-Ubsep+0.866*A3p)/1.5
280 A1p= Ubsep+A2p
290 REM - Number of data points and calendar month number
300 REM - of first month. Enter from disk.
310 REM - Input monthly rainfall and runoff from disk
320 REM - xp(i) is rainfall for month
330 REM = xr(i) is runoff for month
340 OPEN "b:tstdat" FOR INPUT AS #2
350 INPUT #2 Nd:INPUT #2 Xm(1)
360 FOR I = 1 TO Nd
370   INPUT #2 Xp(I)
380   INPUT #2 Xr(I)
390   Xm(I+1)=Xm(I)+1
400   IF Xm(I+1)>12 THEN Xm(I+1)=1
410 NEXT I
420 CLOSE#2
430 Vxpc=EXP(B*Fxpc)
440 Vlim=Vxpc/Fxpc
450 Vlitr=Vlim-(Vlim-Vxpc)*EXP(-B*(1.0-Fxpc))
460 A2=1.0/(Vlitr-1.0)
470 A1=1.0-A2
480 REM - transform precip to v-scale
490 REM - x-scale is original scale of precip
500 REM - little x scale is computed from (x-xpl)/(xpr-xpl)  -  0<xlit<1.0
510 REM  v-scale is linear transform on little v -   1.0<v<2.0
520 FOR I = 1 TO Nd
530   Ang=Xm(I)*0.523599
540   Xpr=Seas(Ang)
550   Xlit=(Xp(I)-Xpl)/(Xpr-Xpl)
560   IF (Xlit<=Fxpc) THEN GOTO 590
570   Vlit=Vlim-(Vlim-Vxpc)*EXP(-B*(Xlit-Fxpc))
580   GOTO 600
590   Vlit=EXP(B*Xlit)
600   Vp(I)=A1+A2*Vlit
610 NEXT I
620 CLEAR (Sc,Swro)
630 REM - Set location of nodes in xp-scale and month scale.
640 Xmnode(1)=1.
650 FOR I=2 TO 4
660   Xmnode(I)=Xmnode(I-1)+3
670 NEXT I
680 FOR I = 1 TO 4
690   Xpnode(I,1)=Xpl
700   Ang=(1+(I-1)*3)*0.523599
710   Xpr=Seas(Ang)
720   Xpnode(I,Nxp)=Xpr
730   Is=Nxp-1
740   Vnode=1.0
750   FOR J= 2 TO Is
760     Vnode=Vnode+1/3
770     Vlit=(Vnode-A1)/A2
780     IF Vlit<=Vxpc THEN GOTO 810
790     Xlit=LOG((Vlit-Vlim)/(Vxpc-Vlim))/(-B)+Fxpc
800     GOTO 820
```

```
810     Xlit=LOG(Vlit)/B
820     Xpnode(I,J)=Xlit*(Xpr-Xpl)+Xpl
830   NEXT J
840 NEXT I
850 PRINT#3 :PRINT#3 "Columns of nodes are located at January,April,July, and October.":PRINT#3
860 PRINT#3 "Location of nodes in precip scale;by months"
870 FOR J= 1 TO 4
880   PRINT#3 J
890   FOR I= 1 TO 4
900     PRINT#3 USING "######"; I;
910     PRINT#3 USING "--####.###";Xpnode(I,J);
920   NEXT I
930   PRINT#3
940 NEXT J
950 PRINT#3
960 REM - Find the x-set and the m-set of nodes and calculate coefficients.
970 OPEN"b:fullc" FOR OUTPUT AS #4
980 FOR I = 1 TO Nd
990   K=1
1000  Xk=K
1010  IF Vp(I)<=(1+Xk/3) THEN GOTO 1040
1020  K=K+1
1030  GOTO 1000
1040  Zp=(Vp(I)-1)*3-K+0.5
1050  M=1
1060  Fm=M
1070  IF Xm(I)<=(1+Fm*3) THEN GOTO 1100
1080  M=M+1
1090  GOTO 1060
1100  Zm=(Xm(I)-1)/3 - Fm +0.5
1110  CLEAR (W,Cw)
1120  Coef Zp
1130  FOR J=1 TO 4
1140    Cp(J)=Coe(J)
1150  NEXT J
1160  Coef Zm
1170  FOR J = 1 TO 4
1180    Cm(J)=Coe(J)
1190  NEXT J
1200  FOR J = 1 TO 4
1210    FOR L = 1 TO 4
1220      Cw(K+J-1,M+L-1)=Cp(J)*Cm(L)
1230    NEXT L
1240  NEXT J
1250  Nxn = 0
1260  FOR J = 1 TO 6
1270    FOR L = 1 TO 7
1280      Nxn= Nxn+1
1290      W(Nxn)=Cw(J,L)
1300    NEXT L
1310  NEXT J
1320  REM - Save m, k, and w(_) coefficients to disk on drive b
1330  PRINT#4 M
1340  PRINT#4 K
1350  FOR J= 1 TO 42
1360    PRINT#4 W(J)
```

```
1370  NEXT J
1380  REM - Collapse the vector and store on disk on drive b
1390  Dw(1)=W(9)+W(13)+2*(W(2)+W(6))
1400  Dw(2)=W(10)+W(14)+2*(W(3)+W(7))
1410  Dw(3)=W(11)+2*W(4)
1420  Dw(4)=W(12)+W(8)+2*(W(5)+W(1))
1430  Dw(5)=W(16)+W(20)-W(2)-W(6)
1440  Dw(6)=W(17)+W(21)-W(3)-W(7)
1450  Dw(7)=W(18)-W(4)
1460  Dw(8)=W(19)+W(15)-W(5)-W(1)
1470  Dw(9)=W(23)+W(27)-W(37)-W(41)
1480  Dw(10)=W(24)+W(28)-W(38)-W(42)
1490  Dw(11)=W(25)-W(39)
1500  Dw(12)=W(26)+W(22)-W(40)-W(36)
1510  Dw(13)=W(30)+W(34)+2*(W(37)+W(41))
1520  Dw(14)=W(31)+W(35)+2*(W(38)+W(42))
1530  Dw(15)=W(32)+2*W(39)
1540  Dw(16)=W(33)+W(29)+2*(W(40)+W(36))
1550  FOR J = 1 TO 16
1560    PRINT #4 Dw(J)
1570  NEXT J
1580  FOR J= 1 TO 16
1590    FOR L= 1 TO 16
1600      Sc(J,L)=Sc(J,L)+Dw(J)*Dw(L)
1610    NEXT L
1620    Swro(J)=Swro(J)+Dw(J)*Xr(I)
1630  NEXT J
1640 NEXT I
1650 CLOSE#4
1660 REM - Square the matrix
1670 FOR J = 1 TO 16
1680   FOR L= 1 TO 16
1690     Sc(L,J)=Sc(J,L)
1700   NEXT L
1710   Sc(J,17)=Swro(J)
1720 NEXT J
1730 Simeqns 16
1740 REM - Construct the full lattice of nodes
1750 FOR I = 1 TO 16
1760   Ir=I+8+((I-1)/4*3)
1770   Yfull(Ir)=Sc(I,17)
1780 NEXT I
1790 FOR I = 2 TO 5
1800   Yfull(I)=2*Yfull(I+7)-Yfull(I+14)
1810 NEXT I
1820 FOR I= 37 TO 40
1830   Yfull(I)=2*Yfull(I-7)-Yfull(I-14)
1840 NEXT I
1850 FOR I = 1 TO 36 STEP 7
1860   Yfull(I)=Yfull(I+4)
1870 NEXT I
1880 FOR I = 6 TO 41 STEP 7
1890   Yfull(I)=Yfull(I-4)
1900   Yfull(I+1)=Yfull(I-3)
1910 NEXT I
1920 PRINT#3"      The full lattice of nodes"
```

```
1930 K=1
1940 FOR I = 1 TO 42
1950   PRINT#3 USING "####";I;
1960   PRINT#3 USING "--####.###";Yfull(I);
1970   IF K=7 THEN PRINT#3:K=1:GOTO 1990
1980   K=K+1
1990 NEXT I
2000 PRINT #3
2010 REM - Convert to 6x7 grid of YNOD
2020 K=0
2030 FOR I = 1 TO 6
2040   FOR J = 1 TO 7
2050     K=K+1
2060     Ynod(I,J)=Yfull(K)
2070   NEXT J
2080 NEXT I
2090 REM - Increment on m and v,interpolate, and transform to v and p scale
2100 REM - This interpolation generates surface for display
2110 FOR Ixm = 1 TO 25
2120   Xmi= Ixm - 1
2130   Xmi = Xmi/2.0+1.0
2140   Ang=Xmi*0.523599
2150   Xpr = Seas(Ang)
2160   M=1
2170   Fm = M
2180   IF Xmi<=(1.+Fm*3.) THEN GOTO 2210
2190   M=M+1
2200   GOTO 2170
2210   Zm=(Xmi-1.0)/3.0-Fm+0.5
2220   Coef Zm
2230   FOR I = 1 TO 4
2240     Cm(I)=Coe(I)
2250   NEXT I
2260   FOR Ivp = 1 TO 20
2270     Fivp = Ivp
2280     Surf(Ivp)=0.0
2290     Vpi=(18.0+Fivp)/19.0
2300     K=1
2310     Fk = K
2320     IF Vpi<=(1.0+Fk/3.0) THEN GOTO 2350
2330     K=K+1
2340     GOTO 2310
2350     Zp = (Vpi-1.0)*3.0-Fk+0.5
2360     Coef Zp
2370     FOR I = 1 TO 4
2380       Cp(I)=Coe(I)
2390     NEXT I
2400     Vlit=(Vpi-A1)/A2
2410     IF Vlit<=Vxpc THEN GOTO 2440
2420     Xlit=LOG((Vlit-Vlim)/(Vxpc-Vlim))/(-B)+Fxpc
2430     GOTO 2450
2440     Xlit=LOG(Vlit)/B
2450     Xpi(Ivp)=Xlit*(Xpr-Xpl)+Xpl
2460     FOR I = 1 TO 4
2470       FOR J = 1 TO 4
2480         Surf(Ivp)=Surf(Ivp)+Cm(I)*Cp(J)*Ynod(J+K-1,I+M-1)
```

```
2490      NEXT J
2500    NEXT I
2510  NEXT Ivp
2520  PRINT#3 "Rainfall and runoff at  ";Xmi;"  in month scale"
2530  K=1
2540  FOR Ivp = 1 TO 20
2550    PRINT#3 USING "####";Ivp;
2560    PRINT#3 USING "--####.###";Xpi(Ivp);Surf(Ivp);
2570    IF K= 5 THEN PRINT#3:K=1: GOTO 2590
2580    K=K+1
2590  NEXT Ivp
2600  PRINT#3
2610 NEXT Ixm
2620 REM - Recall m, k,dw(iw) and w(iw) vectors and interpolate at data
2630 REM - points to get residual errors
2640 CLEAR(Sumw,Sumwe,Sumww)
2650 OPEN "b:fullc" FOR INPUT AS #4
2660 FOR I = 1 TO Nd
2670  INPUT #4 M
2680  INPUT #4 K
2690  FOR Iw = 1 TO 42
2700    INPUT #4 W(Iw)
2710  NEXT Iw
2720  FOR Iw = 1 TO 16
2730    INPUT #4 Dw(Iw)
2740  NEXT Iw
2750  Dsurf = 0.0
2760  FOR Inod = 1 TO 16
2770    Im=(Inod-1)-INT((Inod-1)/4)*4+1
2780    Ik=(Inod-Im)/4+1
2790    Nnu=(Im+M-1)+(Ik+K-2)*7
2800    Dsurf=Dsurf+W(Nnu)*Yfull(Nnu)
2810  NEXT Inod
2820  Surfer=Xr(I)-Dsurf
2830  REM - Store surface error in xp(i) vextor
2840  Xp(I)=Surfer
2850  FOR Nodset = 1 TO 16
2860    Sumw(Nodset)=Sumw(Nodset)+Dw(Nodset)
2870    Sumwe(Nodset)=Sumwe(Nodset)+Surfer*Surfer*Dw(Nodset)*Dw(Nodset)
2880    Sumww(Nodset)=Sumww(Nodset)+Dw(Nodset)*Dw(Nodset)
2890  NEXT Nodset
2900 NEXT I
2910 PRINT#3"     Residual surface errors"
2920 K=1
2930 FOR I = 1 TO Nd
2940  PRINT#3 USING "####";I;
2950  PRINT#3 USING "--###.###";Xp(I);
2960  IF K=12 THEN K=1:PRINT#3: GOTO 2980
2970  K=K+1
2980 NEXT I
2990 PRINT#3:PRINT#3
3000 REM - Compute nodal variance and standard deviation
3010 PRINT#3 "       Variance and standard deviation of nodes"
3020 FOR Nodset = 1 TO 16
3030  Nodvar(Nodset)=Sumwe(Nodset)/Sumww(Nodset)*Nd/(Nd-16)/Sumw(Nodset)
3040 NEXT Nodset
```

```
3050 PRINT #3 "          sum     sum weights*   sum weights            standard"
3060 PRINT #3 "  Node  weights  error sqrd.    squared     variance  deviation"
3070 FOR I = 1 TO 16
3080   PRINT #3 USING "#####";I;
3090   PRINT #3 USING "#####.###";Sumw(I);Sumwe(I);
3100   PRINT #3 USING "#########.###";Sumww(I);
3110   PRINT #3 USING "#######.###";Nodvar(I);
3120   PRINT #3 USING "#######.###";SQR(Nodvar(I))
3130 NEXT I
3140 REM - Compute correlation coefficient
3150 Sige2=0.0
3160 Sigy=0.0
3170 Sigy2=0.0
3180 FOR I=1 TO Nd
3190   Sige2=Sige2+Xp(I)*Xp(I)
3200   Sigy2=Sigy2+Xr(I)*Xr(I)
3210   Sigy=Sigy+Xr(I)
3220 NEXT I
3230 Sigy2=Sigy2-Sigy*Sigy/Nd
3240 Cor=(Sigy2-Sige2)/Sigy2
3250 PRINT#3 "Sum of: ro,adj ro squared,err squared,cor squared and cor."
3260 PRINT#3 USING "#####.####";Sigy;Sigy2;Sige2;Cor;SQR(Cor)
3270 PRINT#3
3280 CLOSE#3
3290 KILL "b:fullc"
3300 CLOSE#4
3310 STOP
3320 END
ENDFILE
```

Program B-6.1

Simulate Random Sample With Segmented Distribution

```
SOURCE
PRECISION 12
OPTION BASE 1
PROCS=1
REAL ARRAY(100): X,Xa
INTEGER: Ns,Ni,M,L,Ip,I,R
REAL: P,A,B,C,Ar

PROCEDURE: Arrange
  INTEGER ARG: Ni
END PROCEDURE

PROCEDURE: Arrange
  EXTERNAL: X,Xa
  INTEGER: I,K,L,Ip
    10 Xa(1)=X(1)
    20 FOR I = 2 TO Ni
    30   FOR K = 1 TO I-1
    40     IF X(I)<Xa(K) THEN EXIT
    50   NEXT K
    60   FOR L = I-1 TO K STEP -1
    70     Xa(L+1)=Xa(L)
    80   NEXT L
    90   Xa(K)=X(I)
   100 NEXT I
   110 Ip = 1
   120 FOR I = Ni TO 1 STEP -1
   130   PRINT#5 USING "####";Ni-I+1;
   140   PRINT#5 USING "--###.###"; Xa(I);
   150   IF Ip = 10 THEN Ip =1:PRINT#5:GOTO 170
   160   Ip = Ip + 1
   170 NEXT I
   180 PRINT#5
END PROCEDURE

'MAIN Program:


   10 REM - Simulate random samples of 4-span segmented linear distribution
   20 REM - Number of items in each sample is ni - not more than 100.
   30 REM - Number of samples is ns
   40 OPEN "prn" FOR OUTPUT AS #5
   50 PRINT#5 " Random samples of 4-span segmented linear distribution"
   60 INPUT "number of samples to generate";Ns
   70 INPUT"number of items to generate in each sample";Ni
   80 INPUT "random number seed";R
   90 RANDOMIZE R
  100 FOR M = 1 TO Ns
  110   PRINT#5 " Sample number  ";M
  120   PRINT#5 " Order of items as generated"
```

```
130  FOR L = 1 TO Ni
140    P= RND(2)
150    IF P <= 0.137 THEN GOTO 190
160    IF P <= 0.5 THEN GOTO 210
170    IF P <= 0.863 THEN GOTO 240
180    GOTO 270
190    X(L)=SQR(16.392*P)
200    GOTO 310
210    A=0.03920:B=0.06543:C=(3.938-5.346)/28.686-P
220    X(L)=(-B+SQR(B*B-4*A*C))/(2*A)
230    GOTO 310
240    A=-0.03920:B=0.53580:C=(14.344-35.99)/28.686-P
250    X(L)=(-B+SQR(B*B-4*A*C))/(2*A)
260    GOTO 310
270    A=-0.06101:B=0.73206:C=(24.748-59.062)/28.686-P
280    Ar=B*B-4*A*C
290    IF Ar<0 THEN Ar=0
300    X(L)=(-B+SQR(Ar))/(2*A)
310  NEXT L
320  Ip=0
330  FOR I = 1 TO Ni
340    Ip=Ip+1
350    PRINT#5 USING "####";I;
360    PRINT#5 USING "--###.###";X(I);
370    IF Ip = 10 THEN Ip=0:PRINT#5
380  NEXT I
390  PRINT#5
400  PRINT#5 " Arranged in order of magnitude"
410  Arrange Ni
420  PRINT#5
430 NEXT M
440 PRINT#5
450 CLOSE#5
460 STOP

ENDFILE
```

Program B-6.2

Simulate Random Samples With Four-Point Sliding Polynomial Distribution

```
SOURCE
PRECISION 12
OPTION BASE 1
PROCS=1
REAL ARRAY(100): X,Xa
REAL ARRAY(10): Nod,Snod,Dnod
STRING: T[100]
INTEGER: I,Ns,Ni,Nr,J
INTEGER: K,Ip
REAL: Vl,Vr,Dv,Lt,Mt,Kt,Rn
REAL: Rp,Il,Sl,Ir,Vm,Z,C1,C2
```

```
REAL: C3,C4,Pm,V,Srn,M,Rnn

PROCEDURE: Arrange
  INTEGER ARG: Ni
END PROCEDURE

PROCEDURE: Arrange
  EXTERNAL: X,Xa
  INTEGER: I,K,L,Ip
    10 Xa(1)=X(1)
    20 FOR I = 2 TO Ni
    30   FOR K = 1 TO I-1
    40     IF X(I)<Xa(K) THEN EXIT
    50   NEXT K
    60   FOR L = I-1 TO K STEP -1
    70     Xa(L+1)=Xa(L)
    80   NEXT L
    90   Xa(K)=X(I)
   100 NEXT I
   110 Ip = 1
   120 FOR I = Ni TO 1 STEP -1
   130   PRINT#6 USING "####";Ni-I+1;
   140   PRINT#6 USING "--###.##"; Xa(I);
   150   IF Ip = 10 THEN Ip =1:PRINT#6:GOTO 170
   160   Ip = Ip + 1
   170 NEXT I
   180 PRINT#6
END PROCEDURE

'MAIN Program:


  10 REM - Simulate ns samples of ni items each.
  20 REM - Assumes 4 sliding polynomial arcs from 00 to 100 % probability.
  30 REM - Uses interval-halving method to do reverse interpolation.
  40 REM - Uses four-point sliding ploynomials on 7 nodes:two are outliers.
  50 REM - Performs a variate de-transform from v to x
  60 REM - Input seven internal nodes.
  70 PRINT " input the nodes"
  80 FOR I = 2 TO 8
  90   PRINT "node number";I-1;:INPUT Nod(I)
 100   Snod(I)=Nod(I)
 110 NEXT I
 120 REM - Input standard deviations of 5 interior nodes.
 130 PRINT " input the standard deviations of the nodes"
 140 FOR I = 3 TO 7
 150   PRINT "node number ";I-1;:INPUT Dnod(I)
 160 NEXT I
 170 INPUT "number of samples to generate";Ns
 180 INPUT "number of items to generate for each sample";Ni
 190 REM - vl is 0.49 for example 6-4.
 200 INPUT "v-value of node number two";Vl
 210 REM - vr is 3.69 for example 6-4.
 220 INPUT "v-value of node number six";Vr
 230 INPUT "random number seed";Rn
 240 Dv=(Vr-Vl)/4
```

```
250 INPUT "problem  title";T
260 OPEN "prn" FOR OUTPUT AS #6
270 PRINT#6 "                    Stochastic simulation"
280 PRINT#6 "          ";T
290 FOR I = 2 TO 8
300   PRINT#6 I;Nod(I);
310 NEXT I
320 PRINT#6
330 PRINT#6 " Left value of v  ";Vl;"          Right value of v  ";Vr
340 PRINT#6
350 REM - Set parameters to transform v to x
360 Lt=0.18
370 Mt=95.5
380 Kt=2
390 REM - Start simulation
400 RANDOMIZE Rn
410 Nr=3
420 FOR I = 1 TO Ns
430   PRINT#6 " Sample number  ";I
440   FOR J = 1 TO Ni
450     Rp = RND(2)*100
460     REM - find the span that rp falls in
470     K=3
480     IF Rp <= Nod(K) THEN GOTO 500
490     K=K+1: GOTO 480
500     Il=Vl+(K-4)*Dv:Sl=Il
510     Ir=Vl+(K-3)*Dv
520     REM - Interpolate at middle of the span found.
530     Vm=(Il+Ir)/2
540     Z=-0.5+(Vm-Sl)/Dv
550     C1=(((-8*Z+4)*Z+2)*Z-1)/16
560     C2=(((24*Z-4)*Z-22)*Z+9)/16
570     C3=(((-24*Z-4)*Z+22)*Z+9)/16
580     C4=(((8*Z+4)*Z-2)*Z-1)/16
590     Pm=C1*Nod(K-2)+C2*Nod(K-1)+C3*Nod(K)+C4*Nod(K+1)
600     IF ABS(Il-Ir) < (Dv/100) THEN GOTO 650
610     IF Pm > Rp THEN Ir=Vm
620     IF Pm <=Rp THEN Il=Vm
630     Vm=(Il+Ir)/2
640     GOTO 540
650     V=(Il+Ir)/2
660     IF V > 2 THEN GOTO 720
670     IF V < 0.0005 THEN GOTO 700
680     X(J)=LOG(V/Kt)/Lt + Mt
690     GOTO 770
700     X(J)=-999
710     GOTO 770
720     IF V > 3.9995 THEN GOTO 750
730     X(J)=Mt-LOG(2-V/Kt)/Lt
740     GOTO 770
750     X(J)=999
760     REM - Randomize an interior node.
770     Srn = 0.0
780     FOR M = 1 TO 12
790       Srn = Srn + RND(2)
800     NEXT M
```

```
810    Rnn = Srn - 6.0
820    Nod(Nr)=Snod(Nr)+Rnn*Dnod(Nr)
830    IF Nr = 7 THEN GOTO 860
840    Nr=Nr+1
850    GOTO 870
860    Nr=3
870  NEXT J
880  REM - One sample simulated - now print it
890  Ip=1
900  FOR J = 1 TO Ni
910    PRINT#6 USING "####";J;
920    PRINT#6 USING "--###.##";X(J);
930    IF Ip =10 THEN GOTO 950
940    Ip=Ip+1: GOTO 960
950    PRINT#6: Ip=1
960  NEXT J
970  PRINT#6:PRINT#6
980  REM - Arrange the sample in order of magnitude and print again.
990  PRINT#6 " Sample arranged in order of magnitude"
1000  Arrange Ni
1010  PRINT#6:PRINT#6
1020 NEXT I
1030 CLOSE#6
1040 STOP
1050 END

ENDFILE
```

Program B-6.3

Fit Four-Point Sliding Polynomial to Probability Distribution

```
SOURCE
PRECISION 12
OPTION BASE 1
PROCS=2
REAL ARRAY(100): H
REAL ARRAY(40): Ch,Ct,Cp,E
REAL ARRAY(40,7): C
REAL ARRAY(6,6): Sc
REAL ARRAY(5): Sd
REAL ARRAY(4): Co
STRING: T[100]
INTEGER: N,I,J,Ip,Jc
INTEGER: K
REAL: Hm,Hh,Sh,Shh,Shhh,Xh,Hb,Ssd
REAL: Sk,Hpm,C1,Vb,F,D,V
REAL ARRAY(10): Xn
REAL: Dp,Z
REAL: Z2s,Z1s,Z0s

PROCEDURE: Coef
```

```
  REAL ARG: Z
END PROCEDURE

PROCEDURE: Simeqns
  INTEGER ARG: N
END PROCEDURE

PROCEDURE: Coef
  EXTERNAL: Co
    10 Co(1)=(((-8*Z+4)*Z+2)*Z-1)/16
    20 Co(2)=(((24*Z-4)*Z-22)*Z+9)/16
    30 Co(3)=(((-24*Z-4)*Z+22)*Z+9)/16
    40 Co(4)=(((8*Z+4)*Z-2)*Z-1)/16
END PROCEDURE

PROCEDURE: Simeqns
  EXTERNAL: Sc
  REAL: X,Y
  INTEGER: J,I,K,M
    10 REM - solve set of n equations
    20 FOR J = 1 TO N
    30   REM - statements 480 to 520 find the first equation with a
    40   REM - non-zero coefficient for the current column.
    50   M=0
    60   FOR I = J TO N
    70     IF Sc(I,J)<>0 THEN GOTO 90
    80     GOTO 120
    90     M=1
   100     K=I
   110     EXIT
   120   NEXT
   130   IF M= 1 THEN GOTO 180
   160   PRINT "no unique solution": STOP
   170   REM  statements 540 to 580 move that equation up to current row.
   180   I=K
   190   FOR K= 1 TO N+1
   200     X=Sc(J,K)
   210     Sc(J,K)=Sc(I,K)
   220     Sc(I,K)=X
   230   NEXT
   240   REM - statements 610 to 640 get a 1 coefficient in the first non-zero
   250   REM - column of the current row
   260   Y=1/Sc(J,J)
   270   FOR K= 1 TO N+1
   280     Sc(J,K)=Y*Sc(J,K)
   290   NEXT
   300   REM - statements 670 to 730 subtract the current equation from
   310   REM - th other rows,
   320   FOR I = 1 TO N
   330     IF I = J THEN GOTO 380
   340     Y=-Sc(I,J)
   350     FOR K = 1 TO N+1
   360       Sc(I,K)=Sc(I,K)+Y*Sc(J,K)
   370     NEXT
   380   NEXT
   390   REM - this process is repeated for all equations.
```

```
    400 NEXT
    410 PRINT #3 " solution of equations"
    420 M = 1
    430 FOR I = 1 TO N
    440   M= M+1
    450   PRINT #3 USING "#####"; I;
    460   PRINT#3 USING "--#####.####";Sc(I,N+1);
    470   IF M=>6 GOTO 500
    480   PRINT #3 "   ";
    490   GOTO 510
    500   PRINT #3: M= 1
    510 NEXT
    520 PRINT #3
END PROCEDURE
```

'MAIN Program:

```
 10 REM - One-dimensional probability analysis.
 20 REM - Uses four spans of the four-point sliding polynomial.
 30 REM - Will handle positive or negative skew.
 40 REM - Smoothing is over 40 classes
 50 OPEN "prn" FOR OUTPUT AS #3
 60 OPEN "b:whtholpr" FOR INPUT AS #4
 70 INPUT "problem title";T
 80 PRINT#3 "              Probability analysis by smoothing over 40 classes"
 90 PRINT#3 T
100 PRINT "input a min h used with 0 or positive skew and a max h used with negative skew"
110 INPUT "minimum value of class limit of h";Hm
120 INPUT "maximum value of class limit of h";Hh
130 PRINT#3 "Minimum value of class limit of h ";Hm
140 PRINT#3 "Maximum value of class limit of h ";Hh
150 INPUT "number of data points ";N
160 FOR I = 1 TO N
170   INPUT #4 H(I)
180 NEXT I
190 CLOSE #4
200 REM - Find first three moments of data
210 Sh=0
220 Shh=0
230 Shhh=0
240 FOR I = 1 TO N
250   Sh=Sh+H(I)
260   Xh=H(I)*H(I)
270   Shh=Shh+Xh
280   Shhh=Shhh+Xh*H(I)
290 NEXT I
300 Hb=Sh/N
310 Ssd=SQR(Shh/N-Hb*Hb)
320 Sk=((Shhh/N)-(3*Hb*Shh/N)+(2*Hb*Hb*Hb))/Ssd/Ssd/Ssd
330 PRINT#3 " Mean ";Hb;" std. dev. ";Ssd;" coef. of skew ";Sk
340 REM - Test for negative or positive skew
350 IF Sk < 0. THEN GOTO 700
360 REM - Positive skew
370 Hpm=(Hm-Hb)/Ssd+Sk/8
380 C1=2.+.375*Sk
```

```
390 Vb=4.05+(C1-2.)
400 F=LOG((4-Vb)/(C1-Vb))/Hpm
410 D=F*(Vb-C1)/C1
420 V=0.
430 PRINT#3 " c ";C1;" vb ";Vb;" f ";F;" d ";D
440 REM - Calculate class limits
450 FOR I = 1 TO 39
460   V=V+.1
470   IF V>C1 THEN GOTO 500
480   Ch(I)=-Ssd*(LOG(V/C1)/D+Sk/8)+Hb
490   GOTO 510
500   Ch(I)=Ssd*(LOG((V-Vb)/(C1-Vb))/F-Sk/8)+Hb
510 NEXT I
520 Ch(40)=Hm
530 REM - Calculate sample probabilities
540 FOR J = 1 TO N
550   FOR I = 1 TO 40
560     IF H(J)=>Ch(I) THEN GOTO 580
570     GOTO 600
580     Ct(I)=Ct(I)+1
590     EXIT
600   NEXT I
610 NEXT J
620 FOR J = 2 TO 40
630   Ct(J)=Ct(J)+Ct(J-1)
640 NEXT J
650 FOR I = 1 TO 40
660   Ct(I)=Ct(I)/N
670 NEXT I
680 GOTO 1020
690 REM - Negative skew
700 Hpm=(Hh-Hb)/Ssd+Sk/8
710 C1=2.+.375*Sk
720 Vb=.05+(2.-C1)
730 D=-LOG(Vb/(C1+Vb))/Hpm
740 F=D*(C1+Vb)/(4-C1)
750 PRINT#3 " c ";C1;" vb ";Vb;" f ";F;" d ";D
760 REM - calculate class limits
770 V=0.
780 FOR I = 2 TO 40
790   V=V+.1
800   IF V > C1 THEN GOTO 830
810   Ch(I)=Ssd*(-LOG((V+Vb)/(C1+Vb))/D-Sk/8)+Hb
820   GOTO 840
830   Ch(I)=Ssd*(LOG((4-V)/(4-C1))/F-Sk/8)+Hb
840 NEXT I
850 Ch(1)=Hh
860 REM - Calculate sample probabilities
870 FOR J = 1 TO N
880   FOR I = 2 TO 40
890     IF H(J)>=Ch(I) THEN GOTO 910
900     GOTO 930
910     Ct(I)=Ct(I)+1
920     EXIT
930   NEXT I
940 NEXT J
```

```
950 FOR J = 2 TO 40
960   Ct(J)=Ct(J)+Ct(J-1)
970 NEXT J
980 FOR I = 2 TO 40
990   Ct(I)=Ct(I)/N
1000 NEXT I
1010 Ct(1)=0.
1020 PRINT#3 "       Class limits and sample probabilities"
1030 J = 1
1040 FOR I = 1 TO 40
1050   PRINT#3 USING "####";I;
1060   PRINT#3 USING "--##.####";Ch(I);Ct(I);
1070   IF J = 3 THEN GOTO 1100
1080   J = J+ 1
1090   GOTO 1120
1100   J = 1
1110   PRINT#3
1120 NEXT I
1130 PRINT#3
1140 REM - Locate seven nodes
1150 Xn(1)=1
1160 FOR I = 2 TO 7
1170   Xn(I)=Xn(I-1)+1.
1180 NEXT I
1190 REM - Calculate sliding polynomial coefficients c(i,j);
1200 REM - seven for each of 40 classes
1210 FOR Ip = 1 TO 40
1220   Jc = 1
1230   IF (Ip*0.1)<=Jc THEN GOTO 1260
1240   Jc = Jc + 1
1250   GOTO 1230
1260   Dp = (Ip*0.1)-Jc+1
1270   Z=-.5+Dp
1280   IF Sk<0. THEN Z = Z - 0.1
1290   Coef Z
1300   C(Ip,Jc) = Co(1)
1310   C(Ip,Jc+1)=Co(2)
1320   C(Ip,Jc+2)=Co(3)
1330   C(Ip,Jc+3)=Co(4)
1340 NEXT Ip
1350 IF Sk < 0. THEN GOTO 1500
1360 REM - If positive skew, put in left boundary zero and
1370 REM - zero slope.  Right boundary is free.
1380 FOR Ip = 1 TO 40
1390   C(Ip,3)=C(Ip,3)+C(Ip,1)
1400 NEXT Ip
1410 REM - Move to left and add class probabilitiy to matrix for least squares
1420 FOR Ip = 1 TO 40
1430   FOR J = 1 TO 5
1440     C(Ip,J)=C(Ip,J+2)
1450   NEXT J
1460   C(Ip,6)=Ct(Ip)
1470 NEXT Ip
1480 GOTO 1600
1490 REM - If negative skew, put in right boundary of 1. and
1500 REM - zero slope.   Left boundary is free
```

```
1510 FOR Ip = 1 TO 40
1520   C(Ip,5)=C(Ip,5)+C(Ip,7)
1530 NEXT Ip
1540 REM - Add class probabilities and node 6 = 1.0 for boundary condition.
1550 REM - Save c(1p,6) in 7
1560 FOR Ip = 1 TO 40
1570   C(Ip,7)=C(Ip,6)
1580   C(Ip,6)=Ct(Ip)-C(Ip,7)
1590 NEXT Ip
1600 REM - Calculate sums of squares matrix
1610 FOR K = 1 TO 40
1620   FOR I = 1 TO 6
1630     FOR J = I TO 6
1640       Sc(I,J)=Sc(I,J)+C(K,I)*C(K,J)
1650     NEXT J
1660   NEXT I
1670 NEXT K
1680 PRINT " sums of squares calculated"
1690 REM - Fill in complete matrix
1700 FOR I = 1 TO 5
1710   FOR J = 1 TO 5
1720     Sc(J,I)=Sc(I,J)
1730   NEXT J
1740 NEXT I
1750 REM - Call procedure to solve simultaneous equations
1760 Simeqns 5
1770 PRINT#3:PRINT#3
1780 REM - Lay in sliding polynomial through nodes
1790 PRINT#3 " Smoothed probabilities over 40 classes"
1800 CLEAR(Cp)
1810 IF Sk < 0. THEN GOTO 2000
1820 FOR I = 1 TO 40
1830   IF I> 30 THEN GOTO 1940
1840   IF I > 20 THEN GOTO 1900
1850   IF I > 10 THEN GOTO 1880
1860   Cp(I)=C(I,1)*Sc(1,6)+C(I,2)*Sc(2,6)
1870   GOTO 1970
1880   Cp(I)=C(I,1)*Sc(1,6)+C(I,2)*Sc(2,6)+C(I,3)*Sc(3,6)
1890   GOTO 1970
1900   FOR J = 1 TO 4
1910     Cp(I)=Cp(I)+C(I,J)*Sc(J,6)
1920   NEXT J
1930   GOTO 1970
1940   FOR J = 2 TO 5
1950     Cp(I)=Cp(I)+C(I,J)*Sc(J,6)
1960   NEXT J
1970 NEXT I
1980 GOTO 2120
1990 REM - For negative skew
2000 FOR I = 1 TO 40
2010   IF I > 30 THEN GOTO 2100
2020   IF I > 20 THEN GOTO 2080
2030   IF I > 10 THEN GOTO 2060
2040   Cp(I)=C(I,1)*Sc(1,6)+C(I,2)*Sc(2,6)+C(I,3)*Sc(3,6)+C(I,4)*Sc(4,6)
2050   GOTO 2110
2060   Cp(I)=C(I,2)*Sc(2,6)+C(I,3)*Sc(3,6)+C(I,4)*Sc(4,6)+C(I,5)*Sc(5,6)
```

```
2070   GOTO 2110
2080   Cp(I)=C(I,3)*Sc(3,6)+C(I,4)*Sc(4,6)+C(I,5)*Sc(5,6)+C(I,7)
2090   GOTO 2110
2100   Cp(I)=C(I,4)*Sc(4,6)+C(I,5)*Sc(5,6)+C(I,7)
2110 NEXT I
2120 J =1
2130 FOR I = 1 TO 40
2140   PRINT#3 USING "####";I;
2150   PRINT#3 USING "--###.###";Cp(I);
2160   IF J = 5 THEN GOTO 2190
2170   J = J+1
2180   GOTO 2210
2190   J = 1
2200   PRINT#3
2210 NEXT I
2220 FOR I = 1 TO 40
2230   E(I)=Cp(I)-Ct(I)
2240 NEXT I
2250 FOR J = 1 TO 5
2260   Z2s=0.
2270   Z1s=0.
2280   Z0s=0.
2290   FOR I = 1 TO 40
2300     Z2s=Z2s+E(I)*E(I)*C(I,J)*C(I,J)
2310     Z1s=Z1s+C(I,J)*C(I,J)
2320     Z0s=Z0s+C(I,J)
2330   NEXT I
2340   IF Z0s<=0 THEN GOTO 2360
2350   GOTO 2380
2360   Sd(J)=99.99
2370   GOTO 2390
2380   Sd(J)=SQR(Z2s/Z1s/Z0s*N/(N-5))
2390 NEXT J
2400 PRINT#3 " Standard deviation of nodes"
2410 FOR J = 1 TO 5
2420   PRINT#3 USING "####";J;
2430   PRINT#3 USING "--##.####";Sd(J);
2440 NEXT J
2450 PRINT#3
2460 PRINT#3 "Problem end"
2470 STOP
2480 END

ENDFILE
```

Program B-6.4

Compute and Store Sliding Polynomial Coefficients for Program B-6.5

```
SOURCE
PRECISION 12
OPTION BASE 1
PROCS=1
```

```
REAL ARRAY(4): Co,Cj,Ci
REAL ARRAY(8,8): Gc
REAL ARRAY(20): Wt
REAL ARRAY(20,20): Sx
INTEGER: I,J,K,L,M,Ic
INTEGER: Js,Is
STRING: T[100]
REAL: U,Z

PROCEDURE: Coef
  REAL ARG: Z
END PROCEDURE

PROCEDURE: Coef
  EXTERNAL: Co
    10 Co(1)=(((-8*Z+4)*Z+2)*Z-1)/16
    20 Co(2)=(((24*Z-4)*Z-22)*Z+9)/16
    30 Co(3)=(((-24*Z-4)*Z+22)*Z+9)/16
    40 Co(4)=(((8*Z+4)*Z-2)*Z-1)/16
END PROCEDURE

'MAIN Program:


  10 REM - Compute sliding ploynomial coefficients for continuous monthly
  20 REM - analysis and store on disk
  30 REM - Also compute sums of squares matrix and store on disk.
  40 PRINT "put formatted disk in drive b"
  50 INPUT "problem title ";T
  60 OPEN "b:spcoefs" FOR OUTPUT AS #5
  70 OPEN "prn" FOR OUTPUT AS #3
  80 CLEAR (Sx)
  90 Ic = 0
 100 U=-0.5
 110 FOR J = 1 TO 12
 120   Js = 1 + INT((J-1)/3)
 130   IF U > .3 THEN U =-.5
 140   Coef U
 150   FOR K = 1 TO 4:Cj(K)=Co(K):NEXT K
 160   Z=-0.5
 170   FOR I = 1 TO 41
 180     CLEAR(Gc)
 190     Ic = Ic + 1
 200     Is= 1 + INT((I-1)/10)
 210     IF Z > .45 THEN Z = -.5
 220     Coef Z
 230     FOR K = 1 TO 4:Ci(K)=Co(K):NEXT K
 240     FOR K = 1 TO 4
 250       FOR L = 1 TO 4
 260         Gc(Js+K-1,Is+L-1)=Cj(K)*Ci(L)
 270       NEXT L
 280     NEXT K
 290     REM - Set boundaries
 300     REM - Left boundary has zero node and zero slope
 310     FOR K = 1 TO 7
 320       Gc(K,3)=Gc(K,3)+Gc(K,1)
```

```
330      NEXT K
340      REM - Top and bottom boundaries wrap around for cylinder function
350      FOR K = 3 TO 7
360        Gc(2,K)=Gc(2,K)+Gc(6,K)
370        Gc(5,K)=Gc(5,K)+Gc(1,K)
380        Gc(3,K)=Gc(3,K)+Gc(7,K)
390      NEXT K
400      REM - Linearize the gc( , ) matrix;save; and take sums of squares
410      M=0
420      FOR K = 2 TO 5
430        FOR L = 3 TO 7
440          M = M + 1
450          Wt(M)=Gc(K,L)
460          PRINT#5 Wt(M)
470        NEXT L
480      NEXT K
490      FOR K = 1 TO M
500        FOR L = K TO M
510          Sx(K,L)=Sx(K,L)+Wt(K)*Wt(L)
520        NEXT L
530      NEXT K
540      Z=Z+.1
550    NEXT I
560    U=U+1/3
570 NEXT J
580 CLOSE #5
590 REM - Fill in full square matrix
600 FOR K = 1 TO M
610    FOR L = K TO M
620      Sx(L,K)=Sx(K,L)
630    NEXT L
640 NEXT K
650 REM - Save sx matrix to disk
660 OPEN "b:ssmat" FOR OUTPUT AS #5
670 FOR K = 1 TO M
680    FOR L = 1 TO M
690      PRINT#5 Sx(K,L)
700    NEXT L
710 NEXT K
720 CLOSE#5
730 STOP
740 END

ENDFILE
```

Program B-6.5

Fit Two-Dimensional Four-Point Sliding Polynomial Surface to Monthly Data

```
SOURCE
PRECISION 12
OPTION BASE 1
PROCS=1
REAL ARRAY(12): Hb,Hsd,Hsk,Hmin
```

```
REAL ARRAY(12,41): Lim,Cf
STRING: T[100]
REAL ARRAY(22,22): Sc
REAL ARRAY(20): Wt,Sy,Lu1,Lu2
REAL ARRAY(20): Lu3,Nvar
REAL ARRAY(492): Py,E
REAL ARRAY(40): Ct,Ch
INTEGER: Ny,I,J,K,L
STRING: CK$[16]
REAL: NY1,KS,SMH,HH,THH,TH,SHH,SHHH
REAL: M2,HB3,HS3,HP,V1,ITT,N,TP
REAL: TP2,IP,T1,SES,T2,SSY,RS
REAL ARRAY(12,100): Hd
REAL: Chk,Cp
REAL: Vr,F,D

PROCEDURE: Simeqns
  INTEGER ARG: N
END PROCEDURE

PROCEDURE: Simeqns
  EXTERNAL: Sc
  REAL: X,Y
  INTEGER: J,I,K,M
    10 REM - solve set of n equations
    20 FOR J = 1 TO N
    30   REM - statements 480 to 520 find the first equation with a
    40   REM - non-zero coefficient for the current column.
    50   M=0
    60   FOR I = J TO N
    70     IF Sc(I,J)<>0 THEN GOTO 90
    80     GOTO 120
    90     M=1
    100    K=I
    110    EXIT
    120  NEXT
    130  IF M= 1 THEN GOTO 180
    160  PRINT "no unique solution": STOP
    170  REM  statements 540 to 580 move that equation up to current row.
    180  I=K
    190  FOR K= 1 TO N+1
    200    X=Sc(J,K)
    210    Sc(J,K)=Sc(I,K)
    220    Sc(I,K)=X
    230  NEXT
    240  REM - statements 610 to 640 get a 1 coefficient in the first non-zero
    250  REM - column of the current row
    260  Y=1/Sc(J,J)
    270  FOR K= 1 TO N+1
    280    Sc(J,K)=Y*Sc(J,K)
    290  NEXT
    300  REM - statements 670 to 730 subtract the current equation from
    310  REM - th other rows,
    320  FOR I = 1 TO N
    330    IF I = J THEN GOTO 380
    340    Y=-Sc(I,J)
```

```
  350    FOR K = 1 TO N+1
  360      Sc(I,K)=Sc(I,K)+Y*Sc(J,K)
  370    NEXT
  380   NEXT
  390   REM - this process is repeated for all equations.
  400 NEXT
  410 PRINT #3 " solution of equations"
  420 M = 1
  430 FOR I = 1 TO N
  440   M= M+1
  450   PRINT #3 USING "#####"; I;
  460   PRINT#3 USING "--#####.####";Sc(I,N+1);
  470   IF M=>6 GOTO 500
  480   PRINT #3 "   ";
  490   GOTO 510
  500   PRINT #3: M= 1
  510 NEXT
  520 PRINT #3
END PROCEDURE
```

'MAIN Program:

```
  10 REM ##################################################
  20 REM  ######### THIS PROGRAM FITS A 2-D SLIDING POLYNOMIAL  ##########
  30 REM  ######### SURFACE TO A SAMPLE MADE UP OF ONE EVENT ###########
  40 REM  ######### PER MONTH. RECORD FOR EACH MONTH IS CONVERTED ######
  50 REM ######### TO A SAMPLE ACCUMULATED PROBABILITY CURVE  ##########
  60 REM ######### ACROSS 40 CLASSES. HEIGHT OF THE SURFACE IS #############
  70 REM  ######### PROBABILITY. DATA ARE INPUT FROM DISK       #############
  80 REM ######### SEQUENTIALLY AND CLASSIFIED IN V-SCALE BY  #############
  90 REM  ######### MONTH. DATA ARE INPUT THRU DRIVE B. SLIDING ##########
 100 REM ######### POLYNOMIAL COEFFICIENTS AND SUMS OF SQUARES ##########
 110 REM  ######### OF THESE COEFFICIENTS ARE INPUT FROM DISK IN #########
 120 REM  #########     DRIVE A.                  #####################
 130 REM ##################################################
 140 PRINT " place data disk in drive b"
 150 PRINT " place coefficient disk in drive a"
 160 OPEN "prn" FOR OUTPUT AS #3
 170 INPUT "problem title";T
 180 PRINT#3 "              Seasonally continuous probability distribution"
 190 PRINT#3 T
 200 INPUT "number of years of record";Ny
 210 OPEN "b:whtholpr" FOR INPUT AS #4
 220 INPUT"can hmin be negative, y or n";CK$
 230 REM - Input data from drive b  store as h(_,_)
 240 FOR I = 1 TO Ny
 250   FOR J = 1 TO 12
 260     INPUT#4 Hd(J,I)
 270   NEXT J
 280 NEXT I
 290 CLOSE #4
 300 NY1=Ny
 310 KS=.2
 320 PRINT#3 " month    mean     std. dev.   coef. skew";
 330 PRINT#3 "    hp       cp       vr        f        d       hmin"
 340 FOR I=1 TO 12
```

```
350 REM ######### FIND FIRST THREE MOMONTS FOR MONTH I
#############
360 PRINT#3 USING "####";I;
370 CLEAR(SMH,SHH,SHHH)
380 CLEAR(Ct)
390 Hmin(I)=99999.
400 FOR J=1 TO Ny
410   Chk=Hd(I,J)
420   IF Chk<Hmin(I) THEN Hmin(I) = Chk
430   SMH=SMH+Hd(I,J)
440   HH=Hd(I,J)*Hd(I,J)
450   THH=THH+HH
460   TH=TH+Hd(I,J)
470   SHH=SHH+HH
480   SHHH=SHHH+HH*Hd(I,J)
490 NEXT J
500 Hb(I)=SMH/NY1
510 M2=(SHH-SMH*SMH/Ny)/NY1
520 Hsd(I)=SQR(M2)
530 HB3=Hb(I)*Hb(I)*Hb(I)
540 HS3=Hsd(I)*Hsd(I)*Hsd(I)
550 Hsk(I)=((SHHH/NY1)-(3*Hb(I)*SHH)/NY1+(2*HB3))/HS3
560 REM ######### KS=0.2 SETS Hmin 20% OF 1 STD DEV BELOW #############

570 REM ######### MINIMUM DATA VALUE.                    #############
580 Hmin(I)=Hmin(I)-KS*Hsd(I)
590 REM ######### MAY BE LIMITED TO ZERO FOR MINIMUM
#############
600 IF CK$="n" THEN GOTO 610 ELSE GOTO 630
610 IF Hmin(I)<0. THEN GOTO 620 ELSE GOTO 630
620 Hmin(I)=0.
630 PRINT#3 USING "--#####.#####";Hb(I);Hsd(I);Hsk(I);
640 REM #########  SET TRANSFORM PARAMETERS       #############
650 HP=(Hmin(I)-Hb(I))/Hsd(I)+Hsk(I)/8
660 Cp=2.+0.375*Hsk(I)
670 Vr=4.05+(Cp-2.)
680 F=LOG((4.-Vr)/(Cp-Vr))/HP
690 D=F*(Vr-Cp)/Cp
700 PRINT#3 USING "--###.####";HP;Cp;Vr;F;D;Hmin(I)
710 V1=0.
720 REM #########  CALCULATE CLASS LIMITS           #############
730 FOR K=1 TO 39
740   V1=V1+.1
750   IF V1>Cp THEN GOTO 800
760   REM #########  LEFT SIDE OF TRANSFORM        #############
770   Ch(K)=-Hsd(I)*(LOG(V1/Cp)/D+Hsk(I)/8)+Hb(I)
780   GOTO 810
790   REM #########  RIGHT SIDE OF TRANSFORM       #############
800   Ch(K)=Hsd(I)*(LOG((V1-Vr)/(Cp-Vr))/F-Hsk(I)/8)+Hb(I)
810 NEXT K
820 Ch(40)=Hmin(I)
830 ITT=1
840 Cf(I,1)=0.
850 Lim(I,1)=99999.99
860 REM #########  CALCULATE SAMPLE CLASS PROBABILITIES   #############
```

```
870  FOR J=1 TO Ny
880   FOR K=ITT TO 40
890    IF Hd(I,J)>=Ch(K) THEN GOTO 910
900    GOTO 930
910    Ct(K)=Ct(K)+1
920    EXIT
930   NEXT K
940  NEXT J
950  REM
960  FOR J=2 TO 40
970   Ct(J)=Ct(J)+Ct(J-1)
980  NEXT J
990  FOR K=ITT TO 40
1000   Ct(K)=Ct(K)/Ny
1010  NEXT K
1020 REM ######### REALIGN LIMITS AND SAMPLE PROBS INTO 41 CLASSES  #######

1030  FOR N=1 TO 40
1040   Cf(I,N-ITT+2)=Ct(N)
1050   Lim(I,N-ITT+2)=Ch(N)
1060  NEXT N
1070  FOR N=1 TO 41
1080   TP=TP+Cf(I,N)
1090   TP2=TP2+Cf(I,N)*Cf(I,N)
1100  NEXT N
1110 NEXT I
1120 REM  #########        END OF MONTH LOOP                 #############
1130 PRINT#3"    SAMPLE  CLASS PROBABILITIES"
1140 PRINT#3 "     ";
1150 FOR I=1 TO 12
1160  PRINT#3 USING "########";I;
1170 NEXT I
1180 PRINT#3
1190 FOR I=1 TO 41
1200  PRINT#3 USING "#####";I-1;
1210  FOR J=1 TO 12
1220   PRINT#3 USING "--####.###";Cf(J,I);
1230  NEXT J
1240  PRINT#3
1250 NEXT I
1260 PRINT#3
1270 PRINT#3"      CLASS LIMITS"
1280 PRINT#3 "     ";
1290 FOR I=1 TO 12
1300  PRINT#3 USING "#########";I;
1310 NEXT I
1320 PRINT#3
1330 FOR I=1 TO 41
1340  PRINT#3 USING "####";I-1;
1350  FOR J=1 TO 12
1360   IF Lim(J,I)=99999.99 THEN GOTO 1370 ELSE GOTO 1390
1370   PRINT#3"    ++++++";
1380   GOTO 1400
1390   PRINT#3 USING "--#####.###";Lim(J,I);
1400  NEXT J
1410  PRINT#3
```

```
1420 NEXT I
1430 PRINT#3
1440 CLOSE#5
1450 REM #########  READ IN WEIGHTS A DATA POINT AT A TIME    #############

1460 REM #########  ACCUMULATE XY VECTOR.                  #############
1470 OPEN "a:spcoefs" FOR INPUT AS #5
1480 FOR I=1 TO 12
1490   PRINT I;
1500   FOR J=1 TO 41
1510     FOR L=1 TO 20
1520       INPUT#5 Wt(L)
1530       Sy(L)=Sy(L)+Wt(L)*Cf(I,J)
1540     NEXT L
1550   NEXT J
1560 NEXT I
1570 CLOSE#5
1580 REM - Read in sums of squares matrix
1590 OPEN "a:ssmat" FOR INPUT AS #6
1600 FOR K = 1 TO 20
1610   FOR L = 1 TO 20
1620     INPUT #6 Sc(K,L)
1630   NEXT L
1640 NEXT K
1650 CLOSE #6
1660 REM - Add probability-product vector to sums of squares matrix
1670 FOR K = 1 TO 20
1680   Sc(K,21)=Sy(K)
1690 NEXT K
1700 Simeqns 20
1710 REM - Lay in sliding polynomial surface through solution nodes
1720 OPEN "a:spcoefs" FOR INPUT AS #5
1730 IP=0
1740 CLEAR(Lu1,Lu2,Lu3)
1750 FOR I = 1 TO 12
1760   FOR J = 1 TO 41
1770     IP = IP + 1
1780     FOR L = 1 TO 20
1790       INPUT#5 Wt(L)
1800       Py(IP)=Py(IP)+Wt(L)*Sc(L,21)
1810     NEXT L
1820     REM - Calculate residuals
1830     E(IP)=Cf(I,J)-Py(IP)
1840     T1=E(IP)*E(IP)
1850     SES=SES+T1
1860     FOR L = 1 TO 20
1870       T2=Wt(L)*Wt(L)
1880       Lu2(L)=Lu2(L)+T1*T2
1890       Lu1(L)=Lu1(L)+T2
1900       Lu3(L) = Lu3(L) + Wt(L)
1910     NEXT L
1920   NEXT J
1930 NEXT I
1940 CLOSE#5
1950 PRINT#3
1960 PRINT#3"                Smoothed probability values"
```

```
1970 PRINT#3 "     ";
1980 FOR I = 1 TO 12
1990   PRINT#3 USING "##########";I;
2000 NEXT I
2010 PRINT#3
2020 FOR J = 1 TO 41
2030   PRINT#3 USING "####";J-1;
2040   FOR I = J TO (451 + J) STEP 41
2050     PRINT#3 USING "--####.####";Py(I);
2060   NEXT I
2070   PRINT#3
2080 NEXT J
2090 PRINT#3 "                          Residual errors"
2100 PRINT#3 "     ";
2110 FOR J = 1 TO 12
2120   PRINT#3 USING "##########";J;
2130 NEXT J
2140 PRINT#3
2150 FOR I = 1 TO 41
2160   PRINT#3 USING "####";I-1;
2170   FOR J = I TO (451+I) STEP 41
2180     PRINT#3 USING "--####.####";E(J);
2190   NEXT J
2200   PRINT#3
2210 NEXT I
2220 FOR L = 1 TO 20
2230   IF Lu3(L)<0. THEN GOTO 2260
2240   Nvar(L)=Lu2(L)/Lu1(L)*Ny*12/(Ny*12-20)/Lu3(L)
2250   GOTO 2270
2260   Nvar(L)=9999.999
2270 NEXT L
2280 PRINT#3"    Nodal variance    Standard deviation      Sum of weights"
2290 FOR J = 1 TO 20
2300   PRINT#3 USING "####";J;
2310   PRINT#3 USING "--#########.####";Nvar(J);SQR(Nvar(J));Lu3(J)
2320 NEXT J
2330 SSY=TP2-TP*TP/492
2340 PRINT#3" Total adjusted sum of squares   "; SSY
2350 PRINT#3" Total residual error squared  ";SES
2360 PRINT#3" Mean residual error ";SQR(SES/492)
2370 RS=(SSY-SES)/SSY
2380 PRINT#3 "Synthesized correlation coefficient  ";SQR(RS)
2390 PRINT#3
2400 CLOSE#3
2410 STOP
2420 END

ENDFILE
```

Program B-6.6

Test Two Integrals For Significant Difference Through Simulation

```
SOURCE
PRECISION 12
OPTION BASE 1
PROCS=1
REAL ARRAY(200): Aa,Ab,Da,Ada
STRING: T[100]
INTEGER: Ns,I,K,L
REAL: Y1,Sd1
REAL: Y2,Sd2,Y3,Sd3,Y4,Sd4,Y5,Sd5
REAL: Y6,Sd6,Y7,Sd7,Y8,Sd8,Rn,Dxa
REAL: Dxb,Y1r,Y2r,Y3r,Y4r,Y5r,Y6r,Y7r
REAL: Y8r,Yab,Yaf,Ybb,Ybf,Suma,Sumb,Sumd
REAL: Sum2a,Sum2b,Sum2d,Ava,Avb,Avd,Sda,Sdb
REAL: Sdd,X

REAL FUNCTION: Rannorm
END FUNCTION

REAL FUNCTION: Rannorm
  REAL: Sum
  INTEGER: J
    10 Sum=0
    20 FOR J = 1 TO 12
    30   Sum = Sum + RND(1)
    40 NEXT J
    50 RESULT = Sum -6.
END FUNCTION

'MAIN Program:


  10 REM - Test the difference between the integrals of two functions.
  20 REM - Each function is approximated by three spans of the
  30 REM - four-point sliding polynomials.
  40 REM - Testing is by random number simulation.
  50 PRINT"input four nodes and standard deviations of function 'a'"
  60 INPUT "1";Y1,Sd1
  70 INPUT "2";Y2,Sd2
  80 INPUT "3";Y3,Sd3
  90 INPUT "4";Y4,Sd4
 100 PRINT"input four nodes and standard deviations of function 'b'"
 110 INPUT "1";Y5,Sd5
 120 INPUT "2";Y6,Sd6
 130 INPUT "3";Y7,Sd7
 140 INPUT "4";Y8,Sd8
 150 INPUT "random number seed";Rn
 160 RANDOMIZE Rn
 170 INPUT "problem title";T
 180 OPEN "prn" FOR OUTPUT AS #3
 190 PRINT#3 "Test the integrals of two functions for significant difference"
```

```
200 PRINT#3 T
210 PRINT#3 " The input nodes and standard deviations"
220 PRINT#3 "Function a";:PRINT#3 USING "--###.####";Y1;Sd1;Y2;Sd2;Y3;Sd3;Y4;Sd4
230 PRINT#3 "Function b";:PRINT#3 USING "--###.####";Y5;Sd5;Y6;Sd6;Y7;Sd7;Y8;Sd8
240 PRINT#3:PRINT#3
250 INPUT "number of simulations";Ns
260 PRINT#3 "Number of simulations  ";Ns:PRINT#3
270 INPUT "distance between nodes for function a";Dxa
280 INPUT "distance between nodes for function b";Dxb
290 REM - Loop through 'ns' simulations
300 FOR I = 1 TO Ns
310   Y1r=Sd1*Rannorm +Y1
320   Y2r=Sd2*Rannorm +Y2
330   Y3r=Sd3*Rannorm +Y3
340   Y4r=Sd4*Rannorm +Y4
350   Y5r=Sd5*Rannorm +Y5
360   Y6r=Sd6*Rannorm +Y6
370   Y7r=Sd7*Rannorm +Y7
380   Y8r=Sd8*Rannorm +Y8
390   REM - Comput left and right outlier nodes by parabolic extrapolation.
400   Yab=3*Y1r-3*Y2r+Y3r
410   Yaf=3*Y4r-3*Y3r+Y2r
420   Ybb=3*Y5r-3*Y6r+Y7r
430   Ybf=3*Y8r-3*Y7r+Y6r
440   Aa(I)=-0.041667*(Yab+Y1r+Y2r+Y3r+Y4r+Yaf)
450   Aa(I)=(Aa(I)+.541667*(Y1r+Y4r+2*(Y2r+Y3r)))*Dxa
460   Ab(I)=-0.041667*(Ybb=Y5r=Y6r+Y7r+Y8r+Ybf)
470   Ab(I)=(Ab(I)+.541667*(Y5r+Y8r+2*(Y6r+Y7r)))*Dxb
480   Da(I)=Aa(I)-Ab(I)
490   PRINT I;
500 NEXT I
510 PRINT#3 "Simulated integrals of function a"
515 K=0
520 FOR I = 1 TO Ns
525   K = K+1
530   PRINT#3 USING "####";I;
540   PRINT#3 USING "--###.###";Aa(I);
550   IF K = 10 THEN PRINT#3:K=0
560 NEXT I
570 PRINT#3:PRINT#3:PRINT#3 "Simulated integrals of function b"
575 K = 0
580 FOR I = 1 TO Ns
590   K = K+1
600   PRINT#3 USING "####";I;
610   PRINT#3 USING "--###.###";Ab(I);
620   IF K = 10 THEN PRINT#3:K=0
630 NEXT I
640 PRINT#3:PRINT#3:PRINT#3"Integral of function a minus integral of function  b"
650 K=0
660 FOR I = 1 TO Ns
670   K = K+1
680   PRINT#3 USING "####";I;
690   PRINT#3 USING "--###.###";Da(I);
700   IF K = 10 THEN PRINT#3:K=0
710 NEXT I
720 PRINT#3:PRINT#3
```

```
730 REM - Compute means and standard deviations of simulated functions and differences
735 CLEAR(Suma,Sumb,Sumd,Sum2a,Sum2b,Sum2d)
740 FOR I = 1 TO Ns
750   Suma=Suma+Aa(I)
760   Sumb=Sumb+Ab(I)
770   Sumd=Sumd+Da(I)
780   Sum2a=Sum2a+Aa(I)*Aa(I)
790   Sum2b=Sum2b+Ab(I)*Ab(I)
800   Sum2d=Sum2d+Da(I)*Da(I)
810 NEXT I
820 Ava=Suma/Ns
830 Avb=Sumb/Ns
840 Avd=Sumd/Ns
850 Sda=SQR((Sum2a-Suma*Suma/Ns)/(Ns-1))
860 Sdb=SQR((Sum2b-Sumb*Sumb/Ns)/(Ns-1))
870 Sdd=SQR((Sum2d-Sumd*Sumd/Ns)/(Ns-1))
880 PRINT#3 "Mean of simulations:-  function a - function b - difference"
890 PRINT#3 USING "--#####.###";Ava;Avb;Avd
900 PRINT#3 "Standard deviations of simulation"
910 PRINT#3 USING "--#####.###";Sda;Sdb;Sdd
920 REM - Arrange differences in order of magnitude and print
930 Ada(1)=Da(1)
940 FOR I = 2 TO Ns
950   X=Da(I)
960   FOR K = 1 TO I-1
970     IF X>Ada(K) THEN EXIT
980   NEXT K
990   FOR L = I-1 TO K STEP -1
1000     Ada(L+1)=Ada(L)
1010   NEXT L
1020   Ada(K)=X
1030 NEXT I
1040 PRINT#3:PRINT#3:PRINT#3 "Integral differences in order of magnitude"
1050 K=0
1060 FOR I = 1 TO Ns
1070   K = K+1
1080   PRINT#3 USING "####";I;
1090   PRINT#3 USING "--###.###";Ada(I);
1100   IF K = 10 THEN PRINT#3:K=0
1110 NEXT I
1120 PRINT#3 :PRINT#3 "End of problem"
1130 STOP
1140 END

ENDFILE
```

Program B-6.7

Simulate A Distribution of Random Slopes

```
SOURCE
PRECISION 12
OPTION BASE 1
PROCS=2
```

```
REAL ARRAY(200): S
REAL ARRAY(4): C,Y,Sd
INTEGER: I,J,K,N
REAL: Sum,Se,Dx
REAL: Z
STRING: T[100]
REAL ARRAY(4): Yr
REAL: Ss,S2,Avs,Ssd

PROCEDURE: Coef
  REAL ARG: Z
END PROCEDURE

PROCEDURE: Rannor
END PROCEDURE

PROCEDURE: Coef
  EXTERNAL: C
    10 C(1)=((-12*Z+4)*Z+1)/8
    20 C(2)=((36*Z-4)*Z-11)/8
    30 C(3)=((-36*Z-4)*Z+11)/8
    40 C(4)=((12*Z+4)*Z-1)/8
END PROCEDURE

PROCEDURE: Rannor
  EXTERNAL: Sum
  INTEGER: J
    10 Sum=0.0
    20 FOR J = 1 TO 12
    30   Sum=Sum+RND(1)
    40 NEXT J
    50 Sum=Sum-6.
END PROCEDURE

'MAIN Program:


  10 REM - Generate a distribution of random slopes at any point in the
  20 REM - central span of four nodes Y(i) with standard deviations sd(i).
  30 REM - Uses the four-point sliding polynomial
  40 OPEN "prn" FOR OUTPUT AS #3
  50 INPUT "problem title";T
  60 PRINT#3  T
  70 PRINT "input four nodes"
  80 FOR I = 1 TO 4
  90   PRINT I;" ";:INPUT Y(I)
 100 NEXT I
 110 PRINT "input the four standard deviations"
 120 FOR I = 1 TO 4
 130   PRINT I;" ";:INPUT Sd(I)
 140 NEXT I
 150 INPUT "random number seed  ";Se
 160 RANDOMIZE Se
 170 INPUT "number of slopes to simulate ";N
 180 INPUT " distance between nodes ";Dx
 190 PRINT "input z-value in central span where slope is to be calculated"
```

```
200 PRINT " use range -0.5< =z< =0.5"
210 INPUT "z ";Z
220 PRINT#3  " The input nodes "
230 FOR I = 1 TO 4:PRINT#3 Y(I);:NEXT I:PRINT#3
240 PRINT#3 "Their standard deviations"
250 FOR I = 1 TO 4:PRINT#3 Sd(I);:NEXT I:PRINT#3
260 PRINT#3 "Dist. between nodes  ";Dx;"    Value of z  ";Z
270 Coef Z
280 CLEAR(S,Ss,S2)
290 FOR J= 1 TO N
300   FOR I = 1 TO 4
310     Rannor
320     Yr(I)=Y(I)+Sd(I)*Sum
330   NEXT I
340   FOR I= 1 TO 4
350     S(J)= S(J)+C(I)*Yr(I)
360   NEXT I
370   S(J)=S(J)/Dx
380   Ss=Ss+S(J)
390   S2=S2+S(J)*S(J)
400 NEXT J
410 PRINT#3:PRINT#3 " The generated slopes"
420 K=0
430 FOR J = 1 TO N
440   K = K+1
450   PRINT#3 USING "####";J;
460   PRINT#3 USING "--##.####";S(J);
470   IF K = 10 THEN K=0:PRINT#3
480 NEXT J
490 Avs=Ss/N
500 Ssd=SQR((S2-Ss/N*Ss)/(N-1))
510 PRINT#3 "Mean of generated slopes ";:PRINT#3 USING "--####.####";Avs;
520 PRINT#3 "          Standard deviation of generated slopes ";
530 PRINT#3 USING "####.####";Ssd
540 PRINT#3
550 CLOSE#3
560 STOP
570 END

ENDFILE
```

Program B-7.1

Solution of y dy/dx = a^2x exp(-2bx)(1-bx) by Nonlinear Least Squares

```
SOURCE
PRECISION 12
OPTION BASE 1
PROCS=5
STRING: Title[60]
REAL ARRAY(50): R,Rhat,Err,Predy
```

```
REAL ARRAY(20): Inode
REAL ARRAY(6): Co,Dco
REAL ARRAY(20,20): Sc
REAL ARRAY(20,50): Dd1
REAL: Xl,Xr,Dx,A
REAL: B,X,Sy,Delx
INTEGER: I,J,Nr,Nrc
REAL: Np,K
REAL ARRAY(20): Sinode,Ddnode

PROCEDURE: Sixpoint
  REAL ARG: Z
END PROCEDURE

PROCEDURE: Tdsixpoint
END PROCEDURE

PROCEDURE: Dsixpoint
  REAL ARG: Z
END PROCEDURE

PROCEDURE: Fhat
END PROCEDURE

PROCEDURE: Simeqns
  INTEGER ARG: N
END PROCEDURE

PROCEDURE: Sixpoint
  EXTERNAL: Co
    10 Co(1)=(((((-160*Z+16)*Z+144)*Z-40)*Z-26)*Z+9)/768
    20 Co(2)=(((((800*Z-48)*Z-848)*Z+312)*Z+162)*Z-75)/768
    30 Co(3)=(((((-1600*Z+32)*Z+1824)*Z-272)*Z-1124)*Z+450)/768
    40 Co(4)=(((((1600*Z+32)*Z-1824)*Z-272)*Z+1124)*Z+450)/768
    50 Co(5)=(((((-800*Z-48)*Z+848)*Z+312)*Z-162)*Z-75)/768
    60 Co(6)=(((((160*Z+16)*Z-144)*Z-40)*Z+26)*Z+9)/768
END PROCEDURE

PROCEDURE: Tdsixpoint
END PROCEDURE

PROCEDURE: Dsixpoint
  EXTERNAL: Dco
    10 Dco(1)=((((-400*Z+32)*Z+216)*Z-40)*Z-13)/384
    20 Dco(2)=((((2000*Z-96)*Z-1272)*Z+312)*Z+81)/384
    30 Dco(3)=((((-4000*Z+64)*Z+2736)*Z-272)*Z-562)/384
    40 Dco(4)=((((4000*Z+64)*Z-2736)*Z-272)*Z+562)/384
    50 Dco(5)=((((-2000*Z-96)*Z+1272)*Z+312)*Z-81)/384
    60 Dco(6)=((((400*Z+32)*Z-216)*Z-40)*Z+13)/384
END PROCEDURE

PROCEDURE: Fhat
  INTEGER: J,I,K,L
  EXTERNAL: Co,Dco,Inode
  REAL: Z,Y,Dy
  EXTERNAL: Sixpoint,Dsixpoint,Predy,Delx
```

```
  440 K=1
  450 FOR I= 1 TO 12
  460   Z=-0.5
  470   FOR J=1 TO 4
  480     K=K+1
  490     Z=Z+0.25
  520     Y=0:Dy=0.
  522     Sixpoint Z
  524     Dsixpoint Z
  530     FOR L= 1 TO 6
  540       Y=Y+Co(L)*Inode(I+L-1)
  550       Dy=Dy+Dco(L)*Inode(I+L-1)
  560     NEXT
  565     Dy=Dy/Delx
  570     Predy(K)=Y*Dy
  590   NEXT
  600 NEXT
END PROCEDURE

PROCEDURE: Simeqns
 EXTERNAL: Sc
 REAL: X,Y
 INTEGER: J,I,K,M
  440 REM - solve set of n equations
  450 FOR J = 1 TO N
  460   REM - statements 480 to 520 find the first equation with a
  470   REM - non-zero coefficient for the current column.
  475   M=0
  480   FOR I = J TO N
  490     IF Sc(I,J)<>0 THEN GOTO 500
  495     GOTO 514
  500     M=1
  505     K=I
  510     EXIT
  514   NEXT
  518   IF M=1 THEN GOTO 535
  520   PRINT "no unique solution": STOP
  530   REM  statements 540 to 580 move that equation up to current row.
  535   I=K
  540   FOR K= 1 TO N+1
  550     X=Sc(J,K)
  560     Sc(J,K)=Sc(I,K)
  570     Sc(I,K)=X
  580   NEXT
  590   REM - statements 610 to 640 get a 1 coefficient in the first non-zero
  600   REM - column of the current row
  610   Y=1/Sc(J,J)
  620   FOR K= 1 TO N+1
  630     Sc(J,K)=Y*Sc(J,K)
  640   NEXT
  650   REM - statements 670 to 730 subtract the current equation from
  660   REM - th other rows,
  670   FOR I = 1 TO N
  680     IF I = J THEN GOTO 730
  690     Y=-Sc(I,J)
  700     FOR K = 1 TO N+1
```

```
710       Sc(I,K)=Sc(I,K)+Y*Sc(J,K)
720      NEXT
730    NEXT
740    REM - this process is repeated for all equations.
750 NEXT
760 PRINT #3 " solution of equations"
770 M = 1
780 FOR I = 1 TO N
785    M= M+1
786    PRINT #3 I;"  ";Sc(I,N+1);
800    IF M=>6 GOTO 820
805    PRINT #3 "   ";
810    GOTO 830
820    PRINT #3: M= 1
830 NEXT
840 PRINT #3
END PROCEDURE
```

'MAIN Program:

```
 10 REM - Solution of non-linear differential equation by non-linear
 20 REM - least squares. Uses six-point sliding polynomial.
 30 REM - Equation: ydy/dx = a*a*x*exp(-2*b*x)*(1-b*x)
 40 REM - 17 nodes are input as initial values. No. 3 is fixed
 50 REM - for boundary condition. Remaining 16 are "at solution"..
 60 OPEN "prn" FOR OUTPUT AS #3
 70 INPUT "title";Title
 80 PRINT#3 Title
 90 INPUT "value of x at node 3";Xl
100 INPUT "value of x at node 15";Xr
110 PRINT#3 "Left boundary in x  ";Xl;"  ";"Right boundary  ";Xr
120 INPUT "print divided difference? yes(1) or no(2)";Np
130 INPUT "number of rounds to try?";Nr
140 Nrc=0
150 Dx=(Xr-Xl)/48
160 Delx=(Xr-Xl)/12
170 PRINT#3 "           Input nodes"
180 PRINT "input 17 nodes"
190 FOR I = 1 TO 17
200    PRINT I;:INPUT Inode(I)
210    PRINT#3 I;Inode(I);"     ";
220 NEXT
230 PRINT#3
240 INPUT " constant 'a'";A
250 INPUT " constant 'b'";B
260 PRINT#3 "a  ";A;"       b  ";B
270 REM - Compute 'r'
280 FOR I = 1 TO 49
290    X=Xl+(I-1)*Dx
300    R(I)=A*A*X*EXP(-2*B*X)*(1-B*X)
310 NEXT
320 REM - Start iteration
330 REM - Compute rhat and err
340 REM - at left boundary
350 Nrc=Nrc+1
```

```
360 Dsixpoint -0.5
370 Sy=0.
380 FOR I=1 TO 6
390   Sy=Sy+Dco(I)*Inode(I)
400 NEXT
410 Sy=Sy/Delx
420 Rhat(1)=Inode(3)*Sy
430 Err(1)=R(1)-Rhat(1)
440 Fhat
450 FOR I = 2 TO 49
460   Rhat(I)=Predy(I)
470   Err(I)=R(I)-Rhat(I)
480 NEXT
490 PRINT#3 "Match-point no.,r,rhat, and error"
500 FOR K = 1 TO 49 STEP 3
510   FOR I = K TO K+2
520     IF I>49 THEN EXIT
530     PRINT#3 I;
540     PRINT#3 USING "--####.###";R(I);Rhat(I);Err(I);
550     PRINT#3 "    ";
560   NEXT
570   PRINT#3
580 NEXT
590 PRINT#3
600 REM - Compute the divided differences
610 K=0
620 FOR I = 1 TO 17
630   IF I= 3  THEN GOTO 840
640   K=K+1
650   Sinode(I)=Inode(I)
660   Inode(I)=Inode(I)*1.02
670   Ddnode(I)=Inode(I)-Sinode(I)
680   IF ABS(Ddnode(I))<.001 THEN GOTO 690 ELSE GOTO 710
690   Ddnode(I)=.001
700   Inode(I)=.001+Sinode(I)
710   Dsixpoint -0.5
720   Sy=0.
730   FOR J=1 TO 6
740     Sy=Sy+Dco(J)*Inode(J)
750   NEXT
760   Sy=Sy/Delx
770   Predy(1)=Inode(3)*Sy
780   Dd1(K,1)=(Predy(1)-Rhat(1))/Ddnode(I)
790   Fhat
800   FOR J= 2 TO 49
810     Dd1(K,J)=(Predy(J)-Rhat(J))/Ddnode(I)
820   NEXT
830   Inode(I)=Sinode(I)
840 NEXT
850 PRINT#3
860 ON Np GOTO 870,1020
870 PRINT#3 "            Divided differences"
880 FOR I= 1 TO 49
890   PRINT#3 "Match-point number  ";I
900   FOR J=1 TO 8
910     PRINT#3 USING "####";J;
```

```
920     PRINT#3 USING "--###.####";Dd1(J,I);
930   NEXT
940   PRINT#3
950   FOR J= 9 TO 16
960     PRINT#3 USING "####";J;
970     PRINT#3 USING "--###.####";Dd1(J,I);
980   NEXT
990   PRINT#3
1000 NEXT
1010 REM - Compute sums of squares matrix
1020 CLEAR (Sc)
1030 FOR I= 1 TO 49
1040   Dd1(17,I)=Err(I)
1050 NEXT
1060 FOR I=1 TO 49
1070   FOR J=1 TO 17
1080     IF Dd1(J,I)=0 THEN GOTO 1130
1090     FOR K=J TO 17
1100       IF Dd1(K,I)=0 THEN GOTO 1120
1110       Sc(J,K)=Sc(J,K)+Dd1(J,I)*Dd1(K,I)
1120     NEXT
1130   NEXT
1140 NEXT
1150 FOR J=1 TO 17
1160   FOR I= J TO 17
1170     Sc(I,J)=Sc(J,I)
1180   NEXT
1190 NEXT
1200 REM - Solve simultaneous equations
1210 J=16
1220 Simeqns J
1230 REM - Update the nodes
1240 K=0
1250 FOR I = 1 TO 17
1260   IF I=3 THEN GOTO 1290
1270   K=K+1
1280   Inode(I)=Inode(I)+Sc(K,17)
1290 NEXT
1300 PRINT#3
1310 PRINT#3 "    Updated set of nodes after round  ";Nrc
1320 FOR J= 1 TO 9
1330   PRINT#3 USING "####";J;
1340   PRINT#3 USING "--###.####";Inode(J);
1350 NEXT
1360 PRINT#3
1370 FOR J= 10 TO 17
1380   PRINT#3 USING "####";J;
1390   PRINT#3 USING "--###.####";Inode(J);
1400 NEXT
1410 PRINT#3:PRINT#3:PRINT#3
1420 IF Nrc>=Nr THEN GOTO 1450
1430 REM - End of iteration
1440 GOTO 350
```

```
1450 PRINT#3
1460 CLOSE#3
1470 STOP
```

ENDFILE

Program B-7.2

Solution of dy/dx + *y* f(y) = a(x) by Nonlinear Least Squares

```
SOURCE
PRECISION 12
OPTION BASE 1
PROCS=5
REAL ARRAY(20): Inode
REAL ARRAY(25): Fnode
REAL ARRAY(6): Co,Dco
REAL ARRAY(20,20): Sc
REAL ARRAY(50): Predy
REAL: Delx,Funy,Xl,Xr,Np,Nr,Nrc
REAL: Dx,I,J,Sy,Y,K
REAL ARRAY(50): R,Rhat,Err
REAL ARRAY(20): Sinode,Ddnode
REAL ARRAY(20,50): Dd1
STRING: Titl[60]

PROCEDURE: Sixpoint
  REAL ARG: Z
END PROCEDURE

PROCEDURE: Dsixpoint
  REAL ARG: Z
END PROCEDURE

PROCEDURE: Fhat
END PROCEDURE

PROCEDURE: Funtny
  REAL ARG: Y
END PROCEDURE

PROCEDURE: Simeqns
  INTEGER ARG: N
END PROCEDURE

PROCEDURE: Sixpoint
  EXTERNAL: Co
    10 Co(1)=(((((-160*Z+16)*Z+144)*Z-40)*Z-26)*Z+9)/768
    20 Co(2)=(((((800*Z-48)*Z-848)*Z+312)*Z+162)*Z-75)/768
    30 Co(3)=(((((-1600*Z+32)*Z+1824)*Z-272)*Z-1124)*Z+450)/768
    40 Co(4)=(((((1600*Z+32)*Z-1824)*Z-272)*Z+1124)*Z+450)/768
```

```
    50 Co(5)=(((((-800*Z-48)*Z+848)*Z+312)*Z-162)*Z-75)/768
    60 Co(6)=(((((160*Z+16)*Z-144)*Z-40)*Z+26)*Z+9)/768
END PROCEDURE

PROCEDURE: Dsixpoint
  EXTERNAL: Dco
    10 Dco(1)=((((-400*Z+32)*Z+216)*Z-40)*Z-13)/384
    20 Dco(2)=((((2000*Z-96)*Z-1272)*Z+312)*Z+81)/384
    30 Dco(3)=((((-4000*Z+64)*Z+2736)*Z-272)*Z-562)/384
    40 Dco(4)=((((4000*Z+64)*Z-2736)*Z-272)*Z+562)/384
    50 Dco(5)=((((-2000*Z-96)*Z+1272)*Z+312)*Z-81)/384
    60 Dco(6)=((((400*Z+32)*Z-216)*Z-40)*Z+13)/384
END PROCEDURE

PROCEDURE: Fhat
  INTEGER: J,I,K,L
  EXTERNAL: Co,Dco,Inode
  REAL: Z,Y,Dy
  EXTERNAL: Sixpoint,Dsixpoint,Predy,Delx,Funtny,Funy
    10 K=1
    20 FOR I= 1 TO 12
    30   Z=-0.5
    40   FOR J=1 TO 4
    50     K=K+1
    60     Z=Z+0.25
    70     Y=0:Dy=0.
    80     Sixpoint Z
    90     Dsixpoint Z
   100     FOR L= 1 TO 6
   110       Y=Y+Co(L)*Inode(I+L-1)
   120       Dy=Dy+Dco(L)*Inode(I+L-1)
   130     NEXT
   140     Funtny Y
   150     Dy=Dy/Delx
   160     Predy(K)=Dy+Funy*Y
   170   NEXT
   180 NEXT
END PROCEDURE

PROCEDURE: Funtny
  EXTERNAL: Funy,Fnode,Co
  INTEGER: I,J
  REAL: Z,Ty
  EXTERNAL: Sixpoint
     5 Funy= 0.
    10 I=1:Ty=1.
    20 IF Y>Ty THEN GOTO 40
    30 GOTO 60
    40 I=I+1:Ty=Ty+1.
    50 GOTO 20
    60 Z=Y-(Ty-1.)-.5
    70 Sixpoint Z
    80 FOR J=1 TO 6
    90   Funy=Funy+Fnode(I+J-1)*Co(J)
   100 NEXT
END PROCEDURE
```

```
PROCEDURE: Simeqns
  EXTERNAL: Sc
  REAL: X,Y
  INTEGER: J,I,K,M
   440 REM - solve set of n equations
   450 FOR J = 1 TO N
   460   REM - statements 480 to 520 find the first equation with a
   470   REM - non-zero coefficient for the current column.
   475   M=0
   480   FOR I = J TO N
   490     IF Sc(I,J)<>0 THEN GOTO 500
   495     GOTO 514
   500     M=1
   505     K=I
   510     EXIT
   514   NEXT
   518   IF M=1 THEN GOTO 535
   520   PRINT "no unique solution": STOP
   530   REM  statements 540 to 580 move that equation up to current row.
   535   I=K
   540   FOR K= 1 TO N+1
   550     X=Sc(J,K)
   560     Sc(J,K)=Sc(I,K)
   570     Sc(I,K)=X
   580   NEXT
   590   REM - statements 610 to 640 get a 1 coefficient in the first non-zero
   600   REM - column of the current row
   610   Y=1/Sc(J,J)
   620   FOR K= 1 TO N+1
   630     Sc(J,K)=Y*Sc(J,K)
   640   NEXT
   650   REM - statements 670 to 730 subtract the current equation from
   660   REM - th other rows,
   670   FOR I = 1 TO N
   680     IF I = J THEN GOTO 730
   690     Y=-Sc(I,J)
   700     FOR K = 1 TO N+1
   710       Sc(I,K)=Sc(I,K)+Y*Sc(J,K)
   720     NEXT
   730   NEXT
   740   REM - this process is repeated for all equations.
   750 NEXT
   760 PRINT #3 " solution of equations"
   770 M = 1
   780 FOR I = 1 TO N
   785   M= M+1
   786   PRINT #3 I;"  ";Sc(I,N+1);
   800   IF M=>6 GOTO 820
   805   PRINT #3 "   ";
   810   GOTO 830
   820   PRINT #3: M= 1
   830 NEXT
   840 PRINT #3
END PROCEDURE
```

'MAIN Program:

```
10 REM - Solution of non-linear differential equation by non-linear
20 REM - least squares. Uses six-point sliding polynomial.
30 REM - Equation:dy/dx+f(y)*y=a(x)
40 REM - y in this version is restricted to 0=<y=<20.
50 REM - f(y) is empirical function input as set of nodes
60 REM - a(x) is discrete function of x. 49 values are input. Can be zero.
70 REM - 17 nodes are input as initial values. No. 3 is fixed
80 REM - for boundary condition. Remaining 16 are "at solution"..
90 OPEN "prn" FOR OUTPUT AS #3
100 INPUT "title";Titl
110 PRINT#3 Titl
120 INPUT "value of x at node 3";Xl
130 INPUT "value of x at node 15";Xr
140 PRINT#3 "Left boundary in x  ";Xl;"  ";"Right boundary  ";Xr
150 INPUT "print divided difference? yes(1) or no(2)";Np
160 INPUT "number of rounds to try?";Nr
170 Nrc=0
180 Dx=(Xr-Xl)/48
190 Delx=(Xr-Xl)/12
200 PRINT#3 "          Input nodes"
210 PRINT "input 17 nodes"
220 FOR I = 1 TO 17
230   PRINT I;:INPUT Inode(I)
240   PRINT#3 I;Inode(I);"   ";
250 NEXT
260 PRINT#3
270 PRINT "input 25 'fnodes' for f(y)"
280 PRINT#3"   Function f(y) nodes"
290 J=0
300 FOR I = 1 TO 25
310   J=J+1
320   PRINT I;:INPUT Fnode(I)
330   PRINT#3 I;Fnode(I);
340   IF J=8 THEN GOTO 370
350   PRINT#3 "  ";
360   GOTO 380
370   J=0:PRINT#3
380 NEXT
390 PRINT#3
400 PRINT "input 49 values of a(x) into space r(i)"
410 PRINT#3 "Discrete function a(x)"
420 J=0
430 FOR I=1 TO 49
440   J=J+1
450   PRINT I;:INPUT R(I)
460   PRINT#3 I;R(I);
470   IF J= 10 THEN GOTO 500
480   PRINT#3 "  ";
490   GOTO 510
500   J=0:PRINT#3
510 NEXT
520 PRINT#3
530 REM - Begin iteration
```

```
540 REM - Compute rhat and err
550 REM -    - -  at left boundary - - for each round.
560 Nrc=Nrc+1
570 Dsixpoint -0.5
580 Sy=0.
590 FOR I=1 TO 6
600   Sy=Sy+Dco(I)*Inode(I)
610 NEXT
620 Y=Inode(3)
630 Funtny Y
640 Sy=Sy/Delx
650 Rhat(1)=Sy+Funy*Y
660 Err(1)=R(1)-Rhat(1)
670 Fhat
680 FOR I = 2 TO 49
690   Rhat(I)=Predy(I)
700   Err(I)=R(I)-Rhat(I)
710 NEXT
720 PRINT#3 "Match-point no.,r,rhat, and error"
730 FOR K = 1 TO 49 STEP 3
740   FOR I = K TO K+2
750     IF I>49 THEN EXIT
760     PRINT#3 USING "###";I;
770     PRINT#3 USING "--####.###";R(I);Rhat(I);Err(I);
780     PRINT#3 "   ";
790   NEXT
800   PRINT#3
810 NEXT
820 PRINT#3
830 REM - Compute the divided differences
840 K=0
850 FOR I = 1 TO 17
860   IF I= 3  THEN GOTO 1090
870   K=K+1
880   Sinode(I)=Inode(I)
890   Inode(I)=Inode(I)*1.02
900   Ddnode(I)=Inode(I)-Sinode(I)
910   IF ABS(Ddnode(I))<.001 THEN GOTO 920 ELSE GOTO 940
920   Ddnode(I)=.001
930   Inode(I)=.001+Sinode(I)
940   Dsixpoint -0.5
950   Sy=0.
960   FOR J=1 TO 6
970     Sy=Sy+Dco(J)*Inode(J)
980   NEXT
990   Y=Inode(3)
1000  Funtny Y
1010  Sy=Sy/Delx
1020  Predy(1)=Sy+Funy*Y
1030  Dd1(K,1)=(Predy(1)-Rhat(1))/Ddnode(I)
1040  Fhat
1050  FOR J= 2 TO 49
1060    Dd1(K,J)=(Predy(J)-Rhat(J))/Ddnode(I)
1070  NEXT
1080  Inode(I)=Sinode(I)
1090 NEXT
```

```
1100 PRINT#3
1110 ON Np GOTO 1120,1270
1120 PRINT#3 "          Divided differences"
1130 FOR I= 1 TO 49
1140   PRINT#3 "Match-point number  ";I
1150   FOR J=1 TO 8
1160     PRINT#3 USING "####";J;
1170     PRINT#3 USING "--###.####";Dd1(J,I);
1180   NEXT
1190   PRINT#3
1200   FOR J= 9 TO 16
1210     PRINT#3 USING "####";J;
1220     PRINT#3 USING "--###.####";Dd1(J,I);
1230   NEXT
1240   PRINT#3
1250 NEXT
1260 REM - Compute sums of squares matrix
1270 CLEAR (Sc)
1280 FOR I= 1 TO 49
1290   Dd1(17,I)=Err(I)
1300 NEXT
1310 FOR I=1 TO 49
1320   FOR J=1 TO 17
1330     IF Dd1(J,I)=0 THEN GOTO 1380
1340     FOR K=J TO 17
1350       IF Dd1(K,I)=0 THEN GOTO 1370
1360       Sc(J,K)=Sc(J,K)+Dd1(J,I)*Dd1(K,I)
1370     NEXT
1380   NEXT
1390 NEXT
1400 FOR J=1 TO 17
1410   FOR I= J TO 17
1420     Sc(I,J)=Sc(J,I)
1430   NEXT
1440 NEXT
1450 REM - Solve simultaneous equations
1460 J=16
1470 Simeqns J
1480 REM - Update the nodes
1490 K=0
1500 FOR I = 1 TO 17
1510   IF I=3 THEN GOTO 1540
1520   K=K+1
1530   Inode(I)=Inode(I)+Sc(K,17)
1540 NEXT
1550 PRINT#3
1560 PRINT#3 "   Updated set of nodes after round  ";Nrc
1570 FOR J= 1 TO 9
1580   PRINT#3 J;FIX(Inode(J)*10000)/10000;"   ";
1590 NEXT
1600 PRINT#3
1610 FOR J= 10 TO 17
1620   PRINT#3 J;FIX(Inode(J)*10000)/10000;"   ";
1630 NEXT
1640 PRINT#3:PRINT#3:PRINT#3
1650 IF Nrc>=Nr THEN GOTO 1680
```

```
1660 REM - End of iteration
1670 GOTO 560
1680 PRINT#3
1690 CLOSE#3
1700 STOP

ENDFILE
```

Program B-7.3

Solution of d^2y/dx^2 + gy dy/dx = asin(bx) exp (-*f*x) by Nonlinear Least Squares

```
SOURCE
PRECISION 12
OPTION BASE 1
PROCS=5
STRING: Title[40]
REAL ARRAY(50): W,What,Err
INTEGER: I,J,K
REAL: Xl,Xr,Dx
REAL ARRAY(17): Inode
REAL: A,B
REAL: X
REAL ARRAY(6): Co,Dco
REAL: Sy,Y
REAL ARRAY(17): Sinode
REAL ARRAY(17): Ddnode
REAL ARRAY(50): Predy
REAL ARRAY(16,16): Sc
REAL ARRAY(16,50): Dd1
INTEGER: Np,Nr,Nrc
REAL ARRAY(6): D2co
REAL: Delx,S2y,F,G
REAL: T1,T2,T3,D1y,D2y

PROCEDURE: Sixpoint
  REAL ARG: Z
END PROCEDURE

PROCEDURE: Dsixpoint
  REAL ARG: Z
END PROCEDURE

PROCEDURE: Yhat
END PROCEDURE

PROCEDURE: Simeqns
  INTEGER ARG: N
END PROCEDURE

PROCEDURE: D2sixpoint
```

```
  REAL ARG: Z
END PROCEDURE

PROCEDURE: Sixpoint
  EXTERNAL: Co
    10 Co(1)=(((((-160*Z+16)*Z+144)*Z-40)*Z-26)*Z+9)/768
    20 Co(2)=(((((800*Z-48)*Z-848)*Z+312)*Z+162)*Z-75)/768
    30 Co(3)=(((((-1600*Z+32)*Z+1824)*Z-272)*Z-1124)*Z+450)/768
    40 Co(4)=(((((1600*Z+32)*Z-1824)*Z-272)*Z+1124)*Z+450)/768
    50 Co(5)=(((((-800*Z-48)*Z+848)*Z+312)*Z-162)*Z-75)/768
    60 Co(6)=(((((160*Z+16)*Z-144)*Z-40)*Z+26)*Z+9)/768
END PROCEDURE

PROCEDURE: Dsixpoint
  EXTERNAL: Dco
    10 Dco(1)=((((-400*Z+32)*Z+216)*Z-40)*Z-13)/384
    20 Dco(2)=((((2000*Z-96)*Z-1272)*Z+312)*Z+81)/384
    30 Dco(3)=((((-4000*Z+64)*Z+2736)*Z-272)*Z-562)/384
    40 Dco(4)=((((4000*Z+64)*Z-2736)*Z-272)*Z+562)/384
    50 Dco(5)=((((-2000*Z-96)*Z+1272)*Z+312)*Z-81)/384
    60 Dco(6)=((((400*Z+32)*Z-216)*Z-40)*Z+13)/384
END PROCEDURE

PROCEDURE: Yhat
  INTEGER: J,I,K,L
  EXTERNAL: Co,Dco,Inode
  REAL: Z,Y,Dy
  EXTERNAL: Sixpoint,Dsixpoint,Predy,D2sixpoint,D2co
  REAL: D2y
  EXTERNAL: Delx,G
    440 K=1
    450 FOR I= 1 TO 12
    460   Z=-0.5
    470   FOR J=1 TO 4
    480     K=K+1
    490     Z=Z+0.25
    520     Y=0:Dy=0.:D2y=0.
    522     Sixpoint Z
    524     Dsixpoint Z
    525     D2sixpoint Z
    530     FOR L= 1 TO 6
    540       Y=Y+Co(L)*Inode(I+L-1)
    550       Dy=Dy+Dco(L)*Inode(I+L-1)
    555       D2y=D2y+D2co(L)*Inode(I+L-1)
    560     NEXT
    561     Dy=Dy/Delx
    562     D2y=D2y/Delx/Delx
    570     Predy(K)=Y*Dy *G+D2y
    590   NEXT
    600 NEXT
END PROCEDURE

PROCEDURE: Simeqns
  EXTERNAL: Sc
  REAL: X,Y
  INTEGER: I,J,K,M
```

```
    440 REM - solve set of n equations
    450 FOR J = 1 TO N
    460   REM - statements 480 to 520 find the first equation with a
    470   REM - non-zero coefficient for the current column.
    475   M=0
    480   FOR I = J TO N
    490     IF Sc(I,J)<>0 THEN GOTO 500
    495     GOTO 514
    500     M=1
    505     K=I
    510     EXIT
    514   NEXT
    518   IF M=1 THEN GOTO 535
    520   PRINT "no unique solution": STOP
    530   REM  statements 540 to 580 move that equation up to current row.
    535   I=K
    540   FOR K= 1 TO N+1
    550     X=Sc(J,K)
    560     Sc(J,K)=Sc(I,K)
    570     Sc(I,K)=X
    580   NEXT
    590   REM - statements 610 to 640 get a 1 coefficient in the first non-zero
    600   REM - column of the current row
    610   Y=1/Sc(J,J)
    620   FOR K= 1 TO N+1
    630     Sc(J,K)=Y*Sc(J,K)
    640   NEXT
    650   REM - statements 670 to 730 subtract the current equation from
    660   REM - th other rows,
    670   FOR I = 1 TO N
    680     IF I = J THEN GOTO 730
    690     Y=-Sc(I,J)
    700     FOR K = 1 TO N+1
    710       Sc(I,K)=Sc(I,K)+Y*Sc(J,K)
    720     NEXT
    730   NEXT
    740   REM - this process is repeated for all equations.
    750 NEXT
    760 PRINT #3 " solution of equations"
    770 M = 1
    780 FOR I = 1 TO N
    785   M= M+1
    786   PRINT #3 I;"  ";Sc(I,N+1);
    800   IF M=>6 GOTO 820
    805   PRINT #3 "    ";
    810   GOTO 830
    820   PRINT #3: M= 1
    830 NEXT
    840 PRINT #3
END PROCEDURE

PROCEDURE: D2sixpoint
  EXTERNAL: D2co
    10 D2co(1)=(((-200*Z+12)*Z+54)*Z-5)/48
    20 D2co(2)=(((1000*Z-36)*Z-318)*Z+39)/48
    30 D2co(3)=(((-2000*Z+24)*Z+684)*Z-34)/48
```

```
    40 D2co(4)=(((2000*Z+24)*Z-684)*Z-34)/48
    50 D2co(5)=(((-1000*Z-36)*Z+318)*Z+39)/48
    60 D2co(6)=(((200*Z+12)*Z-54)*Z-5)/48
END PROCEDURE
```

'MAIN Program:

```
 10 REM - Solution of non-linear differential equation by non-linear
 20 REM - least squares. Uses six-point sliding polynomial.
 30 REM - Equation: d2y/dx2+gydy/dx = F(asin(bx)exp(-fx)
 40 REM - 17 nodes are input as initial values. Nos. 3 and 15 are fixed
 50 REM - for boundary conditions. Remaining 15 are "at solution"..
 60 OPEN "prn" FOR OUTPUT AS #3
 70 INPUT "title";Title
 80 PRINT#3 Title
 90 INPUT "value of x at node 3";Xl
100 INPUT "value of x at node 15";Xr
110 PRINT#3 "Left boundary in x  ";Xl;"  ";"Right boundary  ";Xr
120 INPUT "print divided difference? yes(1) or no(2)";Np
130 INPUT "number of rounds to try?";Nr
140 Nrc=0
150 Dx=(Xr-Xl)/48
160 Delx=(Xr-Xl)/12
170 PRINT#3 "        Input nodes"
180 PRINT "input 17 nodes"
190 FOR I = 1 TO 17
200   PRINT I;:INPUT Inode(I)
210   PRINT#3 I;Inode(I);"    ";
220 NEXT
230 PRINT#3
240 INPUT " constant 'a'";A
250 INPUT " constant 'b'";B
260 INPUT "constant 'f'";F
270 INPUT "constant 'g'";G
280 PRINT#3 "a  ";A;"     b  ";B;"    f  ";J;"     g  ";G
290 REM - Compute 'w'
300 FOR I = 1 TO 49
310   X=Xl+(I-1)*Dx
320   T1=SIN(B*X):T2=COS(B*X):T3=EXP(-F*X)
330   Y=A*T1*T3
340   D1y=A*(-F*T1+B*T2)*T3
350   D2y=A*((F*F-B*B)*T1-2*F*B*T2)*T3
360   W(I)=D2y+G*Y*D1y
370 NEXT
380 REM - Begin iteration
390 REM - Compute what and err
400 REM - at left boundary.
410 Nrc=Nrc+1
420 Dsixpoint -0.5
430 D2sixpoint -0.5
440 Sy=0.:S2y=0.
450 FOR I=1 TO 6
460   Sy=Sy+Dco(I)*Inode(I)
470   S2y=S2y+D2co(I)*Inode(I)
480 NEXT
```

```
490 Sy=Sy/Delx
500 S2y=S2y/Delx/Delx
510 What(1)=Inode(3)*Sy*G+S2y
520 Err(1)=W(1)-What(1)
530 Yhat
540 FOR I = 2 TO 49
550   What(I)=Predy(I)
560   Err(I)=W(I)-What(I)
570 NEXT
580 PRINT#3 "Match-point no.,w,what, and error"
590 FOR K = 1 TO 49 STEP 3
600   FOR I = K TO K+2
610     IF I>49 THEN EXIT
620     PRINT#3 USING "####";I;
630     PRINT#3 USING "--###.####";W(I);What(I);Err(I);
640   NEXT
650   PRINT#3
660 NEXT
670 PRINT#3
680 REM - Compute the divided differences
690 K=0
700 FOR I = 1 TO 17
710   IF I= 3 OR I=15 THEN GOTO 950
720   K=K+1
730   Sinode(I)=Inode(I)
740   Inode(I)=Inode(I)*1.02
750   Ddnode(I)=Inode(I)-Sinode(I)
760   IF ABS(Ddnode(I))<.001 THEN GOTO 770 ELSE GOTO 790
770   Ddnode(I)=0.001
780   Inode(I)=0.001+Sinode(I)
790   Dsixpoint -0.5
800   D2sixpoint -0.5
810   Sy=0.:S2y=0.
820   FOR J=1 TO 6
830     Sy=Sy+Dco(J)*Inode(J)
840     S2y=S2y+D2co(J)*Inode(J)
850   NEXT
860   Sy=Sy/Delx
870   S2y=S2y/Delx/Delx
880   Predy(1)=Inode(3)*Sy*G+S2y
890   Dd1(K,1)=(Predy(1)-What(1))/Ddnode(I)
900   Yhat
910   FOR J= 2 TO 49
920     Dd1(K,J)=(Predy(J)-What(J))/Ddnode(I)
930   NEXT
940   Inode(I)=Sinode(I)
950 NEXT
960 PRINT#3
970 ON Np GOTO 980,1130
980 PRINT#3 "          Divided differences"
990 FOR I= 1 TO 49
1000   PRINT#3 "match-point number  ";I
1010   FOR J=1 TO 8
1020     PRINT#3 USING "####";J;
1030     PRINT#3 USING "--###.####";Dd1(J,I);
1040   NEXT
```

```
1050   PRINT#3
1060   FOR J= 9 TO 15
1070     PRINT#3 USING "####";J;
1080     PRINT#3 USING "--###.####";Dd1(J,I);
1090   NEXT
1100   PRINT#3
1110 NEXT
1120 REM - Compute sums of squares matrix
1130 CLEAR (Sc)
1140 FOR I= 1 TO 49
1150   Dd1(16,I)=Err(I)
1160 NEXT
1170 FOR I=1 TO 49
1180   FOR J=1 TO 16
1190     FOR K=J TO 16
1200       Sc(J,K)=Sc(J,K)+Dd1(J,I)*Dd1(K,I)
1210     NEXT
1220   NEXT
1230 NEXT
1240 FOR J=1 TO 16
1250   FOR I= J TO 16
1260     Sc(I,J)=Sc(J,I)
1270   NEXT
1280 NEXT
1290 REM - Solve simultaneous equations
1300 J=15
1310 Simeqns J
1320 REM - Update the nodes
1330 K=0
1340 FOR I = 1 TO 17
1350   IF I=3 OR I=15 THEN GOTO 1380
1360   K=K+1
1370   Inode(I)=Inode(I)+Sc(K,16)
1380 NEXT
1390 PRINT#3
1400 PRINT#3 "    Updated set of nodes after round  ";Nrc
1410 FOR J= 1 TO 9
1420   PRINT#3 USING "####";J;
1430   PRINT#3 USING "--###.####";Inode(J);
1440 NEXT
1450 PRINT#3
1460 FOR J= 10 TO 17
1470   PRINT#3 USING "####";J;
1480   PRINT#3 USING "--###.####";Inode(J);
1490 NEXT
1500 PRINT#3:PRINT#3:PRINT#3
1510 IF Nrc>=Nr THEN GOTO 1540
1520 REM - End of iteration
1530 GOTO 410
1540 PRINT#3
1550 CLOSE#3
1560 STOP
```

ENDFILE

Program B-7.4

Solution of $d^2y/dx^2 + ay\, dy/dx = bx + c(m)$ by Nonlinear Least Squares

```
SOURCE
PRECISION 12
OPTION BASE 1
PROCS=5
STRING: Title[60]
REAL ARRAY(50): R,Rhat,Err,Con
REAL ARRAY(20): Inode,Sinode,Ddnode
REAL ARRAY(50): Predy
REAL ARRAY(6): Co,Dco,D2co
REAL ARRAY(20,20): Sc
REAL ARRAY(20,50): Dd1
REAL: Xl,Xr,Dx,A,B,X,Sy,S2y
REAL: Delx
INTEGER: I,J,K,Np,Nr,Nrc

PROCEDURE: Sixpoint
  REAL ARG: Z
END PROCEDURE

PROCEDURE: Dsixpoint
  REAL ARG: Z
END PROCEDURE

PROCEDURE: D2sixpoint
  REAL ARG: Z
END PROCEDURE

PROCEDURE: Fhat
END PROCEDURE

PROCEDURE: Simeqns
  INTEGER ARG: N
END PROCEDURE

PROCEDURE: Sixpoint
  EXTERNAL: Co
    10 Co(1)=(((((-160*Z+16)*Z+144)*Z-40)*Z-26)*Z+9)/768
    20 Co(2)=(((((800*Z-48)*Z-848)*Z+312)*Z+162)*Z-75)/768
    30 Co(3)=(((((-1600*Z+32)*Z+1824)*Z-272)*Z-1124)*Z+450)/768
    40 Co(4)=(((((1600*Z+32)*Z-1824)*Z-272)*Z+1124)*Z+450)/768
    50 Co(5)=(((((-800*Z-48)*Z+848)*Z+312)*Z-162)*Z-75)/768
    60 Co(6)=(((((160*Z+16)*Z-144)*Z-40)*Z+26)*Z+9)/768
END PROCEDURE

PROCEDURE: Dsixpoint
  EXTERNAL: Dco
    10 Dco(1)=((((-400*Z+32)*Z+216)*Z-40)*Z-13)/384
    20 Dco(2)=((((2000*Z-96)*Z-1272)*Z+312)*Z+81)/384
```

```
      30 Dco(3)=((((-4000*Z+64)*Z+2736)*Z-272)*Z-562)/384
      40 Dco(4)=((((4000*Z+64)*Z-2736)*Z-272)*Z+562)/384
      50 Dco(5)=((((-2000*Z-96)*Z+1272)*Z+312)*Z-81)/384
      60 Dco(6)=((((400*Z+32)*Z-216)*Z-40)*Z+13)/384
END PROCEDURE

PROCEDURE: D2sixpoint
   EXTERNAL: D2co
      10 D2co(1)=(((-200*Z+12)*Z+54)*Z-5)/48
      20 D2co(2)=(((1000*Z-36)*Z-318)*Z+39)/48
      30 D2co(3)=(((-2000*Z+24)*Z+684)*Z-34)/48
      40 D2co(4)=(((2000*Z+24)*Z-684)*Z-34)/48
      50 D2co(5)=(((-1000*Z-36)*Z+318)*Z+39)/48
      60 D2co(6)=(((200*Z+12)*Z-54)*Z-5)/48
END PROCEDURE

PROCEDURE: Fhat
   INTEGER: J,I,K,L
   EXTERNAL: Co,Dco,Inode
   REAL: Z,Y,Dy
   EXTERNAL: Sixpoint,Dsixpoint,Predy,D2sixpoint,D2co,A
   REAL: D2y
   EXTERNAL: Delx
     440 K=1
     450 FOR I= 1 TO 12
     460   Z=-0.5
     470   FOR J=1 TO 4
     480     K=K+1
     490     Z=Z+0.25
     520     Y=0:Dy=0.:D2y=0.
     522     Sixpoint Z
     524     Dsixpoint Z
     525     D2sixpoint Z
     530     FOR L= 1 TO 6
     540       Y=Y+Co(L)*Inode(I+L-1)
     550       Dy=Dy+Dco(L)*Inode(I+L-1)
     555       D2y=D2y+D2co(L)*Inode(I+L-1)
     560     NEXT
     561     Dy=Dy/Delx
     562     D2y=D2y/Delx/Delx
     570     Predy(K)=Y*Dy *A+D2y
     590   NEXT
     600 NEXT
END PROCEDURE

PROCEDURE: Simeqns
   EXTERNAL: Sc
   REAL: X,Y
   INTEGER: J,I,K,M
     440 REM - solve set of n equations
     450 FOR J = 1 TO N
     460   REM - statements 480 to 520 find the first equation with a
     470   REM - non-zero coefficient for the current column.
     475   M=0
     480   FOR I = J TO N
     490     IF Sc(I,J)<>0 THEN GOTO 500
```

```
495    GOTO 514
500    M=1
505    K=I
510    EXIT
514  NEXT
518  IF M=1 THEN GOTO 535
520  PRINT "no unique solution": STOP
530  REM  statements 540 to 580 move that equation up to current row.
535  I=K
540  FOR K= 1 TO N+1
550    X=Sc(J,K)
560    Sc(J,K)=Sc(I,K)
570    Sc(I,K)=X
580  NEXT
590  REM - statements 610 to 640 get a 1 coefficient in the first non-zero
600  REM - column of the current row
610  Y=1/Sc(J,J)
620  FOR K= 1 TO N+1
630    Sc(J,K)=Y*Sc(J,K)
640  NEXT
650  REM - statements 670 to 730 subtract the current equation from
660  REM - th other rows,
670  FOR I = 1 TO N
680    IF I = J THEN GOTO 730
690    Y=-Sc(I,J)
700    FOR K = 1 TO N+1
710      Sc(I,K)=Sc(I,K)+Y*Sc(J,K)
720    NEXT
730  NEXT
740  REM - this process is repeated for all equations.
750 NEXT
760 PRINT #3 " solution of equations"
770 M = 1
780 FOR I = 1 TO N
785  M= M+1
786  PRINT #3 I;"  ";Sc(I,N+1);
800  IF M=>6 GOTO 820
805  PRINT #3 "  ";
810  GOTO 830
820  PRINT #3: M= 1
830 NEXT
840 PRINT #3
END PROCEDURE
```

'MAIN Program:

```
10 REM - Solution of non-linear differential equation by non-linear
20 REM - least squares. Uses six-point sliding polynomial.
30 REM - Equation: d2y/dx2+a*dy/dx*y=bx+c(m)
40 REM - "c(m)" in the equation can vary for each "match-point"
50 REM - 17 nodes are input as initial values. Nos. 3 and 15 are fixed
60 REM - for boundary conditions. Remaining 15 are "at solution"..
70 OPEN "prn" FOR OUTPUT AS #3
80 INPUT "title";Title
90 PRINT#3 Title
```

```
100 INPUT "value of x at node 3";Xl
110 INPUT "value of x at node 15";Xr
120 PRINT#3 "Left boundary in x  ";Xl;"  ";"Right boundary  ";Xr
130 INPUT "print divided difference? yes(1) or no(2)";Np
140 INPUT "number of rounds to try?";Nr
150 Nrc=0
160 Dx=(Xr-Xl)/48
170 Delx=(Xr-Xl)/12
180 PRINT#3 "          Input nodes"
190 PRINT "input 17 nodes"
200 FOR I = 1 TO 17
210   PRINT I;:INPUT Inode(I)
220   PRINT#3 I;Inode(I);"    ";
230 NEXT
240 PRINT#3
250 PRINT#3 "Values of 'c'"
260 PRINT " input 49 values of 'c'"
270 FOR I = 1 TO 49
280   PRINT I;:INPUT Con(I)
290   PRINT#3 I;Con(I);"   ";
300 NEXT
310 PRINT#3
320 INPUT " constant 'a'";A
330 INPUT " constant 'b'";B
340 PRINT#3 "a  ";A;"      b  ";B
350 REM - Compute 'r'
360 FOR I = 1 TO 49
370   X=Xl+(I-1)*Dx
380   R(I)=B*X+Con(I)
390 NEXT
400 REM - Begin iteration
410 REM - Compute rhat and err
420 REM - at left boundary.
430 Nrc=Nrc+1
440 Dsixpoint -0.5
450 D2sixpoint -0.5
460 Sy=0.:S2y=0.
470 FOR I=1 TO 6
480   Sy=Sy+Dco(I)*Inode(I)
490   S2y=S2y+D2co(I)*Inode(I)
500 NEXT
510 Sy=Sy/Delx
520 S2y=S2y/Delx/Delx
530 Rhat(1)=Inode(3)*Sy*A+S2y
540 Err(1)=R(1)-Rhat(1)
550 Fhat
560 FOR I = 2 TO 49
570   Rhat(I)=Predy(I)
580   Err(I)=R(I)-Rhat(I)
590 NEXT
600 PRINT#3 "Match-point no.,r,rhat, and error"
610 FOR K = 1 TO 49 STEP 3
620   FOR I = K TO K+2
630     IF I>49 THEN EXIT
640     PRINT#3 USING "#####";I;
650     PRINT#3 USING "--####.####";R(I);Rhat(I);Err(I);
```

```
660   NEXT
670   PRINT#3
680 NEXT
690 PRINT#3
700 REM - Compute the divided differences
710 K=0
720 FOR I = 1 TO 17
730   IF I= 3 OR I=15 THEN GOTO 970
740   K=K+1
750   Sinode(I)=Inode(I)
760   Inode(I)=Inode(I)*1.02
770   Ddnode(I)=Inode(I)-Sinode(I)
780   IF ABS(Ddnode(I))<.001 THEN GOTO 790 ELSE GOTO 810
790   Ddnode(I)=0.001
800   Inode(I)=0.001+Sinode(I)
810   Dsixpoint -0.5
820   D2sixpoint -0.5
830   Sy=0.:S2y=0.
840   FOR J=1 TO 6
850     Sy=Sy+Dco(J)*Inode(J)
860     S2y=S2y+D2co(J)*Inode(J)
870   NEXT
880   Sy=Sy/Delx
890   S2y=S2y/Delx/Delx
900   Predy(1)=Inode(3)*Sy*A+S2y
910   Dd1(K,1)=(Predy(1)-Rhat(1))/Ddnode(I)
920   Fhat
930   FOR J= 2 TO 49
940     Dd1(K,J)=(Predy(J)-Rhat(J))/Ddnode(I)
950   NEXT
960   Inode(I)=Sinode(I)
970 NEXT
980 PRINT#3
990 ON Np GOTO 1000,1150
1000 PRINT#3 "         Divided differences"
1010 FOR I= 1 TO 49
1020   PRINT#3 "Match-point number  ";I
1030   FOR J=1 TO 8
1040     PRINT#3 USING"####";J;
1050     PRINT#3 USING "--###.####";Dd1(J,I);
1060   NEXT
1070   PRINT#3
1080   FOR J= 9 TO 15
1090     PRINT#3 USING "####";J;
1100     PRINT#3 USING "--###.####";Dd1(J,I);
1110   NEXT
1120   PRINT#3
1130 NEXT
1140 REM - Calculate sums of squares matrix
1150 CLEAR (Sc)
1160 FOR I= 1 TO 49
1170   Dd1(16,I)=Err(I)
1180 NEXT
1190 FOR I=1 TO 49
1200   FOR J=1 TO 16
1210     FOR K=J TO 16
```

```
1220       Sc(J,K)=Sc(J,K)+Dd1(J,I)*Dd1(K,I)
1230     NEXT
1240   NEXT
1250 NEXT
1260 FOR J=1 TO 16
1270   FOR I= J TO 16
1280     Sc(I,J)=Sc(J,I)
1290   NEXT
1300 NEXT
1310 REM - Solve the simultaneous equations
1320 J=15
1330 Simeqns J
1340 REM - Update the nodes
1350 K=0
1360 FOR I = 1 TO 17
1370   IF I=3 OR I=15 THEN GOTO 1400
1380   K=K+1
1390   Inode(I)=Inode(I)+Sc(K,16)
1400 NEXT
1410 PRINT#3
1420 PRINT#3 "    Updated set of nodes after round  ";Nrc
1430 FOR J= 1 TO 9
1440   PRINT#3 USING "####";J;
1450   PRINT#3 USING "--###.####";Inode(J);
1460 NEXT
1470 PRINT#3
1480 FOR J= 10 TO 17
1490   PRINT#3 USING "####";J;
1500   PRINT#3 USING"--###.####";Inode(J);
1510 NEXT
1520 PRINT#3:PRINT#3:PRINT#3
1530 IF Nrc>=Nr THEN GOTO 1560
1540 REM - End iteration
1550 GOTO 430
1560 PRINT#3
1570 CLOSE#3
1580 STOP

ENDFILE
```

Program B-8.1

Finite Element Solution of Dispersion Equation

```
SOURCE
PRECISION 12
OPTION BASE 1
PROCS=0
REAL ARRAY(8,8): Ct,Cz,B
REAL ARRAY(11,22): Cs
REAL ARRAY(11): Rh,Cb
```

```
REAL ARRAY(8): Cq,S,Q
REAL ARRAY(170): Xc,Mc
STRING: Ti[100]
INTEGER: J,I
INTEGER: Nc,M,L,K,O
REAL: Eb
INTEGER: Nr
REAL: D,R,Ss,T
REAL: Yp,V,Aa,Bb,Cc,Dd,Z
INTEGER: I1,Ix

'MAIN Program:

 10 REM - Solution of two dimensional differential dispersion equation
 20 REM - Solution covers eight spans
 30 REM - Nodes are non-uniformly spaced using exponential transform
 40 REM - Other configurations can be used.
 50 REM - Regard this as elemental programming for an elemental dispersion model
 60 REM - Note reversal of signs due to reversal of time direction.
 70 OPEN "prn" FOR OUTPUT AS #3
 80 INPUT "problem title  ";Ti
 90 PRINT#3 "            ";Ti
100 REM - Bring in sliding polynomial/finite element coefficients
110 OPEN "b:febf" FOR INPUT AS #4
120 OPEN "b:febh" FOR INPUT AS #5
130 OPEN "b:feba" FOR INPUT AS #6
140 CLEAR (Ct,Cz,Cs,Rh,Cb,B,Cq,S,Q,Xc,Mc)
150 J = 1
160 INPUT#6 Cq(J)
170 I = 1
180 INPUT #4 Ct(I,J)
190 INPUT #5 Cz(I,J)
200 IF I = 8 THEN GOTO 230
210 I = I + 1
220 GOTO 180
230 IF J = 8 THEN GOTO 260
240 J = J + 1
250 GOTO 160
260 CLOSE#4
270 CLOSE#5
280 CLOSE#6
290 Eb=0.00575364145
300 INPUT "number of rounds to calculate  ";Nr
310 Nc = 1
320 INPUT "dispersion coefficient d  ";D
330 DATA .0000166,.0000828,.0002152,.0004138,.0004138,.0002152,.0000828,.0000166
340 I = 1
350 READ S(I): IF I = 8 THEN GOTO 370
360 I = I + 1: GOTO 350
370 PRINT "  values of material inputs for spans 1 thru 8."
380 I = 1
390 PRINT I;:INPUT " q(i)  ";Q(I):IF I = 8 THEN GOTO 410
400 I = I + 1: GOTO 390
410 M = 1
420 L = 1
```

```
430 K = 1
440 J = M + K - 1
450 I = L + K - 1
460 O = L + K + 10
470 Cs(J,I)=Cs(J,I)+Ct(M,L)+D*S(K)*Cz(M,L)
480 Cs(J,O)=Cs(J,O)+Ct(M,L+4)+D*S(K)*Cz(M,L+4)
490 IF K = 8 THEN GOTO 520
500 K = K + 1
510 GOTO 440
520 IF L = 4 THEN GOTO 550
530 L = L + 1
540 GOTO 430
550 IF M = 4 THEN GOTO 580
560 M = M + 1
570 GOTO 420
580 PRINT#3:PRINT#3 " Dispersion coefficient is  ";D
590 PRINT#3:PRINT #3 "Material input by spans  "
600 M = 1
610 PRINT#3 USING "###";M;:PRINT#3 USING "--##.####";Q(M);
620 IF M = 8 THEN GOTO 640
630 M = M + 1: GOTO 610
640 PRINT#3
650 PRINT " Input eleven initial value ordinates"
660 PRINT "   Value 1 must equal value 3 and value 11 must equal value 9"
670 PRINT "   Values 2 and 10 must be 0"
680 K = 1
690 PRINT K;:INPUT Cb(K)
700 IF K = 11 THEN GOTO 730
710 K = K + 1
720 GOTO 690
730 M = 1
740 Rh(M)= 0.0
750 L = 1
760 Rh(M)=Rh(M)+Cb(L)*Cs(M,L+11)
770 IF L = 11 THEN GOTO 800
780 L = L + 1
790 GOTO 760
800 IF M = 11 THEN GOTO 840
810 M = M + 1
820 GOTO 740
830 REM - Augment right hand by span input constants
840 M = 1
850 K = 1
860 J = M + K - 1
870 Rh(J)=Rh(J)-Q(K)*Cq(M)
880 IF K = 8 THEN GOTO 910
890 K = K + 1
900 GOTO 860
910 IF M = 4 THEN GOTO 980
920 M = M + 1
930 GOTO 850
940 REM - Put in left and right boundary conditions
950 REM - First make node(1)=node(3) and node(9) = node(11) for zero slope
960 REM - Set node(2) and node(10) equal to zero
970 REM - Rewrite to a new 7 x 7 matrix
980 J = 1
```

```
990 Cs(J,1)=Cs(J+2,1)+Cs(J+2,3)
1000 I = 2
1010 Cs(J,I)=Cs(J+2,I+2)
1020 IF I = 6 THEN GOTO 1050
1030 I = I + 1
1040 GOTO 1010
1050 Cs(J,7)=Cs(J+2,9)+Cs(J+2,11)
1060 IF J = 7 THEN GOTO 1090
1070 J = J + 1
1080 GOTO 990
1090 IF Nc > 1 THEN GOTO 1630
1100 R = 7
1110 I = 1
1120 B(I,I)=1.
1130 IF I = R THEN GOTO 1160
1140 I = I + 1
1150 GOTO 1120
1160 J = 1
1170 I = J
1180 IF Cs(I,J) <> 0 THEN GOTO 1240
1190 IF I = R GOTO 1220
1200 I = I + 1
1210 GOTO 1180
1220 PRINT "singular matrix"
1230 GOTO 1820
1240 K = 1
1250 Ss = Cs(J,K)
1260 Cs(J,K)=Cs(I,K)
1270 Cs(I,K)=Ss
1280 Ss=B(J,K)
1290 B(J,K)=B(I,K)
1300 B(I,K)=Ss
1310 IF K = R THEN GOTO 1340
1320 K = K + 1
1330 GOTO 1250
1340 T = 1/Cs(J,J)
1350 K = 1
1360 Cs(J,K)=T*Cs(J,K)
1370 B(J,K)=T*B(J,K)
1380 IF K = R THEN GOTO 1410
1390 K = K + 1
1400 GOTO 1360
1410 L = 1
1420 IF L = J THEN GOTO 1500
1430 T = -Cs(L,J)
1440 K = 1
1450 Cs(L,K)=Cs(L,K)+T*Cs(J,K)
1460 B(L,K)=B(L,K)+T*B(J,K)
1470 IF K = R THEN GOTO 1500
1480 K = K + 1
1490 GOTO 1450
1500 IF L = R THEN GOTO 1530
1510 L = L + 1
1520 GOTO 1420
1530 IF J = R THEN GOTO 1560
1540 J = J + 1
```

```
1550 GOTO 1170
1560 PRINT#3 "    Initial ordinates"
1570 I = 1
1580 PRINT#3 USING "###";I;
1590 PRINT#3 USING "--###.###";Cb(I);
1600 IF I = 11 THEN GOTO 1630
1610 I = I + 1
1620 GOTO 1580
1630 PRINT#3
1640 PRINT#3:PRINT#3 "      Round  ";Nc
1650 PRINT#3 "      Solution of equations"
1660 I = 1
1670 Yp = 0.0
1680 J = 1
1690 Yp = Yp + B(I,J)*Rh(J+2)
1700 IF J = R THEN GOTO 1730
1710 J = J + 1
1720 GOTO 1690
1730 PRINT#3 USING "###";I;
1740 PRINT#3 USING "--###.####";Yp;
1750 Cb(I+2)=Yp
1760 IF I = R THEN GOTO 1790
1770 I = I + 1
1780 GOTO 1670
1790 PRINT#3
1800 Cb(1)=Cb(3)
1810 Cb(2)=0.0
1820 Cb(10)=0.0
1830 Cb(11)=Cb(9)
1840 PRINT#3 "      Full curves by interpolation for round ";Nc;".  Values of x and C"
1850 V = -4.
1860 I = 1
1870 Aa = (9*(Cb(I+1)+Cb(I+2))-Cb(I)-Cb(I+3))/16
1880 Bb = (11*(Cb(I+2)-Cb(I+1))+Cb(I)-Cb(I+3))/8
1890 Cc = (Cb(I)-Cb(I+1)-Cb(I+2)+Cb(I+3))/4
1900 Dd = (3*(Cb(I+1)-Cb(I+2))-Cb(I)+Cb(I+3))/2
1910 Z = -0.5
1920 K = 1
1930 V = V + 0.05
1940 Z = Z + 0.05
1950 I1 = (I-1)*20+K
1960 Mc(I1)=((Dd*Z+Cc)*Z+Bb)*Z+Aa
1970 IF V<=0.0 THEN GOTO 2010
1980 IF V = 4. THEN GOTO 2080
1990 Xc(I1)=LOG((4.-V)/4.)/-Eb
2000 GOTO 2020
2010 Xc(I1)=LOG((4.+V)/4.)/Eb
2020 IF K = 20 THEN GOTO 2050
2030 K = K + 1
2040 GOTO 1930
2050 IF I = 8 THEN GOTO 2080
2060 I = I + 1
2070 GOTO 1870
2080 J = 1
2090 I = (J-1)*6
2100 Ix = I+1
```

```
2110 PRINT#3 USING "--###.####";Xc(Ix);Mc(Ix);
2120 IF Ix = 159 THEN GOTO 2200
2130 IF Ix = I + 6 THEN GOTO 2170
2140 PRINT#3 "   ";
2150 Ix = Ix + 1
2160 GOTO 2110
2170 PRINT#3
2180 J = J + 1
2190 GOTO 2090
2200 PRINT#3:PRINT#3" Solution completed for round  ";Nc
2210 IF Nc = Nr THEN GOTO 2240
2220 Nc = Nc + 1
2230 GOTO 730
2240 PRINT#3 " All rounds completed"
2250 CLOSE#3
2260 STOP
2270 END

ENDFILE
^ZSOURCE
PRECISION 12
OPTION BASE 1
PROCS=0
REAL ARRAY(8,8): Ct,Cz,B
REAL ARRAY(11,22): Cs
REAL ARRAY(11): Rh,Cb
REAL ARRAY(8): Cq,S,Q
REAL ARRAY(170): Xc,Mc
STRING: Ti[100]
INTEGER: J,I
INTEGER: Nc,M,L,K,O
REAL: Eb
INTEGER: Nr
REAL: D,R,Ss,T
REAL: Yp,V,Aa,Bb,Cc,Dd,Z
INTEGER: I1,Ix

'MAIN Program:


 10 REM - Solution of two dimensional differential dispersion equation
 20 REM - Solution covers eight spans
 30 REM - Nodes are non-uniformly spaced using exponential transform
 40 REM - Other configurations can be used.
 50 REM - Regard this as elemental programming for an elemental dispersion model
 60 REM - Note reversal of signs due to reversal of time direction.
 70 OPEN "prn" FOR OUTPUT AS #3
 80 INPUT "problem title  ";Ti
 90 PRINT#3 "            ";Ti
100 REM - Bring in sliding polynomial/finite element coefficients
110 OPEN "b:febf" FOR INPUT AS #4
120 OPEN "b:febh" FOR INPUT AS #5
130 OPEN "b:feba" FOR INPUT AS #6
140 CLEAR (Ct,Cz,Cs,Rh,Cb,B,Cq,S,Q,Xc,Mc)
150 J = 1
160 INPUT#6 Cq(J)
```

```
170 I = 1
180 INPUT #4 Ct(I,J)
190 INPUT #5 Cz(I,J)
200 IF I = 8 THEN GOTO 230
210 I = I + 1
220 GOTO 180
230 IF J = 8 THEN GOTO 260
240 J = J + 1
250 GOTO 160
260 CLOSE#4
270 CLOSE#5
280 CLOSE#6
290 Eb=0.00575364145
300 INPUT "number of rounds to calculate ";Nr
310 Nc = 1
320 INPUT "dispersion coefficient d ";D
330 DATA .0000166,.0000828,.0002152,.0004138,.0004138,.0002152,.0000828,.0000166
340 I = 1
350 READ S(I): IF I = 8 THEN GOTO 370
360 I = I + 1: GOTO 350
370 PRINT " values of material inputs for spans 1 thru 8."
380 I = 1
390 PRINT I;:INPUT " q(i) ";Q(I):IF I = 8 THEN GOTO 410
400 I = I + 1: GOTO 390
410 M = 1
420 L = 1
430 K = 1
440 J = M + K - 1
450 I = L + K - 1
460 O = L + K + 10
470 Cs(J,I)=Cs(J,I)+Ct(M,L)+D*S(K)*Cz(M,L)
480 Cs(J,O)=Cs(J,O)+Ct(M,L+4)+D*S(K)*Cz(M,L+4)
490 IF K = 8 THEN GOTO 520
500 K = K + 1
510 GOTO 440
520 IF L = 4 THEN GOTO 550
530 L = L + 1
540 GOTO 430
550 IF M = 4 THEN GOTO 580
560 M = M + 1
570 GOTO 420
580 PRINT#3:PRINT#3 " Dispersion coefficient is ";D
590 PRINT#3:PRINT #3 "Material input by spans "
600 M = 1
610 PRINT#3 USING "###";M;:PRINT#3 USING "--##.####";Q(M);
620 IF M = 8 THEN GOTO 640
630 M = M + 1: GOTO 610
640 PRINT#3
650 PRINT " Input eleven initial value ordinates"
660 PRINT "   Value 1 must equal value 3 and value 11 must equal value 9"
670 PRINT "   Values 2 and 10 must be 0"
680 K = 1
690 PRINT K;:INPUT Cb(K)
700 IF K = 11 THEN GOTO 730
710 K = K + 1
720 GOTO 690
```

```
730 M = 1
740 Rh(M)= 0.0
750 L = 1
760 Rh(M)=Rh(M)+Cb(L)*Cs(M,L+11)
770 IF L = 11 THEN GOTO 800
780 L = L + 1
790 GOTO 760
800 IF M = 11 THEN GOTO 840
810 M = M + 1
820 GOTO 740
830 REM - Augment right hand by span input constants
840 M = 1
850 K = 1
860 J = M + K - 1
870 Rh(J)=Rh(J)-Q(K)*Cq(M)
880 IF K = 8 THEN GOTO 910
890 K = K + 1
900 GOTO 860
910 IF M = 4 THEN GOTO 980
920 M = M + 1
930 GOTO 850
940 REM - Put in left and right boundary conditions
950 REM - First make node(1)=node(3) and node(9) = node(11) for zero slope
960 REM - Set node(2) and node(10) equal to zero
970 REM - Rewrite to a new 7 x 7 matrix
980 J = 1
990 Cs(J,1)=Cs(J+2,1)+Cs(J+2,3)
1000 I = 2
1010 Cs(J,I)=Cs(J+2,I+2)
1020 IF I = 6 THEN GOTO 1050
1030 I = I + 1
1040 GOTO 1010
1050 Cs(J,7)=Cs(J+2,9)+Cs(J+2,11)
1060 IF J = 7 THEN GOTO 1090
1070 J = J + 1
1080 GOTO 990
1090 IF Nc > 1 THEN GOTO 1630
1100 R = 7
1110 I = 1
1120 B(I,I)=1.
1130 IF I = R THEN GOTO 1160
1140 I = I + 1
1150 GOTO 1120
1160 J = 1
1170 I = J
1180 IF Cs(I,J) <> 0 THEN GOTO 1240
1190 IF I = R GOTO 1220
1200 I = I + 1
1210 GOTO 1180
1220 PRINT "singular matrix"
1230 GOTO 1820
1240 K = 1
1250 Ss = Cs(J,K)
1260 Cs(J,K)=Cs(I,K)
1270 Cs(I,K)=Ss
1280 Ss=B(J,K)
```

```
1290 B(J,K)=B(I,K)
1300 B(I,K)=Ss
1310 IF K = R THEN GOTO 1340
1320 K = K + 1
1330 GOTO 1250
1340 T = 1/Cs(J,J)
1350 K = 1
1360 Cs(J,K)=T*Cs(J,K)
1370 B(J,K)=T*B(J,K)
1380 IF K = R THEN GOTO 1410
1390 K = K + 1
1400 GOTO 1360
1410 L = 1
1420 IF L = J THEN GOTO 1500
1430 T = -Cs(L,J)
1440 K = 1
1450 Cs(L,K)=Cs(L,K)+T*Cs(J,K)
1460 B(L,K)=B(L,K)+T*B(J,K)
1470 IF K = R THEN GOTO 1500
1480 K = K + 1
1490 GOTO 1450
1500 IF L = R THEN GOTO 1530
1510 L = L + 1
1520 GOTO 1420
1530 IF J = R THEN GOTO 1560
1540 J = J + 1
1550 GOTO 1170
1560 PRINT#3 "    Initial ordinates"
1570 I = 1
1580 PRINT#3 USING "###";I;
1590 PRINT#3 USING "--###.###";Cb(I);
1600 IF I = 11 THEN GOTO 1630
1610 I = I + 1
1620 GOTO 1580
1630 PRINT#3
1640 PRINT#3:PRINT#3 "      Round  ";Nc
1650 PRINT#3 "       Solution of equations"
1660 I = 1
1670 Yp = 0.0
1680 J = 1
1690 Yp = Yp + B(I,J)*Rh(J+2)
1700 IF J = R THEN GOTO 1730
1710 J = J + 1
1720 GOTO 1690
1730 PRINT#3 USING "###";I;
1740 PRINT#3 USING "--###.####";Yp;
1750 Cb(I+2)=Yp
1760 IF I = R THEN GOTO 1790
1770 I = I + 1
1780 GOTO 1670
1790 PRINT#3
1800 Cb(1)=Cb(3)
1810 Cb(2)=0.0
1820 Cb(10)=0.0
1830 Cb(11)=Cb(9)
1840 PRINT#3 "      Full curves by interpolation for round ";Nc;".  Values of x and C"
```

```
1850 V = -4.
1860 I = 1
1870 Aa = (9*(Cb(I+1)+Cb(I+2))-Cb(I)-Cb(I+3))/16
1880 Bb = (11*(Cb(I+2)-Cb(I+1))+Cb(I)-Cb(I+3))/8
1890 Cc = (Cb(I)-Cb(I+1)-Cb(I+2)+Cb(I+3))/4
1900 Dd = (3*(Cb(I+1)-Cb(I+2))-Cb(I)+Cb(I+3))/2
1910 Z = -0.5
1920 K = 1
1930 V = V + 0.05
1940 Z = Z + 0.05
1950 I1 = (I-1)*20+K
1960 Mc(I1)=((Dd*Z+Cc)*Z+Bb)*Z+Aa
1970 IF V<=0.0 THEN GOTO 2010
1980 IF V = 4. THEN GOTO 2080
1990 Xc(I1)=LOG((4.-V)/4.)/-Eb
2000 GOTO 2020
2010 Xc(I1)=LOG((4.+V)/4.)/Eb
2020 IF K = 20 THEN GOTO 2050
2030 K = K + 1
2040 GOTO 1930
2050 IF I = 8 THEN GOTO 2080
2060 I = I + 1
2070 GOTO 1870
2080 J = 1
2090 I = (J-1)*6
2100 Ix = I+1
2110 PRINT#3 USING "--###.####";Xc(Ix);Mc(Ix);
2120 IF Ix = 159 THEN GOTO 2200
2130 IF Ix = I + 6 THEN GOTO 2170
2140 PRINT#3 "   ";
2150 Ix = Ix + 1
2160 GOTO 2110
2170 PRINT#3
2180 J = J + 1
2190 GOTO 2090
2200 PRINT#3:PRINT#3" Solution completed for round  ";Nc
2210 IF Nc = Nr THEN GOTO 2240
2220 Nc = Nc + 1
2230 GOTO 730
2240 PRINT#3 " All rounds completed"
2250 CLOSE#3
2260 STOP
2270 END

ENDFILE
```

Program B-8.2

Finite Element Solution of Simple Mass Continuity

```
SOURCE
PRECISION 12
OPTION BASE 1
PROCS=3
REAL ARRAY(26,27): B
REAL ARRAY(8,8): A
REAL ARRAY(13): Ivnod,Bv
REAL ARRAY(10,10): Lo
REAL ARRAY(8,8): Cs
STRING: Ti[100]
INTEGER: R,Ns,T,Nt,I,J
REAL: V,Dx
REAL: Dt,N

PROCEDURE: Simeqns
  INTEGER ARG: N
END PROCEDURE

PROCEDURE: Interpol
END PROCEDURE

PROCEDURE: Joining
END PROCEDURE

PROCEDURE: Simeqns
  REAL: X,Y
  INTEGER: J,I,K,M
  EXTERNAL: B
    10 REM - solve set of n equations
    20 FOR J = 1 TO N
    30  REM - statements 480 to 520 find the first equation with a
    40  REM - non-zero coefficient for the current column.
    50  M=0
    60  FOR I = J TO N
    70    IF B(I,J)<>0 THEN GOTO 90
    80    GOTO 120
    90    M=1
   100    K=I
   110    EXIT
   120  NEXT
   130  IF M= 1 THEN GOTO 180
   160  PRINT "no unique solution": STOP
   170  REM  statements 540 to 580 move that equation up to current row.
   180  I=K
   190  FOR K= 1 TO N+1
   200    X=B(J,K)
   210    B(J,K)=B(I,K)
   220    B(I,K)=X
   230  NEXT
   240  REM - statements 610 to 640 get a 1 coefficient in the first non-zero
```

```
    250  REM - column of the current row
    260  Y=1/B(J,J)
    270  FOR K= 1 TO N+1
    280    B(J,K)=Y*B(J,K)
    290  NEXT
    300  REM - statements 670 to 730 subtract the current equation from
    310  REM - th other rows,
    320  FOR I = 1 TO N
    330    IF I = J THEN GOTO 380
    340    Y=-B(I,J)
    350    FOR K = 1 TO N+1
    360      B(I,K)=B(I,K)+Y*B(J,K)
    370    NEXT
    380  NEXT
    390  REM - this process is repeated for all equations.
    400 NEXT
    410 PRINT #3 " solution of equations"
    420 M = 1
    430 FOR I = 1 TO N
    440   M= M+1
    450   PRINT #3 USING "#####"; I;
    460   PRINT#3 USING "--#####.####";B(I,N+1);
    470   IF M=>6 GOTO 500
    480   PRINT #3 "   ";
    490   GOTO 510
    500   PRINT #3: M= 1
    510 NEXT
    520 PRINT #3
END PROCEDURE

PROCEDURE: Interpol
  EXTERNAL: Ivnod
  INTEGER: I,J
  REAL: Z,Aa,Bb,Cc,Dd,Y
    10 PRINT#3 "     Interpolated curve at tenth-points of spans"
    20 PRINT#3 "     Left boundary node is  ";Ivnod(2)
    30 FOR I = 1 TO 10
    40   PRINT#3 "  Span ";I
    50   Z=-0.5
    60   Aa=(9*(Ivnod(I+1)+Ivnod(I+2))-Ivnod(I)-Ivnod(I+3))/16
    70   Bb=(11*(Ivnod(I+2)-Ivnod(I+1))+Ivnod(I)-Ivnod(I+3))/8
    80   Cc=(Ivnod(I)-Ivnod(I+1)-Ivnod(I+2)+Ivnod(I+3))/4
    90   Dd=(3*(Ivnod(I+1)-Ivnod(I+2))-Ivnod(I)+Ivnod(I+3))/2
   100   FOR J = 1 TO 10
   110     Z=Z+0.1
   120     Y=((Dd*Z+Cc)*Z+Bb)*Z+Aa
   130     PRINT#3 USING "###";J;
   140     PRINT#3 USING "--###.####";Y;
   150   NEXT J
   160   PRINT#3
   170 NEXT I
   180 PRINT#3
END PROCEDURE

PROCEDURE: Joining
  EXTERNAL: Ns,B,Cs
```

```
INTEGER: N,J,I
  10 REM - joining spans along x with one span along t
  20 FOR N = 1 TO Ns
  30   FOR J = 1 TO 4
  40     FOR I = 1 TO 4
  50       B(J+N-1,I+N-1)=B(J+N-1,I+N-1)+Cs(J,I)
  60     NEXT I
  70     FOR I = 5 TO 8
  80       B(J+N-1,I+Ns+N-2)=B(J+N-1,I+Ns+N-2)+Cs(J,I)
  90     NEXT I
 100   NEXT J
 110   FOR J = 5 TO 8
 120     FOR I = 1 TO 4
 130       B(J+Ns+N-2,I+N-1)=B(J+Ns+N-2,I+N-1)+Cs(J,I)
 140     NEXT I
 150     FOR I = 5 TO 8
 160       B(J+Ns+N-2,I+Ns+N-2)=B(J+Ns+N-2,I+Ns+N-2) +Cs(J,I)
 170     NEXT I
 180   NEXT J
 190 NEXT N
END PROCEDURE
```

'MAIN Program:

```
 10 REM - Finite Element solution of simple mass continuity problem.
 20 REM - Uses 10 spans of sliding polynomial in x (along stream).
 30 REM - Uses one span in t and iterates for time steps.
 40 OPEN "prn" FOR OUTPUT AS #3
 50 R = 11
 60 Ns = 10
 70 T = 1
 80 Nt = 5
 90 INPUT "problem title";Ti
100 PRINT#3 Ti
110 INPUT "stream velocity in kilometers per hour";V
120 PRINT " Input material loadings in grams per hour by spans and time steps"
130 FOR I = 1 TO 10
140   PRINT "span";I
150   FOR J = 1 TO 5
160     PRINT "time step";J;
170     INPUT Lo(I,J)
180   NEXT J
190 NEXT I
200 INPUT "Length of span in kilometers";Dx
210 INPUT "Time-step in hours";Dt
220 PRINT#3 "Velocity ";V;" Span length ";Dx;"  Time step  ";Dt
230 PRINT#3 "Material loadings by time-steps for each span"
240 FOR I = 1 TO 10
250   PRINT#3 "Span   ";I
260   FOR J = 1 TO 5
270     PRINT#3 J; Lo(I,J); "   ";
280   NEXT J
290   PRINT#3
300 NEXT I
310 PRINT "  Input initial value nodes"
```

```
320 FOR I = 1 TO 13
330   PRINT I;
340   INPUT Ivnod(I)
350 NEXT I
360 PRINT#3 "   Initial value of nodes"
370 FOR I = 1 TO 7
380   PRINT#3 I;Ivnod(I);"    ";
390 NEXT I
400 PRINT#3
410 FOR I = 8 TO 13
420   PRINT#3 I;Ivnod(I);"    ";
430 NEXT I
440 PRINT#3
450 CLEAR(A,B,Cs)
460 PRINT "  Input a boundary value for each time step"
470 FOR J = 1 TO 5
480   PRINT J;" ";
490   INPUT Bv(J)
500 NEXT J
510 PRINT#3 "   Boundary values"
520 FOR J = 1 TO 5
530   PRINT#3 J;Bv(J);"    ";
540 NEXT J
550 PRINT#3
560 REM - Bring in finite element coefficients for 1st partial derivative with z
570 OPEN "b:febg" FOR INPUT AS #4
580 FOR J = 1 TO 8
590   FOR I = 1 TO 8
600     INPUT #4 A(J,I)
610   NEXT I
620 NEXT J
630 CLOSE#4
640 FOR J = 1 TO 8
650   FOR I = 1 TO 8
660     Cs(J,I)=V*A(J,I)*Dx
670   NEXT I
680 NEXT J
690 REM - Bring in finite element coefficients for 1st partial derivative with t
700 OPEN "b:febf" FOR INPUT AS #4
710 FOR J = 1 TO 8
720   FOR I = 1 TO 8
730     INPUT#4 A(J,I)
740   NEXT I
750 NEXT J
760 CLOSE#4
770 FOR J = 1 TO 8
780   FOR I = 1 TO 8
790     Cs(J,I)=Cs(J,I)+A(J,I)*Dt
800   NEXT I
810 NEXT J
820 REM - Bring in the coefficients for the constant of the equation
830 OPEN "b:feba" FOR INPUT AS #4
840 FOR I = 1 TO 8
850   INPUT#4 A(1,I)
860 NEXT I
870 CLOSE#4
```

```
880 Joining
890 REM - Basic left-hand matrix constructed, now build right-hand vector.
900 REM - Construct part of vector with time and span loadings.
910 FOR N = 1 TO Ns
920   FOR I = 1 TO 4
930     B(I+N-1,27)=B(I+N-1,27)+A(1,I)*Lo(N,T)
940   NEXT I
950   FOR I = 5 TO 8
960     B(I+Ns+N-2,27)=B(I+Ns+N-2,27)+A(1,I)*Lo(N,T)
970   NEXT I
980 NEXT N
990 REM - Add scaled initial value nodes to right-hand vector
1000 REM - Limit r-h vector elements to residual set of equations
1010 REM - Initial values are in ivnod(i)
1020 FOR J = 16 TO 26
1030   FOR I = 1 TO 13
1040     B(J,27)=B(J,27)-B(J,I)*Ivnod(I)
1050   NEXT I
1060 NEXT J
1070 REM - Node 15 is an input boundary value varying by time steps
1080 REM - Node 14 is set to node 16 to give boundary slope of zero
1090 FOR J = 16 TO 26
1100   B(J,27)=B(J,27)-B(J,15)*Bv(T)
1110   B(J,16)=B(J,16)+B(J,14)
1120 NEXT J
1130 REM - Move matrix to beginning index 1,1 for solution of sim. eqns.
1140 FOR J = 16 TO 26
1150   FOR I = 16 TO 27
1160     B(J-15,I-15)=B(J,I)
1170   NEXT I
1180 NEXT J
1190 PRINT#3
1200 PRINT#3 "  Solution for time step number  ";T
1210 PRINT#3 "   Boundary node at head of reach  ";Bv(T)
1220 Simeqns R
1230 REM - Move solution of last time step to initial values for next step.
1240 FOR I =3 TO 13
1250   Ivnod(I)=B(I-2,R+1)
1260 NEXT I
1270 Ivnod(2)=Bv(T)
1280 Ivnod(1)=Ivnod(3)
1290 REM - Fill in tenth-points by sliding polynomial interpolation
1300 Interpol
1310 IF T = Nt THEN GOTO 1380
1320 T = T+1
1330 REM - Clear right-hand vector - b(i,27)
1340 FOR I = 1 TO 26
1350   B(I,27)=0.0
1360 NEXT I
1370 GOTO 910
1380 STOP
```

ENDFILE

Program B-8.3

Finite Element Solution of BOD Concentration

```
SOURCE
PRECISION 12
OPTION BASE 1
PROCS=3
REAL ARRAY(26,27): B
REAL ARRAY(13): Ivnod
REAL ARRAY(8,8): Cs
INTEGER: Ns
REAL ARRAY(8,8): A
REAL ARRAY(10,10): Lo
STRING: Ti[100]
INTEGER: R,T,Nt,I,J,N
REAL: E,U,K1
REAL: K3,Dx,Dt
REAL ARRAY(5): No15

PROCEDURE: Simeqns
  INTEGER ARG: N
END PROCEDURE

PROCEDURE: Interpol
END PROCEDURE

PROCEDURE: Joining
END PROCEDURE

PROCEDURE: Simeqns
  REAL: X,Y
  INTEGER: J,I,K,M
  EXTERNAL: B
    10 REM - solve set of n equations
    20 FOR J = 1 TO N
    30  REM - statements 480 to 520 find the first equation with a
    40  REM - non-zero coefficient for the current column.
    50  M=0
    60  FOR I = J TO N
    70   IF B(I,J)<>0 THEN GOTO 90
    80   GOTO 120
    90   M=1
    100   K=I
    110   EXIT
    120  NEXT
    130  IF M= 1 THEN GOTO 180
    160  PRINT "no unique solution": STOP
    170  REM  statements 540 to 580 move that equation up to current row.
    180  I=K
    190  FOR K= 1 TO N+1
    200   X=B(J,K)
    210   B(J,K)=B(I,K)
```

```
    220     B(I,K)=X
    230   NEXT
    240   REM - statements 610 to 640 get a 1 coefficient in the first non-zero
    250   REM - column of the current row
    260   Y=1/B(J,J)
    270   FOR K= 1 TO N+1
    280     B(J,K)=Y*B(J,K)
    290   NEXT
    300   REM - statements 670 to 730 subtract the current equation from
    310   REM - th other rows,
    320   FOR I = 1 TO N
    330     IF I = J THEN GOTO 380
    340     Y=-B(I,J)
    350     FOR K = 1 TO N+1
    360       B(I,K)=B(I,K)+Y*B(J,K)
    370     NEXT
    380   NEXT
    390   REM - this process is repeated for all equations.
    400 NEXT
    410 PRINT #3 " solution of equations"
    420 M = 1
    430 FOR I = 1 TO N
    440   M= M+1
    450   PRINT #3 USING "#####"; I;
    460   PRINT#3 USING "--#####.####";B(I,N+1);
    470   IF M=>6 GOTO 500
    480   PRINT #3 "   ";
    490   GOTO 510
    500   PRINT #3: M= 1
    510 NEXT
    520 PRINT #3
END PROCEDURE

PROCEDURE: Interpol
  EXTERNAL: Ivnod
  INTEGER: I,J
  REAL: Z,Aa,Bb,Cc,Dd,Y
    10 PRINT#3 "     Interpolated curve at tenth-points of spans"
    20 PRINT#3 "     Left boundary node is  ";Ivnod(2)
    30 FOR I = 1 TO 10
    40   PRINT#3 " Span ";I
    50   Z=-0.5
    60   Aa=(9*(Ivnod(I+1)+Ivnod(I+2))-Ivnod(I)-Ivnod(I+3))/16
    70   Bb=(11*(Ivnod(I+2)-Ivnod(I+1))+Ivnod(I)-Ivnod(I+3))/8
    80   Cc=(Ivnod(I)-Ivnod(I+1)-Ivnod(I+2)+Ivnod(I+3))/4
    90   Dd=(3*(Ivnod(I+1)-Ivnod(I+2))-Ivnod(I)+Ivnod(I+3))/2
   100   FOR J = 1 TO 10
   110     Z=Z+0.1
   120     Y=((Dd*Z+Cc)*Z+Bb)*Z+Aa
   130     PRINT#3 USING "###";J;
   140     PRINT#3 USING "--###.####";Y;
   150   NEXT J
   160   PRINT#3
   170 NEXT I
   180 PRINT#3
END PROCEDURE
```

```
PROCEDURE: Joining
  EXTERNAL: Ns,B,Cs
  INTEGER: N,J,I
    10 REM - joining spans along x with one span along t
    20 FOR N = 1 TO Ns
    30   FOR J = 1 TO 4
    40     FOR I = 1 TO 4
    50       B(J+N-1,I+N-1)=B(J+N-1,I+N-1)+Cs(J,I)
    60     NEXT I
    70     FOR I = 5 TO 8
    80       B(J+N-1,I+Ns+N-2)=B(J+N-1,I+Ns+N-2)+Cs(J,I)
    90     NEXT I
   100   NEXT J
   110   FOR J = 5 TO 8
   120     FOR I = 1 TO 4
   130       B(J+Ns+N-2,I+N-1)=B(J+Ns+N-2,I+N-1)+Cs(J,I)
   140     NEXT I
   150     FOR I = 5 TO 8
   160       B(J+Ns+N-2,I+Ns+N-2)=B(J+Ns+N-2,I+Ns+N-2) +Cs(J,I)
   170     NEXT I
   180   NEXT J
   190 NEXT N
END PROCEDURE

'MAIN Program:


  10 REM - Sliding polynomial/finite element solution for BOD concentration
  20 REM - Uses 10 spans in x along stream
  30 REM - Uses one span in t and iterates for 5 periods
  40 CLEAR(B,A,Cs)
  50 R=11
  60 Ns=10
  70 T=1
  75 Nt=5
  80 INPUT "Problem title";Ti
  90 OPEN "prn" FOR OUTPUT AS #3
 100 PRINT#3 Ti
 105 PRINT "Input equation parameters as called"
 110 INPUT "e";E
 120 INPUT "u";U
 130 INPUT "k1";K1
 140 INPUT "k3";K3
 150 PRINT "Input  10 x 5 matrix of BOD loadings"
 160 FOR I = 1 TO 10
 170   PRINT "span";I
 180   FOR J = 1 TO 5
 190     PRINT "time step";J
 200     INPUT Lo(I,J)
 210   NEXT J
 220 NEXT I
 230 INPUT "Input span length in miles";Dx
 240 INPUT "Time step in days";Dt
 245 PRINT#3 "          Equation parameters"
 250 PRINT#3 "  e  ";E;"  u  ";U
```

```
260 PRINT#3 "  k1  ";K1;"  k3  ";K3
270 PRINT#3 "  Span length  ";Dx;"  Time step  ";Dt
280 PRINT#3 "     BOD loadings by time step for each span"
290 FOR I = 1 TO 10
300   PRINT#3 "  Span  ";I
310   FOR J = 1 TO 5
320     PRINT#3 USING "###";J;
325     PRINT#3 USING "--##.####";Lo(I,J);
330   NEXT J
340   PRINT#3
350 NEXT I
360 PRINT " input 13 initial value nodes"
370 FOR I = 1 TO 13
380   PRINT I;
390   INPUT Ivnod(I)
400 NEXT I
410 PRINT#3 "  Initial value nodes"
420 FOR I = 1 TO 7
430   PRINT#3 USING "###";I;
435   PRINT#3 USING"--###.####";Ivnod(I);
440 NEXT I
450 PRINT#3
460 FOR I = 8 TO 13
470   PRINT#3 USING"###";I;
475   PRINT#3 USING "--###.####";Ivnod(I);
480 NEXT I
490 PRINT#3
491 PRINT "  input head-of-reach  boundary nodes - one for each time step"
492 FOR J = 1 TO 5
493   PRINT J;:INPUT No15(J)
494 NEXT J
500 REM - Bring in sliding polynomial/finite element coefficients
505 REM - Coefficients for second partial derivative with z
510 OPEN "b:febh" FOR INPUT AS #4
520 FOR J= 1 TO 8
530   FOR I = 1 TO 8
540     INPUT#4 A(J,I)
550   NEXT I
560 NEXT J
570 FOR J = 1 TO 8
580   FOR I = 1 TO 8
590     Cs(J,I)=A(J,I)*(-E)*Dx*Dx
600   NEXT I
610 NEXT J
620 CLOSE#4
630 REM - Bring in coefficients for first partial derivative with z
640 OPEN "b:febg" FOR INPUT AS #4
650 FOR J = 1 TO 8
660   FOR I = 1 TO 8
670     INPUT#4 A(J,I)
680   NEXT I
690 NEXT J
700 CLOSE#4
710 FOR J = 1 TO 8
720   FOR I = 1 TO 8
730     Cs(J,I)=Cs(J,I)+U*A(J,I)*Dx
```

```
740   NEXT I
750 NEXT J
760 REM - Bring in coefficients for first partial derivative with t
770 OPEN "b:febf" FOR INPUT AS #4
780 FOR J = 1 TO 8
790   FOR I = 1 TO 8
800     INPUT#4 A(J,I)
810   NEXT I
820 NEXT J
830 CLOSE#4
840 FOR J = 1 TO 8
850   FOR I = 1 TO 8
860     Cs(J,I)=Cs(J,I)+A(J,I)*Dt
870   NEXT I
880 NEXT J
890 REM - Bring in coefficients for l
900 OPEN "b:febe" FOR INPUT AS #4
910 FOR J = 1 TO 8
920   FOR I = 1 TO 8
930     INPUT#4 A(J,I)
940   NEXT I
950 NEXT J
960 CLOSE#4
970 FOR J = 1 TO 8
980   FOR I = 1 TO 8
990     Cs(J,I)=Cs(J,I)+(K1+K3)*A(J,I)
1000   NEXT I
1010 NEXT J
1020 REM - Bring in coefficients for constant
1030 OPEN "b:feba" FOR INPUT AS #4
1040 FOR I = 1 TO 8
1050   INPUT#4 A(1,I)
1060 NEXT I
1070 CLOSE#4
1080 Joining
1090 REM - Basic left-hand matrix constructed, now build right-hand vector
1100 REM - Construct the part of right-hand vector with constant
1110 FOR N = 1 TO Ns
1120   FOR I = 1 TO 4
1130     B(I+N-1,27)=B(I+N-1,27)+A(1,I)*Lo(N,T)
1140   NEXT I
1150   FOR I = 5 TO 8
1160     B(I+Ns+N-2,27)=B(I+Ns+N-2,27)+A(1,I)*Lo(N,T)
1170   NEXT I
1175 NEXT N
1180 REM - Add initial value nodes to right-hand vector
1190 REM - Limit right-hand vector elements to residual set of equations
1200 REM - Initial value nodes are in ivnod(i)
1210 FOR J = 16 TO 26
1220   FOR I = 1 TO 13
1230     B(J,27)=B(J,27)-B(J,I)*Ivnod(I)
1240   NEXT I
1250 NEXT J
1260 REM - Add boundary node to vector
1280 REM - Node 14 is set equal to node 16 to give boundary slope of zero
1290 FOR J = 16 TO 26
```

```
1300   B(J,27)=B(J,27)-B(J,15)*No15(T)
1310   B(J,16)=B(J,16)+B(J,14)
1320 NEXT J
1330 REM - Move matrix to beginning index 1,1 for solution of simultaneous eqns
1340 FOR J = 16 TO 26
1350   FOR I = 16 TO 27
1360     B(J-15,I-15)=B(J,I)
1370   NEXT I
1380 NEXT J
1505 PRINT#3
1510 PRINT#3 "     Solution for round    ";T
1530 Simeqns R
1560 REM -  Move solution of last time step to initial values for next step
1570 FOR I = 3 TO 13
1580   Ivnod(I)=B(I-2,R+1)
1590 NEXT I
1600 Ivnod(2)=No15(T)
1610 Ivnod(1)=Ivnod(3)
1614 REM- Fill in tenth points of curve by sliding polynomial interpolation
1615 Interpol
1616 IF T = Nt THEN GOTO 1670
1617 T = T + 1
1620 REM - Clear right hand vector for next round
1630 FOR I = 1 TO 26
1640   B(I,27)=0.0
1650 NEXT I
1660 GOTO 1110
1670 STOP

ENDFILE
```

INDEX